Feedback Control of MEMS to Atoms

Jason J. Gorman • Benjamin Shapiro
Editors

Feedback Control of MEMS to Atoms

Springer

Editors
Jason J. Gorman
National Institute of Standards
& Technology (NIST)
Intelligent Systems Division
100 Bureau Drive
Stop 8230 Gaithersburg
MD 20899
USA
gorman@nist.gov

Benjamin Shapiro
University of Maryland
2330 Kim Building
College Park
MD 20742
USA
benshap@umd.edu

ISBN 978-1-4419-5831-0 e-ISBN 978-1-4419-5832-7
DOI 10.1007/978-1-4419-5832-7
Springer New York Dordrecht Heidelberg London

Library of Congress Control Number: 2011937573

Printed on acid-free paper

Springer is part of Springer Science+Business Media (www.springer.com)

Preface

This book explores the control of systems on small length scales. Research and development for micro- and nanoscale science and technology has grown quickly over the last decade, particularly in the areas of microelectromechanical systems (MEMS), microfluidics, nanoelectronics, bio-nanotechnologies, nanofabrication, and nanomaterials. However, to date, control theory has played only a small role in the advancement of this research. As we know from the technical progression of macroscale intelligent systems, such as assembly robots and fly-by-wire aircraft, control systems can maximize system performance and, in many cases, enable capabilities that would otherwise not be possible. We expect that control systems will play a similar enabling role in the development of the next generation of micro- and nanoscale devices, as well as in the precision instrumentation that will be used to fabricate and measure these devices. In support of this, each chapter of this book provides an introduction to an application of micro- and nanotechnologies in which control systems have already been shown to be critical to its success. Through these examples, we aim to provide insight into the unique challenges in controlling systems at small length scales and to highlight the benefits in merging control systems and micro- and nanotechnologies.

We conceived of this book because we saw a strong need to bring the control systems and micro- and nanosystems communities closer together. In our view, the intersection between these two groups is still very small, impeding the advancement of active, precise, and robust micro- and nanoscale systems that can meet the demanding requirements for commercial, military, medical, and consumer products. As an example, we attend conferences for both the control systems and micro- and nanoscale science and technology communities and have found the overlap between attendees to be marginal; maybe in the tens of people. Our hope is that this book will be a step toward rectifying this situation by bridging the gap between these two communities and demonstrating that concrete benefits for both fields can be achieved through collaborative research. We also hope to motivate the next generation of young engineers and scientists to pursue a career at this intersection, which offers all of the excitement, frustration, and eventual big rewards that an aspiring researcher could want.

This book is targeted toward both control systems researchers interested in pursuing new application in the micro- and nanoscales domains, and researchers developing micro- and nanosystems who are interested in learning how control systems can benefit their work. For the former, we hope these chapters will show the serious effort required to demonstrate control in a new application area. All of the contributing authors have acquired expertise in at least one new scientific area in addition to control theory (e.g., atomic force microscopy, optics, microfluidics) in order to pursue their area of research. Acquiring dual expertise can take years of effort, but the payoff can be high by providing results that no expert in a single domain can accomplish. Additionally, it can result in fascinating work (we hope some of the challenges and excitement are conveyed). For researchers in micro- and nanoscale science and technology, this book contains concrete examples of the benefits that control can provide. These range from better control of particle size distribution during synthesis, to high-bandwidth and reliable nanoscale positioning and imaging of objects, to optimal control of the spin dynamics of quantum systems. We also hope this book will be of use to those who are not yet experts in either control systems or micro- and nanoscale systems but are interested in both. We believe it will provide a useful and instructive introduction to the breadth of research being performed at the intersection of these two fields.

The topics covered in this book were selected to represent the entire length scale of miniaturized systems, ranging from hundreds of micrometers down to a fraction of a nanometer (hence our title, *Feedback Control of MEMS to Atoms*). They were also selected to cover a broad range of physical systems that will likely provide new material to most readers.

Acknowledgments We would like to express our deepest appreciation to all of the researchers who contributed to this book. Without them this project would not have been possible. It was a pleasure to have the opportunity to work with them. We would also like to thank the staff at Springer and in particular, Steven Elliot, who provided us with outstanding guidance and motivation throughout the process.

Gaithersburg Jason J. Gorman
College Park Benjamin Shapiro

Contents

1 **Introduction** 1
Jason J. Gorman and Benjamin Shapiro

2 **Feedback Control of Particle Size Distribution in Nanoparticle Synthesis and Processing** 7
Mingheng Li and Panagiotis D. Christofides

3 **In Situ Optical Sensing and State Estimation for Control of Surface Processing** 45
Rentian Xiong and Martha A. Grover

4 **Automated Tip-Based 2-D Mechanical Assembly of Micro/Nanoparticles** 69
Cagdas D. Onal, Onur Ozcan, and Metin Sitti

5 **Atomic Force Microscopy: Principles and Systems Viewpoint Enabled Methods** 109
Srinivasa Salapaka and Murti Salapaka

6 **Feedback Control of Optically Trapped Particles** 141
Jason J. Gorman, Arvind Balijepalli, and Thomas W. LeBrun

7 **Position Control of MEMS** 179
Michael S.-C. Lu

8 **Dissecting Tuned MEMS Vibratory Gyros** 211
Dennis Kim and Robert T. M'Closkey

9 **Feedback Control of Microflows** 269
Mike Armani, Zach Cummins, Jian Gong, Pramod Mathai, Roland Probst, Chad Ropp, Edo Waks, Shawn Walker, and Benjamin Shapiro

10 **Problems in Control of Quantum Systems** 321
Navin Khaneja

11 Common Threads and Technical Challenges in Controlling Micro- and Nanoscale Systems 365
Benjamin Shapiro and Jason J. Gorman

Index 377

Chapter 1
Introduction

Jason J. Gorman and Benjamin Shapiro

The goal of this book is to illustrate how control tools can be successfully applied to micro- and nano-scale systems. The book partially explores the wide variety of applications where control can have a significant impact at the micro- and nanoscale, and identifies key challenges and common approaches. This first chapter briefly outlines the range of subjects within micro and nano control and introduces topics that recur throughout the book.

1.1 Controlling Micro- and Nanoscale Systems

Microelectromechanical systems (MEMS) emerged, at the beginning of the 1980s, as a cost effective and highly sensitive solution for many sensor applications, including pressure, force, and acceleration measurements. Since then, MEMS has grown into a $6 billion industry and a number of other microtechnologies have followed, including microfluidics, microrobotics, and micromachining. Simultaneously, nanotechnology has become one of the largest areas of scientific and engineering research, with over $12 billion invested over the last decade by the U.S. Government alone. This research has resulted in a new set of materials and devices that offer unique physical and chemical properties due to their nanoscale dimensions, which are expected to yield better products and services.

J.J. Gorman (✉)
Intelligent Systems Division, Engineering Laboratory, National Institute of Standards and Technology, Gaithersburg, MD 20899, USA
e-mail: gorman@nist.gov

B. Shapiro
Fischell Department of Bioengineering, Institute for Systems Research (ISR), University of Maryland, College Park, MD 20742, USA
e-mail: benshap@umd.edu

J.J. Gorman and B. Shapiro (eds.), *Feedback Control of MEMS to Atoms*,
DOI 10.1007/978-1-4419-5832-7_1, © Springer Science+Business Media, LLC 2012

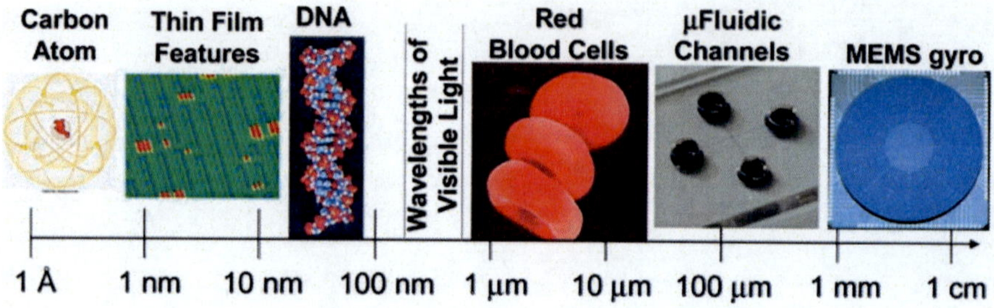

Fig. 1.1 A sense of scale: the sizes of things from a single carbon atom to an integrated MEMS gyro (Images used with permission. Copyrights Denis Kunkel Microscopy, Inc. and Springer)

Micro- and nanotechnology integrated systems refer to a combination of components that provide enhanced functionality that would not be possible with each component alone. Familiar examples of systems at the macroscale include robots, aircraft, automobiles, and information networks, where each system is composed of actuators, sensors, and computational logic that allow for complex and controlled behavior. Micro- and nano-systems only differ from their macroscale counterparts in that essential system behavior occurs at minute length scales. In some cases, micro- and nanosystems are large in size but are dependent on micro- or nanoscale phenomena (e.g., a scanning probe microscope), whereas in other cases the entire system is miniaturized (e.g., a MEMS accelerometer). Other examples of micro- and nanosystems include nanomechanical resonators, cell micromanipulators, and nanofabrication tools. Clearly this is a diverse group of systems, but as will be seen in this book, there are a variety of common threads for the integration and control of such systems, as well as common principles to address these threads.

Feedback control is necessary at small length scales for the same reasons that it is needed in macroscale applications: to correct for errors in system variables in real-time to improve performance, to provide robust operation in the face of unknown or uncertain conditions, and to enable new system capabilities. This book explores emerging efforts to apply control systems to micro- and nanoscale systems in order to realize these benefits and, as a result, accelerate the utility and adoption of these technologies.

Going down in length scales, to micrometers and nanometers (Fig. 1.1), opens up a wide set of technologies, opportunities, and challenges. Sensors and actuators at small scales can directly access and manipulate microscopic and nanoscopic objects, and are thus being used to study surfaces with atomic resolution and to process individual cells. The associated system tasks are new (e.g., manipulate nanoscopic objects), there are additional physical effects to be considered and understood (e.g., molecule to molecule interactions and atomic spins), and previously small phenomena can now dominate (e.g., surface effects, like surface tension, now surpass bulk phenomena, like gravity or momentum). Thus system control techniques must be modified or newly developed, as must the modeling, sensing, actuation, and real-time computation that supports them. As a result, the merging of control systems with micro- and nanoscale systems represents a new field that we expect to grow considerably over the coming decade. The intention of this book is to provide an introductory survey.

1.2 Critical Application Areas

There is a lot of diversity in the micro and nanoscale systems where control is playing a role. However, the majority of the applications fit within at least one of the five following groups: micro- and nanomanufacturing, instruments for nanoscale research, MEMS/NEMS, micro/nanofluidics, and quantum systems. This taxonomy has influenced the structure of this book and provides a starting point for finding the most important applications to pursue. Some of the applications and devices where control can play an important role are listed below. Only a small percentage of these applications have seen a concerted control implementation research effort. Therefore, there remain considerable opportunities for control practitioners to make important contributions to this field in the near and long term.

Micro- and Nanomanufacturing: Nanolithography including scanning probe and nanoimprint techniques, micro- and nanoassembly, directed self-assembly, nanoscale material deposition processes, nanoparticle growth, and formation of composite nanomaterials.

Instruments for Nanoscale Research: Scanning probe microscopy including atomic force microscopy, scanning tunneling microscopy, and near-field scanning optical microscopy. Particle trapping includes optical and magnetic trapping, particle tracking and localization.

MEMS/NEMS: MEMS/NEMS (micro/nano-electro-mechanical-systems) including inertial devices such as accelerometers and gyroscopes; micromirrors and other optomechanical components; filters, switches, and resonators for radiofrequency (RF) communications; probe-based data storage and hard drive read heads; biochemical sensors for medical diagnostics and threat detection; and micro- and nanorobots.

Micro/Nanofluidics: Micro/Nanofluidics include lab-on-a-chip technologies, inexpensive medical diagnostics, embedded drug delivery systems, and inkjet valves for high-volume printing.

Quantum Systems: Quantum systems include quantum computing, quantum communication and encryption, nuclear magnetic resonance imaging, and atom trapping and cooling.

1.3 Overview of the Chapters

The contributors to this book were chosen because they are leaders in their respective research areas and have either been able to demonstrate significant experimental results, or are well on their way towards experiments. It can take a long time to get from an initial control concept to an experimental demonstration: all the chosen contributors have been working on control of small systems for at least 5 to 10 years. Given that the field of micro/nanoscale systems is itself still fairly new (30+ years in the case of MEMS) and that there was a delay between the inception of the field and the subsequent entry of control researchers, in this sense, these contributors are at the leading edge. Of course, we could not include every major researcher at the intersection of controls and micro- and nanosystems, but we believe that we have chosen a representative sampling across a diverse set of applications that demonstrate how control is beginning to be applied on small length scales. We expect that there will be many more researchers in the future with many more exciting, and needed, applications and results.

The chapters have been organized along the lines of the application areas listed in the previous section: micro/nanomanufacturing, instruments for nanoscale research, MEMS, microfluidics, and quantum systems. Chapters 2 to 4 explore controlled manufacturing. Control of nanoparticle size during synthesis is presented in Chap. 2, and Chap. 3 discusses the estimation of nanoscale surface properties during manufacturing using optical measurements and Kalman filtering techniques. Automated assembly of two-dimensional structures composed of micro- and nanoparticles is presented in Chap. 4. The control of instruments for nanoscale research is discussed in Chaps. 5 and 6. Improving the imaging performance of atomic force microscopes using robust control is covered in Chap. 5 and the control of optically trapped particles is discussed in Chap. 6. The control of MEMS and microfluidic systems is the subject of Chaps. 7 through 9. Position control of MEMS actuators is presented in Chap. 7, closed-loop operation of precision MEMS gyroscopes is covered in Chap. 8, and the control of particle motion within a microfluidic system is presented in Chap. 9. Finally, quantum control is presented in Chap. 10 with an emphasis on controlling spin dynamics in quantum mechanical systems. In the final chapter, a review of some of the common challenges encountered throughout the book is presented along with prospects for future research in controlling micro- and nanoscale systems.

1.4 Notes for the Reader

This book was written for scientists and engineers in the fields of both micro/nanotechnologies and control systems with the intention of bridging the gap between the two. For the former group, it shows how control is being applied to miniaturized systems and highlights the benefits of feedback control for these systems. There is also an emphasis on the importance of interdisciplinary collaboration, physical modeling, control design mathematics, and experimentation in realizing these benefits. For the latter group, it provides an introduction to how control is currently being applied, extended, and developed for miniaturized systems. It also documents the need to fully understand the capabilities, requirements, and bottlenecks in new application areas before approaching the control design problem. It is our intention that this book provide an impetus for each group to better learn the technical language of the other – a requirement for successful collaboration between the two. But above all, our greatest hope is that it will spark new ideas and insights to enable better interactions between these two fields and result in significant advances in micro- and nanoscale systems.

Due to the multidisciplinary nature of this book, some background reading may be helpful. Readers who are not familiar with control theory can find an introduction written for a broad audience in [1] and practical 'fast-track' advice for implementing linear feedback controllers in [2]. More rigorous treatments of control theory are found in [3], a concise book that crucially describes not only what control can achieve for any given system but also what it cannot. Control theory is usually introduced in a linear system setting, where strong and comprehensive results are available, but there are also more advanced books that deal with control for nonlinear systems [4]. Nonlinear methods require a higher level of mathematical sophistication, but are needed in many real-world situations where nonlinearities cannot be neglected, as seen in several chapters in this book.

Readers not familiar with micro- and nanoscale systems can find an excellent introduction to microelectromechanical systems (MEMS) in [5–7]. The first of these reference includes, as its first chapter, the classic 1959 Feynman lecture 'There is Plenty of Room at the Bottom' [8]. There are also a number of books that introduce nanoscale science (e.g., [9, 10]) and nanotechnology [11, 12]. Texts relevant to the physics of micro- and nanoscale systems span the spectrum from optics and electronics to mechanics, fluid dynamics, and chemistry and biology. When faced with diving into a new field of physics and learning the basics, the Feynman lectures [13] are a fantastic resource. Each lecture provides a brilliant, concise, and accurate introduction to an entire field.

Finally, for both the controls and micro/nanoreaders, four fairly recent reports provide context for how control methods apply to novel systems in the areas of atomic force microscopy and nanorobotic manipulation [14]: MEMS, biological, chemical, and nanoscale systems [15, 16]; and networks of large and small systems, including aerospace, transportation, information technology, robotics, biology, medicine, and materials [17]. Many of the recommendations made in these reports are mirrored in the research and approaches described in this book.

References

1. R.M. Murray and K.J. Åström, *Feedback systems: An introduction for scientists and engineers*, Princeton University Press, Princeton, NJ, 2008.
2. A. Abramovici and J. Chapsky. *Feedback control systems: A fast-track guide for scientists and engineerings*, Kluwer, Norwell, MA, 2000.
3. J.C. Doyle, B.A. Francis, and A.R. Tannenbaum. *Feedback control theory*, Macmillan, New York, 1992.
4. A. Isidori. *Nonlinear control systems*, Springer, London, 1995.
5. W.S. Trimmer (editor). *Micromechanics and MEMS: Classic and seminal papers to 1990*, IEEE Press, New York, 1997.
6. N. Maluf. *An introduction to microelectromechanical systems engineering*, Artech House, Boston, MA, 2000.
7. C. Liu. *Foundations of MEMS*, Prentice-Hall, Englewood Cliffs, NJ, 2011.
8. R. Feynman. *There's plenty of room at the bottom. Caltech engineering and science magazine*, 23, 1960.
9. E.L. Wolf. *Nanophysics and nanotechnology: An introduction to modern concepts in nanoscience*, Wiley, Weinheim, Germany, 2006.
10. S. Lindsay. *Introduction to nanoscience*, Oxford University Press, New York, 2009.
11. B. Bhushan. *Springer handbook of nanotechnology*, Springer, New York, 2010.
12. A. Busnaina. *Nanomanufacturing handbook*, CRC Press, Boca Raton, FL, 2006.
13. R.P. Feynman, R.B. Leighton, and M. Sands. *The Feynman lectures on physics*, Addison-Wesley, Boston, MA , 1964.
14. M. Sitti. *NSF workshop on future directions in nano-scale systems, dynamics and control*, final report, 2003.
15. B. Shapiro. *NSF workshop on control and system integration of micro- and nano-scale systems*, final report, 2004. Available: http://www.isr.umd.edu/CMN-NSFwkshp/.
16. B. Shapiro. Workshop on control of micro- and nano-scale systems, *IEEE control systems magazine*, 25:82–88, 2005.
17. R.M. Murray (editor). *Control in an information rich world: Report of the panel on future directions in control, dynamics, and systems*. SIAM, Philadelphia, PA. 2003. Available: http://www.cds.caltech.edu/~murray/cdspanel.

Chapter 2
Feedback Control of Particle Size Distribution in Nanoparticle Synthesis and Processing

Mingheng Li and Panagiotis D. Christofides

2.1 Introduction

Particulate processes (also known as dispersed-phase processes) are characterized by the co-presence of and strong interaction between a continuous (gas or liquid) phase and a particulate (dispersed) phase and are essential in making many high-value industrial products. Particulate processes play a prominent role in a number of process industries since about 60% of the products in the chemical industry are manufactured as particulates with an additional 20% using powders as ingredients. Representative examples of particulate processes for micro- and nano-particle synthesis and processing include the crystallization of proteins for pharmaceutical applications [2], the emulsion polymerization of nano-sized latex particles [50], the aerosol synthesis of nanocrystalline catalysts [64], and thermal spray processing of nanostructured functional thermal barrier coatings to protect turbine blades [1]. The industrial importance of particulate processes and the realization that the physicochemical and mechanical properties of materials made with particulates depend heavily on the characteristics of the underlying particle-size distribution (PSD) have motivated significant research attention over the last ten years on model-based control of particulate processes. These efforts have also been complemented by recent and ongoing developments in measurement technology which allow the accurate and fast online measurement of key process variables including important characteristics of PSDs (e.g., [37,55,56]). The recent efforts on model-based control

M. Li (✉)
Department of Chemical and Materials Engineering, California State Polytechnic University, Pomona, CA 91768, USA
e-mail: minghengli@csupomona.edu

P.D. Christofides
Department of Chemical and Biomolecular Engineering, University of California, Los Angeles, CA 90095, USA
e-mail: pdc@seas.ucla.edu

J.J. Gorman and B. Shapiro (eds.), *Feedback Control of MEMS to Atoms*,
DOI 10.1007/978-1-4419-5832-7_2, © Springer Science+Business Media, LLC 2012

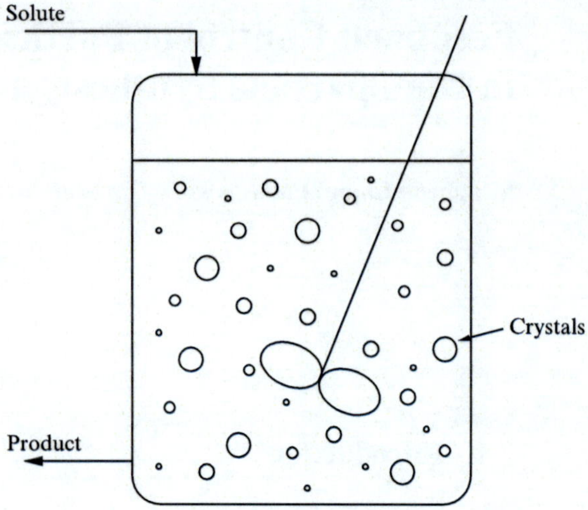

Fig. 2.1 Schematic of a continuous crystallizer

of particulate processes have also been motivated by significant advances in the physical modeling of highly coupled reaction-transport phenomena in particulate processes that cannot be easily captured through empirical modeling. Specifically, population balances have provided a natural framework for the mathematical modeling of PSDs in broad classes of particulate processes (see, for example, the tutorial article [30] and the review article [54]), and have been successfully used to describe PSDs in emulsion polymerization reactors (e.g., [13, 15]), crystallizers (e.g., [4, 55]), aerosol reactors (e.g., [23]), and cell cultures (e.g., [12]). To illustrate the structure of the mathematical models that arise in the modeling and control of particulate processes, we focus on three representative examples: continuous crystallization, batch crystallization, and aerosol synthesis.

2.1.1 *Continuous Crystallization*

Crystallization is a particulate process, which is widely used in industry for the production of many micro- or nano-sized products including fertilizers, proteins, and pesticides. A typical continuous crystallization process is shown in Fig. 2.1. Under the assumptions of isothermal operation, constant volume, well-mixed suspension, nucleation of crystals of infinitesimal size and mixed product removal, a dynamic model for the crystallizer can be derived from a population balance for the particle phase and a mass balance for the solute concentration and has the following mathematical form [32, 39]:

$$\begin{aligned} \frac{\partial n(r,t)}{\partial t} &= -\frac{\partial (R(t)n(r,t))}{\partial r} - \frac{n(r,t)}{\tau} + \delta(r-0)Q(t), \\ \frac{\mathrm{d}c(t)}{\mathrm{d}t} &= \frac{(c_0-\rho)}{\varepsilon(t)\tau} + \frac{(\rho - c(t))}{\tau} + \frac{(\rho - c(t))}{\varepsilon(t)}\frac{\mathrm{d}\varepsilon(t)}{\mathrm{d}t}, \end{aligned} \tag{2.1}$$

where $n(r,t)dr$ is the number of crystals in the size range of $[r, r+dr]$ at time t per unit volume of suspension, τ is the residence time, ρ is the density of the crystal, $c(t)$ is the solute concentration in the crystallizer, c_0 is the solute concentration in the feed, and

$$\varepsilon(t) = 1 - \int_0^\infty n(r,t)\frac{4}{3}\pi r^3 \mathrm{d}r$$

is the volume of liquid per unit volume of suspension. $R(t)$ is the crystal growth rate, $\delta(r-0)$ is the standard Dirac function, and $Q(t)$ is the crystal nucleation rate. The term $\delta(r-0)Q(t)$ accounts for the production of crystals of infinitesimal (zero) size via nucleation. An example of expressions of $R(t)$ and $Q(t)$ is the following:

$$R(t) = k_1(c(t) - c_s), \quad Q(t) = \varepsilon(t)k_2 \mathrm{e}^{-\frac{k_3}{(c(t)/c_s - 1)^2}}, \tag{2.2}$$

where k_1, k_2, and k_3 are constants and c_s is the concentration of solute at saturation. For a variety of operating conditions (see [6] for model parameters and detailed studies), the continuous crystallizer model of (2.1) exhibits highly oscillatory behavior (the main reason for this behavior is that the nucleation rate is much more sensitive to supersaturation relative to the growth rate – i.e., compare the dependence of $R(t)$ and $Q(t)$ on the values of $c(t)$ and c_s), which suggests the use of feedback control to ensure stable operation and attain a crystal size distribution (CSD) with desired characteristics. To achieve this control objective, the inlet solute concentration can be used as the manipulated input and the crystal concentration as the controlled and measured output.

2.1.2 Batch Protein Crystallization

Batch crystallization plays an important role in the pharmaceutical industry. We consider a batch crystallizer, which is used to produce tetragonal HEW (hen-egg-white) lysozyme crystals from a supersaturated solution [62]. A schematic of the batch crystallizer is shown in Fig. 2.2. Applying population, mass and energy balances to the process, the following mathematical model is obtained:

$$\begin{aligned}
&\frac{\partial n(r,t)}{\partial t} + G(t)\frac{\partial n(r,t)}{\partial r} = 0, \quad n(0,t) = \frac{B(t)}{G(t)},\\
&\frac{\mathrm{d}C(t)}{\mathrm{d}t} = -24\rho k_v G(t)\mu_2(t),\\
&\frac{\mathrm{d}T(t)}{\mathrm{d}t} = -\frac{UA}{MC_\mathrm{p}}(T(t) - T_j(t)),
\end{aligned} \tag{2.3}$$

Fig. 2.2 Schematic of a batch cooling crystallizer

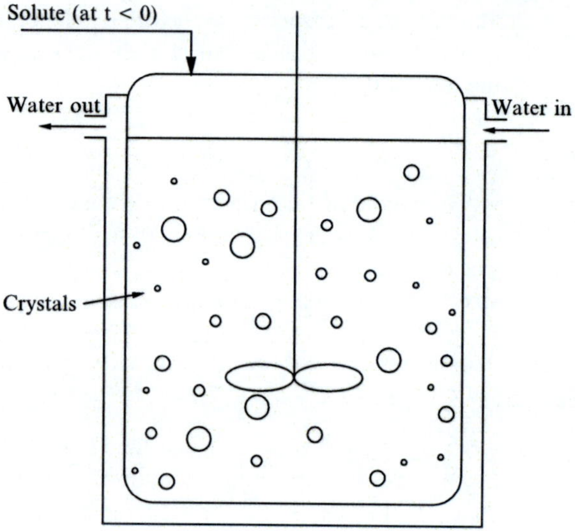

where $n(r,t)$ is the CSD, $B(t)$ is the nucleation rate, $G(t)$ is the growth rate, $C(t)$ is the solute concentration, $T(t)$ is the crystallizer temperature, $T_j(t)$ is the jacket temperature, ρ is the density of crystals, k_v is the volumetric shape factor, U is the overall heat-transfer coefficient, A is the total heat-transfer surface area, M is the mass of solvent in the crystallizer, C_p is the heat capacity of the solution, and $\mu_2(t) = \int_0^{\infty} r^2 n(r,t)dr$ is the second moment of the CSD. The nucleation rate, $B(t)$, and the growth rate, $G(t)$, are given by [62]:

$$B(t) = k_a C(t) \exp\left(-\frac{k_b}{\sigma^2(t)}\right), \quad G(t) = k_g \sigma^g(t), \tag{2.4}$$

where $\sigma(t)$, the supersaturation, is a dimensionless variable and is defined as $\sigma(t) = \ln(C(t)/C_s(T(t)))$, $C(t)$ is the solute concentration, g is the exponent relating growth rate to the supersaturation, and $C_s(T)$ is the saturation concentration of the solute, which is a nonlinear function of the temperature of the form:

$$C_s(T) = 1.0036 \times 10^{-3} T^3 + 1.4059 \times 10^{-2} T^2 - 0.12835 T + 3.4613. \tag{2.5}$$

The existing experimental results [68] show that the growth condition of tetragonal HEW lysozyme crystal is significantly affected by the supersaturation. Low supersaturation will lead to the cessation of the crystal growth. On the other hand, rather than forming tetragonal crystals, large amount of needle crystals will form when the supersaturation is too high. Therefore, a proper range of supersaturation is necessary to guarantee the product's quality. The jacket temperature, T_j, is manipulated to achieve the desired crystal shape and size distribution.

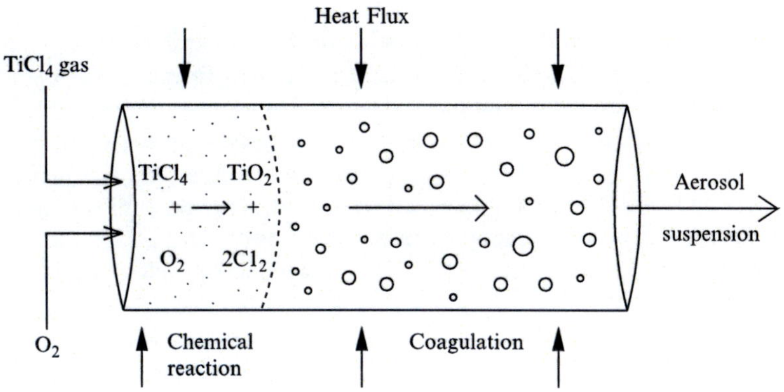

Fig. 2.3 Schematic of a titania aerosol reactor

2.1.3 Aerosol Synthesis

Aerosol processes are increasingly being used for the large-scale production of nano- and micron-sized particles. A typical aerosol flow reactor for the synthesis of titania aerosol with simultaneous chemical reaction, nucleation, condensation, coagulation, and convective transport is shown in Fig. 2.3. A general mathematical model, which describes the spatiotemporal evolution of the particle size distribution in such aerosol processes can be obtained from a population balance and consists of the following nonlinear partial integro-differential equation [33, 34]:

$$\frac{\partial n(v,z,t)}{\partial t} + v_z \frac{\partial n(v,z,t)}{\partial z} + \frac{\partial (G(\bar{x},v,z)n(v,z,t))}{\partial v} - I(v^*)\delta(v-v^*)$$
$$= \frac{1}{2}\int_0^v \beta(v-\bar{v},\bar{v},\bar{x})n(v-\bar{v},t)n(\bar{v},z,t)\mathrm{d}\bar{v} - n(v,z,t)\int_0^\infty \beta(v,\bar{v},\bar{x})n(\bar{v},z,t)\mathrm{d}\bar{v}, \quad (2.6)$$

where $n(v,z,t)$ denotes the particle size distribution function, v is the particle volume, t is the time, $z \in [0,L]$ is the spatial coordinate, L is the length scale of the process, v^* is the size of the nucleated aerosol particles, v_z is the velocity of the fluid, $\bar{x}$ is the vector of the state variables of the continuous phase, $G(\cdot,\cdot,\cdot), I(\cdot), \beta(\cdot,\cdot,\cdot)$ are nonlinear scalar functions which represent the growth, nucleation, and coagulation rates and $\delta(\cdot)$ is the standard Dirac function. The model of (2.6) is coupled with a mathematical model, which describes the spatiotemporal evolution of the concentrations of species and temperature of the gas phase $(\bar{x})$ that can be obtained from mass and energy balances. The control problem is to regulate process variables such as inlet flow rates and wall temperature to produce aerosol products with desired size distribution characteristics.

The mathematical models of (2.1), (2.3) and (2.6) demonstrate that particulate process models are nonlinear and distributed parameter in nature. These properties have motivated extensive research on the development of efficient numerical methods for the accurate computation of their solution (see, for example, [12, 23, 25, 38, 48, 54, 63]). However, in spite of the rich literature on population balance modeling, numerical solution, and dynamical analysis of particulate processes, up to about ten years ago, research on model-based control of particulate processes had been very limited. Specifically, early research efforts had mainly focused on the understanding of fundamental control-theoretic properties (controllability and observability) of population balance models [58] and the application of conventional control schemes (such as proportional-integral and proportional-integral-derivative control, self-tuning control) to crystallizers and emulsion polymerization processes (see, for example, [13, 57, 59] and the references therein). The main difficulty in synthesizing nonlinear model-based feedback controllers for particulate processes is the distributed parameter nature of the population balance models, which does not allow their direct use for the synthesis of low-order (and therefore, practically implementable) model-based feedback controllers. Furthermore, a direct application of the aforementioned solution methods to particulate process models leads to finite dimensional approximations of the population balance models (i.e., nonlinear ordinary differential equation (ODE) systems in time) which are of very high order, and thus inappropriate for the synthesis of model-based feedback controllers that can be implemented in realtime. This limitation had been the bottleneck for model-based synthesis and real-time implementation of model-based feedback controllers on particulate processes.

2.2 Model-Based Control of Particulate Processes

2.2.1 *Overview*

Motivated by the lack of population balance-based control methods for particulate processes and the need to achieve tight size distribution control in many particulate processes, we developed, over the last ten years, a general framework for the synthesis of nonlinear, robust, and predictive controllers for particulate processes based on population balance models [6–9, 16, 33, 35, 60, 62]. Specifically, within the developed framework, nonlinear low-order approximations of the particulate process models are initially derived using order reduction techniques and are used for controller synthesis. Subsequently, the infinite-dimensional closed-loop system stability, performance and robustness properties were precisely characterized in terms of the accuracy of the approximation of the low-order models. Furthermore, controller designs were proposed that deal directly with the key practical issues of uncertainty in model parameters, unmodeled actuator/sensor dynamics and constraints in the capacity of control actuators and the magnitude of the process state variables. It is also important to note that owing to the low-dimensional structure of the controllers, the computation of the control action involves the solution of a small set of ODEs, and thus, the developed controllers can be readily

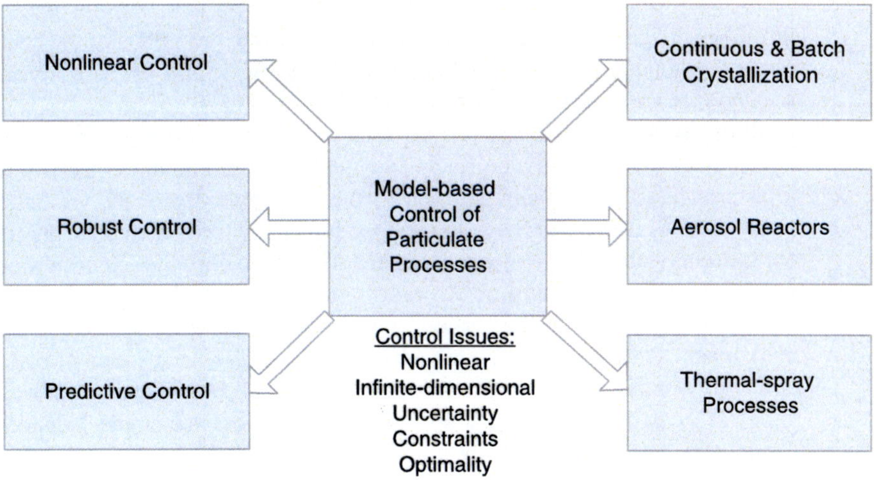

Fig. 2.4 Summary of our research on model-based control of particulate processes

implemented in realtime with reasonable computing power, thereby resolving the main issue on model-based control of particulate processes. In addition to theoretical developments, we also successfully demonstrated the application of the proposed methods to size distribution control in continuous and batch crystallization, aerosol, and thermal spray processes and documented their effectiveness and advantages with respect to conventional control methods. Figure 2.4 summarizes these efforts. The reader may refer to [4, 12, 15] for recent reviews of results on simulation and control of particulate processes.

2.2.2 *Particulate Process Model*

To present the main elements of our approach to model-based control of particulate processes, we focus on a general class of spatially homogeneous particulate processes with simultaneous particle growth, nucleation, agglomeration, and breakage. Examples of such processes have been introduced in the previous section. Assuming that particle size is the only internal particle coordinate and applying a dynamic material balance on the number of particles of size r to $r+dr$ (population balance), we obtain the following general nonlinear partial integro-differential equation, which describes the rate of change of the PSD, $n(r,t)$:

$$\frac{\partial n}{\partial t} = -\frac{\partial (G(x,r)n)}{\partial r} + w(n,x,r), \tag{2.7}$$

where $n(r,t)$ is the particle number size distribution, $r \in [0, r_{\max}]$ is the particle size, and $r_{\max}$ is the maximum particle size (which may be infinity), t is the time and

$x \in \mathbb{R}^n$ is the vector of state variables, which describe properties of the continuous phase (for example, solute concentration, temperature, and pH in a crystallizer); see (2.8) for the system that describes the dynamics of x. $G(x,r)$ and $w(n,x,r)$ are nonlinear scalar functions whose physical meaning can be explained as follows: $G(x,r)$ accounts for particle growth through condensation and is usually referred to as growth rate. It usually depends on the concentrations of the various species present in the continuous phase, the temperature of the process, and the particle size. On the other hand, $w(n,x,r)$ represents the net rate of introduction of new particles into the system. It includes all the means by which particles appear or disappear within the system including particle agglomeration (merging of two particles into one), breakage (division of one particle to two) as well as nucleation of particles of size $r \geq 0$ and particle feed and removal. The rate of change of the continuous-phase variables x can be derived by a direct application of mass and energy balances to the continuous phase and is given by a nonlinear integro-differential equation system of the general form:

$$\dot{x} = f(x) + g(x)u(t) + A\int_0^{r_{\max}} a(n,r,x)\mathrm{d}r, \tag{2.8}$$

where $f(x)$ and $a(n,r,x)$ are nonlinear vector functions, $g(x)$ is a nonlinear matrix function, A is a constant matrix and $u(t) = [u_1\ u_2\ \cdots\ u_m] \in \mathbb{R}^m$ is the vector of manipulated inputs. The term $A\int_0^{r_{\max}} a(n,r,x)\mathrm{d}r$ accounts for mass and heat transfer from the continuous phase to all the particles in the population (see [8] for details).

2.2.3 *Model Reduction of Particulate Process Models*

While the population balance models are infinite dimensional systems, the dominant dynamic behavior of many particulate process models has been shown to be low dimensional. Manifestations of this fundamental property include the occurrence of oscillatory behavior in continuous crystallizers [32] and the ability to capture the long-term behavior of aerosol systems with self-similar solutions [23]. Motivated by this, we introduced a general methodology for deriving low-order ODE systems that accurately reproduce the dominant dynamics of the nonlinear integro-differential equation system of (2.7) and (2.8) [6]. The proposed model reduction methodology exploits the low-dimensional behavior of the dominant dynamics of the system of (2.7) and (2.8) and is based on a combination of the method of weighted residuals with the concept of approximate inertial manifolds.

Specifically, the proposed approach initially employs the method of weighted residuals (see [54] for a comprehensive review of results on the use of this method for solving population balance equations) to construct a nonlinear, possibly high-order, ODE system that accurately reproduces the solutions and dynamics of the distributed parameter system of (2.7) and (2.8). We first consider an orthogonal set of basis functions $\phi_k(r)$, where $r \in [0, r_{\max})$, $k = 1, \ldots, \infty$, and expand the particle size distribution function $n(r,t)$ in an infinite series in terms of $\phi_k(r)$ as follows:

$$n(r,t) = \sum_{k=1}^{\infty} a_k(t)\phi_k(r), \tag{2.9}$$

where $a_k(t)$ are time-varying coefficients. In order to approximate the system of (2.7) and (2.8) with a finite set of ODEs, we obtain a set of N equations by substituting (2.9) into (2.7) and (2.8), multiplying the population balance with N different weighting functions $\psi_\nu(r)$ (that is, $\nu = 1, \ldots, N$), and integrating over the entire particle size spectrum. In order to obtain a finite dimensional model, the series expansion of $n(r,t)$ is truncated up to order N. The infinite dimensional system of (2.7) reduces to the following finite set of ODEs:

$$\int_0^{r_{\max}} \psi_\nu(r) \sum_{k=1}^{N} \phi_k(r) \frac{\partial a_{kN}(t)}{\partial t} \mathrm{d}r = \sum_{k=1}^{N} a_{kN}(t) \int_0^{r_{\max}} \psi_\nu(r) \frac{\partial (G(x_N, r)\phi_k(r))}{\partial r} \mathrm{d}r,$$

$$+ \int_0^{r_{\max}} \psi_\nu(r) w\left(\sum_{k=1}^{N} a_{kN}(t)\phi_k(r), x_N, r \right) \mathrm{d}r, \;\; \nu = 1, \ldots, N$$

$$\dot{x}_N = f(x_N) + g(x_N)u(t) + A \int_0^{r_{\max}} a\left(\sum_{k=1}^{N} a_{kN}(t)\phi_k(r), r, x_N \right) \mathrm{d}r, \tag{2.10}$$

where x_N and a_{kN} are the approximations of x and a_k obtained by an N-th order truncation. From (2.10), it is clear that the form of the ODEs that describe the rate of change of $a_{kN}(t)$ depends on the choice of the basis and weighting functions, as well as on N. The system of (2.10) was obtained from a direct application of the method of weighted residuals (with arbitrary basis functions) to the system of (2.7) and (2.8), and thus, may be of very high order in order to provide an accurate description of the dominant dynamics of the particulate process model. High-dimensionality of the system of (2.10) leads to complex controller design and high-order controllers, which cannot be readily implemented in practice. To circumvent these problems, we exploited the low-dimensional behavior of the dominant dynamics of particulate processes and proposed an approach based on the concept of inertial manifolds to derive low-order ODE systems that accurately describe the dominant dynamics of the system of (2.10) [6]. This order reduction technique initially employs singular perturbation techniques to construct nonlinear approximations of the modes neglected in the derivation of the finite dimensional model of (2.10) (i.e., modes of order $N+1$ and higher) in terms of the first N modes. Subsequently, these steady-state expressions for the modes of order $N+1$ and higher (truncated up to appropriate order) are used in the model of (2.10) (instead of setting them to zero) and significantly improve the accuracy of the model of (2.10) without increasing its dimension; details on this procedure can be found in [6].

It is important to note that the method of weighted residuals reduces to the method of moments when the basis functions are chosen to be Laguerre polynomials and the weighting functions are chosen as $\psi_\nu = r^\nu$. The moments of the particle size distribution are defined as:

$$\mu_\nu = \int_0^\infty r^\nu n(r,t)\mathrm{d}r, \quad \nu = 0,\ldots,\infty \tag{2.11}$$

and the moment equations can be directly generated from the population balance model by multiplying it by r^ν, $\nu = 0,\ldots,\infty$ and integrating from 0 to ∞. The procedure of forming moments of the population balance equation very often leads to terms that may not reduce to moments, terms that include fractional moments, or to an unclosed set of moment equations. To overcome this problem, the particle size distribution may be expanded in terms of Laguerre polynomials defined in $L_2[0,\infty)$ and the series solution using a finite number of terms may be used to close the set of moment equations (this procedure has been successfully used for models of crystallizers with fine traps used to remove small crystals [7]).

2.2.4 *Model-Based Control Using Low-Order Models*

2.2.4.1 Nonlinear Control

Low-order models can be constructed using the techniques described in the previous section. We describe an application to the continuous crystallization process of Sect. 2.1.1. First, the method of moments is used to derive the following infinite-order dimensionless system from (2.1) for the continuous crystallization process:

$$\begin{aligned}
\frac{\mathrm{d}\tilde{x}_0}{\mathrm{d}t} &= -\tilde{x}_0 + (1-\tilde{x}_3)Da e^{-F/\tilde{y}^2},\\
\frac{\mathrm{d}\tilde{x}_1}{\mathrm{d}t} &= -\tilde{x}_1 + \tilde{y}\tilde{x}_0,\\
\frac{\mathrm{d}\tilde{x}_2}{\mathrm{d}t} &= -\tilde{x}_2 + \tilde{y}\tilde{x}_1,\\
\frac{\mathrm{d}\tilde{x}_3}{\mathrm{d}t} &= -\tilde{x}_3 + \tilde{y}\tilde{x}_2,\\
\frac{\mathrm{d}\tilde{x}_\nu}{\mathrm{d}t} &= -\tilde{x}_\nu + \tilde{y}\tilde{x}_{\nu-1}, \quad \nu = 4,5,6\ldots,\\
\frac{\mathrm{d}\tilde{y}}{\mathrm{d}t} &= \frac{1-\tilde{y}-(\alpha-\tilde{y})\tilde{y}\tilde{x}_2}{1-\tilde{x}_3},
\end{aligned} \tag{2.12}$$

where $\tilde{x}_i$ and $\tilde{y}$ are the dimensionless i-th moment and solute concentration, respectively, and Da and F are dimensionless parameters [6]. On the basis of the system of (2.12), it is clear that the moments of order four and higher do not affect those of order three and lower, and moreover, the state of the infinite dimensional system:

$$\frac{\mathrm{d}\tilde{x}_\nu}{\mathrm{d}t} = -\tilde{x}_\nu + \tilde{y}\tilde{x}_{\nu-1}, \ \nu = 4, \ldots, \tag{2.13}$$

is bounded when x_3 and y are bounded, and it converges to a globally exponentially stable equilibrium point when $\lim_{t\to\infty} x_3 = c_1$ and $\lim_{t\to\infty} \tilde{y} = c_2$, where c_1 and c_2 are constants. This implies that the dominant dynamics (that is, dynamics associated with eigenvalues that are close to the imaginary axis) of the process of (2.1) can be adequately captured by the following fifth-order moment model:

$$\begin{aligned}
\frac{\mathrm{d}\tilde{x}_0}{\mathrm{d}t} &= -\tilde{x}_0 + (1-\tilde{x}_3)Da\mathrm{e}^{-F/\tilde{y}^2}, \\
\frac{\mathrm{d}\tilde{x}_1}{\mathrm{d}t} &= -\tilde{x}_1 + \tilde{y}\tilde{x}_0, \\
\frac{\mathrm{d}\tilde{x}_2}{\mathrm{d}t} &= -\tilde{x}_2 + \tilde{y}\tilde{x}_1, \\
\frac{\mathrm{d}\tilde{x}_3}{\mathrm{d}t} &= -\tilde{x}_3 + \tilde{y}\tilde{x}_2, \\
\frac{\mathrm{d}\tilde{y}}{\mathrm{d}t} &= \frac{1-\tilde{y}-(\alpha-\tilde{y})\tilde{y}\tilde{x}_2}{1-\tilde{x}_3}.
\end{aligned} \tag{2.14}$$

The ability of the above fifth-order moment model to reproduce the dynamics, and to some extent the solutions, of the distributed parameter model of (2.1) is shown in Fig. 2.5, where the profiles of the total particle concentration generated by the two models are compared (both models start from the same initial conditions). Even though the discrepancy of the total particle concentration profiles predicted by the two models increases with time (this is expected due to the open-loop instability of the process), it is clear that the fifth-order moment model of (2.14) provides a very good approximation of the distributed parameter model of (2.1), thereby establishing that the dominant dynamics of the system of (2.1) are low dimensional and motivating the use of the moment model for nonlinear controller design.

For the batch crystallization process, the following low-order model can be derived from (2.3) using the method of moments:

$$\begin{aligned}
\frac{\mathrm{d}\mu_0}{\mathrm{d}t} &= \left(1-\frac{4}{3}\pi\mu_3\right)k_2\mathrm{e}^{-\frac{k_3}{(c/c_s-1)^2}}\mathrm{e}^{-\frac{E_b}{RT}}, \\
\frac{\mathrm{d}\mu_1}{\mathrm{d}t} &= k_1(c-c_s)\mathrm{e}^{-\frac{E_g}{RT}}\mu_0, \\
\frac{\mathrm{d}\mu_2}{\mathrm{d}t} &= 2k_1(c-c_s)\mathrm{e}^{-\frac{E_g}{RT}}\mu_1,
\end{aligned}$$

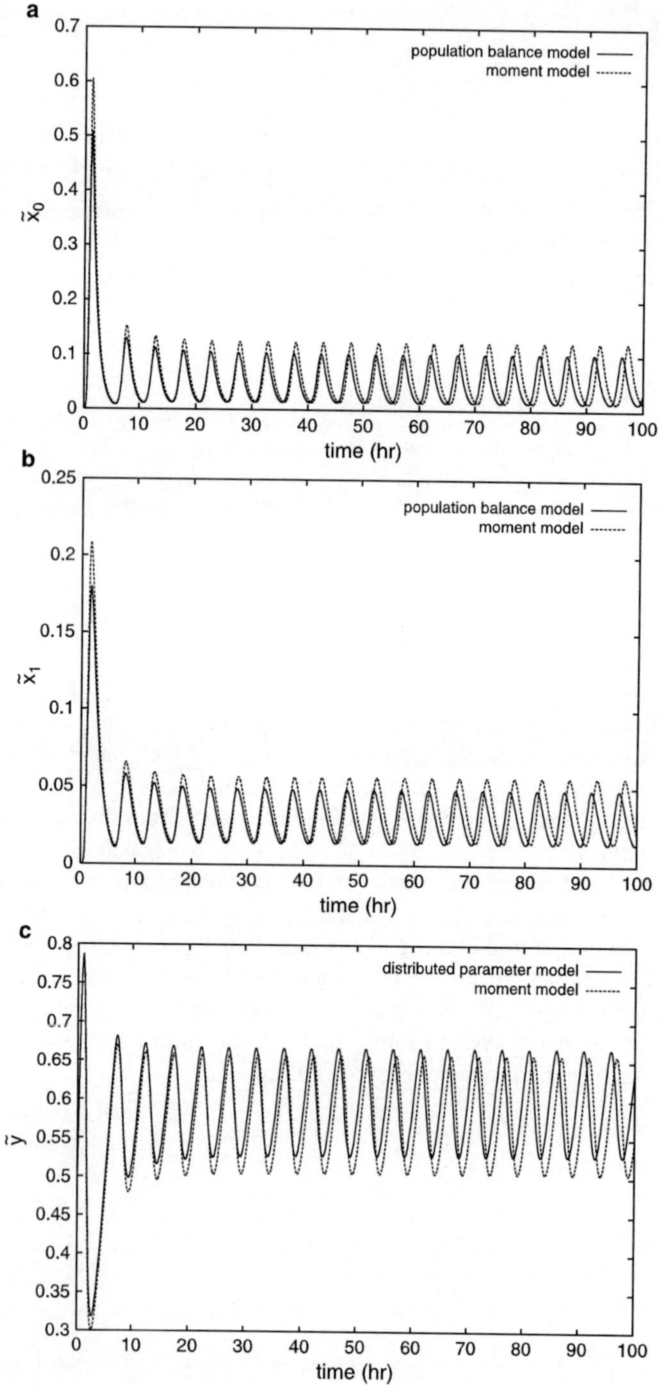

Fig. 2.5 Comparison of open-loop profiles of (**a**) crystal concentration, (**b**) total crystal size, and (**c**) solute concentration obtained from the distributed parameter model and the moment model

$$\frac{\mathrm{d}\mu_3}{\mathrm{d}t} = 3k_1(c-c_s)\mathrm{e}^{-\frac{E_g}{RT}}\mu_2,$$
$$\frac{\mathrm{d}c}{\mathrm{d}t} = \frac{-4\pi(c-c_s)\mu_2(\rho-c)}{(1-\frac{4}{3}\pi\mu_3)},$$
$$\frac{\mathrm{d}T}{\mathrm{d}t} = -\frac{\rho_c\Delta H_c}{\rho C_\mathrm{p}}4\pi k_1(c-c_s)\mathrm{e}^{-\frac{E_g}{RT}}\mu_2 - \frac{UA_c}{\rho C_\mathrm{p}V}(T-T_c), \tag{2.15}$$

where E_g and E_b denote the activation energies for growth and nucleation, respectively. The objective is to control the interplay between the particle nucleation and growth rates such that a CSD with a larger average particle size is obtained at the end of the batch run by manipulating the cooling water temperature.

Based on the low-order models, nonlinear finite-dimensional state and output feedback controllers have been synthesized that guarantee stability and enforce output tracking in the closed-loop finite dimensional system. It has also been established that these controllers exponentially stabilize the closed-loop particulate process model. The output feedback controller is constructed through a standard combination of the state feedback controller with a state observer. Specifically, in the case of the continuous crystallization example, the nonlinear output feedback controller has the following form:

$$\frac{\mathrm{d}\omega_0}{\mathrm{d}t} = -\omega_0 + (1-\omega_3)Da\mathrm{e}^{-F/\omega_4^2} + L_0(\tilde{h}(\tilde{x}) - \tilde{h}(\omega)),$$
$$\frac{\mathrm{d}\omega_1}{\mathrm{d}t} = -\omega_1 + \omega_4\omega_0 + L_1(\tilde{h}(\tilde{x}) - \tilde{h}(\omega)),$$
$$\frac{\mathrm{d}\omega_2}{\mathrm{d}t} = -\omega_2 + \omega_4\omega_1 + L_2(\tilde{h}(\tilde{x}) - \tilde{h}(\omega)),$$
$$\frac{\mathrm{d}\omega_3}{\mathrm{d}t} = -\omega_3 + \omega_4\omega_2 + L_3(\tilde{h}(\tilde{x}) - \tilde{h}(\omega)),$$
$$\frac{\mathrm{d}\omega_4}{\mathrm{d}t} = \frac{1-\omega_4-(\alpha-\omega_4)\omega_4\omega_2}{1-\omega_3} + L_4(\tilde{h}(\tilde{x}) - \tilde{h}(\omega))$$
$$+\frac{[\beta_2 L_{\tilde{g}} L_{\tilde{f}}\tilde{h}(\omega)]^{-1}\left\{v - \beta_0\tilde{h}(\omega) - \beta_1 L_{\tilde{f}}\tilde{h}(\omega) - \beta_2 L_{\tilde{f}}^2\tilde{h}(\omega)\right\}}{1-\omega_3},$$
$$\bar{u}(t) = [\beta_2 L_{\tilde{g}} L_{\tilde{f}}\tilde{h}(\omega)]^{-1}\left\{v - \beta_0\tilde{h}(\omega) - \beta_1 L_{\tilde{f}}\tilde{h}(\omega) - \beta_2 L_{\tilde{f}}^2\tilde{h}(\omega)\right\}, \tag{2.16}$$

where v is the set-point, β_0, β_1, β_2 and $L = [L_0\ L_1\ L_2\ L_3\ L_4]^T$ are controller parameters and $\tilde{h}(\omega) = \omega_0$ or $\tilde{h}(\omega) = \omega_1$.

The nonlinear controller of (2.16) was also combined with a PI controller (that is, the term $v - \beta_0\tilde{h}(\omega)$ was substituted by $v - \beta_0\tilde{h}(\tilde{x}) + \frac{1}{\tau_i'}\xi$, where $\dot{\xi} = v - \tilde{h}(\tilde{x})$, $\xi(0) = 0$ and τ_i' is the integral time constant) to ensure offsetless tracking in the presence of constant uncertainty in process parameters. The practical implementation of the nonlinear controller of (2.16) requires online measurements of the controlled outputs $\tilde{x}_0$ or $\tilde{x}_1$; in practice, such measurements can be obtained by using, for example, light scattering [3, 55]. In (2.16), the feedback controller is synthesized via geometric control methods and the state observer is an extended Luenberger-type observer [6].

Several simulations have been performed in the context of the continuous crystallizer process model presented before to evaluate the performance and robustness properties of the nonlinear controllers designed based on the reduced order models, and to compare them with the ones of a PI controller. In all the simulation runs, the initial condition:

$$n(r,0) = 0.0,\ c(0) = 990.0\ \text{kg/m}^3$$

was used for the process model of (2.1) and (2.2) and the finite difference method with 1,000 discretization points was used for its simulation. The crystal concentration, $\tilde{x}_0$, was considered to be the controlled output and the inlet solute concentration was chosen to be the manipulated input. Initially, the set-point tracking capability of the nonlinear controller was evaluated under nominal conditions for a 0.5 increase in the value of the set-point.

Figure 2.6 shows the closed-loop output (left plot) and manipulated input (right plot) profiles obtained by using the nonlinear controller (solid lines). For the sake of comparison, the corresponding profiles under proportional-integral (PI) control are also included (dashed lines); the PI controller was tuned so that the closed-loop output response exhibits the same level of overshoot to the one of the closed-loop output under non-linear control. Clearly, the nonlinear controller drives the controlled output to its new set-point value in a significantly shorter time than the one required by the PI controller, while both controlled outputs exhibit very similar overshoot. For the same simulation run, the evolution of the closed-loop profile and the final steady-state profile of the CSD are shown in Fig. 2.7. An exponentially decaying CSD is obtained at the steady state. The reader may refer to [6] for extensive simulation results.

2.2.4.2 Hybrid Predictive Control

In addition to handling nonlinear behavior, an important control problem is to stabilize the crystallizer at an unstable steady-state (which corresponds to a desired PSD) using constrained control action. Currently, the achievement of high performance, under control and state constraints, relies to a large extent on the use of model predictive control (MPC) policies. In this approach, a model of the process is used to make predictions of the future process evolution and compute control actions, through repeated solution of constrained optimization problems, which ensure that the process state variables satisfy the imposed limitations. However, the ability of the available model predictive controllers to guarantee closed-loop stability and enforce constraint satisfaction is dependent on the assumption of

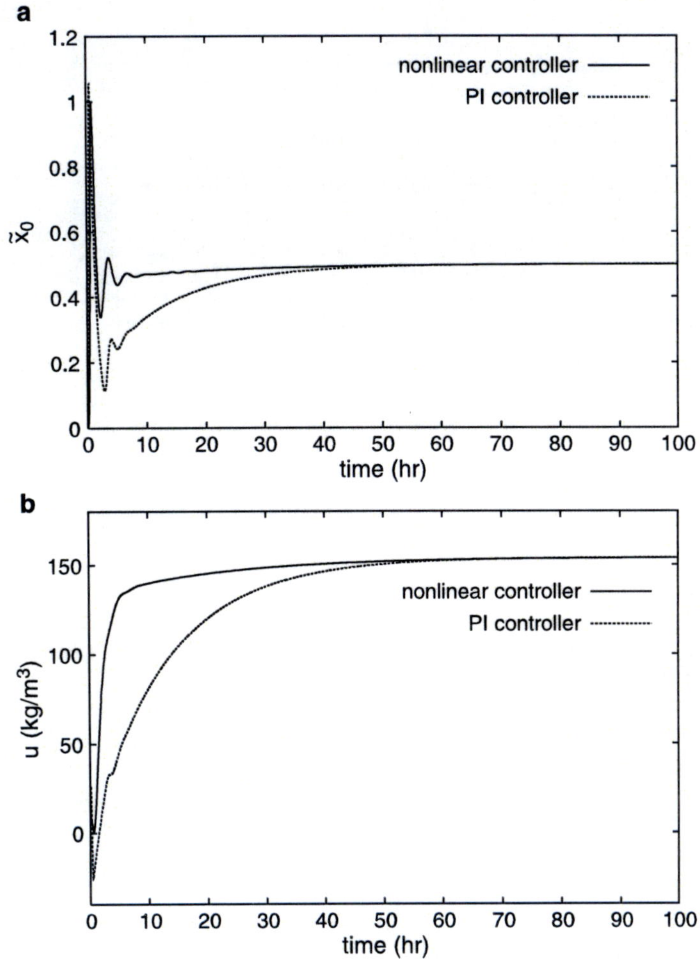

Fig. 2.6 (**a**) Closed-loop output and (**b**) manipulated input profiles under nonlinear and PI control, for a 0.5 increase in the set-point ($\tilde{x}_0$ is the controlled output) [6]

feasibility (i.e., existence of a solution) of the constrained optimization problem. This limitation strongly impacts the practical implementation of the MPC policies and limits the a priori (i.e., before controller implementation) characterization of the set of initial conditions starting from where the constrained optimization problem is feasible and closed-loop stability is guaranteed. This problem typically results in the need for extensive closed-loop simulations and software verification (before online implementation) to search over the whole set of possible initial operating conditions that guarantee stability. This in turn can lead to prolonged periods for plant commissioning. Alternatively, the lack of a priori knowledge of the stabilizing initial conditions may necessitate limiting process operation within a

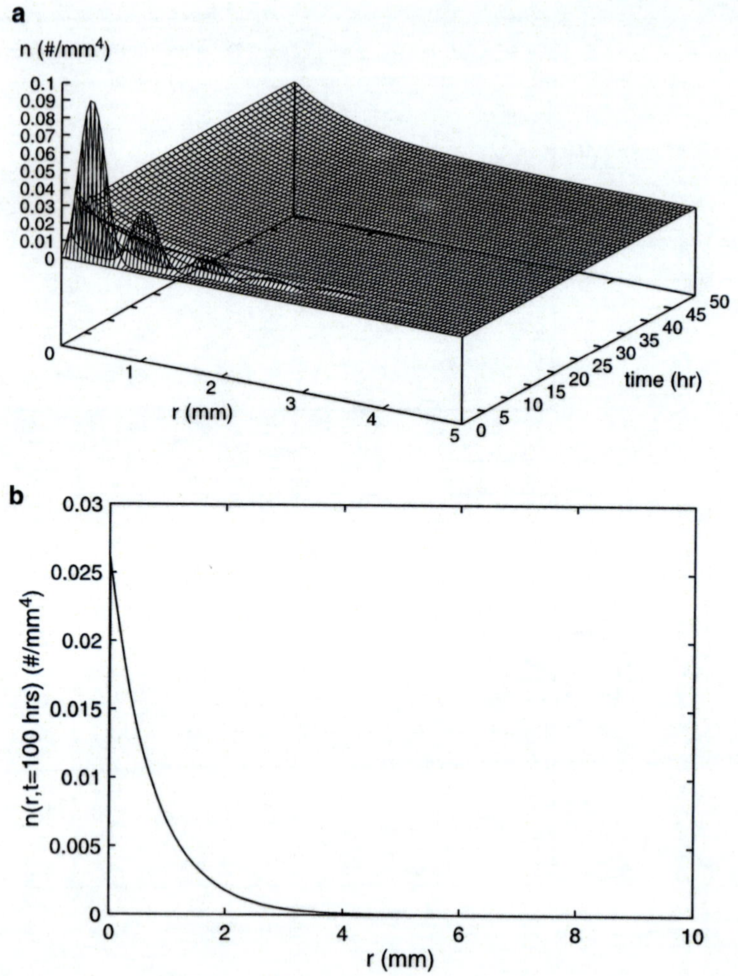

Fig. 2.7 Profile of evolution of crystal size distribution (*top*) and final steady-state crystal size distribution (*bottom*) under nonlinear control ($\tilde{x}_0$ is the controlled output) [6]

small conservative neighborhood of the desired set-point in order to avoid extensive testing and simulations. Given the tight product quality specifications, however, both of these two remedies can impact negatively on the efficiency and profitability of the process by limiting its operational flexibility. Lyapunov-based analytical control designs allow for an explicit characterization of the constrained stability region [17, 18, 47]; however, their closed-loop performance properties cannot be transparently characterized.

To overcome these difficulties, we recently developed [20] a hybrid predictive control structure that provides a safety net for the implementation of predictive control algorithms. The central idea is to embed the implementation of MPC within the stability region of a bounded controller and devise a set of switching rules that orchestrate the transition from MPC to the bounded controller in the event that MPC is unable to achieve closed-loop stability (e.g., due to inappropriate choice of the horizon length, infeasibility, or computational difficulties). Switching between the two controllers allows reconciling the tasks of optimal stabilization of the constrained closed-loop system (through MPC) with that of computing a priori the set of initial conditions for which closed-loop stability is guaranteed (through Lyapunov-based [17, 18] bounded nonlinear control).

We demonstrated the application of the hybrid predictive control strategy to the continuous crystallizer of (2.1) and (2.2). The control objective was to suppress the oscillatory behavior of the crystallizer and stabilize it at an unstable steady state that corresponds to a desired PSD by manipulating the inlet solute concentration. To achieve this objective, measurements or estimates of the first four moments and of the solute concentration are assumed to be available. Subsequently, the proposed methodology was employed for the design of the controllers using a low-order model constructed by using the method of moments. We compared the hybrid predictive control scheme, with an MPC controller designed with a set of stabilizing constraints and a Lyapunov-based nonlinear controller.

In the first set of simulation runs, we tested the ability of the MPC controller with the stability constraints to stabilize the crystallizer starting from the initial condition $x(0) = [0.066\ 0.041\ 0.025\ 0.015\ 0.560]'$, corresponding to the dimensionless moments of the CSD as well as the dimensionless concentration of the solute in the crystallizer [60]. The result is shown by the solid lines in Fig. 2.8a–e where it is seen that the predictive controller, with a horizon length of $T = 0.25$, is able to stabilize the closed-loop system at the desired equilibrium point. Starting from the initial condition $x(0) = [0.033\ 0.020\ 0.013\ 0.0075\ 0.570]'$, however, the MPC controller with the stability constraints yields no feasible solution. If the stability constraints are relaxed to make the MPC feasible, we see from the dashed lines in Fig. 2.8a–e that the resulting control action cannot stabilize the closed-loop system, and leads to a stable limit cycle. On the other hand, the bounded controller is able to stabilize the system from both initial conditions (this was guaranteed because both initial conditions lied inside the stability region of the controller). The state trajectory starting from $x(0) = [0.033\ 0.020\ 0.013\ 0.0075\ 0.570]'$ is shown in Fig. 2.8a–e with the dotted profile. This trajectory, although stable, presents slow convergence to the equilibrium as well as a damped oscillatory behavior that the MPC does not show when it is able to stabilize the system.

When the hybrid predictive controller is implemented from the initial condition $x(0) = [0.033\ 0.020\ 0.013\ 0.0075\ 0.570]'$, the supervisor detects initial infeasibility of MPC and implements the bounded controller in the closed loop. As the closed-loop states evolve under the bounded controller and get closer to the desired steady state, the supervisor finds (at $t = 5.8$ h) that the MPC becomes feasible and, therefore, implements it for all future times. Note that despite the "jump" in the control action profile as we switch from the bounded controller to MPC at $t = 5.8$ h, (see the difference between dotted and dash-dotted profiles in Fig. 2.8f), the

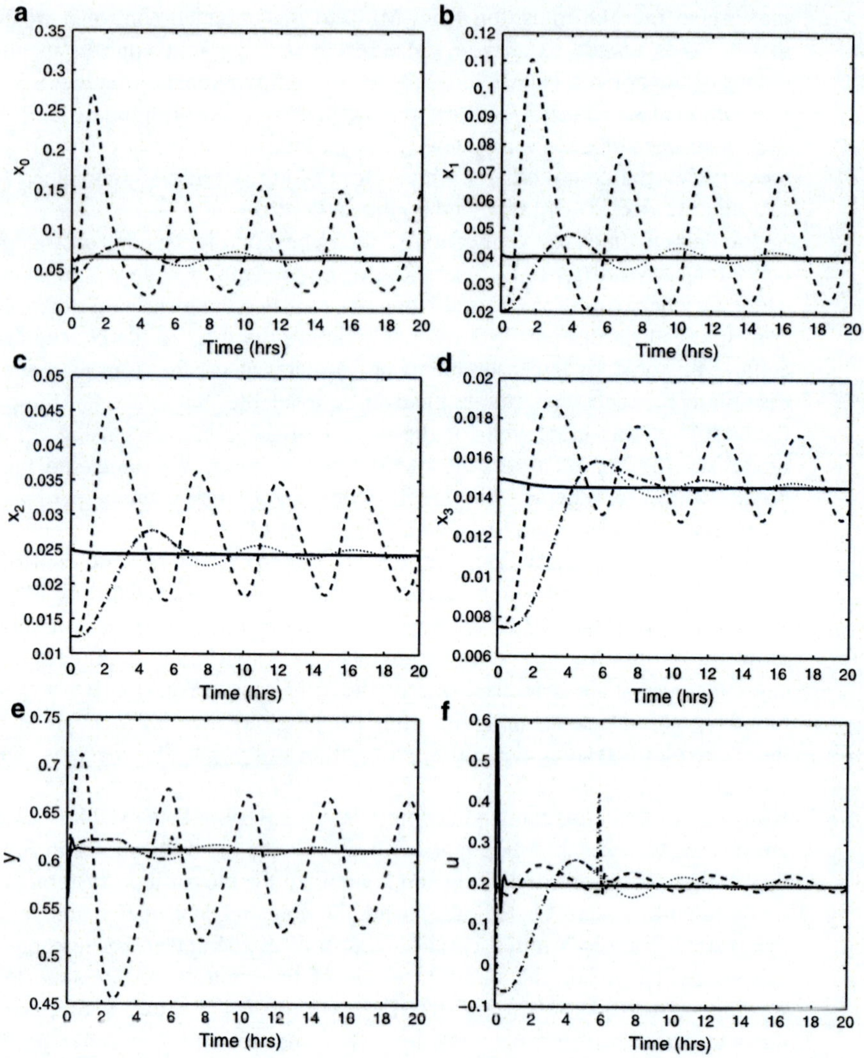

Fig. 2.8 Continuous crystallizer example: closed-loop profiles of the dimensionless crystallizer moments (**a–d**), the solute concentration in the crystallizer (**e**) and the manipulated input (**f**) under MPC with stability constraints (*solid lines*), under MPC without stability constraints (*dashed lines*), under the bounded controller (*dotted lines*), and using the hybrid predictive controller (*dash-dotted lines*) [60]. Note the different initial states

moments of the PSD in the crystallizer continue to evolve smoothly (dash-dotted lines in Fig. 2.8a–e). The supervisor finds that MPC continues to be feasible and is implemented in closed-loop to stabilize the closed-loop system at the desired steady state. Compared with the simulation results under the bounded controller, the hybrid predictive controller (dash-dotted lines) stabilizes the system much faster, and achieves a better performance, reflected in a lower value of the performance index (0.1282 *vs* 0.1308). The manipulated input profiles for the three scenarios are shown in Fig. 2.8f.

2.2.4.3 Predictive Control of Size Distribution in a Batch Protein Crystallizer

In batch crystallization, the main objective is to achieve a desired particle size distribution at the end of the batch and to satisfy state and control constraints during the whole batch run. Significant previous work has focused on CSD control in batch crystallizers, e.g., [55,70]. In [52], a method was developed for assessing parameter uncertainty and studied its effects on the open-loop optimal control strategy, which maximized the weight mean size of the product. To improve the product quality expressed in terms of the mean size and the width of the distribution, an online optimal control methodology was developed for a seeded batch cooling crystallizer [72]. In these previous works, most efforts were focused on the open-loop optimal control of the batch crystallizer, i.e., the optimal operating condition was calculated offline based on mathematical models. The successful application of such a control strategy relies, to a large extent, on the accuracy of the models. Furthermore, an open-loop control strategy may not be able to manipulate the system to follow the optimal trajectory because of the ubiquitous existence of modeling error. Motivated by this, we developed a predictive feedback control system to maximize the volume-averaged tetragonal lysozyme crystal size (i.e., μ_4/μ_3 where μ_3, μ_4 are the third and fourth moments of the CSD; see (2.11)) by manipulating the jacket temperature, T_j [60]. The principle moments are calculated from the online measured CSD, n, which can be obtained by measurement techniques such as the laser light scattering method. The concentration and crystallizer temperature are also assumed to be measured in real time. In the closed-loop control structure, a reduced-order moments model was used within the predictive controller for the purpose of prediction. The main idea is to use this model to obtain a prediction of the state of the process at the end of the batch operation, t_f, from the current measurement at time t. Using this prediction, a cost function that depends on this value is minimized subject to a set of operating constraints. Manipulation input limitations and constraints on supersaturation and crystallizer temperature are incorporated as input and state constraints on the optimization problem. The optimization algorithm computes the profile of the manipulated input T_j from the current time until the end of the batch operation interval, then the current value of the computed input is implemented on the process, and the optimization problem is resolved and the input is updated each time a new measurement is available (receding horizon control strategy). The optimization problem that is solved at each sampling instant takes the following form:

$$\min_{T_j} \; -\frac{\mu_4(t_f)}{\mu_3(t_f)}$$

$$\text{such that} \quad \frac{\mathrm{d}\mu_0}{\mathrm{d}t} = k_a C \exp\left(-\frac{k_b}{\sigma^2}\right),$$

$$
\begin{aligned}
&\frac{\mathrm{d}\mu_i}{\mathrm{d}t} = i k_g \sigma^g \mu_{i-1}(t), \quad i = 1,\ldots,4,\\
&\frac{\mathrm{d}C}{\mathrm{d}t} = -24\rho k_v k_g \sigma^g \mu_2(t),\\
&\frac{\mathrm{d}T}{\mathrm{d}t} = -\frac{UA}{MC_\mathrm{p}}(T - T_j),\\
&T_{\min} \le T \le T_{\max},\\
&T_{j\,\min} \le T_j \le T_{j\,\max},\\
&\sigma_{\min} \le \sigma \le \sigma_{\max},\\
&\left\|\frac{\mathrm{d}C_\mathrm{s}}{\mathrm{d}t}\right\| \le k_1,
\end{aligned}
\tag{2.17}
$$

$$
n(0,t) \le n_{\mathrm{fine}}, \forall\ t \ge t_\mathrm{f}/2, \tag{2.18}
$$

where $T_{\min}$ and $T_{\max}$ are the constraints on the crystallizer temperature, T, and are specified as 4°C and 22°C, respectively. $T_{j\,\min}$ and $T_{j\,\max}$ are the constraints on the manipulated variable, T_j, and are specified as 3°C and 22°C, respectively. The constraints on the supersaturation σ are $\sigma_{\min} = 1.73$ and $\sigma_{\max} = 2.89$. The constant, k_1, (chosen to be 0.065mg/ml·min) specifies the maximum rate of change of the saturation concentration C_s. n_{fine} is the largest allowable number of nuclei at any time instant during the second half of the batch run, and is set to $5/\mu$ m/ml. In the simulation, the sampling time is 5 min, while the batch process time t_f is 24 h. The optimization problem is solved using sequential quadratic programming (SQP). A second-order accurate finite difference scheme with 3,000 discretization points is used to obtain the solution of the population balance model of (2.3). Referring to the predictive control formulation of (2.17) and (2.18), it is important to note that previous work has shown that the objective of maximizing the volume-averaged crystal size can result in a large number of fines (crystals whose size is very small compared to the mean crystal size) in the final product [49]. To enhance the ability of the predictive control strategy to maximize the performance objective while avoiding the formation of a large number of fines in the final product, the predictive controller of (2.17) and (2.18) includes a constraint (2.18) on the number of fines present in the final product. Specifically, the constraint of (2.18), by restricting the number of nuclei formed at any time instant during the second half of the batch run limits the fines in the final product. Note that predictive control without a constraint on fines can result in a product with a large number of fines (see Fig. 2.9a), which is undesirable. The implementation of the predictive controller with the constraint of (2.18), designed to reduce the fines in the product, results in a product with much less fines while still maximizing the volume-averaged crystal size (see Fig. 2.9b). The reader may refer to [60, 62] for further results on the performance of the predictive controller and comparisons with the performance of two other open-loop control strategies, Constant Temperature Control (CTC) and Constant Supersaturation Control (CSC).

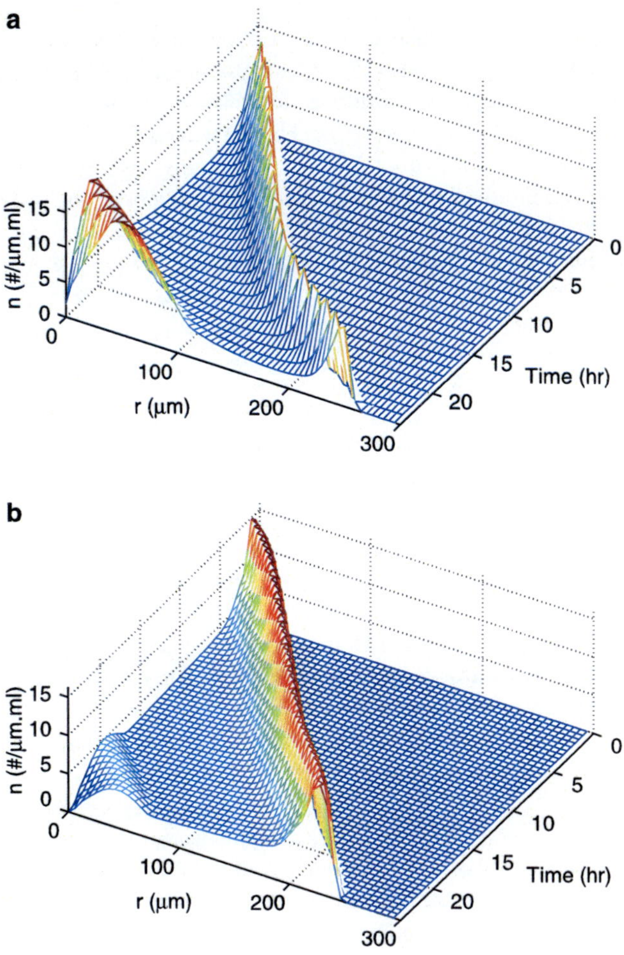

Fig. 2.9 Evolution of particle size distribution under (**a**) predictive control without a constraint on fines, and (**b**) predictive control with a constraint on fines [62]

2.2.4.4 Fault-Tolerant Control of Particulate Processes

Compared with the significant and growing body of research work on feedback control of particulate processes, the problem of designing fault-tolerant control systems for particulate processes has not received much attention. This is an important problem given the vulnerability of automatic control systems to faults (e.g., malfunctions in the control actuators, measurement sensors, or process equipment), and the detrimental effects that such faults can have on the process operating efficiency and product quality. Given that particulate processes play a key role in a wide range of industries (e.g., chemical, food, and pharmaceutical) where the ability to consistently meet stringent product specifications is critical to the product utility, it is imperative that systematic methods for the timely diagnosis and handling of faults be developed to minimize production losses that could result from operational failures. Motivated by these considerations, recent research efforts have started to tackle this problem by bringing together tools from model-based control, infinite-dimensional systems, fault diagnosis, and hybrid systems theory. For particulate processes modeled by population balance equations with control constraints, actuator faults, and a limited number of process measurements, a fault-tolerant control architecture that integrates model-based fault detection, feedback and supervisory control has recently been developed in [19]. The architecture, which is based on reduced-order models that capture the dominant dynamics of the particulate process, consists of a family of control configurations, together with a fault detection filter and a supervisor. For each configuration, a stabilizing output feedback controller with well-characterized stability properties is designed through a combination of a state feedback controller and a state observer that uses the available measurements of the principal moments of the PSD and the continuous-phase variables to provide appropriate state estimates. A fault detection filter that simulates the behavior of the fault-free, reduced-order model is then designed, and its discrepancy from the behavior of the actual process state estimates is used as a residual for fault detection. Finally, a switching law based on the stability regions of the constituent control configurations is derived to reconfigure the control system in a way that preserves closed-loop stability in the event of fault detection. Appropriate fault detection thresholds and control reconfiguration criteria that account for model reduction and state estimation errors were derived for the implementation of the control architecture on the particulate process. The methodology was successfully applied to a continuous crystallizer example using computer simulations where the control objective was to stabilize an unstable steady state and achieve a desired CSD in the presence of constraints and actuator faults.

In addition to the synthesis of actuator fault-tolerant control systems for particulate processes, recent research efforts have also investigated the problem of preserving closed-loop stability and performance of particulate processes in the presence of sensor data losses [24]. Typical sources of sensor data losses include measurement sampling losses, intermittent failures associated with measurement techniques, as well as data packet losses over transmission lines. In this work, two representative particulate process examples – a continuous crystallizer and a batch protein crystallizer – were considered. In both examples, feedback control systems were first designed on the basis of low-order models and applied to the population balance models to enforce closed-loop stability and constraint satisfaction. Subsequently, the robustness of the control systems in the presence of sensor data losses was investigated using a stochastic formulation developed in [51] that models sensor failures as a random Poisson process. In the case of the continuous crystallizer, a Lyapunov-based nonlinear output feedback controller was designed and shown to stabilize an open-loop unstable steady state of the population balance model in the presence of input constraints. Analysis of the closed-loop system under sensor malfunctions showed that the controller is robust with respect to significant sensor data losses, but cannot maintain closed-loop stability when the rate of data losses exceeds a certain threshold. In the case of the batch crystallizer, a predictive controller was designed to obtain a desired CSD at the end of the batch while satisfying state and input constraints. Simulation results showed how constraint modification in the predictive controller formulation can assist in achieving constraint satisfaction under sensor data losses.

2.2.4.5 Nonlinear Control of Aerosol Reactors

The crystallization process examples discussed in the previous section share the common characteristic of having two independent variables (time and particle size). In such a case, order reduction, for example with the method of moments, leads to a set of ODEs in time as a reduced-order model. This is not the case, however, when three or more independent variables (time, particle size, and space) are used in the process model. An example of such a process is the aerosol flow reactor presented in the Introduction section. The complexity of the partial integro-differential equation model of (2.6) does not allow its direct use for the synthesis of a practically implementable nonlinear model-based feedback controller for spatially inhomogeneous aerosol processes. Therefore, we developed [33–35] a model-based controller design method for spatially inhomogeneous aerosol processes, which is based on the experimental observation that many aerosol size distributions can be adequately approximated by lognormal functions. The proposed control method can be summarized as follows:

1. Initially, the aerosol size distribution is assumed to be described by a lognormal function and the method of moments is applied to the aerosol population balance model of (2.6) to compute a hyperbolic partial differential equation (PDE) system (where the independent variables are time and space) that describes the spatiotemporal behavior of the three leading moments needed to exactly describe the evolution of the lognormal aerosol size distribution.
2. Then nonlinear geometric control methods for hyperbolic PDEs [10] are applied to the resulting system to synthesize nonlinear distributed output feedback controllers that use process measurements at different locations along the length of the process to adjust the manipulated input (typically, wall temperature), in order to achieve an aerosol size distribution with desired characteristics (e.g., geometric average particle volume).

We carried out an application of this nonlinear control method to an aerosol flow reactor, including nucleation, condensation, and coagulation, used to produce NH_4Cl particles [33] and a titania aerosol reactor [34]. Specifically, for an aerosol flow reactor used to produce NH_4Cl particles, the following chemical reaction takes place $NH_3 + HCl \rightarrow NH_4Cl$ where NH_3, HCl are the reactant species and NH_4Cl is the monomer product species. Under the assumption of lognormal aerosol size distribution, the mathematical model that describes the evolution of the first three moments of the distribution, together with the monomer (NH_4Cl) and reactant (NH_3, HCl) concentrations and reactor temperature takes the form:

$$
\begin{aligned}
\frac{\partial N}{\partial \theta} &= -v_{zl}\frac{\partial N}{\partial \bar{z}} + I' - \xi N^2, \\
\frac{\partial V}{\partial \theta} &= -v_{zl}\frac{\partial V}{\partial \bar{z}} + I'k^* + \eta(S-1)N, \\
\frac{\partial V_2}{\partial \theta} &= -v_{zl}\frac{\partial V_2}{\partial \bar{z}} + I'k^{*2} + 2\varepsilon(S-1)V + 2\zeta V^2, \\
\frac{\partial S}{\partial \theta} &= -v_{zl}\frac{\partial S}{\partial \bar{z}} + C\bar{C}_1\bar{C}_2 - I'k^* - \eta(S-1)N, \\
\frac{\partial \bar{C}_1}{\partial \theta} &= -v_{zl}\frac{\partial \bar{C}_1}{\partial \bar{z}} - A_1\bar{C}_1\bar{C}_2, \\
\frac{\partial \bar{C}_2}{\partial \theta} &= -v_{zl}\frac{\partial \bar{C}_2}{\partial \bar{z}} - A_2\bar{C}_1\bar{C}_2, \\
\frac{\partial \bar{T}}{\partial \theta} &= -v_{zl}\frac{\partial \bar{T}}{\partial \bar{z}} + B\bar{C}_1\bar{C}_2\bar{T} + E\bar{T}(\bar{T}_{\mathrm{w}} - \bar{T}),
\end{aligned} \tag{2.19}
$$

where θ is the dimensionless time, $\bar{z}$ is the dimensionless length, v_{zl} is the dimensionless velocity, I' is the dimensionless nucleation rate, S is the saturation ratio, $\bar{C}_1$ and $\bar{C}_2$ are the dimensionless concentrations of NH_3 and HCl, respectively, $\bar{T}, \bar{T}_{\mathrm{w}}$ are the dimensionless reactor and wall temperatures, respectively, and A_1, A_2, B, C, E are dimensionless quantities [33]. The controlled output is the geometric average particle volume in the outlet of the reactor, and the manipulated input is the wall temperature.

Figure 2.10 displays the steady-state profile of the dimensionless total particle concentration, N, as a function of reactor length. As expected, N increases very fast close to the inlet of the reactor (approximately, the first 3% of the reactor) due to a nucleation burst, and then, it slowly decreases in the remaining part of the reactor due to coagulation. Even though coagulation decreases the total number of particles, it leads to the formation of bigger particles, and thus, it increases the geometric average particle volume, v_{g}. We formulate the control problem as the one of controlling the geometric average particle volume in the outlet of the reactor, $v_{\mathrm{g}}(1,\theta)$, ($v_{\mathrm{g}}(1,\theta)$ is directly related to the geometric average particle diameter, and hence, it is a key product characteristic of industrial aerosol processes) by manipulating the wall temperature, i.e.:

$$
y(\theta) = \mathscr{C} v_{\mathrm{g}} = v_{\mathrm{g}}(1,\theta), \;\; u(\theta) = \bar{T}_{\mathrm{w}}(\theta) - \bar{T}_{ws}, \tag{2.20}
$$

where $\mathscr{C}(\cdot) = \int_0^1 \delta(\bar{z}-1)(\cdot)\mathrm{d}z$ and $\bar{T}_{ws} = T_{ws}/T_o = 1$. Since coagulation is the main mechanism that determines the size of the aerosol particles, we focus on controlling the part of the reactor where coagulation occurs. Therefore, the wall temperature is

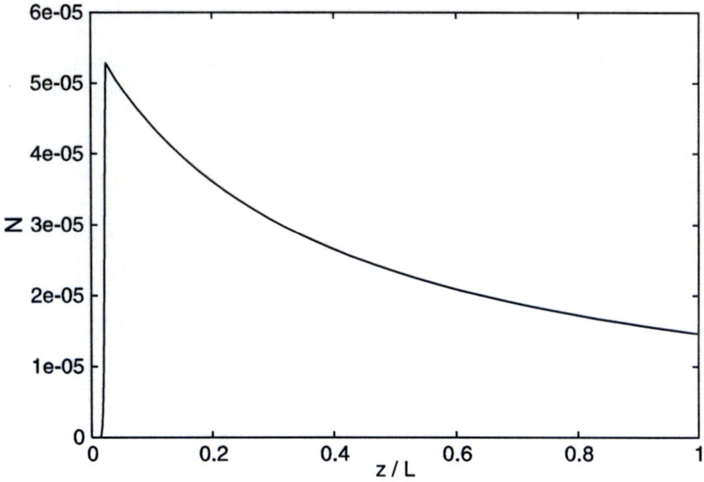

Fig. 2.10 Steady-state profile of dimensionless particle concentration

assumed to be equal to its steady-state value in the first 3.5% of the reactor (where nucleation mainly occurs), and it is adjusted by the controller in the remaining part of the reactor (where coagulation takes place).

The model of (2.19) was used as the basis for the synthesis of a nonlinear controller utilizing the above-mentioned control method. For this model, σ (geometric standard deviation of particle number distribution) was found to be equal to 2 and the necessary controller was synthesized using the nonlinear distributed state feedback formula developed in [10] and is of the form:

$$u = \left[\mathscr{C}\gamma_\sigma L_g \left(\sum_{j=1}^{n} \frac{\partial x_j}{\partial \bar{z}} L_{a_j} + L_f \right) h(x) b(\bar{z}) \right]^{-1}$$

$$\left\{ y_{sp} - \mathscr{C}h(x) - \sum_{\nu=1}^{2} \mathscr{C}\gamma_\nu \left(\sum_{j=1}^{n} \frac{\partial x_j}{\partial \bar{z}} L_{a_j} + L_f \right)^{\nu} h(x) \right\}, \qquad (2.21)$$

where $\gamma_1 = 580$ and $\gamma_2 = 1.6 \times 10^5$ to enforce a slightly underdamped response.

Two simulation runs were performed to evaluate the set-point tracking capabilities of the nonlinear controller and compare its performance with a proportional-integral controller. In both simulation runs, the aerosol reactor was initially assumed to be at steady-state and a 5% increase in the set-point value of $v_g(1,0)$ was imposed at $t = 0$ s (i.e., $y_{sp} = 1.05 v_g(1,0)$). Figure 2.11 (top plot – solid line) shows the profile of the controlled output which is the mean particle volume at the outlet of the reactor $v_g(1,t)$, while Fig. 2.11 (bottom plot – solid line) displays the corresponding profile of the manipulated input which is the wall temperature. The nonlinear controller

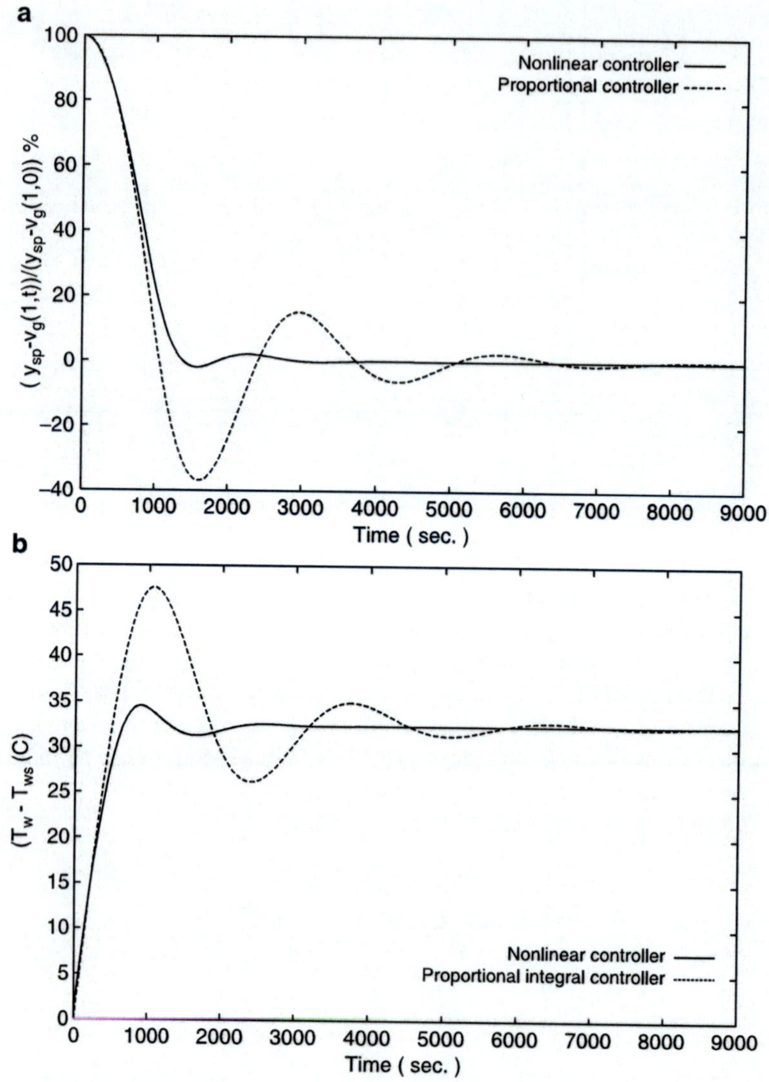

Fig. 2.11 (**a**) Closed-loop profiles of scaled mean particle volume in the outlet of the reactor under proportional-integral and nonlinear controllers. (**b**) Manipulated input profiles for proportional-integral and nonlinear controllers [33]

of (2.21) regulates $v_g(1,t)$ successfully to its new set-point value. For the sake of comparison, we also implemented on the process a proportional-integral controller; this controller was tuned so that the time at which the closed-loop output needs to reach the final steady state is the same as for the closed-loop output under nonlinear control. The profiles of the controlled output and manipulated input are shown in

Fig. 2.11 (dashed lines show the corresponding profiles for the proportional-integral controller). It is clear that the nonlinear controller outperforms the proportional-integral controller.

2.3 Multiscale Modeling and Control of HVOF Thermal Spray Coating Processes

2.3.1 Multiscale Modeling of Coating Microstructure

The past decade has witnessed a shift of synthesis to processing in nanotechnology research, i.e. the manufacture of functional coatings and bulk structures using nanostructured powders [5]. One example is HVOF thermal spray processing of functional coatings from nanostructured agglomerate powders. The nanostructured coatings prepared by HVOF are extensively used in many industries as thermal-barrier and wear-resistant surface layers to extend product life, increase performance, and reduce production and maintenance costs. Thermal spray has also been a molding method for the fabrication of micro-components [69].

A representative diagram of the HVOF thermal spray process is shown in Fig. 2.12. The high-pressure combustion of a fuel (typically hydrogen, propane, or kerosene) with oxygen generates a supersonic jet, which propels and heats up the powder of particles added to the gas stream. The powder particles are accelerated, softened in the gas stream, and deformed on the substrate, forming a dense coating.

Because the highly coupled transport phenomena of the HVOF thermal spray cannot be fully revealed by experimental studies, mathematical modeling has been an excellent complement in order to provide system-level understanding of the underlying physics of the HVOF thermal spray process to guide optimal system design and operation [14, 53]. Moreover, to fabricate coatings of a consistent quality, the compensation of feed disturbances and process variability during real-time process

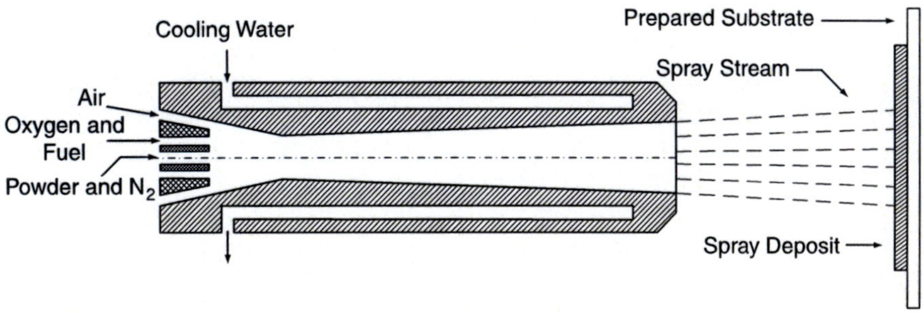

Fig. 2.12 Schematic of a representative HVOF thermal spray process

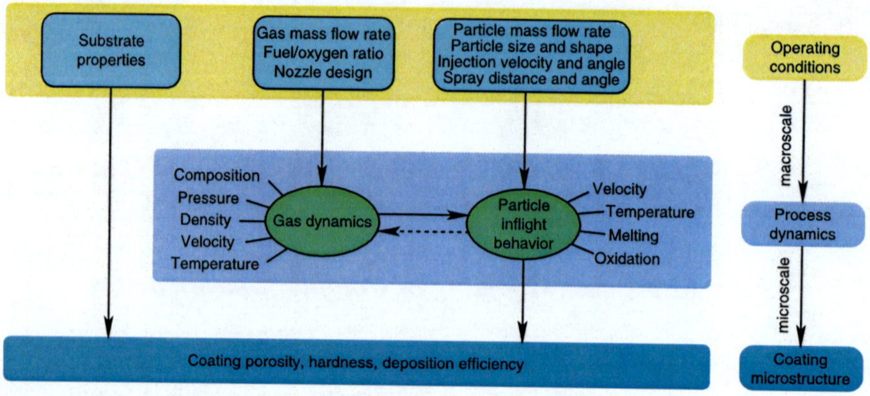

Fig. 2.13 Multiscale character of the HVOF thermal spray process [44]

operation becomes essential. This motivates the development and implementation of real-time control systems in the HVOF thermal spray process to suppress variations in the particle characteristics at the point of impact on the substrate. The major challenge in this problem is the development of multiscale models linking the macroscopic scale process behavior (i.e., gas dynamics and particle inflight behavior) and the microscopic scale process characteristics (evolution of coating microstructure), and the integration of models, measurements, and control theory to develop measurement/model-based control strategies. The multiscale character of the HVOF thermal spray process is shown in Fig. 2.13. The microstructure of HVOF-sprayed coatings results from the deformation, solidification, and sintering of the deposited particles, which are dependent on the substrate properties (e.g., substrate temperature) as well as the physical and chemical state (e.g., temperature, velocity, melting ratio, and oxidant content) of the particles at the point of impact on the substrate. On the other hand, the particle inflight characteristics are coupled with the gas dynamics, which can be manipulated by adjusting operating conditions such as the gas flow rates of fuel and oxygen. While the macroscopic thermal/flow field can be readily described by continuum type differential equations governing the compressible two-phase flow, the process of particle deposition is stochastic and discrete in nature, and thus, it can be best described by stochastic simulation methods [36]. By manipulating macro-scale operating conditions such as gas feed flow rates, one can control the coating microstructure which determines the coating mechanical and physical properties.

In the past several years, we developed a multiscale computational framework for the HVOF thermal spray processing of nanostructured coatings [40–46, 61]. The multiscale process model encompasses gas dynamics of the supersonic reacting flow, evolution of particle velocity, temperature and molten state during flight, and stochastic growth of coating microstructure, as shown in Fig. 2.14. The modeling work demonstrates that the coating microstructure, porosity, and roughness, as well as the deposition efficiency, are highly dependent on the particle

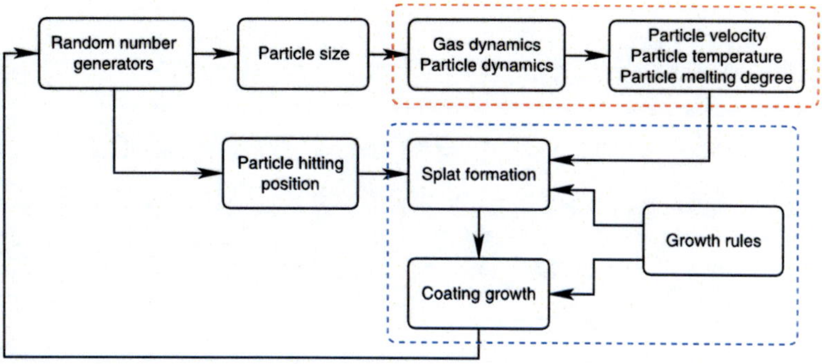

Fig. 2.14 Multiscale modeling of the HVOF thermal spray process (based on [44,61])

characteristics (primarily velocity, temperature, and molten state), which is consistent with experimental observations [26,28,29]. For example, the effect of particle melting degree on the coating microstructure is shown in Fig. 2.15 [42]. When all the particles are fully melted, which is typical in a plasma spray, an ideal lamellar microstructure is formed. However, under normal HVOF processing conditions, many particles might be partially melted or even unmelted [44,71]. When a partially melted particle lands on the substrate and deforms, the resulting splat typically has a "fried-egg" shape which features a nearly hemispherical core located in the center of a thin disk [31]. As a result, a different coating microstructure is formed which deviates from the ideal lamellar microstructure. The fact that the particles are not necessarily fully melted to form a coating with excellent microstructure is very important in the processing of nanostructured coatings because the nanostructure in the powder particles could be destroyed if the particles are heated too much and go through a phase change during flight. However, if the particle melting degree is very low, unmelted particles may bounce off the substrate, resulting in a high-porosity coating with a low deposition efficiency. In addition to particle melting degree, the model also predicts that the higher the particle impact velocity, the higher the flattening ratio. As a result, the coating porosity is lower and the coating is denser.

2.3.2 Control of Particle Velocity and Temperature in HVOF Thermal Spray

Based on the above analysis, one should suppress the variation in the particle characteristics upon impact on the substrate to enhance the consistency of the coating quality. Both modeling and experimental studies [44, 65, 67] reveal that the particle velocity and temperature (or melting degree) at impact with the substrate can be almost independently adjusted by manipulating the pressure in the combustion chamber and the fuel/oxygen ratio. As shown in Fig. 2.16, when the combustion pressure increases from 5 bar to 15 bar with a fixed equivalence ratio

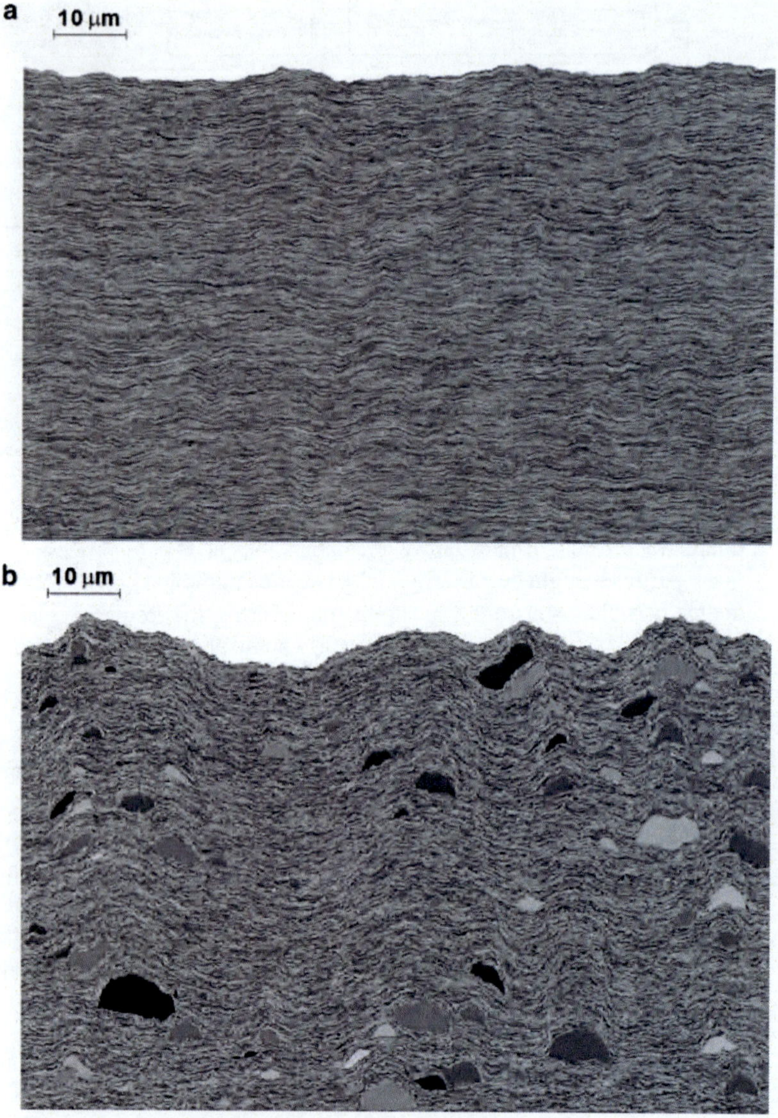

Fig. 2.15 Simulated microstructure of coatings formed by fully melted particles and particles with mixed molten states [42]

(or the fuel/oxygen ratio divided by its stoichiometric value), the gas momentum flux (ρv_g^2), which is roughly proportional to the drag force for particle motion, is almost tripled. However, the gas temperature increases by about 4% only. When the equivalence ratio varies from 0.5 to 1.5 with a fixed chamber pressure, the gas temperature varies about 12% from its lowest value to the peak occurring at an equivalence ratio around 1.2. However, the gas momentum flux remains almost the

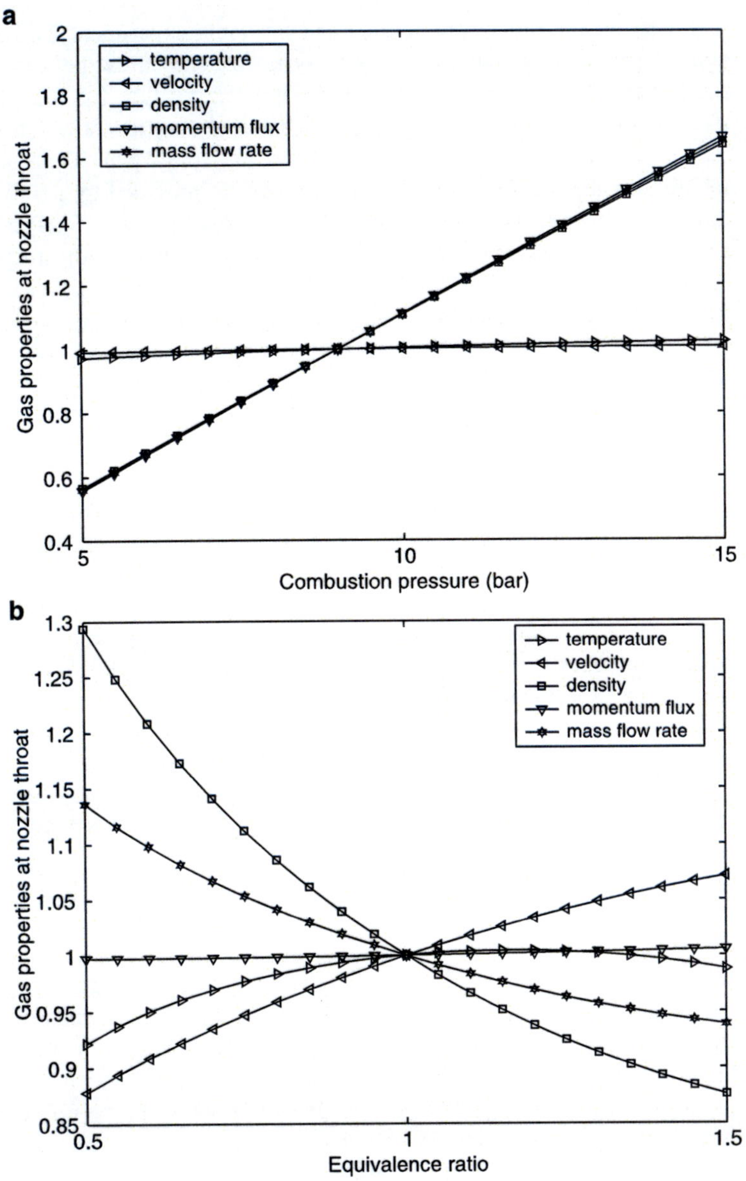

Fig. 2.16 Influence of pressure and fuel/oxygen ratio on gas momentum flux and gas temperature [44]

same in the entire range. It is worth noting that the window for particle temperature control in the HVOF thermal spray is narrower than in the plasma spray where the particle temperature can be adjusted in a wider range by manipulating the torch current [21].

Based on the model predictions and available experimental observations, the control problem for the HVOF process is formulated as the one of regulating the volume-based averages of velocity and temperature (or melting degree) of particles at impact on the substrate by manipulating the flow rates of fuel and oxygen at the entrance of the HVOF thermal spray gun. The particle sensing including temperature, velocity, and size can be provided by a variety of online diagnostic techniques developed by different groups [22, 27, 66]. The manipulation of combustion pressure and equivalence ratio is realized by adjusting the flow rate of fuel, $u_1(t)$, and oxygen, $u_2(t)$ (see (2.23) below). Note that the chamber pressure is dependent on the flow rates of fuel and oxygen as follows:

$$\dot{m} = \frac{p_0}{\sqrt{T_0}} A_{\mathrm{th}} \sqrt{\frac{\gamma \bar{M}_{\mathrm{pr}}}{R_{\mathrm{g}}} \left(\frac{2}{\gamma+1} \right)^{\frac{\gamma+1}{\gamma-1}}}, \tag{2.22}$$

where $\dot{m}$ is the total mass flow rate, A_{th} is the cross-sectional area at the throat (where the area is the minimum), R_{g} is the molecular gas constant, $\bar{M}_{\mathrm{pr}}$ is the average molecular weight of the combustion products, and T_0 and p_0 are the stagnation temperature and stagnation pressure in the combustion chamber, respectively.

Owing to the almost decoupled nature of the manipulated input/controlled output pairs, two proportional-integral (PI) controllers were proposed in [41,44] to regulate the process. Specifically, the controllers have the following form:

$$\begin{aligned} \dot{\zeta}_i &= y_{\mathrm{sp}_i} - y_i, \quad \zeta_i(0) = 0, \quad i = 1,2 \\ u'_i &= K_{\mathrm{c}_i} \left[(y_{\mathrm{sp}_i} - y_i) + \frac{1}{\tau_{\mathrm{c}_i}} \zeta_i \right] + u'_{0_i}, \quad i = 1,2 \\ \{u_1, u_2\} &= f(u'_1, u'_2), \end{aligned} \tag{2.23}$$

where y_{sp_i} is the desired set-point value and y_i is the value of the output obtained from the measurement system (y_1 is the volume-based average of particle velocity and y_2 is the volume-based average of particle temperature or melting degree), u'_1 is the combustion pressure and u'_2 is the equivalence ratio. f is the mapping between the flow rates and the chamber pressure as well as the equivalence ratio. K_{c_i} is the proportional gain and τ_{c_i} is the integral time constant. If the gas phase measurement is available, a model-based scheme can be used to estimate the particle properties through the dynamic particle-inflight model [45].

Closed-loop simulations under the control scheme of (2.23) have been carried out to demonstrate the effectiveness of the proposed control formulation [46]. It is assumed in the computer simulations that the responses of gas and particle dynamics to the change of gas flow rates are very fast, which is reasonable for such a supersonic flow. With this simplification, it has been demonstrated that the feedback controllers are very effective with respect to set-point changes in both particle velocity and temperature (i.e., 5% increase in both particle velocity and melting degree). As seen in Fig. 2.17, both the flow rates of oxygen and fuel increase

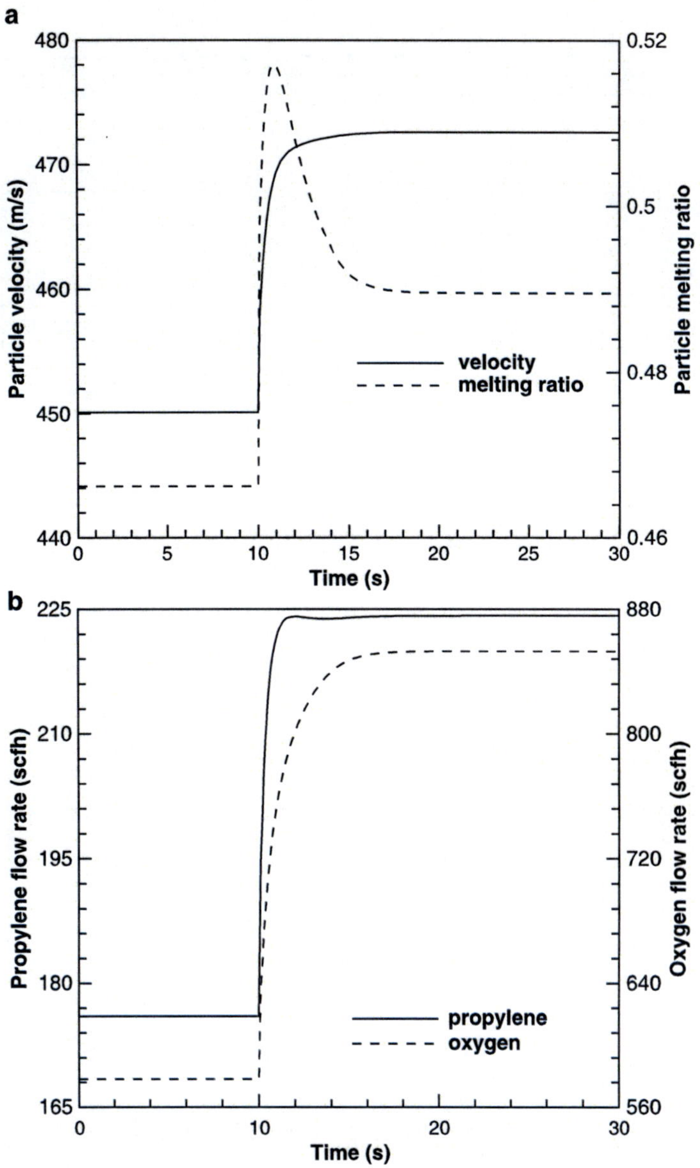

Fig. 2.17 Profiles of (**a**) controlled outputs (average particle velocity and melting ratio) and (**b**) manipulated inputs (flow rates of propylene and oxygen) under the request of 5% increase in particle velocity and 5% increase in melting ratio [46]

in order to have a higher particle velocity. However, the temperature increases and exceeds its desired value due to the increased chamber pressure. As a result, the rate of change of oxygen flow becomes slower than the one of fuel after a short period of

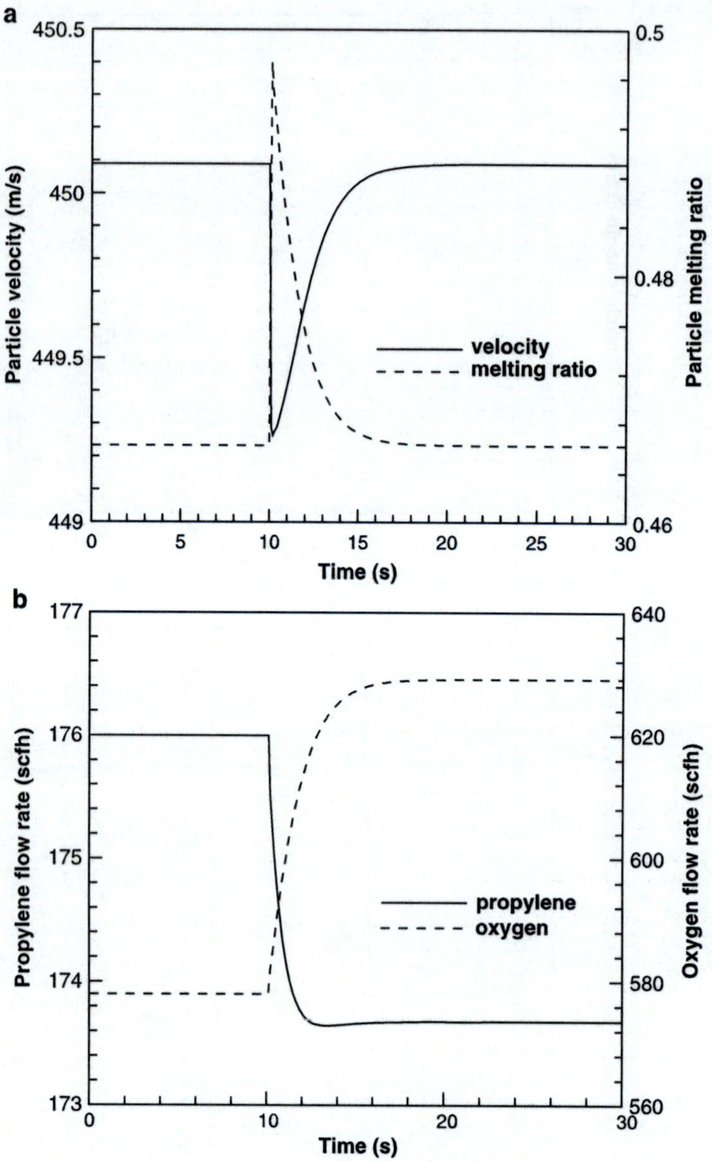

Fig. 2.18 Profiles of (**a**) controlled outputs (average particle velocity and melting ratio) and (**b**) manipulated inputs (flow rates of propylene and oxygen) in the presence of 10% decrease in spray distance [46]

time, which lowers the equivalence ratio and drives the temperature down to its set point. Figure 2.18 demonstrates the response of the feedback controller in order to maintain the same particle velocity and temperature levels in the presence of a 10%

decrease in the spray distance (process disturbance). The particle velocity does not change much while the particle temperature increases significantly. Under feedback control, the manipulated inputs adjust to drive the process outputs to their original steady-state values in 10 s, which demonstrates the robustness of the controller.

To the best knowledge of the authors, no experimental implementation of HVOF thermal spray control has been reported. Feedback control of average particle temperature and velocity in plasma spray has been studied by Fincke et al. [21]. With the development of fast and reliable online gas and particle sensing and diagnostic tools by companies and institutions (e.g., Idaho National Laboratory, Tecnar Automation, Canada, and Oseir Ltd., Finland), the demonstration of HVOF spray control should be expected in the near future.

2.4 Conclusions

Control of particulate processes systems is a cross-disciplinary and rapidly growing research area that brings together fundamental modeling, numerical simulation, nonlinear dynamics, and control theory. This chapter presents recent advances in systematic methods for the design of easy-to-implement nonlinear feedback controllers for broad classes of particulate processes. It is expected that feedback control will play an important role in the synthesis and processing of nano- and micro-size particles with the ever-increasing research and development in advanced materials and semiconductor manufacturing, nanotechnology, and biotechnology. The reader may refer to [11] for a detailed discussion on future problems on control of particulate processes.

Acknowledgements Financial support from the NSF (ITR), CTS-0325246, and the Office of Naval Research (2001 Young Investigator Award) is gratefully acknowledged. The authors note that all the figures included in this Chapter are original and have permission to include them in the present chapter.

References

1. L. Ajdelsztajn, F. Tang, J.M. Schoenung, G.E. Kim, and V. Provenzano. Synthesis and oxidation behavior of nanocrystalline MCrAlY bond coatings. *J. Thermal Spray Technol.*, 14:23–30, 2005.
2. A. Azarani. Automated high-throughput protein crystallization. In *The proteomics protocols handbook,* Walker, J. M. (Ed.), pages 955–966. Humana Press, Totowa, New Jersey, 2005.
3. C.F. Bohren, and D.R. Huffman. *Absorption and scattering of light by small particles.* Wiley, New York, 1983.
4. R.D. Braatz, and S. Hasebe. Particle size and shape control in crystallization processes. In *AIChE Symposium Series: Proceedings of 6th international conference on chemical process control,* Rawlings, J. B. *et al.* (Eds.), pages 307–327, 2002.

5. D. Cheng, G. Trapaga, J.W. McKelliget, and E.J. Lavernia. Mathematical modelling of high velocity oxygen fuel thermal spraying of nanocrystalline materials: an overview. *Modell. Simul. Mater. Sci. Eng.*, 11:R1–R31, 2003.
6. T. Chiu, and P.D. Christofides. Nonlinear control of particulate processes. *AIChE J.*, 45: 1279–1297, 1999.
7. T. Chiu, and P.D. Christofides. Robust control of particulate processes using uncertain population balances. *AIChE J.*, 46:266–280, 2000.
8. P.D. Christofides. *Model-based control of particulate processes.* Kluwer Academic Publishers, Particle Technology Series, Netherlands, 2002.
9. P.D. Christofides, and T. Chiu. Nonlinear control of particulate processes. In *AIChE annual meeting, paper 196a, Los Angeles, CA*, 1997.
10. P.D. Christofides, and P. Daoutidis. Feedback control of hyperbolic PDE systems. *AIChE J.*, 42:3063–3086, 1996.
11. P.D. Christofides, M. Li, and L. Mädler. Control of particulate processes: Recent results and future challenges. *Powder Technol.*, 175:1–7, 2007.
12. P. Daoutidis, and M. Henson. Dynamics and control of cell populations. In *Proceedings of 6th international conference on chemical process control*, pages 308–325, Tucson, AZ, 2001.
13. J. Dimitratos, G. Elicabe, and C. Georgakis. Control of emulsion polymerization reactors. *AIChE J.*, 40:1993–2021, 1994.
14. E. Dongmo, M. Wenzelburger, and R. Gadow. Analysis and optimization of the HVOF process by combined experimental and numerical approaches. *Surf. Coat. Technol.*, 202:4470–4478, 2008.
15. F.J. Doyle, M. Soroush, and C. Cordeiro. Control of product quality in polymerization processes. In *AIChE symposium series: Proceedings of 6th international conference on chemical process control,* rawlings, J. B. *et al.* (Eds.), pages 290–306, 2002.
16. N.H. El-Farra, T. Chiu, and P.D. Christofides. Analysis and control of particulate processes with input constraints. *AIChE J.*, 47:1849–1865, 2001.
17. N.H. El-Farra, and P.D. Christofides. Integrating robustness, optimality, and constraints in control of nonlinear processes. *Chem. Eng. Sci.*, 56:1–28, 2001.
18. N.H. El-Farra, and P.D. Christofides. Bounded robust control of constrained multivariable nonlinear processes. *Chem. Eng. Sci.*, 58:3025–3047, 2003.
19. N.H. El-Farra, and A. Giridhar. Detection and management of actuator faults in controlled particulate processes using population balance models. *Chem. Eng. Sci.*, 63:1185–1204, 2008.
20. N.H. El-Farra, P. Mhaskar, and P.D. Christofides. Hybrid predictive control of nonlinear systems: Method and applications to chemical processes. *Int. J. Robust Nonlinear Control*, 14:199–225, 2004.
21. J.R. Fincke, W.D. Swank, R.L. Bewley, D.C. Haggard, M. Gevelber, and D. Wroblewski. Diagnostics and control in the thermal spray process. *Surf. Coat. Technol.*, 146-147:537–543, 2001.
22. J.R. Fincke, W.D. Swank, and C.L. Jeffrey. Simultaneous measurement of particle size, velocity and temperature in thermal plasmas. *IEEE Trans. Plasma Sci.*, 18:948–957, 1990.
23. S.K. Friendlander. *Smoke, dust and haze: Fundamentals of aerosol dynamics (2nd Ed.).* Oxford University Press, New York, USA, 2000.
24. A. Gani, P. Mhaskar, and P.D. Christofides. Handling sensor malfunctions in control of particulate processes. *Chem. Eng. Sci.*, 63:1217–1229, 2008.
25. F. Gelbard, and J.H. Seinfeld. Numerical solution of the dynamic equation for particulate processes. *J. Comput. Phys.*, 28:357–375, 1978.
26. L. Gil, and M.H. Staia. Influence of HVOF parameters on the corrosion resistance of NiWCrBSi coatings. *Thin Solid Films*, 420–421:446–454, 2002.
27. E. Hamalainen, J. Vattulainen, T. Alahautala, R. Hernberg, P. Vuoristo, and T. Mantyla. Imaging diagnostics in thermal spraying. "spraywatch" system. In *Thermal spray: Surface engineering via applied research, Proceedings of the international thermal spray conference*, pages 79–83, Montreal, QC, Canada, 2000.

28. T.C. Hanson, and G.S. Settles. Particle temperature and velocity effects on the porosity and oxidation of an HVOF corrosion-control coating. *J. Therm. Spray Technol.*, 12:403–415, 2003.
29. J.A. Hearley, J.A. Little, and A.J. Sturgeon. The effect of spray parameters on the properties of high velocity oxy-fuel NiAl intermetallic coatings. *Surf. Coat. Technol.*, 123:210–218, 2000.
30. H.M. Hulburt, and S. Katz. Some problems in particle technology: A statistical mechanical formulation. *Chem. Eng. Sci.*, 19:555–574, 1964.
31. M. Ivosevic, R.A. Cairncross, and R. Knight. 3D predictions of thermally sprayed polymer splats: Modeling particle acceleration, heating and deformation on impact with a flat substrate. *Int. J. Heat Mass Transfer*, 49:3285–3297, 2006.
32. G.R. Jerauld, Y. Vasatis, and M.F. Doherty. Simple conditions for the appearance of sustained oscillations in continuous crystallizers. *Chem. Eng. Sci.*, 38:1675–1681, 1983.
33. A. Kalani, and P.D. Christofides. Nonlinear control of spatially-inhomogeneous aerosol processes. *Chem. Eng. Sci.*, 54:2669–2678, 1999.
34. A. Kalani, and P.D. Christofides. Modeling and control of a titania aerosol reactor. *Aerosol Sci. Technol.*, 32:369–391, 2000.
35. A. Kalani, and P.D. Christofides. Simulation, estimation and control of size distribution in aerosol processes with simultaneous reaction, nucleation, condensation and coagulation. *Comput. Chem. Eng.*, 26:1153–1169, 2002.
36. O. Knotek, and R. Elsing. Monte carlo simulation of the lamellar structure of thermally sprayed coatings. *Surf. Coat. Technol.*, 32:261–271, 1987.
37. P.A. Larsen, J.B. Rawlings, and N.J. Ferrier. An algorithm for analyzing noisy, in situ images of high-aspect-ratio crystals to monitor particle size distribution. *Chem. Eng. Sci.*, 61: 5236–5248, 2006.
38. K. Lee, and T. Matsoukas. Simultaneous coagulation and break-up using constant-n monte carlo. *Powder Technol.*, 110:82–89, 2000.
39. S.J. Lei, R. Shinnar, and S. Katz. The stability and dynamic behavior of a continuous crystallizer with a fines trap. *AIChE J.*, 17:1459–1470, 1971.
40. M. Li, and P.D. Christofides. Modeling and analysis of HVOF thermal spray process accounting for powder size distribution. *Chem. Eng. Sci.*, 58:849–857, 2003.
41. M. Li, and P.D. Christofides. Feedback control of HVOF thremal spray process accounting for powder size distribution. *J. Therm. Spray Technol.*, 13:108–120, 2004.
42. M. Li, and P.D. Christofides. Multi-scale modeling and analysis of HVOF thermal spray process. y *Chem. Eng. Sci.*, 60:3649–3669, 2005.
43. M. Li, and P.D. Christofides. Computational study of particle in-flight behavior in the HVOF thermal spray process. *Chem. Eng. Sci.*, 61:6540–6552, 2006.
44. M. Li, D. Shi, and P.D. Christofides. Diamond jet hybrid HVOF thermal spray: Gas-phase and particle behavior modeling and feedback control design. *Ind. Eng. Chem. Res.*, 43:3632–3652, 2004.
45. M. Li, D. Shi, and P.D. Christofides. Model-based estimation and control of particle velocity and melting in HVOF thermal spray. *Chem. Eng. Sci.*, 59:5647–5656, 2004.
46. M. Li, D. Shi, and P.D. Christofides. Modeling and control of HVOF thermal spray processing of WC-Co coatings. *Powder Technol.*, 156:177–194, 2005.
47. Y. Lin, and E.D. Sontag. A universal formula for stabilization with bounded controls. *Syst. Contr. Lett.*, 16:393–397, 1991.
48. Y.L. Lin, K. Lee, and T. Matsoukas. Solution of the population balance equation using constant-number Monte Carlo. *Chem. Eng. Sci.*, 57:2241–2252, 2002.
49. D.L. Ma, D.K. Tafti, and R.D. Braatz. Optimal control and simulation of multidimensional crystallization processes. *Comput. Chem. Eng.*, 26:1103–1116, 2002.
50. A. Martinez, C. Gonzalez, M. Porras, and J.M. Gutierrez. Nano-sized latex particles obtained by emulsion polymerization using an amphiphilic block copolymer as surfactant. *Colloids and Surfaces A: Physicochemical and Engineering Aspects*, 270–271:67–71, 2005.
51. P. Mhaskar, A. Gani, C. McFall, P.D. Christofides, and J.F. Davis. Fault-tolerant control of nonlinear process systems subject to sensor faults. *AIChE J.*, 53:654–668, 2007.

52. S.M. Miller, and J.B. Rawlings. Model identification and control strategies for batch cooling crystallizers. *AIChE J.*, 40:1312–1327, 1994.
53. J. Mostaghimi, S. Chandra, R. Ghafouri-Azar, and A. Dolatabadi. Modeling thermal spray coating processes: a powerful tool in design and optimization. *Surf. Coat. Technol.*, 163-164: 1–11, 2003.
54. D. Ramkrishna. The status of population balances. *Rev. Chem. Eng.*, 3:49–95, 1985.
55. J.B. Rawlings, S.M. Miller, and W.R. Witkowski. Model identification and control of solution crystallizatin process – a review. *Ind. Eng. Chem. Res.*, 32:1275–1296, 1993.
56. J.B. Rawlings, C.W. Sink, and S.M. Miller. Control of crystallization processes. In *Industrial crystallization - theory and practice*, pages 179–207, Butterworth, Boston, 1992.
57. S. Rohani, and J.R. Bourne. Self-tuning control of crystal size distribution in a cooling batch crystallizer. *Chem. Eng. Sci.*, 12:3457–3466, 1990.
58. D. Semino, and W.H. Ray. Control of systems described by population balance equations-I. controllability analysis. *Chem. Eng. Sci.*, 50:1805–1824, 1995.
59. D. Semino, and W.H. Ray. Control of systems described by population balance equations-II. emulsion polymerization with constrained control action. *Chem. Eng. Sci.*, 50:1825–1839, 1995.
60. D. Shi, N.H. El-Farra, M. Li, P. Mhaskar, and P.D. Christofides. Predictive control of particle size distribution in particulate processes. *Chem. Eng. Sci.*, 61:268–281, 2006.
61. D. Shi, M. Li, and P.D. Christofides. Diamond jet hybrid HVOF thermal spray: Rule-based modeling of coating microstructure. *Ind. Eng. Chem. Res.*, 43:3653–3665, 2004.
62. D. Shi, P. Mhaskar, N.H. El-Farra, and P.D. Christofides. Predictive control of crystal size distribution in protein crystallization. *Nanotechnology*, 16:S562–S574, 2005.
63. T. Smith, and T. Matsoukas. Constant-number Monte Carlo simulation of population balances. *Chem. Eng. Sci.*, 53:1777–1786, 1998.
64. W.J. Stark, A. Baiker, and S.E. Pratsinis. Nanoparticle opportunities: pilot-scale flame synthesis of vanadia/titania catalysts. *Part. Part. Syst. Charact.*, 19:306–311, 2002.
65. W.D. Swank, J.R. Fincke, D.C. Haggard, G. Irons, and R. Bullock. HVOF particle flow field characteristics. In *Thermal spray industrial applications, Proceedings of the 7th national thermal spray conference*, pages 319–324, Boston, Massachusetts, 1994.
66. Tecnar Automation. *DPV-2000 Reference Manual.*
67. E. Turunen, T. Varis, S.-P. Hannula, A. Vaidya, A. Kulkarni, J. Gutleber, S. Sampath, and H. Herman. On the role of particle state and deposition procedure on mechanical, tribological and dielectric response of high velocity oxy-fuel sprayed alumina coatings. *Mat. Sci. Eng. A*, 415:1–11, 2006.
68. P.G. Vekilov, and F. Rosenberger. Dependence of lysozyme growth kinetics on step sources and impurities. *J. Cryst. Growth*, 158:540–551, 1996.
69. J. Wilden, J.P. Bergmann, and T. Luhn. Aspects of thermal spray molding of micro components. In *Thermal spray: Science, innovation, and application, Proceedings of the 2006 international thermal spray conference*, pages 1243–1246, Seattle, WA, 2006.
70. W. Xie, S. Rohani, and A. Phoenix. Dynamic modeling and operation of a seeded batch cooling crystallizer. *Chem. Eng. Comm.*, 187:229–249, 2001.
71. D. Zhang, S.J. Harris, and D.G. McCartney. Microstructure formation and corrosion behaviour in HVOF-sprayed Inconel 625 coatings. *Mat. Sci. Eng. A*, 344:45–56, 2003.
72. G.P. Zhang, and S. Rohani. On-line optimal control of a seeded batch cooling crystallizer. *Chem. Eng. Sci.*, 58:1887–1896, 2003.

Chapter 3
In Situ Optical Sensing and State Estimation for Control of Surface Processing

Rentian Xiong and Martha A. Grover

3.1 Introduction

Measuring surface properties during modification is challenging, since the surface cannot be directly contacted, as that would disrupt the surface being modified. Calibration and pilot wafers can be used to characterize a process, by performing measurements during [2] or after [1] surface modification. However, in a manufacturing setting this creates a loss of productivity, since this wafer is not used to produce any product. In contrast, optical sensors have found common use in measuring thin film properties during surface modification [2, 4, 5], because they do not directly contact the surface and are thus referred to as noninvasive. The general concept is shown in Fig. 3.1.

A beam of light with known properties is generated by a light source (e.g., a lamp or laser), and is then directed at the surface of interest. During in situ optical measurements, this beam is typically passed through a window, to separate and protect the light source and the processing environment from each other. Unless the surface is completely black and at absolute zero temperature, such that it absorbs all the incident radiation and emits none, outgoing light will be produced by the interaction of the incoming beam with the surface. The properties of the light that is produced are dependent upon the properties of the surface, so by measuring the outgoing beam, one can infer properties about the surface as it evolves during processing.

The details of the incoming beam depend upon the exact technique being used [3, 4]. Due to the potential for variations in this light source, the incoming beam is also typically measured by a separate detector. In *spectroscopic* methods, the

R. Xiong • M.A. Grover (✉)
School of Chemical & Biomolecular Engineering, Georgia Institute of Technology, Atlanta, GA 30332, USA
e-mail: xiongrt@gmail.com; martha.grover@chbe.gatech.edu

J.J. Gorman and B. Shapiro (eds.), *Feedback Control of MEMS to Atoms*,
DOI 10.1007/978-1-4419-5832-7_3,

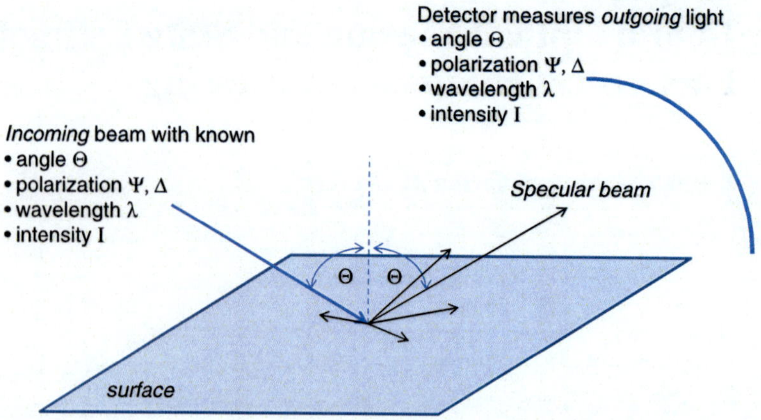

Fig. 3.1 Schematic of an optical surface measurement

incoming light has a continuous distribution of wavelengths. A detector also measures the intensity of the outgoing light as a function of wavelength. *Ellipsometry* measurements refer to the use of a polarized incoming beam, with the detector used to measure the change in polarization that occurs due to interaction with the surface [5]. Spectroscopic ellipsometry measurements can be performed, in which both wavelength and polarization are used to infer the surface structure. Raman techniques rely on the use of multiple wavelengths, but in this case a single wavelength is used for the incoming beam, while the detector measures the small amount of light that is emitted at different wavelengths.

Depending on the technique being used, the detector may measure the outgoing beam at one or at multiple angles. The detector angle may be adjusted to match the incidence angle as it is intentionally varied, to measure the *specular* beam, or the detector may also measure off-specular intensity, such as light scattered by a rough surface, or diffraction patterns from a periodic surface structure.

Real-time optical measurements can be used either for detailed scientific studies such as reaction mechanisms or thin film growth mechanisms; or they can be used as a process monitoring tool for detection and control of growth rate or surface roughness. The former task is referred to as an "optical diagnostic," while the latter is termed an "optical sensor." This latter task of monitoring and control with optical sensors is what we are concerned with here.

3.1.1 Pyrometry and Temperature Control

One commonly used optical sensor is the pyrometer [6, 7]. In this case, there is no actual incoming beam, but instead the detector measures the light emitted by a hot surface. This radiation, which is primarily in the infrared wavelength range, is then measured at normal incidence by a photodetector. The intensity of the light at one or more wavelengths is then correlated to the temperature. Like most optical measurement techniques, the pyrometer must be calibrated. The amount of radiation emitted by the surface is a function not only of the surface temperature but also the emissivity of the surface and the geometry of the measurement system, including the size of the photodetector and the distance between the surface and the detector. The emissivity is a material-dependent surface property that can be measured *ex situ*, while the geometric factors are specific to the particular pyrometer system being used. Temperature control in surface processing has been successfully implemented under the name rapid thermal processing [8, 9]. Continuing challenges in implementation include the difficulty of temperature control across an entire wafer, and implementation of advanced control strategies for tracking more aggressive temperature profiles [10].

One limitation of the pyrometer in surface processing is that the surface emissivity may be changing as the surface is being modified. Thus, it is not possible to infer the temperature based on an *ex situ* measured surface emissivity. One solution that has been proposed is to additionally *estimate* the surface emissivity in real-time, using a normal incidence reflection measurement. This combination is referred to as the emissivity correcting pyrometer (ECP) [11] and has been successfully used for closed-loop temperature control of a chemical vapor deposition process [12]. The ECP measurement relies on a reflection measurement of the surface, such that a single-wavelength unpolarized beam of light is directed at the surface at normal incidence. The normal incidence reflected beam can be measured by the same detector that is used for the pyrometer: a rotating chopper is put in front of the source beam so that the reflected light for the reflection measurement can be measured independently from the emitted light for pyrometry.

There are a number of reasons why the surface emissivity will change during surface modification, and this change will affect the actual surface temperature, as well as the signal measured by the pyrometer. A changing chemical composition will lead to a change in both the surface emissivity and the corresponding temperature dynamics. An endpoint control strategy was recently implemented for composition control in solar cell materials, by relating the emissivity to the thin film copper composition [13].

An important optical feature of thin films is the ability to create constructive and destructive interference. When the film is at least partially transparent, then the multiple internal reflections within the film can lead to interference of light. The ratio of the film thickness to the wavelength of the light is needed to predict the interference, and thus the interference effect can even be used to infer film thickness. This effect is illustrated in Fig. 3.2, for an incoming beam of light. However, even with no incoming beam, the emission of radiation from the underlying substrate can also cause constructive and destructive interference [14], and, therefore, also an oscillation in the emission of radiation from the surface. This oscillation due to emission has been used for film thickness control via endpoint detection [15]. The surface roughness of a thin film can also further alter the emissivity [16], and emission measurements have even been used to infer surface roughness during

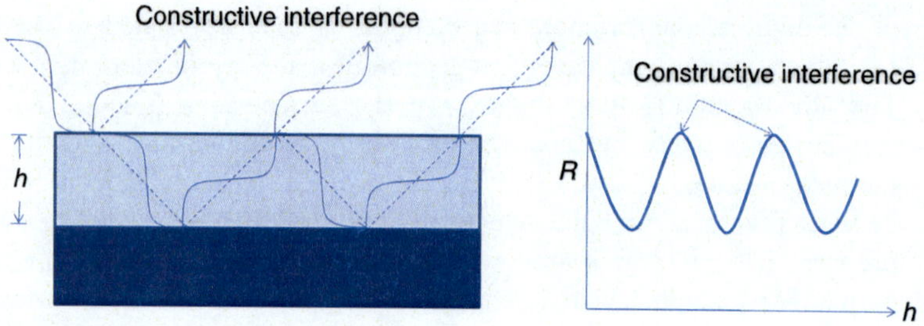

Fig. 3.2 Illustration of constructive interference in a thin transparent film, and the resulting dependence of film reflectance R on the film thickness h

deposition [17]. It is also understood that patterned surfaces, such as those created by lithographic patterning for microelectronics applications, will alter the emissivity of a surface [18].

3.1.2 Reflectometry and Film Thickness Control

The reflectometer alone, independent from the ECP system, has also been successfully used for in situ measurement and control of surface properties, especially in the deposition and removal (etching) of thin films. When a beam of light having a single wavelength is directed at a transparent thin film, some light will reflect from the top surface of the film, but also some of the light will be transmitted into the film. When the light reaches the interface between the thin film and the substrate, a portion may be transmitted into the substrate bulk, while the remainder will be reflected back into the film. Depending upon the thickness of the film and the wavelength of the light, the multiple beams of light that are transmitted back out of the film toward the detector will interfere either constructively or destructively, as shown in Fig. 3.2. During surface modification a film may either become thicker during deposition or thinner during etching, and this film thickness h can be monitored using a reflectometer, if the index of refraction of the film is known. In this case, the numbers of peaks in the reflectance measurement can be directly mapped to the thickness h.

In most thin film deposition and etching processes, the nominal desired growth rate of the film is constant throughout the process. Therefore, the film thickness h is often determined by first estimating a constant growth rate G. Once G has been determined, the time-varying film thickness h is simply the time-integral of G. To relate the measured reflection to the film thickness, the refractive index n and extinction coefficient k (absorption) of the film must be known. When the film is partially transparent, then the optical properties of the underlying substrate, n_s and k_s, must also be known. Breiland and Killeen considered the inverse problem: how to simultaneously estimate five constant parameters (G, n, k, n_s, and k_s) using a single-wavelength normal incidence reflection measurement [19]. One challenge articulated by this study was that these five parameters to be estimated are highly correlated with each other in the reflectance measurement.

They demonstrated in simulation and then experimentally that all five quantities can be estimated during film deposition, as long as at least $1/4$ of an oscillation is present. The parameter estimation was performed by minimizing the error between the model and the measurements, with no consideration of the correlations between the parameters. Breiland and Killeen also suggested the possibility of real-time control of the growth rate. However, in their parameter fitting method they assume that all four optical constants and the film growth rate are constant, which would not be a good assumption if the chemical composition or growth rate is the quantity to be controlled.

There have been several reports of the use of reflectometry for real-time estimation of film thickness. The most common estimation strategy is the extended Kalman filter (EKF), which has been applied to both film deposition by chemical vapor deposition [20, 21] and to etching for film removal [22]. Recently we have applied moving horizon estimation (MHE) [23] to estimate film thickness in situ [24, 25] and have compared the performance and robustness of MHE to EKF in our chemical vapor deposition system [26]. In fact, MHE is a generalization of EKF, and also a generalization of the least squares fitting performed by Breiland and Killeen [19]. This work will be discussed in detail in Sect. 3.2.

Reflectometry has also been used for real-time control of film thickness. Vincent et al. used the EKF with a proportional–integral controller in a silicon etch process to reduce final thickness variation between runs [27]. Lee et al. used least squares estimation along with a model-based multivariable controller to achieve uniform film thickness across a wafer, by manipulating the heater power in various zones [28].

The previous discussion and analysis in this section on reflectometry does not take into account any inhomogeneities in the surface or film. During most surface processing, the surface of the film is not atomically smooth, and the interior of the film is not a single perfect crystal. Polycrystalline thin films are often deposited when a film of one material is deposited on a substrate of another material. The individual crystals, or grains, may have preferred (low-energy) crystal facets, leading to a nonsmooth surface, with grain boundaries inside the film that alter the effective refractive index and extinction coefficient. Lithographically patterned surfaces also alter the reflectivity. Both cases are illustrated by Fig. 3.3.

As with emissivity and the pyrometry temperature measurement, it is possible to model the reflection behavior of microscopically rough thin films [4, 24–27]. A few studies have gone further, and considered the problem of estimating the surface roughness with a reflectometry measurement [29–33]. In the work of Zuiker, the thickness, surface roughness, and extinction coefficient were simultaneously estimated, although the results were mixed, with negative (unphysical) absorption and noisy estimates [29]. Recently we applied moving horizon estimation to the problem of estimating thickness, optical constants, and roughness in our chemical vapor deposition process [24], which we discuss in Sect. 3.2. One challenge in estimating surface roughness is in selecting a sufficiently accurate optical model.

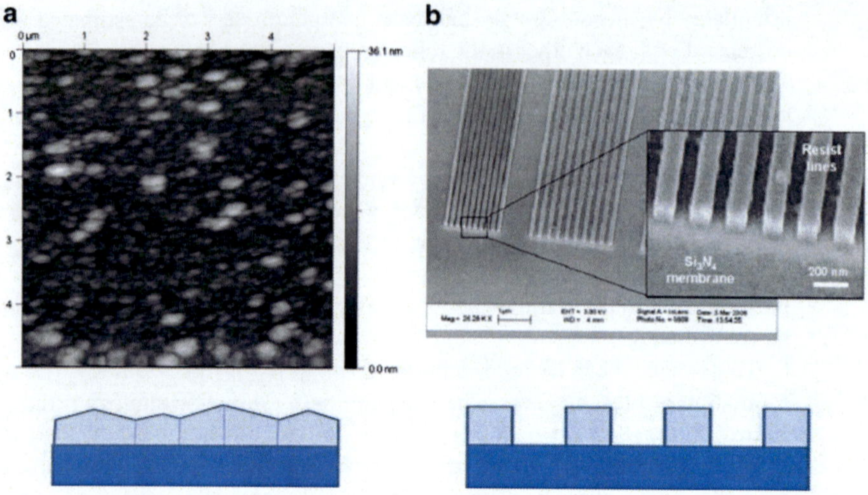

Fig. 3.3 Examples of nonsmooth surfaces in thin film deposition: **(a)** atomic force microscopy (AFM) image of a polycrystalline yttrium oxide thin film (Grover lab). The lateral scale is in micrometers, while the vertical scale is in nanometers; the schematic underneath represents the crosssection of the film and substrate. **(b)** Electron-beam (e-beam) image of a lithographic resist. The vertical and horizontal features are measured in nanometers. Line edge roughness is visible along the lines (image courtesy of Cliff Henderson and Richard Lawson)

For large scale features, geometric optics can be used [16, 34], while for smaller scale features, scalar scattering theory [35] or the effective medium model [36] may be more appropriate.

3.1.3 Spectroscopic Ellipsometry for Control of Thickness and Composition

Compared to reflectometry, the measurements provided by ellipsometry can provide additional information about the surface, by monitoring the polarization change of the incident light due to the surface [5]. However, despite the additional information provided, many of the same challenges seen in reflectometry also apply here. Approximate inversion relations have been proposed for certain limiting cases [37], as also done in reflectometry [19], but in general the inversion requires consideration of the full nonlinear model. The parameters to be estimated again have significant correlations in the optical model [38, 39]. The surface roughness of a thin film can also be estimated using ellipsometry measurements; in fact ellipsometry has been used extensively in the development and validation of optical models, such as the effective medium theory [5, 40].

Despite the challenges of the estimation problem, ellipsometry has been used for real-time monitoring and control of thin film composition and thickness. The original work of Aspnes established the feasibility and utility of feedback control for nanoscale graded structures comprised of compound III–V semiconductors [41]. The measured optical properties were used to adjust the aluminum precursor flowrate in real-time to track the desired trajectory of film stoichiometry. During the past decade, more advanced control analysis has been applied. Model-based controller design for III–V layered structures [42, 43] has been used to control the thickness and chemical composition. The ellipsometer's vendor-supplied software was used to obtain estimates of these two quantities, and then two single-input, single-output linear controllers were implemented. In the deposition of silicon–germanium alloy films, the extended Kalman filter was used for state estimation of thickness and composition [44], with model predictive control [45] implemented to enable the simultaneous control of thickness and composition. Model predictive control (MPC) requires an online optimization after each measurement to compute the new control action, and thus requires significantly more online computation than a precomputed feedback law such as a proportional–integral controller. However, MPC also holds the promise to improve tracking performance.

3.1.4 Scatterometry for Lithography Control

The extensive use of nanoscale patterning in the microelectronics industry creates an additional challenge for in situ optical sensing. The International Technology Roadmap for Semiconductors outlines the short-term and long-term challenges for the industry, and metrology continues to be an area of critical need [46]. Feedback control in semiconductor processing often takes the form of run-to-run control, in which a wafer is examined after the process is completed, so that the recipe can be adjusted for subsequent wafers [1]. In situ sensing and control is less common, due to the difficulty of both the measurement and the adjustment, but it also holds greater promise for reducing measurement delays and correcting the process sooner, which is desired due to the high cost of each wafer that is processed.

To monitor and control patterned features in microelectronics fabrication, periodic test structures are created on a small portion of each wafer. By monitoring properties of these regular structures, changes in the overall process can be detected and corrected. The critical dimension (CD), which is the characteristic length scale in a transistor, must be tightly controlled, and additionally the line edge roughness (LER) of these features must be minimized. As shown in Fig. 3.3b [47, 48], lithographically patterned features are measured in nanometers. The length scale of the CD is measured in tens of nanometers, while LER is measured in nanometers.

Current research in metrology technology is largely focused on CD-SEM and scatterometry (or optical CD, or OCD) [49]. CD-SEM is fundamentally suited toward run-to-run control, since the surface cannot be imaged in the SEM (scanning electron microscope) while it is being processed. In contrast, scatterometry is based on an optical reflection measurement, and can be used for either run-to-run or in situ control. The detector in a scatterometer actually measures a diffraction pattern from the surface, since a regular array of nanoscale lines will cause the incoming light to be diffracted. This diffraction pattern is very sensitive to the dimensions and shape of the pattern, and thus can be used for monitoring and for control [50]. Fundamental optical models are used for predicting the diffraction pattern of different surface profiles, and table lookup or linear regression models are used to invert this database in order to estimate the surface profile from the measured diffraction pattern. As with the other optical measurements discussed in this chapter, a unique inverse does not always exist, due to correlations in the parameters to be estimated [50].

3.2 Moving Horizon Estimation in Chemical Vapor Deposition Using In Situ Reflectometry Measurements

Because the models of optical response are nonlinear and can be high-dimensional, the inversion of these models – to obtain surface structure from optical measurements – is challenging. Significant correlations in the fitted parameters are known to occur, yet these correlations are rarely used in the actual estimation process. Formal estimation techniques, such as the extended Kalman filter (EKF) and moving horizon estimation (MHE), directly predict and then use these correlations, thus providing the opportunity for improved estimation [26].

In most reported inversion studies from the surface physics community, a different approach is employed to obtain accurate estimates of multiple film properties – the sum squared error between the measurements and the model predictions is minimized, over some predefined window of past data. The additional information provided by using multiple measurements enables a successful inversion to estimate the film properties. However, the window of data that is required may be long. In the reflectometry study by Breiland and Killeen, it was reported that the window must be at least one-fourth of an oscillation of the reflectance measurement [19]. In the case of very thin films, this may be a significant portion of the entire deposition, eliminating the possibility for real-time correction of the process. Moreover, in this method of simple least squares fitting to the optical measurement, the variables to be estimated are constrained to be constant parameters, even though they may be varying over the time span of the window.

3.2.1 Experimental Apparatus

We have applied several estimation methods to reflectometry measurements of a chemical vapor deposition (CVD) process [25, 51]. It is a thermal CVD process, such that the source material, or *precursor*, undergoes a chemical reaction that is

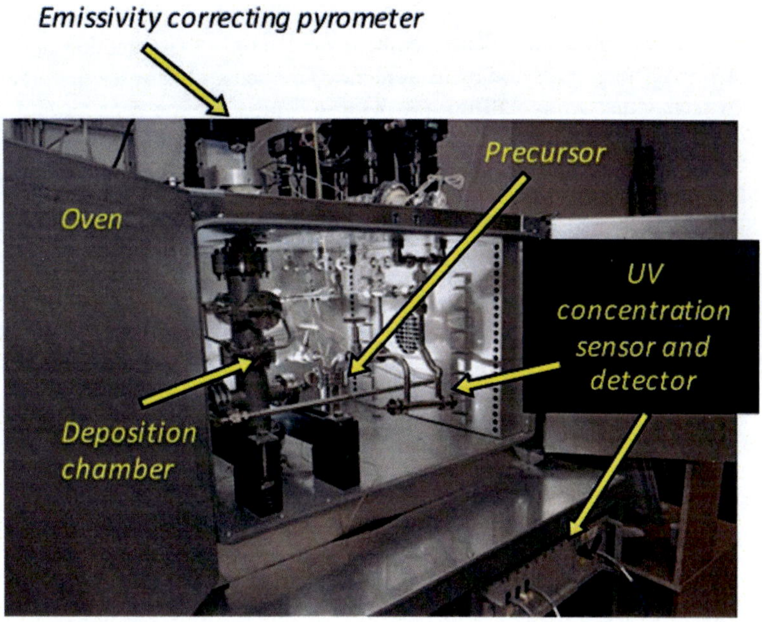

Fig. 3.4 Experimental apparatus for chemical vapor deposition of metal oxide thin films

induced by the high temperature (~700°C) of the substrate. In our studies this substrate is a silicon wafer. The metal organic precursor material (yttrium-2,2,6,6-tetramethyl-3,5-heptanedionate) is a solid-phase powder at room temperature, but at elevated temperatures near 150°C, the precursor sublimates into the gas phase. In our deposition system we flow an inert gas (argon) through the precursor canister and into the deposition chamber. When the precursor reaches the vicinity of the hot substrate, it breaks down to form a solid thin film on the substrate. A photograph of the CVD system is shown in Fig. 3.4.

The deposition chamber is constructed from two four-way stainless steel crosses, and is on the left side of the oven in Fig. 3.4. The right side of the oven houses the canister containing the precursor material, as well as an ultraviolet sensor used for monitoring and control of the inlet precursor concentration [52]. All of the lines from the precursor canister through the deposition chamber are housed in the oven so that the precursor does not re-condense in the lines or on the precursor walls. The oven is held between 150 and 200°C throughout the deposition process.

A vacuum pump and pressure control valve outside the oven are connected to the outlet of the deposition chamber, enabling the regulation of a setpoint pressure in the range of 1–10 torr. Mass flow controllers regulate the flow of argon and oxygen into the oven. The oxygen is needed to promote the chemical reaction and to provide the oxygen for the yttrium oxide (Y_2O_3) thin film. Our reflectometer is mounted above the reactor and oven, as shown in Fig. 3.4. This sensor is actually an emissivity-correcting pyrometer (purchased from SVT Associates), although in the estimation studies described here we use only the reflection measurement (not the pyrometer emission measurement). This sensor measures the reflectance at two wavelengths: $\lambda_1 = 950\,\text{nm}$ and $\lambda_2 = 470\,\text{nm}$. All the sensors and actuators are connected to a LabView program, which enables monitoring, data acquisition, and real-time control. More detail on the apparatus and experimental procedure can be found in our previous publications [21, 24, 25].

3.2.2 State Space Model

For this estimation problem we consider the general nonlinear autonomous discrete-time state space system

$$x_{j+1} = f(x_j) + w_j \tag{3.1}$$

$$y_j = g(x_j) + v_j \tag{3.2}$$

where j is the discrete-time index, x_j is the internal state at time j, and y_j is the measurement at a time j. Noise variables w_j and v_j are usually assumed to be zero mean, independent, and normally distributed for convenience, although in practice they might also represent the effects of unmodeled dynamics or correlated random disturbances.

In our work, the process model f is a simple linear function representing parameters that may drift, with additional integrating states that represent the thickness of the film, h, and the thickness h_e of the "roughness layer" in the effective medium model. The G's are growth rates of the two layers.

$$\begin{bmatrix} h \\ G \\ h_e \\ G_e \\ n_1 \\ k_1 \\ n_2 \\ k_2 \end{bmatrix}_{j+1} = \begin{bmatrix} 1 & \Delta t & 0 & 0 & 0 & 0 & 0 & 0 \\ 0 & 1 & 0 & 0 & 0 & 0 & 0 & 0 \\ 0 & 0 & 1 & \Delta t & 0 & 0 & 0 & 0 \\ 0 & 0 & 0 & 1 & 0 & 0 & 0 & 0 \\ 0 & 0 & 0 & 0 & 1 & 0 & 0 & 0 \\ 0 & 0 & 0 & 0 & 0 & 1 & 0 & 0 \\ 0 & 0 & 0 & 0 & 0 & 0 & 1 & 0 \\ 1 & 0 & 0 & 0 & 0 & 0 & 0 & 1 \end{bmatrix} \begin{bmatrix} h \\ G \\ h_e \\ G_e \\ n_1 \\ k_1 \\ n_2 \\ k_2 \end{bmatrix}_{j} \tag{3.3}$$

The refractive index n and extinction coefficient k are estimated at each of the two measurement wavelengths λ_1 and λ_2. In contrast to the simple process model, the sensor model g must be very accurate, if the sensor is to be useful. It predicts the reflectance R, which is the fraction of incident light that is reflected back to the detector at normal incidence. In our work on estimating roughness [24], we used the equation for multilayer reflectance response [5, 53–55], including the roughness as a separate layer according to effective medium theory and the Bruggeman approximation [5, 40].

$$y = R + v = \left| \frac{r_{12} + r_{23}\mathbf{e}^{-i2\delta_2} + r_{34}\mathbf{e}^{-i2(\delta_2+\delta_3)} + r_{12}r_{23}r_{34}\mathbf{e}^{-i2\delta_3}}{1 + r_{12}r_{23}\mathbf{e}^{-2\delta_2} + r_{12}r_{34}\mathbf{e}^{-2(\delta_2+\delta_3)} + r_{23}r_{34}\mathbf{e}^{-2\delta_3}} \right|^2 + v \qquad (3.4)$$

Here, medium 1 is the surrounding vacuum, medium 2 is the roughness layer, medium 3 is the film, and medium 4 is the substrate. Thus, r_{ab} is the interface reflectivity between media a and b. A phase change δ_a occurs as the light passes through medium a. Thus, the overall reflectance of the film-substrate system depends on r_{ab} and δ_a. Both quantities depend upon the refractive indices and extinction coefficients of the film and substrate. Specifically,

$$\begin{aligned} r_{ab} &= \frac{\widehat{n}_a - \widehat{n}_b}{\widehat{n}_a + \widehat{n}_b} \\ \delta_a &= \frac{2\pi \widehat{n}_a h_a}{\lambda} \\ \widehat{n}_a &= n_a - ik_a \end{aligned} \qquad (3.5)$$

Further detail on our model and assumptions can be found in [24].

3.2.3 Moving Horizon Estimation

Moving horizon estimation (Fig. 3.5) [23,26] is based on the concept of least squares fitting. A model is used to compare the measured data to the model-predicted data, from (3.2), over a window of the m most recent time intervals. Ideally these two quantities would match identically, but in practice there will always be some nonideal behavior. The idea in MHE is to include additional sources of information to estimate the true value of the state, including the expected relationship between states from (3.1), and the expected value of the state at the beginning of the window. Expressed formally, one arrives at the following minimization problem:

$$\min_{x_{j-m+1},\ldots,x_j} \left[\left(x^e_{j-m+1}\right)^T P^{-1}_{j-m+1|j-m} x^e_{j-m+1} + \sum_{l=j-m+1}^{j} v_l^T R^{-1} v_l + \sum_{l=j-m+1}^{j-1} w_l^T Q^{-1} w_l \right] \qquad (3.6)$$

$$x^e_{j-m+1} = x_{j-m+1} - x_{j-m+1|j-m} \qquad (3.7)$$

$$v_l = y_l - g(x_l) \qquad (3.8)$$

$$w_l = x_{l+1} - f(x_l) \qquad (3.9)$$

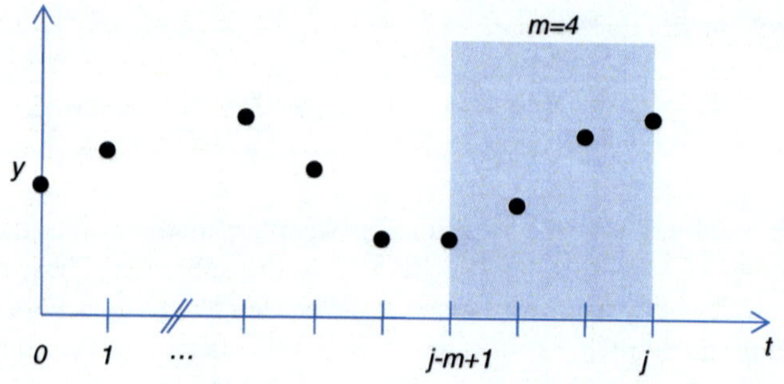

Fig. 3.5 Illustration of moving horizon estimation. Only the most recent m measurements are directly included in the estimation. As each new measurement is acquired, the window of data (shaded in the figure) moves to the right. Earlier measurements are included via the a priori state estimate at $k = j - m + 1$

In this minimization problem, the state at each point in the window is varied. The quantity to be minimized has three terms, each representing a squared error in the estimate. The first term represents the uncertainty in the state estimate at the beginning of the window, $t = (j - m + 1)\Delta t$. It is computed using an extended Kalman filter, based on the measurements only up to time $j - m$. The covariance matrix of this state estimate, P, is also calculated by the extended Kalman filter, and acts as a weighting matrix. The second term in the minimization is the sum squared error associated with the measurement, v_j. The weighting matrix for this term comes from R, which is the covariance matrix on the sensor error. The third term is the deviation from the process model at each point in the window, w_j, with its corresponding covariance matrix Q.

In addition to applying the general MHE method to our reflectometry measurements, we apply several limiting cases. The first is the simple least squares fitting approach over a window, as applied by Breiland and Killeen [19]. In this case, no initial estimate is used at the beginning of the window, which is equivalent to setting $P^{-1} = 0$ in the minimization problem. Additionally, the process model is used as a constraint instead of as a term in the objective function, which is equivalent to setting $Q = 0$. Thus, the only term retained in (3.6) is the second term, which corresponds to the difference between the measurements and the corresponding model predictions. With these simplifications, only the initial state at the beginning of the window is varied during the optimization, since the remaining states are computed using the process model. This greatly reduces the dimension of the minimization problem.

We also applied an intermediate approach, in which the estimate at the beginning of the window is included in the objective function, while the process model is constrained to be deterministic such that $Q = 0$. We find that this approach provides good state estimates compared to MHE, due to the slowly drifting nature of our disturbances, but that the computation associated with the optimization is significantly reduced [25]. In other words, the use of the state estimate at the beginning of the window (computed with the EKF) is helpful in avoiding overfitting of noisy data in the window, by including a priori knowledge of the state based on the earlier measurements.

As a further comparison, we also apply the EKF alone, which is conceptually equivalent to MHE with $m = 1$. In fact, the two are not exactly equivalent due to a modification of the EKF used in the MHE algorithm [23]. Regardless, the EKF does not achieve the same robustness as the estimation methods based on a longer window of data, which can better account for the nonlinearities in the sensor model.

Implicit in this discussion is that our nonlinear state space system will be observable, i.e., that it is theoretically possible to infer the states from the measurements [56], although there is no guarantee that would be the case. Significant correlations exist in the states to be estimated. In particular, the product hn is directly related to the period of the oscillation via the phase change δ, but h and n can be more difficult to independently estimate [19]. Moreover, both the extinction coefficient k and the surface roughness h_e cause a decay in the amplitude of the reflectance oscillation, so it may also be difficult to separate out these two effects using the measurements.

A common test of local observability is based on the linearization of the system about a single operating point, but for any single operating point, the observability matrix for our system is not full rank, suggesting that the state is not observable. However, this is not a necessary condition for nonlinear observability [57]. Here we use the notion of strong local observability [58], meaning that any two trajectories that start nearby can be distinguished from each other, given a sufficiently large window of measurements. While this concept of observability is consistent with our "window-based" estimation methods, it is unclear how well the extended Kalman filter will work, since it only processes one measurement at a time, based on the linearization of the model around the current estimate. We find in our simulation studies that the EKF can converge under near-ideal conditions, but that the other three methods using a *window* of data at each iteration are much more robust to realistic noise levels and disturbances.

3.3 Results

The surfaces of our polycrystalline yttria thin films are rough, as shown in Fig. 3.3. The root-mean-square roughness, as measured by atomic force microscopy (AFM), is typically in the range of 20–30 nm, while the lateral grain size is near 1 μm. Even though this surface roughness is an order of magnitude less than the wavelength of light, the surface roughness can still affect the normal incidence surface reflectance R. We have confirmed this point using ex situ ellipsometry measurements, after removing the films from the deposition chamber. The ellipsometry software uses the effective medium model to fit a roughness layer that is approximately 50 nm. The thickness of the effective rough layer predicted by the effective models has previously been correlated to be twice the root-mean-square roughness [35]. Since our effective layer is 50 nm and our root-mean-square roughness is 20–30 nm (half

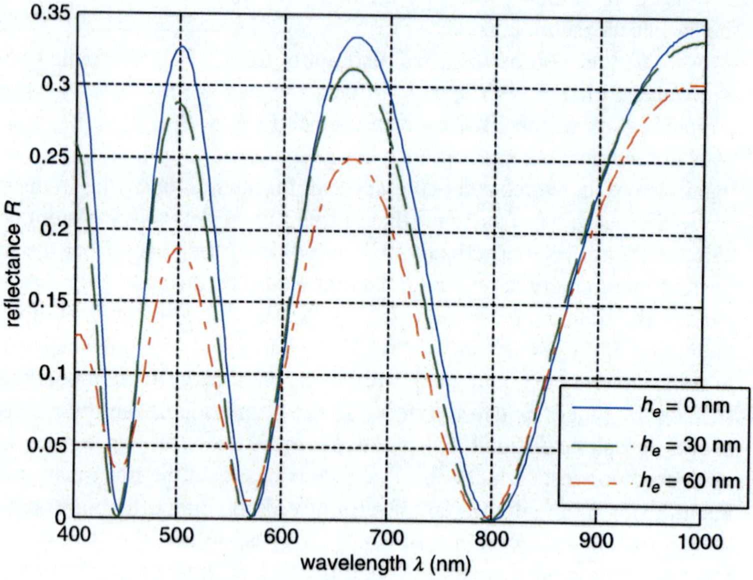

Fig. 3.6 A simulated spectral reflectometry measurement. Comparison of thin film reflectance R with and without the effective roughness layer, for various levels of roughness

of 50 nm), it may be possible to estimate the actual surface roughness measured by AFM from our in situ reflectometry measurements.

Since roughness does in fact alter the reflectometry measurement, we can ask two specific questions: (1) by neglecting the effect of roughness on the measurement, can we still accurately estimate the film thickness? (2) by including roughness in our optical model, can we simultaneously estimate thickness and roughness. We consider both questions in this chapter. We addressed the first question in [25], where we did not include a roughness optical model in the moving horizon estimation. Instead, we viewed the roughness effect as an unmodeled disturbance, and showed that the modified moving horizon estimates of film thickness and refractive index are significantly more accurate, compared to the simple least squares method and to the extended Kalman filter. The window of data together with the state estimate at the beginning of the window dramatically improved the robustness of the estimator, particularly when there were significant noise levels and/or unmodeled effects.

We addressed the second question in a separate publication [24], using our nominal sensor model in (3.4) to quantify the relationship between surface roughness and reflectivity. Before actually implementing the estimator, we consider what effect the roughness would have on the reflectivity, based on the effective medium approximation. This prediction is shown in Fig. 3.6, in which the reflectance R is plotted as a function of wavelength. This plot is a simulation of an ex situ spectroscopic reflectometry measurement, for a film with a thickness of 500 nm, which is typical for our CVD thin films. Similarly, the optical constants of the substrate are chosen to represent silicon ($n = 4.0$ and $k = 0.1$) [59] and the film to represent yttria ($n = 2.0$ and $k = 0.01$) [24, 60]. Note that the two wavelengths of our in situ sensor are 470 nm and 950 nm, which are both contained within the wavelength range on the plot. During film growth, we observe oscillations due to the changing thickness at a constant wavelength, while in Fig. 3.6 the interference oscillations are due to a changing wavelength at a constant thickness. In fact, it is the ratio of the thickness and wavelength that is critical for creating the oscillations, as seen in the expression for δ_a in (3.5).

Clearly the effect of surface roughness is significant in the effective medium model, even at roughness values much less than the measurement wavelength. Another important trend in Fig. 3.6 is that the roughness effect is more significant at the shorter measurement wavelengths. For example, at 470 nm we should expect our measured reflectance to be more altered than at 950 nm. For a roughness layer of 30 nm, and a measurement at 470 nm, R should be altered only near the maximum in the oscillation (constructive interference), while at a higher roughness of 60 nm, R will be altered at all phases of the interference.

We explicitly modeled and estimated this surface roughness to simultaneously estimate the optical constants, film thickness, and roughness [24]. Typical results are shown in Figs. 3.7 and 3.8. Figure 3.7 shows the estimates of all eight states from (3.3) using a simple least squares fitting of the optical model. The estimate of the film thickness is quite reasonable up until the end of the deposition, while the surface roughness and film extinction coefficient are highly variable and appear to be overfitting nonideal aspects of the experimental measurements. In contrast, Fig. 3.8 illustrates the results of moving horizon estimation with the deterministic process model. Although the computational demands are similar to the least squares fitting, the results are much more consistent with our understanding of the CVD process. The period of oscillation in the reflectance is increasing throughout the deposition, indicating that the film growth rate is decreasing. This can also be seen in the estimates of h and G. Overall the estimates are much less sensitive to the choice of horizon lengths, compared to Fig. 3.7. The final film roughness and refractive index are near the nominal values expected for this process. The estimates of k may be less reliable, perhaps due to the overall low absorbance of yttria.

3.4 Discussion, Perspectives, and Conclusions

The use of optical sensors to estimate and control surface properties is not easy or straightforward. However, the use of optical diagnostics in surface science is extremely common and widespread, due to the wealth of qualitative and quantitative information that can be inferred about a surface.

A number of feedback control strategies based on optical sensors have been successfully demonstrated. They range in complexity from a simple endpoint detection [13] or a proportional–integral controller [27], which requires little

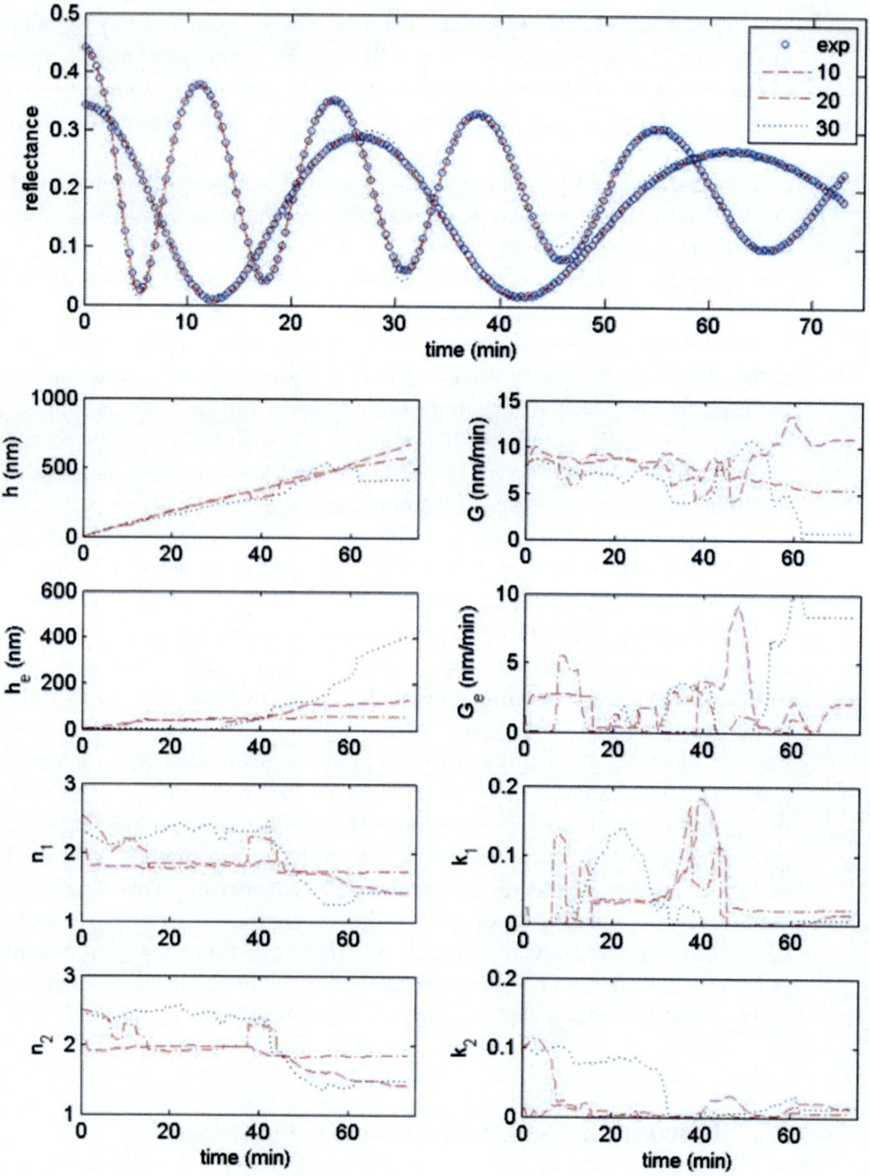

Fig. 3.7 Application of simple least squares fitting to experimental CVD data, for three horizon lengths

real-time computation, up through multivariable model predictive control, which requires an on-line optimization after each measurement [45]. Other model-based control strategies have been proposed in the literature, including methods based on stochastic atomic-scale models of thin film growth [61]. The best controller

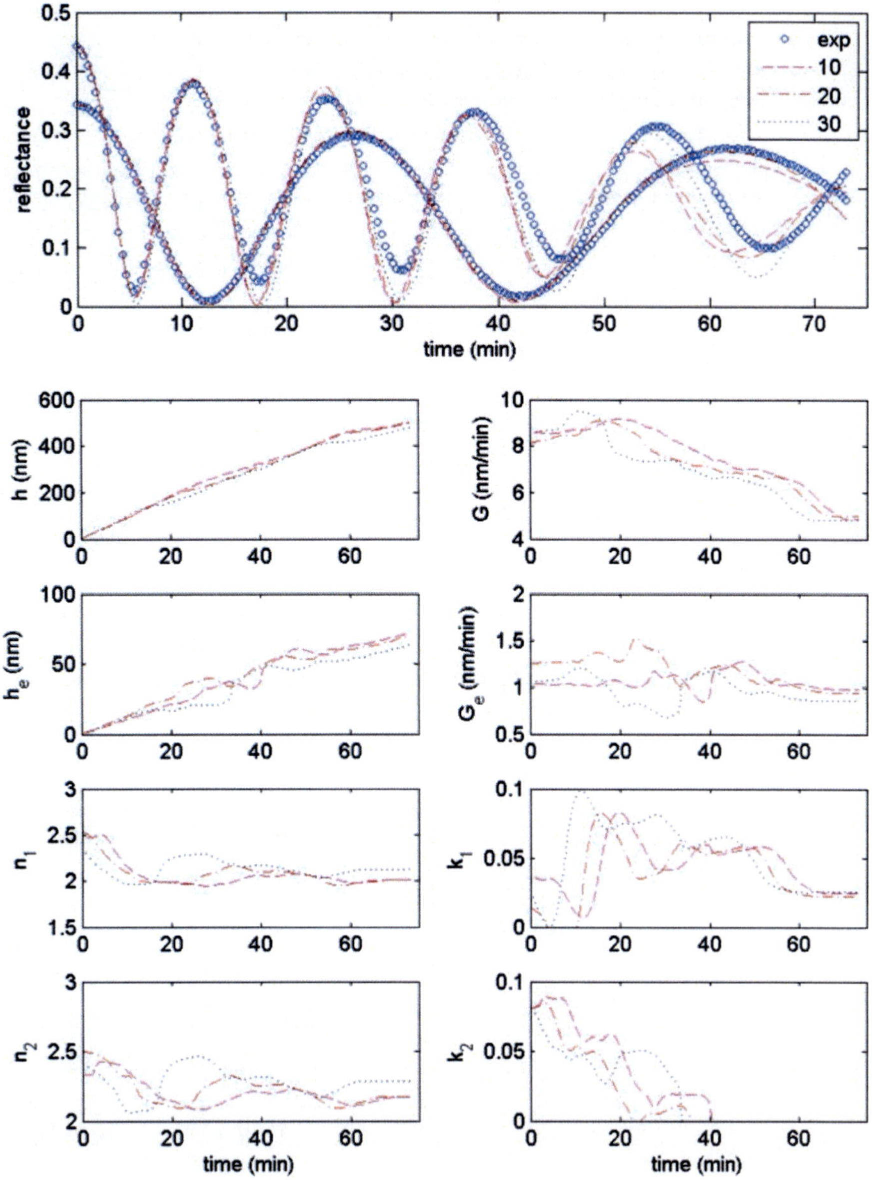

Fig. 3.8 Application of the modified moving horizon estimation to experimental CVD data, for three horizon lengths

structure for a particular application will depend on the dynamics of the system, the estimated quantities available for feedback, and the trajectory to be tracked. However, in many cases a simple controller will suffice to correct for the slow drift that often characterizes disturbances to deposition and etching processes – once an accurate state estimate has been obtained.

The difficulty of the sensor modeling task – computing the optical response of a known surface structure – has not been the focus of this book chapter, but of course it is essential when using any optical diagnostics or sensors. Due to the importance of optical diagnostics in scientific research, the modeling of optical measurements is quite advanced. Of course, any such model should be validated against ex situ experimental data before actually implementing any real-time estimation and process control.

One unique challenge for optical sensors and process control, relative to optical diagnostics for scientific inquiry, is the need for real-time inversion of the optical model. Depending upon the speed of the process and the sampling time of the sensor, the allowable computation time for the inversion may be measured in seconds or as long as minutes. Of the inversion and estimation methods discussed here, the extended Kalman filter has the lowest computation, while moving horizon estimation has a computational demand that scales with the length of the window [25]. The use of the deterministic process model in MHE enables a significant reduction of the computation to near that of the EKF and simple least squares methods. All of the estimation methods discussed here were implemented using a local optimization algorithm, so if multiple minima exist, the global minimum may not be found. However, because the optimization methods can be implemented with a "warm start" based on the previous estimate, it is less likely that an incorrect local minimum would be identified as the estimate.

Library-based methods and regression models used in scatterometry provide a conceptually different approach in which the dynamics of the process are not used. These methods are more likely to converge to the wrong local minimum, because there is no inclusion of dynamics. These methods have been developed for scatterometry as an ex situ diagnostic, but estimation methods like MHE that specifically incorporate a process dynamic model could be applied to in situ scatterometry, yielding more robust estimates than for the ex situ case, assuming that some a priori knowledge of the initial surface structure exists.

An additional challenge of in situ sensing and control (relative to ex situ optical diagnostics) is that the sensor measurement may have significant noise. A photodetector measures the number of photons incident on its surface, and converts this intensity into a voltage measurement. Because the surface is not changing in an optical diagnostics application, the measurement can be integrated over a long period of time, such that the noise level is low. However, the integration time in a control application must be shorter than the typical time scale of the process dynamics, so the noise level in the sensor may be higher.

An advantage of estimation methods like EKF and MHE is that the covariances of the estimates and the noise sources are modeled and predicted. Due to the significance of unmodeled dynamics and disturbances, this explicit consideration of uncertainty and correlations can improve the accuracy and robustness of the estimates. However, this advantage comes with a price. Certain aspects of the uncertainty must be specified before the estimator can be used. In our work on EKF and MHE, this includes the noise covariance matrices Q and R, the initial state estimate x_0, and the error covariance of this initial estimate P_0. If these matrices are incorrectly specified, the resulting estimates might be even worse than if a simpler inversion method was used. Fortunately, our work suggests that the exact values of these matrices are not critical [24]; what is most important is that the largest sources of uncertainty are assigned the highest values of the covariance. In any case, the need to specify weighting matrices in EKF and MHE makes their implementation more complicated.

An additional modeling step that has not been discussed at length here is the selection of the states or parameters to be estimated. This will be of critical importance in any inversion from optical measurements to surface properties. The surface being measured can potentially be characterized by an infinitely large number of states, via a Fourier transform, or even by quantifying the location of each atom on the surface. Of course this would not be practical due to measurement limitations, and would also not be relevant for most engineering and control applications. In any case, the user must make a decision about which metrics to use in characterizing the surface. In the case of polycrystalline thin films, this set may contain the average film thickness, the average width of grains, and the average height of grains (surface roughness). However, due to insensitivity of the measurement technique to the lateral width of grains, this quantity may not be included in the inversion of the optical model. In our two recent papers we considered the cases of estimating surface roughness [24] or leaving it out [25], even though the two films under consideration were similarly rough.

In general, as one adds more metrics to be estimated, the simultaneous estimation of all parameters becomes more difficult, and this will be reflected in larger covariance matrices for the prediction error. In the case of patterned surfaces, similar decisions must be made. Even though test sections are intentionally designed to be easily characterized, one must parameterize the deviation from the desired structure, using metrics such as line edge roughness and critical dimension. Ideally, the metrics that have been neglected will either not vary or else they will not be significant to the measured signal. This situation could be studied by treating the surface as an infinite dimensional system, and considering the invertibility and null space based on a particular optical measurement.

In conclusion, there is widespread use of optical measurements in industry and academia to quantify microscale and nanoscale surface structure and in some cases to also control it. Nonlinear models relate the infinite-dimensional surface to the noisy optical measurements, and significant challenges remain in the estimation of surface structure. Better approaches for inverting this relationship are needed. Due to the importance of optical measurements in nanoscale surface structure and process control, there should be significant future interest and development in optical measurement technology. Correspondingly, significant future opportunities exist for control researchers and systems engineers to join with the domain experts to tackle practical needs on a case-by-case basis. Systems engineers must bring and synthesize their expertise in modeling, analysis, statistics, and control to solve critical problems in nanoscale processing, which may also further elucidate the need for new systems theories and tools.

Acknowledgments This research was supported by a grant from the National Science Foundation, entitled "CAREER: A Systems Approach to Materials Processing." The research described in this chapter was performed with graduate student Rentian Xiong, with additional support on the experimental work by Paul Wissmann. Helpful discussions with Georgia Tech Professors Jay Lee and Ian Ferguson are appreciated. The generous contribution of the e-beam image of a lithographically patterned surface by Georgia Tech Professor Cliff Henderson and graduate student Richard Lawson is also gratefully acknowledged.

References

1. T.F. Edgar, S.W. Butler, W.J. Campbell, C. Pfeiffer, C. Bode, S.B. Hwang, B.S. Balakrishnan, and J. Hahn, "Automatic control in microelectronics manufacturing: Practices, challenges, and possibilities," *Automatica*, vol. 36, pp. 1567–1603, 2000.
2. M. Freed, M. Kruger, C.J. Spanos, and K. Poolla, "Autonomous on-wafer sensors for process modeling, diagnosis, and control," *IEEE Transactions on Semiconductor Manufacturing,* vol. 14, pp. 255–264, 2001.
3. I.P. Herman, *Optical diagnostics for thin film processing*. San Diego, CA, Academic Press, 1996.
4. O. Auciello and A.R. Krauss, *In-situ real-time characterization of thin films*. New York, NY, John Wiley and Sons, 2001.
5. H.G. Tompkins and E.A. Irene, *Handbook of ellipsometry.* Heidelberg, Springer, 2005.
6. *Theory and practice of radiation thermometry.* New York, NY, John Wiley and Sons, 1988.
7. Z.M. Zhang, "Surface temperature measurement using optical techniques," *Annual Review of Heat Transfer,* vol. 11, pp. 351–411, 2000.
8. F. Roozeboom, "Rapid thermal-processing systems - A review with emphasis on temperature control," *Journal of Vacuum Science & Technology B,* vol. 8, pp. 1249–1259, 1990.
9. F.Y. Sorrell, S. Yu, and W.J. Kiether, "Applied RTP optical modeling: an argument for model-based control," *IEEE Transactions on Semiconductor Manufacturing,* vol. 7, pp. 454–459, 1994.
10. A. Emami-Naeini, J.L. Ebert, D. deRoover, R.L. Kosut, M. Dettori, L.M.L. Porter, and S. Ghosal, "Modeling and control of distributed thermal systems," *IEEE Transactions on Control Systems Technology,* vol. 11, pp. 668–683, 2003.
11. W.G. Breiland, "Reflectance-correcting pyrometry in thin film deposition applications," Albuquerque, NM 871852003.
12. J.R. Creighton, W.G. Breiland, D.D. Koleske, G. Thaler, and M.H. Crawford, "Emissivity-correcting mid-infrared pyrometry for group-III nitride MOCVD temperature measurement and control," *Journal of Crystal Growth,* vol. 310, pp. 1062–1068, 2008.
13. C. Chityuttakan, P. Chinvetkitvanich, K. Yoodee, and S. Chatraphorn, "In situ monitoring of the growth of Cu(In,Ga)Se2 thin films," *Solar Energy Materials and Solar Cells,* vol. 90, pp. 3124–3129, 2006.
14. K.A. Snail and C.M. Marks, "In situ diamond growth rate measurement using emission interferometry," *Applied Physics Letters,* vol. 60, pp. 3135–3137, 1992.
15. Z.H. Zhou and R. Reif, "Epi-film thickness measurements using emission Fourier transform infrared spectroscopy-Part II: Real-time in situ process monitoring and control," *IEEE Transactions on Semiconductor Manufacturing,* vol. 8, pp. 340–345, 1995.

16. Y.H. Zhou and Z.M. Zhang, "Radiative properties of semitransparent silicon wafers with rough surfaces," *Transactions of the ASME,* vol. 125, pp. 462–470, 2003.
17. Z.L. Akkerman, Y. Song, Z. Yin, and R.W. Smith, "In situ determination of the surface roughness of diamond films using optical pyrometry," *Applied Physics Letters,* vol. 72, pp. 903–905, 1998.
18. J.P. Hebb, K.F. Jensen, and J. Thomas, "The effect of surface roughness on the radiative properties of patterned silicon wafers," *IEEE Transactions on Semiconductor Manufacturing,* vol. 11, pp. 607–614, 1998.
19. W.G. Breiland and K.P. Killeen, "A virtual interface method for extracting growth rates and high temperature optical constants from thin semiconductor films using in situ normal incidence reflectance," *Journal of Applied Physics,* vol. 78, pp. 6726–6736, 1995.
20. W.W. Woo, S.A. Svoronos, H.O. Sankur, J. Bajaj, and S.J.C. Irvine, "In situ estimation of MOCVD growth rate via a modified Kalman filter," *AIChE Journal,* vol. 42, pp. 1319–1325, 1996.
21. R. Xiong, P.J. Wissmann, and M.A. Gallivan, "An extended Kalman filter for in situ sensing of yttria-stabilized zirconia in chemical vapor deposition," *Computers & Chemical Engineering,* vol. 30, pp. 1657–1669, 2006.
22. T.L. Vincent, P.P. Khargonekar, and F.L. Terry, "An extended Kalman filtering-based method of processing reflectometry data for fast in-situ etch rate measurements," *IEEE Transactions on Semiconductor Manufacturing,* vol. 10, pp. 42–51, 1997.
23. D.G. Robertson and J.H. Lee, "A moving horizon-based approach for least-squares estimation," *AIChE Journal,* vol. 42, pp. 2209–2224, 1996.
24. R. Xiong and M.A. Grover, "In situ estimation of thin film growth rate, complex refractive index, and roughness during chemical vapor deposition using a modified moving horizon estimator," *Journal of Applied Physics,* vol. 103, p. 124901, 2008.
25. R. Xiong and M.A. Grover, "A modified moving horizon estimator for in situ sensing of a chemical vapor deposition process," *IEEE Transactions on Control Systems Technology,* vol. 17, pp. 1228–1235, 2009.
26. E.L. Haseltine and J.B. Rawlings, "Critical evaluation of extended Kalman filtering and moving-horizon estimation," *Industrial and Engineering Chemistry Research,* vol. 44, pp. 2451–2460, 2005.
27. T.L. Vincent, P.P. Khargonekar, and F.L. Terry, "End point and etch rate control using dual-wavelength laser reflectometry with a nonlinear estimator," *Journal of the Electrochemical Society,* vol. 144, pp. 2467–2472, 1997.
28. L.L. Lee, C.D. Schaper, and W.K. Ho, "Real-time predictive control of photoresist film thickness uniformity," *IEEE Transactions on Semiconductor Manufacturing,* vol. 15, pp. 51–59, 2002.
29. C.D. Zuiker, D.M. Gruen, and A.R. Krauss, "In situ laser relectance interferometry measurement of diamond film growth," *Journal of Applied Physics,* vol. 79, pp. 3541–3547, 1996.
30. Z. Yin, H.S. Tan, and F.W. Smith, "Determination of the optical constants of diamond films with a rough growth surface," *Diamond and Related Materials,* vol. 5, pp. 1490–1496, 1996.
31. J.L. Luo, X.T. Ting, P.N. Wang, and L.Y. Chen, "Study on the growth of CVD diamond thin films by in situ reflectivity measurement," *Diamond and Related Materials,* vol. 11, pp. 1871–1875, 2002.
32. G. Comina, J. Rodriquez, J.L. Solis, and W. Estrada, "In situ laser reflectometry measurements of pyrolytic ZnO film growth," *Measurement Science and Technology,* vol. 16, pp. 685–690, 2005.
33. R.S. Balmer, C. Pickering, A.J. Pidduck, and T. Martin, "Modelling the high temperature optical constants and surface roughness evolution during MOVPE growth of GaN using in-situ spectral reflectometry," *Journal of Crystal Growth,* vol. 245, pp. 198–206, 2002.
34. K. Tang, P.A. Kawka, and R.O. Buckius, "Geometric optics applied to rough surfaces coated with an absorbing thin film," *Journal of Thermophysics and Heat Transfer,* vol. 13, pp. 169–176, 1999.

35. C.K. Carniglia, "Scalar scattering theory for multilayer optical coatings," *Optical Engineering,* vol. 18, pp. 104–115, 1979.
36. H. Fujiwara, J. Koh, P.I. Rovira, and R.W. Collins, "Assessment of effective-medium theories in the analysis of nucleation and microscopic surface roughness evolution for semiconductor thin films," *Physical Review B,* vol. 61, pp. 10832–10844, 2000.
37. D.E. Aspnes, "Minimal-data approaches for determining outer-layer dielectric responses of films from kinetic reflectometric and ellipsometric measurements," *Applied Physics Letters,* vol. 62, pp. 343–345, 1993.
38. I.K. Kim and D.E. Aspnes, "Analytic determination of n, k, and d of an absorbing film from polarimetric data in the thin-film limit," *Journal of Applied Physics,* vol. 101, p. 033109, 2007.
39. K. Flock, "The simultaneous determination of n, k, and t from polarimetric data," *Thin Solid Films,* vol. 455–456, pp. 349–355, 2004.
40. D.E. Aspnes, J.B. Theeten, and F. Hottier, "Investigation of effective-medium models of microscopic roughness by spectroscopic ellipsometry," *Physical Review B,* vol. 20, pp. 3292–3302, 1979.
41. D.E. Aspnes, W.E. Quinn, M.C. Tamargo, S. Gregory, S.A. Schwarz, M.A.A. Pudensi, M.J.S.P. Brasil, and R.E. Nahory, "Closed-loop control of growth of semiconductor materials and structures by spectroellipsometry," *Journal of Vacuum Science and Technology A,* vol. 10, pp. 1840–1841, 1992.
42. S.C. Warnick and M.A. Dahleh, "Feedback control of MOCVD growth of submicron compound semiconductor films," *IEEE Transactions on Control Systems Technology,* vol. 6, pp. 62–71, 1998.
43. A. Kussmaul, S. Vernon, P.C. Colter, R. Sudharsanan, A. Mastrovito, K.J. Linden, N.H. Karam, S.C. Warnick, and M.A. Dahleh, "In-situ monitoring and control for MOCVD growth of AlGaAs and InGaAs," *Journal of Electronic Materials,* vol. 26, pp. 1145–1153, 1997.
44. S.A. Middlebrooks and J.B. Rawlings, "State estimation approach for determining composition and growth rate of Si(1-x)Ge(x) chemical vapor deposition utilizing real-time ellipsometric measurements," *Applied Optics,* vol. 45, pp. 7043–7055, 2006.
45. S.A. Middlebrooks and J.B. Rawlings, "Model predictive control of Si(1-x)Ge(x) thin film chemical-vapor deposition," *IEEE Transactions on Semiconductor Manufacturing,* vol. 20, pp. 114–125, 2007.
46. "The international technology roadmap for semiconductors," 2007.
47. C.T. Lee, R.A. Lawson, and C.L. Henderson, "Understanding the effects of photoacid distribution homogeneity and diffusivity on critical dimension control and line edge roughness in chemically amplified resists," *Journal of Vacuum Science & Technology B,* vol. 26, pp. 2276–2280, 2008.
48. R.A. Lawson, C.T. Lee, W. Yueh, L. Tolbert, and C.L. Henderson, "Epoxide functionalized molecular resists for high resolution electron-beam lithography," *Microelectronic Engineering,* vol. 85, pp. 959–962, 2008.
49. M. Adel, D. Kandel, V. Levinski, J. Seligson, and A. Kuniavsky, "Diffraction order control in overlay metrology - a review of the roadmap options," in *Proceedings of the SPIE: Metrology, Inspection, and Process Control in Microlithography XXII* San Jose, CA, 2008, pp. 692202–692201.
50. D.C. Wack, J. Hench, L. Poslavsky, J. Fielden, V. Zhuang, W. Mieher, and T. Dziura, "Opportunities and challenges for optical CD metrology in double patterning process control," in *Proceedings of the SPIE: Metrology, Inspection and Process Control for Microlithography XXII.* vol. 6922, J. Allgair and C. Raymond, Eds. San Jose, CA, 2008, p. 69221N.
51. K.L. Choy, "Chemical vapour deposition of coatings," *Progress in Materials Science,* vol. 48, pp. 57–170, 2003.
52. W.J. Desisto and B.J. Rappoli, "Ultraviolet absorption sensors for precursor delivery rate control for metalorganic chemical vapor deposition of multiple component oxide thin films," *Journal of Crystal Growth,* vol. 191, pp. 290–293, 1998.
53. M. Born and E.R. Wolf, *Principles of optics: Electromagnetic theory of propagation, interference, and diffraction of light.* Oxford, Pergamon, 1980.

54. A.W. Crook, "The reflection and transmission of light by any system of parallel isotropic films," *Journal of the Optical Society of America,* vol. 38, p. 954963, 1948.
55. O. Heavens, *Optical properties of thin solid films.* New York, Academic Press, 1955.
56. R.F. Stengel, *Optimal control and estimation.* Mineola, NY, Dover Publications, 1994.
57. R. Hermann and A.J. Krener, "Nonlinear controllability and observability," *IEEE Transactions on Automatic Control,* vol. 22, pp. 728–740, 1977.
58. H. Nijmeijer, "Observability of autonomous discrete time non-linear systems: a geometric approach," *International Journal of Control,* vol. 36, pp. 867–874, 1982.
59. G.E. Jellison and F.A. Modine, "Optical functions of silicon at elevated temperatures," *Journal of Applied Physics,* vol. 76, pp. 3758–3761, 1994.
60. *E.D. Palik Handbook of the optical constants of solids.* Orlando, Academic, 1985.
61. G. Hu, Y. Lou, and P.D. Christofides, "Model parameter estimation and feedback control of surface roughness in a sputtering process," *Chemical Engineering Science,* vol. 63, pp. 1800–1816, 2008.

Chapter 4
Automated Tip-Based 2-D Mechanical Assembly of Micro/Nanoparticles

Cagdas D. Onal, Onur Ozcan, and Metin Sitti

4.1 Introduction

Fabricating micro/nanostructures has been investigated for decades. Electromechanical systems on these small scales are becoming more and more of a necessity in scientific explorations in physics, biology, and chemistry, since current systems have many limitations in their interactions with small-scale phenomena. The construction of miniature agents has the potential to enable massively parallel operations such as distributed sensing and actuation [1], fast manipulation of small-scale materials for repair operations [2], and smart materials that can change their own shape or physical properties [3].

Among other methods for the fabrication of micro/nanodevices, one possibility is to build robotic systems to manipulate matter. The physics at these small scales is very different from what one is used to on the macroscale. Forces scale differently as dimensions are decreased. Inertial (volumetric) forces, such as weight, that dominate on the macroscale are negligible with respect to areal (adhesive) and peripheral (capillary) forces, especially at scales smaller than $10\,\mu$m [4–6]. Thus, to be able to control manipulation, the first step is to theoretically model and experimentally investigate this new environment.

Micro/nanomanipulation implies precise interactions with micro/nanoscale objects by pulling, pushing, cutting, indenting, picking, and placing. In this context, submicron- or nanometer-scale resolution positioning is imperative. The atomic

C.D. Onal (✉)
Computer Science and Artificial Intelligence Laboratory,
Massachusetts Institute of Technology, Cambridge, MA, USA
e-mail: cagdas@mit.edu

O. Ozcan • M. Sitti
Department of Mechanical Engineering,
Carnegie Mellon University, Pittsburgh, PA, USA
e-mail: oozcan@andrew.cmu.edu; sitti@cmu.edu

J.J. Gorman and B. Shapiro (eds.), *Feedback Control of MEMS to Atoms*,
DOI 10.1007/978-1-4419-5832-7_4, © Springer Science+Business Media, LLC 2012

force microscope (AFM) is a suitable tool for micro/nanomanipulation. It offers the benefit of locally and directly interacting with small-scale phenomena on the order of a few nanometers. This ability provides precision and repeatability in micro/nanorobotics. Using an AFM for manipulation has its drawbacks as well. Since manipulating matter from the bottom up with a single end-effector is a serial process, it mainly suffers from a limitation on speed. Moreover, since imaging is generally limited by intensity and the wavelength of light, real-time visual feedback on the nanoscale is challenging. Further, current AFM manipulation systems are generally manually controlled by an operator; future tools must be automated for increased speed, repeatability, and ease of use.

One method of producing micro/nanostructures is to utilize micro/nanoelectromechanical systems (MEMS/NEMS) monolithic fabrication processes, which have a top-down approach. These processes typically start with a block of material and move down to achieve the final product with generally some chemical and/or optical lithography routines. The most important drawback of this approach is that unfortunately, not every three-dimensional geometry can be achieved with it. This fact is summarized in the definition of MEMS/NEMS fabrication processes, which are said to be $2\frac{1}{2}$-D processes. These processes typically yield structures that are extrusions of a 2-D sketch with minimal undercuts, whereas one can potentially achieve almost any 3-D design, with a resolution comparable to the size of subunits, using a layer-by-layer manipulation and assembly process. Also, materials that can be used with MEMS/NEMS processes are limited. More recently, 2-photon polymerization methods have also emerged as a way to create nanoscale 3-D structures in plastics [7–9].

In contrast, micro/nanomanipulation techniques are generally understood to mean moving subunits, such as molecules, particles, and rods, to achieve the final product as an assembly with a bottom-up approach. Potentially, micro/nanomanipulation could achieve any 2-D or 3-D structure imaginable, using layer-by-layer operation similar to the 3-D printers in larger scales, with subunit size resolution. This potential, however, has yet to be investigated. Also, it does not necessarily have limits on the materials to be used, although some material combinations admittedly work better than others due to the interplay between the respective adhesive forces. Current contact manipulation methods suffer from lack of automation [10] and control. The lack of understanding of contact mechanics at the micro/nanoscale is the biggest limiting factor. In addition, a lack of even the simplest sensory feedback makes controlled manipulation problematic.

There have been many 2-D and 3-D micro/nanomanipulation methods proposed in the literature: pushing using a nanoprobe [11, 12], pick and place with a gripper [13, 14], a single finger [15–17], and certain noncontact methods such as optical tweezers [18], magnetic tweezers [19], (di)electrophoresis [20–22], and electroosmosis [23]. While automation has been demonstrated for some of these studies, generally automated utilization of real-time sensory feedback to close the loop is still lacking, especially for nanomanipulation. We believe that control systems will be integral parts of micro/nanomanipulation in the future as precision positioning, repeatability, and performance optimization are able to be handled effectively with feedback control using force and/or visual information.

This work focuses on developing 2-D automated micro/nanomanipulation control systems using mechanical pushing or pulling with an AFM tip. Using robotic systems of relatively inexpensive components, the study aims to bring a powerful alternative for automated micro/nanoparticle manipulation, with control over the success of each operation. Our manipulation algorithms are similar at both micro- and nanoscales. Particle positions are determined by image processing, and particle–target pairs are made by a task planner to minimize obstacles. Particles are pushed in piecewise linear trajectories, and contact loss is checked during manipulation. When the error between particle and target positions is below a certain threshold, manipulation is completed. Using spherical particles simplifies the problem, since it removes the orientation degree of freedom. Nevertheless, the work is generalizable to different materials, sizes, and geometries.

Microparticle manipulation is relatively easier than nanoparticle manipulation, since the positions of microparticles can be detected from visual feedback. Our microparticle manipulation system is detailed in Sect. 4.2. Here, we define microparticles as objects that have characteristic dimensions in the micron size and are visible under a conventional optical microscope without any modifications.

For nanoparticles, conventional visual feedback from an optical microscope is challenging, since such particles are smaller than the diffraction limit of light. Some advanced subpixel averaging algorithms help detect particle positions below this size limit [24], but they may have trouble distinguishing particles that are disordered and closer to each other than a certain threshold, which may limit generality. AFM gives an alternative way to measure nanoparticle positions from an initial AFM scan, called the "reference image." A similar manipulation process can be devised for nanoparticles using force feedback, provided that particle positions can be estimated using force. Therefore, Sect. 4.3 builds on the results of Sect. 4.2 for the case of force-based nanoparticle manipulation.

Spherical micro/nanoparticles deposited on flat substrates are manipulated in linear trajectories by the AFM tip. These trajectories are selected by an automated task planner to achieve obstacle avoidance and to minimize blockages to the linear trajectories and hence maximize speed. Furthermore, to achieve successful assembly of multiple particles, an assembly algorithm is used in addition to the task planner. Automated pattern formation and assembly of multiple micro/nanoparticles are described in Sect. 4.4.

4.2 Vision-Based 2-D Microparticle Manipulation

In this section, an atomic force microscope (AFM) probe tip is utilized to push and pull spherical polystyrene (PS) particles of diameter 4.5 μm on a flat glass substrate in 2-D [25]. Automated contact manipulation of microparticles under an optical microscope requires fast image-processing algorithms to detect the particles in real time, a means to convert the pixel dimensions in image coordinates to distance in world coordinates, and an algorithm to realize closed-loop manipulation.

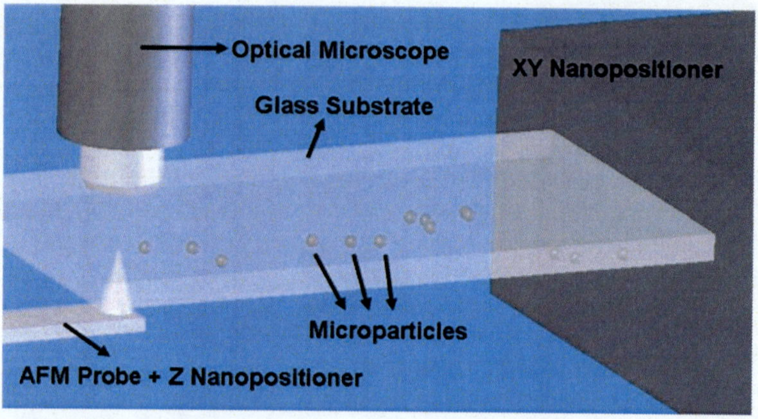

Fig. 4.1 Structure of an automated micromanipulation system using an AFM probe for pushing/pulling microparticles on a glass substrate using top-view optical microscope visual feedback. Reprinted with permission. © 2007 IEEE

The experimental setup is designed as shown in Fig. 4.1 for automated pushing or pulling of microparticles using an AFM probe tip. A glass slide carrying the particles on the bottom side is attached to a nanometer precision XY piezoelectric stage (Queensgate NPS-XYZ-100A, with effective x-y-z range of 100 μm × 100 μm × 15 μm with ±5-nm closed-loop precision). The stage has its own closed-loop PID controller using capacitive sensors and driver that takes position inputs from a PC with a D/A card and moves to those positions rapidly. The z-axis is extracted from the piezoelectric stage and attached to another, manual, stage. An AFM probe is attached to this z-stage upside-down, so it can touch the bottom side of the substrate, and hence the particles, on command. This inverted probe setup is required to avoid obstructing the images of the microparticles during micromanipulation for real-time visual servoing control. This setup is placed under a top-view reflective-type optical microscope (Nikon Eclipse L200) with a video camera (Dage-MTI DC-330), connected to a frame-grabber on the PC to close the loop.

A constraint in this inverted configuration is the necessity of a transparent substrate. Particles are deposited on the glass slide in a deionized water solution that is passively evaporated. The setup can easily be used in an inverted microscope system by depositing the particles on the top of a glass slide and touching with the probe from the top.

As mentioned briefly above, the dominance of inertial and gravitational forces diminishes as objects scale down to the microscale. Therefore, adhesive forces begin to play the most important role in the manipulation process. Since these forces are contact geometry dependent, a pick-and-place approach is hard to achieve due to release problems. It is much more likely for the particle to get stuck when one is using grippers, which necessarily have larger surface areas than that of a

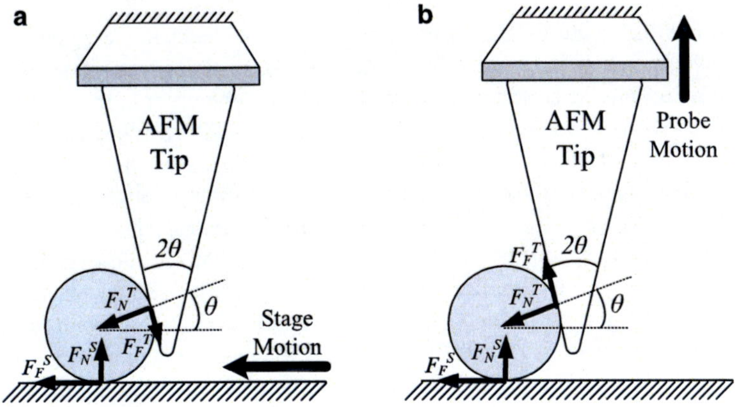

Fig. 4.2 Effective forces during (**a**) particle pulling with the substrate motion and (**b**) particle release with the probe motion. A front view of the conic AFM probe is depicted in this figure, while manipulation is invariant with respect to the relative orientation

single-finger AFM tip. If one uses an AFM probe with a very sharp tip (Ultrasharp NSC12/50 noncontact probe made by Micromasch Inc.), adhesive forces are unable to cause permanent stiction.

In theory, adhesive forces between the substrate and particles should be much larger than those at the tip–particle interface. However, this theory assumes that contact with the object is conducted with only the tip, which is not the case in practice. Instead, particles contact the side of the tip, and the area of contact is much larger. Still, for the selected tip, it has been experimentally verified that repeatable release can be achieved for a vertical motion, and adhesion can stick the particle to the tip for slow horizontal motions in a quasistatic equilibrium. This enables pushing and pulling of particles by the tip, while no permanent stiction occurs. The effective forces during the manipulation of particles horizontally or when they are released and come into contact with particles vertically for the success of a manipulation task are investigated in Sect. 4.2.1 and depicted in Fig. 4.2. The relationship of the net horizontal force to the cone half angle θ and particle radius, while pulling the particles, is shown in Fig. 4.4.

To visually detect microparticle positions, a fast image processing algorithm is utilized. This gradient-based circle-detection algorithm is described in Sect. 4.2.2. Using the particle-detection algorithm, we then devised an iterative sliding mode observer to estimate the parameters of the transformation between image coordinates and world coordinates. An observer is similar to but more versatile than a least-squares estimation, since it can adapt to changes in the transformation matrix over time.

Regarding the design of the particle-control algorithm, we have chosen to use sliding mode control (SMC). SMC is a robust control technique with many important features including insensitivity to matched parameter variations and disturbance rejection. SMC works by constraining the plant on a predefined manifold in the state space, therefore reducing the order of the motion [26–28]. The key to SMC is selecting a manifold such that the control objectives are realized when motion is confined to this manifold. The inclusion of the sliding mode estimator enables one to convert references in image coordinates to world coordinates and also to extrapolate the positions between the two frames for a larger bandwidth and hence smoother and faster motions without overshoot. This is addressed in Sect. 4.2.3.

Finally, an accurate and fast position controller should be designed to move the AFM probe tip relative to the glass slide and interact with particles to move them into their target positions. The simple position controller used is explained in Sect. 4.2.4. The internal controller of the piezostage cannot perform positioning in different axes simultaneously; it has a queue to store commands and performs them sequentially with typical rise times of about 10 ms. To combat this, our control commands a smooth velocity in an arbitrary direction, with negligible zigzags.

4.2.1 Force Modeling for Tip-Based Microparticle Manipulation

Since the AFM probe tip radius is tiny (around 20 nm) with respect to the size of the particles to be manipulated by pulling/pushing, the contact area with the particle is much smaller than the particle substrate contact. This leads to a small adhesive force, which means that a permanent stiction of the particle on the tip can be avoided by a vertical release motion of the tip. When a particle is being pulled horizontally on the glass slide, the adhesive force on the tip–particle interface is opposed by the particle–substrate friction, which may not be enough to prevent the particle from moving. Moreover, contact with the side of the probe (which has a length of about 20–25 μm) is likely, and this further increases the pull forces compared to adhesive forces on the surface.

A theoretical analysis on using pulling for the manipulation of microparticles is discussed in this section. Since the friction for a particle spinning around its vertical axis (spinning friction) is much lower than either rolling or sliding friction for such a particle, even a small alignment error of tip to particle center results in spinning of the particle [12]. This causes an instability and makes it harder to control a push-only manipulation. In contrast, pulling provides a stable alternative. Hence, this force analysis has two objectives. First, it will provide us with an understanding of contact manipulation on the microscale. And second, it will investigate the possibility of pulling as a means of microparticle manipulation.

Effective forces in the system for horizontal (pushing or pulling) and vertical (approaching or retracting) tip motions are depicted in Fig. 4.2. Even for future high-yield production systems, in which the tip will scan as fast as possible (a few hundred hertz), the motion's speed will still be small compared to the adhesive dynamics at the microscale. Therefore, the system can be considered to be in quasistatic equilibrium, so that

$$-F_{\mathrm{N}}^{\mathrm{T}} \cos\theta + F_{\mathrm{F}}^{\mathrm{T}} \sin\theta = F_{\mathrm{F}}^{\mathrm{S}}, \tag{4.1}$$

$$F_{\mathrm{N}}^{\mathrm{T}} \sin\theta - F_{\mathrm{F}}^{\mathrm{T}} \cos\theta = F_{\mathrm{N}}^{\mathrm{S}}, \tag{4.2}$$

where the subscripts denote normal (N) and frictional (F) forces and the superscripts indicate forces at the interface of the particle with the tip (T) and substrate (S). Note that the first equation is written for the pulling case, while the second equation is for particle release. The angle θ is the angle of the tip shape with respect to the vertical axis. For a conical tip, this angle is invariant with respect to the direction.

According to this geometric interpretation, for particles to be pulled into their target positions, the net horizontal force exerted on the particle by the probe should be larger than the frictional force exerted by the glass slide. For the particles to adhere to the tip, the vector sum of the horizontal components of normal and frictional forces on the tip should be larger than the maximum frictional force on the glass slide.

Also, for the particle to be released from the tip after successful positioning, the vector sum of the vertical components of the normal and frictional forces on the tip should be smaller than the adhesive force on the glass slide. This can be further explained by the following inequalities:

$$-F_{\mathrm{N}}^{\mathrm{T}} \cos\theta + F_{\mathrm{F}}^{\mathrm{T^{max}}} \sin\theta > F_{\mathrm{F}}^{\mathrm{S^{max}}}, \tag{4.3}$$

$$-F_{\mathrm{N}}^{\mathrm{T}} \sin\theta + F_{\mathrm{F}}^{\mathrm{T^{max}}} \cos\theta < -F_{\mathrm{N}}^{\mathrm{S}}. \tag{4.4}$$

The maximum negative value (that is, the minimum value) of the normal force in contact mechanics is the adhesive force (i.e., $-F_{\mathrm{N}}^{\mathrm{T}} = F_{\mathrm{A}}^{\mathrm{T}}$ and $-F_{\mathrm{N}}^{\mathrm{S}} = F_{\mathrm{A}}^{\mathrm{S}}$ in the inequalities above).

Adhesive force (pull-off force) is related to the equivalent radius R between two contacting spheres with radii r_1, r_2 and the relative surface energy $\Delta\gamma$ according to the Johnson–Kendall–Roberts (JKR) elastic contact mechanics model by [12,29,30]

$$F_{\mathrm{A}} = \frac{3}{2}\pi\Delta\gamma R. \tag{4.5}$$

For the PS–SiO_2 interface, $\Delta\gamma \approx 2\sqrt{\gamma_{\mathrm{PS}}\gamma_{\mathrm{SiO_2}}}$. Material properties are given in Table 4.1. The equivalent radius can be found from

$$\frac{1}{R} = \frac{1}{r_1} + \frac{1}{r_2}. \tag{4.6}$$

Here, since the tip–particle interaction is a cone–sphere elastic contact, the cone is approximated as a sphere with radius equal to the radius of the level curve (circle) at the contact with the particle:

$$r_{\mathrm{c}} = r_{\mathrm{t}} + (r_{\mathrm{p}}(1+\sin\theta) - L)\tan\theta, \tag{4.7}$$

Table 4.1 Material properties for the theoretical analysis

	E (GPa)	ν	γ (mJ/m^2)
Polystyrene	3.8	0.34	50
Glass (SiO_2)	73	0.17	160

where r_c is the cone radius at the contact, $r_t = 20\,\text{nm}$ is the radius of the tip, $r_p = 2.25\,\mu\text{m}$ is the particle radius, and L is the tip sample separation. Also at the particle–substrate interaction, the flat surface could be considered a sphere of infinite radius to yield an equivalent radius equal to the particle radius.

We define the maximum frictional forces simply as

$$F_F^{max} = \tau A, \tag{4.8}$$

where τ is the interfacial shear strength and $A = \pi a^2$ is the contact area with a as the contact radius. The value of τ is found to be directly proportional to the effective shear modulus [12, 31], given as

$$G = 2G_1G_2/(G_1 + G_2), \tag{4.9}$$

where $G_i = \frac{E_i}{2(1+\nu_i)}$ is the shear modulus, E_i is the Young's modulus, and ν_i is the Poisson ratio for $i = 1,2$.

The ratio τ/G, however, is strongly dependent on the contact radius [32, 33]. For a small contact area ($a < 20\,\text{nm}$), this value approaches the strength of a perfect crystal (1/30), while as the contact radius increases ($a > 50\,\mu\text{m}$), the ratio decreases to a smaller value of about 1/1,286 (Peierls stress) [33]. Using the scale-dependence given in [32] and assuming a single dislocation in sliding friction for simplicity, we are using the relation shown in Fig. 4.3 to calculate the interfacial shear strength in our analysis.

For calculating A, its radius can be determined from the JKR contact radius equation [29]

$$a = \left[\frac{R}{K}\left(N + 3\pi R\Delta\gamma + \sqrt{6\pi R\Delta\gamma N + (3\pi R\Delta\gamma)^2}\right)\right]^{1/3}, \tag{4.10}$$

where N is the net normal force and K is the equivalent Young's modulus of the two spheres, which is defined as

$$K = \left[\frac{3}{4}\left(\frac{1-\nu_1^2}{E_1} + \frac{1-\nu_2^2}{E_2}\right)\right]^{-1}. \tag{4.11}$$

In attempting to pull an object moving horizontally, the vertical (normal) force F_N^s on the particle by the glass slide could be calculated as in (4.2). Putting this force in (4.8), the maximum frictional force on the slide can be determined. Graphing

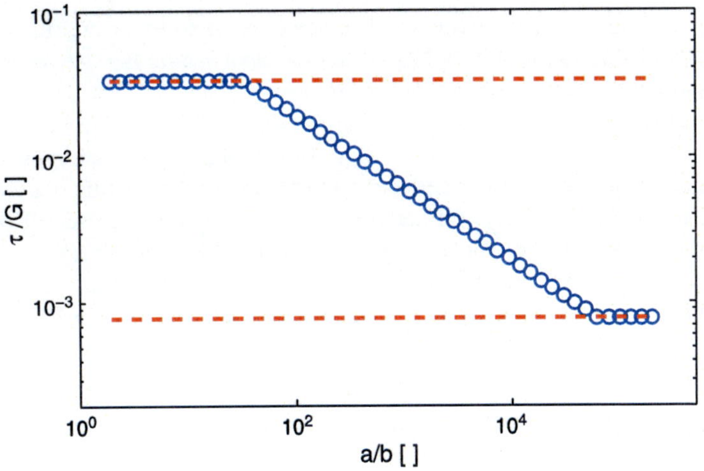

Fig. 4.3 The dependence of the interfacial shear strength on the contact radius normalized by the Burgers vector $b = 0.5$ nm. The *dashed lines* indicate the two limits on the value of τ/G. For small contact radii, interfacial shear strength approaches the strength of a perfect crystal, whereas for large contact area, it converges to the Peierls stress

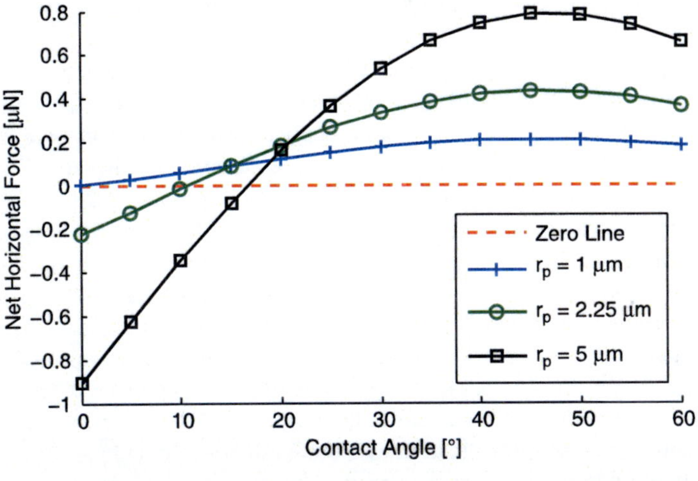

Fig. 4.4 Theoretical net horizontal forces in pulling PS microparticles of different sizes, according to tip-cone half-angle θ. For successful pulling, horizontal forces should be positive (i.e., in the pulling direction). Here $L = 0.5\,\mu$m

$F_{\mathrm{A}}^{\mathrm{T}} \cos\theta + F_{\mathrm{F}}^{\mathrm{T^{max}}} \sin\theta - F_{\mathrm{F}}^{\mathrm{S^{max}}}$ with respect to θ for different particle radii, we investigate theoretically in Fig. 4.4 the effects of the tip-cone half-angle on pulling PS microparticles.

In this figure, the net horizontal value needs to be greater than zero according to (4.3) for successful pulling. A similar analysis on the net vertical force on the particle on being released according to (4.4) showed that this relation is always satisfied.

This analysis showed that a conic AFM probe can be used to pull PS microparticles into position using side contact manipulation. It is obvious that smaller particles are pulled more easily. Furthermore, there is an optimal contact (tip half cone) angle to maximize the horizontal force in pulling direction, which is due to the vector sum of the adhesive and friction forces in the tip–particle interface. Therefore, stable manipulation of microparticles can be achieved by not only pushing but also pulling.

4.2.2 *Image Processing*

The first challenge and the starting point of the experimental work is the real-time detection of particles using optical microscope images. Since an image-based controller will be utilized to ensure the mechanical manipulation of particles to their respective target positions by pushing and pulling with an AFM probe tip, it is imperative to process each frame as it is fed to the frame-grabber and update positions accordingly. The OpenCV library is used to handle all image-processing tasks.

Spherical PS particles (Banglabs Inc.) 4.5 μm in diameter were used for all experiments. PS particles were chosen to be manipulated because they can be used in optical microdevice prototyping applications; are hydrophobic, which removes the effects of capillary forces; and are commercially available in a wide range of sizes. Under the microscope, any spherical object, and hence these particles, appears circular. Many circle-detection schemes have been proposed in the image-processing literature. Among them, the Hough transform is the most commonly used algorithm that is known to be robust under noisy conditions [34]. In general, it is possible to detect any arbitrary shape that can be quantized with parameters [35]. Although it has many favorable properties, the most critical problem of the Hough transform is the fact that it is slow and hard to include in a real-time process. There have been some improvements to the generalized Hough transform, such as adding gradient-direction information and probabilistic calculations [36]. However, even with these improvements, the speed of the process is still inadequate for processing every frame repeatedly.

Utilizing the gradient directions of a gray-scale image, Rad et al. [37] proposed a new algorithm to detect circles quickly. The algorithm is based on the two reasonable assumptions that for each circle:

1. There will be pairs of gradient vectors with opposite directions.
2. The slope of the line that connects the two base points of these vectors will be about the same as the slope of the first vector.

The details of this particle-detection algorithm are given in Algorithm 1.

Algorithm 1 Detection of Particles

1: Calculate the gradient image.

- Calculate the x and y gradients using the Sobel operator.
- Calculate the magnitude and direction of gradients at each pixel.

2: Utilize an adaptive threshold Th according to the maximum gradient magnitude $\max(gr)$ such that $Th = K\max(gr)$, $K \in [0,1]$.
3: Search for pairs of gradient vectors with the two properties mentioned above.
4: Utilize a voting mechanism to avoid false positives.
5: Search for circles with the predefined particle radius.
6: Disregard circles closer to each other than a specified threshold value.

This algorithm is robust in detecting particles, with the exception of the case in which multiple particles are in contact. Application of the algorithm takes about 0.03 s for a frame of 320×240 pixels, which allows every frame to be processed (15 frames per second in our system). We are using this algorithm as a component of our controller due to its above-mentioned benefits.

4.2.3 Parameter Estimation

Before focusing on pushing/pulling of particles, it is necessary to discuss calibration of the camera. One necessity of visually servoed micromanipulation is the calibration of the length scale between the image frame and the real world. A transformation or even a simple lookup table can be used to quantify the mapping between the image and real-world frames. Since using or generating lookup tables is time- and resource-consuming, and using purely mathematical procedures might yield inaccurate results, an iterative discrete sliding-mode parameter observer is proposed to converge to the solution in this study.

The mapping between the two frames is a linear transformation. This mapping can represent any 3-D orientation of the real-world frame as projected onto the image frame. Assuming that a piezostage is fastened carefully under the microscope, the elements of the transformation matrix $\mathbf{A} \in \Re^{2\times 2}$ and the translation vector $\mathbf{b} \in \Re^{2\times 1}$ would be constant. Here, $\mathbf{A} = \begin{bmatrix} a_{11} & a_{12} \\ a_{21} & a_{22} \end{bmatrix}$ and $\mathbf{b} = \begin{pmatrix} b_1 & b_2 \end{pmatrix}^T$. The resulting transformation operation could be described in parameter space as

$$\mathbf{x}_i = \underbrace{\begin{bmatrix} x^r & y^r & 0 & 0 & 1 & 0 \\ 0 & 0 & x^r & y^r & 0 & 1 \end{bmatrix}}_{\mathbf{K}} \underbrace{\begin{pmatrix} a_{11} \\ a_{12} \\ a_{21} \\ a_{22} \\ b_1 \\ b_2 \end{pmatrix}}_{\mathbf{p}}, \tag{4.12}$$

where $\mathbf{x}_i \in \Re^{2\times 1}$ is the state (position) vector in the image frame, x^r, y^r are the positions in real coordinates, and $\mathbf{p} \in \Re^{6\times 1}$ is the parameter vector, to be estimated by the observer. One important requirement for the matrix $\mathbf{K}$ is that rank($\mathbf{K}$) = $m = 6$ for the proposed sliding-mode estimator to converge to a solution. This means that the number of states should be equal to at least the number of parameters, which is not true in its current definition. Assuming that the transformation is constant (for at least two iterations), this matrix is redefined below to become a nonsingular (square) matrix.

The parameter estimator is, as mentioned, an iterative one. This means that discrete values are obtained for the input and output vectors iteratively with a similar approach to the controller previously designed in [27] and [38]. Simply combining the current values with the two previous values, (4.12) becomes

$$\mathbf{x}_i = \underbrace{\begin{bmatrix} x^r_k & y^r_k & 0 & 0 & 1 & 0 \\ 0 & 0 & x^r_k & y^r_k & 0 & 1 \\ x^r_{k-1} & y^r_{k-1} & 0 & 0 & 1 & 0 \\ 0 & 0 & x^r_{k-1} & y^r_{k-1} & 0 & 1 \\ x^r_{k-2} & y^r_{k-2} & 0 & 0 & 1 & 0 \\ 0 & 0 & x^r_{k-2} & y^r_{k-2} & 0 & 1 \end{bmatrix}}_{\mathbf{K}} \underbrace{\begin{pmatrix} a_{11} \\ a_{12} \\ a_{21} \\ a_{22} \\ b_1 \\ b_2 \end{pmatrix}}_{\mathbf{p}} \tag{4.13}$$

redefining the image state vector as $\mathbf{x}_i = \left(x^i_k\; y^i_k\; x^i_{k-1}\; y^i_{k-1}\; x^i_{k-2}\; y^i_{k-2}\right)^T$. Defining the estimated parameters as $\mathbf{u} \in \Re^{6\times 1}$ and the image state vector according to this estimation as $\hat{\mathbf{x}}_i = \mathbf{K}\mathbf{u}$, the estimation error (and the sliding mode variable) becomes

$$\sigma = \mathbf{e} = \mathbf{K}(\mathbf{p} - \mathbf{u}). \tag{4.14}$$

If det($\mathbf{K}$) $\neq 0$, then for stability, a positive definite Lyapunov function of the form $v(\sigma) = \frac{\sigma^T\sigma}{2}$ is used, with derivative $\dot{v}(\sigma) = \sigma^T\dot{\sigma}$. If the control function is designed such that

$$\dot{\sigma} + \mathbf{D}\sigma = 0, \tag{4.15}$$

for a positive definite symmetric matrix $\mathbf{D} \in \Re^{m\times m}$, the Lyapunov function derivative becomes a negative definite function $\dot{v}(\sigma) = -\sigma^T\mathbf{D}\sigma$, which satisfies the Lyapunov stability criterion.

From (4.14), the sliding mode variable is seen to be

$$\sigma = \mathbf{K}(\underbrace{\mathbf{p}}_{\mathbf{u}_{\mathrm{eq}}} - \mathbf{u}). \tag{4.16}$$

Here $\mathbf{u}_{\mathrm{eq}} \in \Re^m$ is the equivalent control defined by Utkin [26], which makes $\sigma = 0$. Solving (4.16) for $\mathbf{u}_{\mathrm{eq}}$ yields

$$\mathbf{u}_{\mathrm{eq}}(t) = \mathbf{u}(t) + \mathbf{K}^{-1}\sigma. \tag{4.17}$$

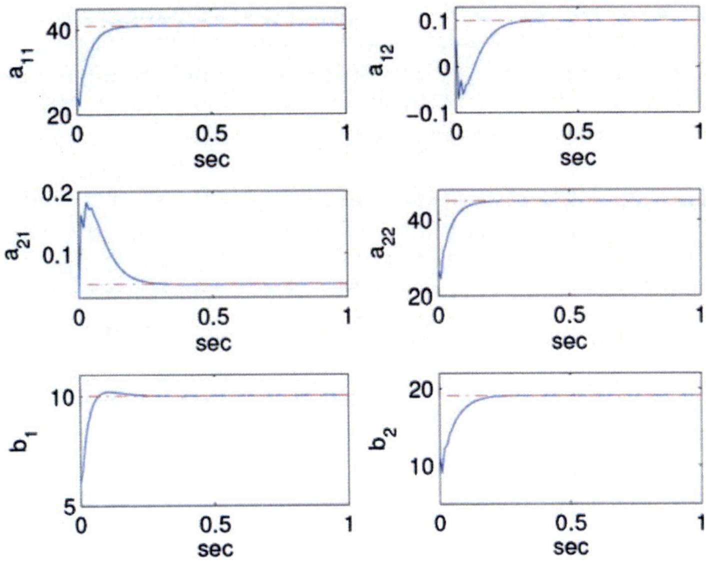

Fig. 4.5 Simulation results for the sliding-mode iterative parameter estimator. Reprinted with permission. © 2007 IEEE

Putting (4.16) in (4.15), we obtain

$$\dot{\sigma} + \mathbf{DK}(\mathbf{u}_{\mathrm{eq}}(t) - \mathbf{u}(t)) = 0, \tag{4.18}$$

and the only unknown that prevents the calculation of $\mathbf{u}$ is $\mathbf{u}_{\mathrm{eq}}$, which is hard to calculate. However, since it is a smooth function, an approximation could be made using the previous time-step value of $\mathbf{u}$ in (4.17) such that $\mathbf{u}_{\mathrm{eq}}(t) \approx \mathbf{u}(t - \Delta t) + \mathbf{K}^{-1}\sigma$. Using this approximation in (4.18) and solving for current $\mathbf{u}$ gives $\mathbf{u}(t) = \mathbf{u}(t - \Delta t) + [\mathbf{DK}]^{-1}(\dot{\sigma} + \mathbf{D}\sigma)|_{t-\Delta t}$, or in discrete time,

$$\mathbf{u}_k = \mathbf{u}_{k-1} + [\mathbf{DK}]^{-1}(\dot{\sigma} + \mathbf{D}\sigma)|_{k-1}. \tag{4.19}$$

The effectiveness of the proposed observer is demonstrated on simulation results in Fig. 4.5. The experimental results are in line with these results, since similar convergence curves are achieved.

4.2.4 Automated Single-Microparticle Manipulation

The piezo stage receives commands of target positions and moves to those specified positions using its own controller. These target positions can, however, be considered control inputs for moving the particles to pixel positions in image coordinates.

There are two initial issues in positioning the particles accurately and moving them with the probe. First, even though the controller can run at a bandwidth of 3–5 kHz, the images are updated at only 15 Hz, which, in turn, defines the speed of the feedback loop. This causes either large rise times or overshoots in positioning, both of which are unacceptable. To solve this problem, the parameter estimator designed in the previous section is utilized to extrapolate the data between the two frames and provide a rough estimate of the real-time position of the particle of interest.

The second problem is due to the nature of the setup. Since the whole glass slide moves with the stage, target positions of the particle are not stationary, and they should be updated by the program. One solution to this problem is to use a reference on the glass slide (another particle) inside the frame that is untouched by the probe. Initial experiments used this approach, which simplifies the updates. Figure 4.6 gives an example in which a stationary particle is picked as the reference.

However, this approach adds a constraint, namely that there needs to be another particle present in the scene during manipulation. One other solution employs the estimated transformation between the world frame and the image frame. There are two sources of error in this approach. First is the error in estimation, which is less than 2 pixels (0.6 μm). Second is the error caused by the pixel discretization, which is by definition smaller than a pixel, but it accumulates throughout the manipulation to result in large errors if it is not accounted for. However, in this work, pixel discretization was taken into account to remove that accumulation, and the constraint of having another particle on the screen all the time is lifted. This approach was used to obtain the results discussed in the final section and demonstrated in Fig. 4.23.

Also, having to have another particle detected in the scene all the time caused the manipulation to be slower than what would have been possible had one not been concerned with not losing the reference particle outside the region of interest described in Sect. 4.2.2. An individual manipulation took a little more than 1 min with this approach. However, with the second method, this time was reduced to 15 s on average. The former method also eliminated the possibility of an assembly operation, since once two particles contacted each other, the gradient image was affected, and a particle that had already been assembled could not be detected as easily and would not be able to be used as a reference. The assembly of particle patterns using the estimated transformation is demonstrated in Sect. 4.4.

With these necessary additions to the manipulation algorithm, a simple controlled trajectory-generation procedure can be devised to realize successful positioning of particles. As mentioned above, motion trajectories of the tip relative to the glass slide are almost linear. First, two z-positions of the probe that induce and eliminate contact with the particles are determined. The vertical distance between the two positions is more than the particle diameter to ensure that tip–particle contact can occur only when intended for manipulation. The algorithm decides whether to move the probe up or down depending on whether it wishes to push or pull the particle.

For manipulation, the substrate is moved such that the tip is right behind the particle according to the line that connects the particle to its target position. Then,

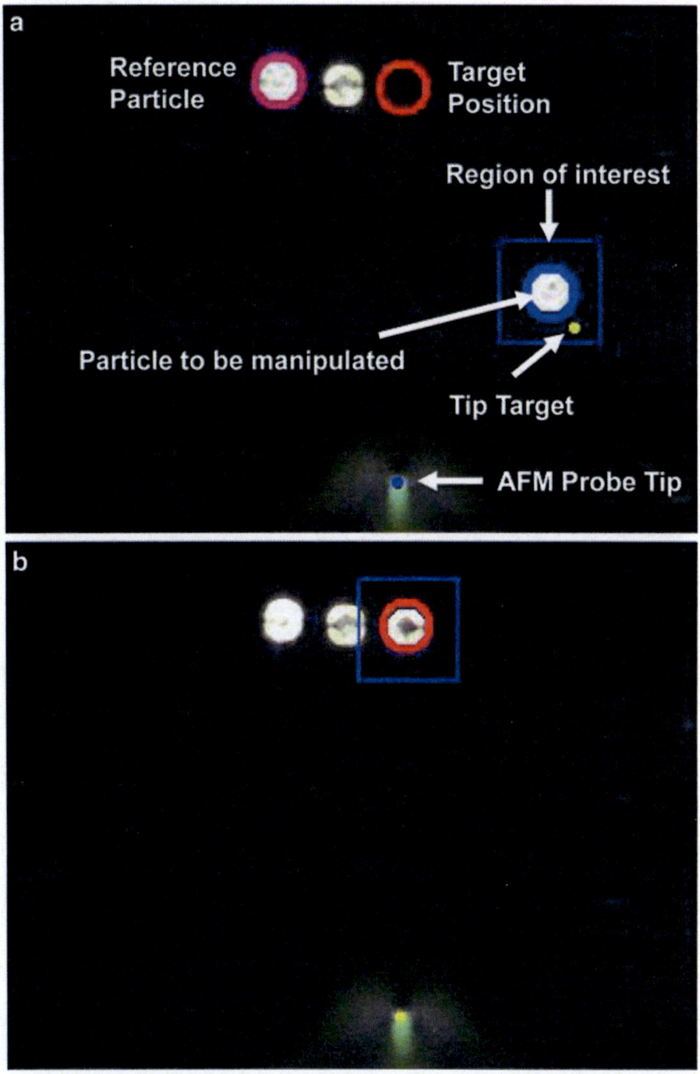

Fig. 4.6 Optical microscope top-view images of (**a**) before and (**b**) after controlled manipulation of a single PS particle of diameter 4.5 μm to a target position autonomously. Although the microparticles are spherical in shape, a cross shape appears on top of them due to optical artifacts. Reprinted with permission. © 2007 IEEE

the probe is moved up, and the particle is pushed forward. If the particle spins around the tip, then it continues to be pulled toward the target. Whenever the tip loses contact (i.e., distance between the two is larger than the radius), the whole process repeats. The algorithm terminates when the position error is less than 2 pixels (0.6 μm). It should be noted that the controller behaves the same for pushing and for pulling particles.

Finally, to realize manipulation it is imperative to move the slide accurately for the tip to reach its target stably. To ensure this, a velocity that varies linearly as a fraction of the pixel distance between the current and target positions of the tip is calculated so that it satisfies the Lyapunov stability criterion

$$\mathbf{v} = \dot{\mathbf{e}}_p = -\mathbf{D}\mathbf{e}_p \tag{4.20}$$

for positive definite $\mathbf{D}$, where $\mathbf{e}_p$ is the position error vector. To eliminate large tip speeds for small position errors, the calculated Lyapunov velocity is saturated to a small maximum value. From the saturated velocity in image coordinates, the x and y components of the next time step are extracted and converted to world coordinates using the transformation given in Sect. 4.2.3. This motion, in very small increments, is run at around 3–5 kHz until the errors are less than 2 pixels (0.6 μm).

This vision-based algorithm is implemented on a PC. Initially, the user is asked for the tip position, the particle to be moved, the target position, and the reference particle, as shown in Fig. 4.6. Then it runs the parameter estimator for 15–30 s to converge to a solution, while the stage is moved in a circular trajectory of radius 10 μm to eliminate any singularities. To reduce any sort of disturbance or discretization error, 10–15 data points (current and previous positions of a particle) are used. Then the process explained above is run until the particle is moved to its target position with an error smaller than 0.6 μm. No obstacle avoidance is implemented at this point.

To test for repeatability and to analyze the errors of positioning, a single particle is pushed/pulled random distances to user-specified random target positions 59 times. The average error and standard deviation in these successive manipulations are 0.49 and 0.14 μm, respectively. With the described method, we encountered no failure in the manipulation of microparticles in our experiments.

4.3 Force-Based 2-D Nanoparticle Manipulation

AFM nanomanipulation systems are mostly utilized for two-dimensional (2-D) manipulation tasks such as lithography (as in writing a pattern on a flat surface) [39–41], dissection [42, 43] and particle positioning. Several groups have worked on AFM based particle manipulation to show its feasibility [39, 40, 44–51]. The techniques used in these studies usually involved turning servo feedback off or decreasing the voltage set point of the signals, which are normally used to scan a surface, in order to decrease the distance between the substrate and the AFM tip, so that particles can be mechanically manipulated rather than having the tip of the AFM probe jump over the particles. These applications all can be referred to as push-and-look manipulation examples. This discussion suggests three of the most important problems in AFM-based nanomanipulation: reliability, speed and precision.

Due to the lack of real-time visual feedback during nanomanipulation, automated manipulation with AFMs has been a strong need in the field. Only one group [52,53] has recently demonstrated automated AFM-based nanoparticle manipulation using a drift-compensation method to position 28 nanoparticles with 15 nm diameter in an operation time of approximately 40 min. The images of manipulated particles were an impressive indication of the reliability of the system; on the other hand, a quantitative analysis of success rates of manipulation attempts is not included, and force feedback of the AFM probe is not utilized during nanomanipulation for control.

To increase the speed of AFM-based nanomanipulation and to demonstrate the success-rate issues in detail, this section focuses on developing a 2-D automated nanomanipulation system using force feedback. The AFM tip is utilized to push gold nanoparticles of diameter 100 nm on a flat mica substrate covered with a thin layer of Poly-l-lysine (PLL) in 2-D. The sample is prepared following the procedure in [46]. The manipulation principle for nanoparticles is very similar to that for microparticles, though with some important additional issues that need to be addressed. First and foremost of these issues is the fact that visual feedback from an optical microscope is no longer useful at the nanoscale. Even though some methods exist that utilize fluorescence or digital image correlation [24], these methods can require additional image-processing capabilities.

Particle position values are more difficult to find in real time than after postprocessing the images. Since AFM was originally an imaging tool, high-resolution 3-D topographic images of the substrate containing the particles can be taken before manipulation to detect the initial positions of the particles. However, it is very challenging to use AFM simultaneously in both imaging and manipulation modes.

A possible method to simultaneously detect and control the position of single particles by AFM is to use force feedback, which may allow the detection of the force exerted by the particle on the tip and an indirect estimation of the particle position. This task, however, has two main issues. First, any nonphysical crosstalk effects on the normal and lateral force signals of the AFM must be compensated. A crosstalk compensation procedure is proposed in [54] to reduce these effects. Second, and more importantly, since there are only two deflection measurements of the 3-D force information, these signals are inherently a coupled representation of the full force vector. This renders force information in 3-D not readily available.

If all three components of the force vector acting on the tip could somehow be available, one could roughly deduce the nanoparticle's position from the projected angle of this vector to the horizontal plane parallel to the flat substrate. Since this information is unavailable due to the unavoidable coupling of horizontal and vertical force components in AFM measurements, the next-best thing is to use the available information to ensure that the tip is in contact with the particle. This force-based contact-loss algorithm provides some control over the success of each individual push (or pull) operation, since it enables the automated system to determine whether it is currently in contact with a particle.

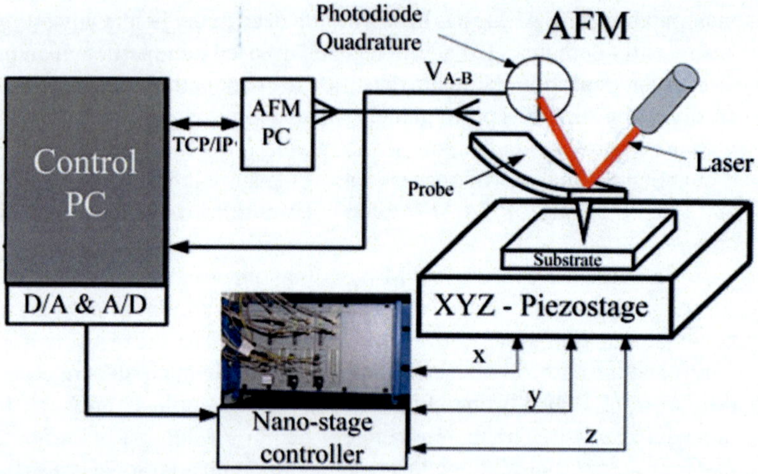

Fig. 4.7 Overall system structure of the automated nanoparticle manipulation setup. Reprinted with permission. © 2009 IEEE

As the most important contribution of this study, the performance of each automated manipulation operation is improved using a contact-loss algorithm that continuously tracks the real-time force feedback of the AFM probe, in contrast to the traditional blind push-and-look approach. The contact-loss algorithm dramatically changes the reliability of the system, because if contact loss can be detected during manipulation, it is easier to detect errors in the positioning and pushing operation. The performance of the system is investigated through a statistical study to form a quantifiable basis of development and comparison. The work can be generalized to different materials and geometries.

Another problem arises due to drift in the AFM system itself. Due to thermal and piezoelectric drift (creep), the particle positions will change over time, and the positions extracted from the reference image will no longer be valid for long operations of multiple-particle manipulation that take more than about 15 min. This problem is discussed in [55], and a compensation procedure that has a small processing burden on the main manipulation task is devised. This procedure allows manipulation operations to take place as if no drift existed. Therefore, drift problems will be neglected for the remainder of the chapter.

4.3.1 System Description

Figure 4.7 displays the overall layout for the automated nanomanipulation system. An AFM (Veeco, Autoprobe M5) with an AFM cantilever (Veeco, BESP, $k_n = 3.2$ N/m, calibrated via Sader's method [56]) is used as the nanomanipulator, which is accessed by a Windows 95-based PC (AFM PC). A client–server program is created on the AFM PC that allows an external PC to connect to the AFM through TCP/IP Ethernet (the Veeco SPMAPI is used to interface with the AFM). The main control PC uses real-time Linux (RTAI 3.3, Ubuntu Linux 2.6.15) and an interface program written in C++. It interfaces with the AFM PC through a direct Ethernet connection. A 3-DOF piezoelectric nanopositioning stage (Physik Instrumente P-753, 12-μm range, $<$ 1nm precision) is used as the positioner in the AFM, which allows for a faster control bandwidth compared to actuating the AFM's nanopositioner through TCP/IP. The three axes of the piezo stage are controlled through its dedicated controller (Physik Instrumente E-612) by the main control PC using three D/A outputs (Adlink PCI-6208). Positions are read from the amplifier using an A/D converter (National Instruments PCI-6024E). In addition, the AFM's normal-deflection signal (A-B voltage signal) is read directly from the AFM with the A/D converter. The total control bandwidth is 1 kHz, which is limited by the data-acquisition system.

For operation, a sample is placed on the custom nanopositioner, and the AFM probe is automatically positioned on the sample a short distance from the surface. The AFM noncontact mode (NCM) is used in order not to move particles inadvertently during imaging. By default, the AFM servo positions the probe with an offset over the surface to maintain a constant vibration amplitude on the substrate. The software in the main control PC has the capabilities of turning the servo feedback control on and off, positioning the tip with high precision, changing the set point of NCM tip-vibration magnitude, taking noncontact images of the substrate, and making single line scans or line travels on a given line with a given length.

An issue for nanoparticle manipulation, especially for a large number of particles, is thermal drift. Due to their thermal properties, all AFM components change size under small changes in temperature, which leads to the AFM tip translating in 3-D space. This motion is generally very slow, but it can be detrimental to the success of long experiments that deal with objects on the order of a few nanometers. To limit problems due to drift, we wait for the drift velocities to go below 5 nm/min before beginning the experiments, and we limit ourselves to proof-of-concept demonstrations that involve the manipulation of fewer than ten nanoparticles. Higher-volume manipulation can be achieved by utilizing a recent drift-compensation method [55], which utilizes a particle-filter algorithm and adds minimal overhead to the manipulation operation.

The 100-nm-diameter gold colloid nanoparticle samples are prepared with the procedure stated by Baur et al. in [46]. Commercially available nanoparticle samples, often used for SPM calibration tasks, are used. The overall procedure consisted in adsorbing 20 μl of 0.1% Poly-L-Lysine (PLL) onto freshly cleaved mica for 20–60 s, rinsing with deionized water, and drying with nitrogen. Immediately after drying, 20 μl of gold colloidal solution was adsorbed onto the treated mica for 5 min or more, depending on the surface concentration needed. The sample was then rinsed again with deionized water. After drying with nitrogen, it was finally incubated in a 60 °C oven for at least 1 h. The positive charge of PLL and the negative charge of Au nanoparticles create an electrostatic bond between the mica surface and the particles, temporarily fixing them on the surface for imaging.

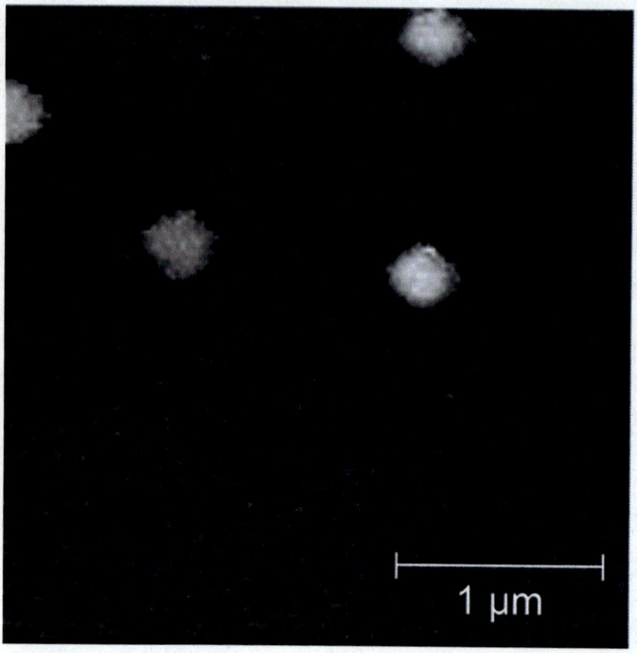

Fig. 4.8 A sample 128 × 128 pixels^2, 3 μm × 3 μm AFM non-contact-mode image of 100-nm-diameter gold nanoparticles

4.3.2 *Imaging*

Since the nanoparticles are loosely held to the substrate by electrostatic forces [46], small contact forces might disturb and move the particles if the conventional contact-mode imaging is attempted with the AFM. To eliminate this possibility, non-contact-mode scans are used for particle imaging. In the contact mode, the repulsive contact forces are used to trace the topography of the surface, while in the noncontact mode, the long-range attractive forces of the sample are used. In the noncontact mode, the cantilever is oscillated with a certain amplitude near its resonant frequency. As the tip interacts with the surface, the natural frequency of the cantilever shifts, and the oscillation amplitude changes. Using a feedback loop to regulate these changes, the tip can trace the surface without ever touching it (but with an associated loss of resolution compared to the contact mode).

A sample 3 μm × 3 μm AFM image of the gold nanoparticles is shown in Fig. 4.8. As seen in this image, nanoparticles look bigger than they really are, due to tip-convolution effects.

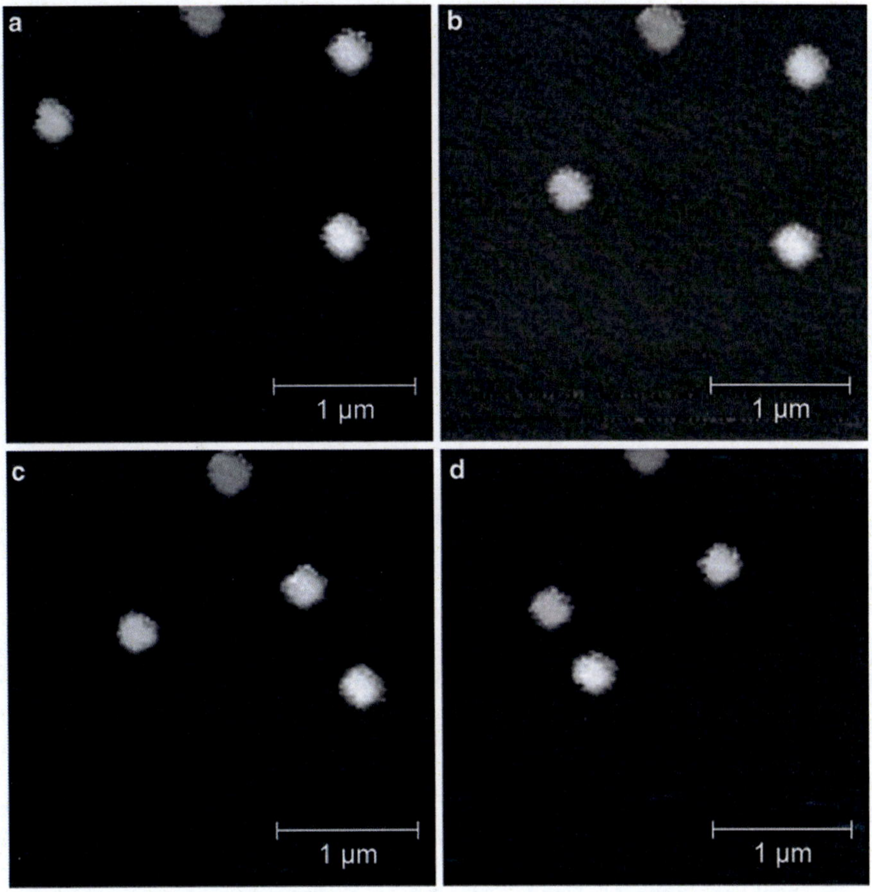

Fig. 4.9 3 μm × 3 μm AFM non-contact-mode images of a sequence of manual manipulation tasks on 100-nm gold nanoparticles

4.3.3 Manual Nanoparticle Manipulation

As mentioned, the method of manipulation for placing the nanoparticles in their respective target positions is mechanical pushing (or pulling) with the AFM tip. A sequence of manipulation operations, performed by manual tip-positioning commands, is given in Fig. 4.9. A simple, and blind, manipulation procedure is to take a "before" image, position the tip behind the particle, turn the z-servo off, and move the tip on a straight line passing through the center of the particle. After this tip motion, the z-servo is turned on, and an "after" image is taken.

Even when an experienced user manually controls the AFM, not all attempts at particle manipulation are successful. The most prominent source of error is the tip trajectory not passing through the particle center. The amount of offset is critical for the success of this operation. Even a little offset can cause the particle to spin around the tip or make it easier for the tip to slide over the particle, resulting in a failure in manipulation.

As mentioned previously, thermal drift is a major concern. Figure 4.9 provides important evidence on the detrimental effects of drift over nanoparticle manipulation. Subsequent scans on the same area show a slight translation in particle positions.

4.3.4 *Force Modeling for Tip-Based Nanoparticle Manipulation*

This section investigates the forces applied to the particle during manipulation by the surface and the tip [57]. These forces were modeled previously for tribological characterization [12, 58], and manipulation experiments [51, 59, 60] with different focuses. Our investigation involves an analysis of pushing and pulling cases in sliding manipulation.

Since the particle diameter in our experiments is 100 nm, a rolling type of motion is not expected, because the rolling resistance moment at this scale is larger than the sliding friction [58]. Another assumption we make is that the center of the particle is found robustly before any manipulation attempt. Our particle-center detection algorithm, which will be described in Sect. 4.3.5, has a convergence limit of 5 nm, which should eliminate any failure due to the particle spinning off of the tip.

The main variable in our analysis is the normal stiffness of the cantilever. The stiffness value of an AFM cantilever can be chosen almost freely. Our experimental results show a preferred stiffness range for successful particle manipulation. One goal here is to understand and explain this preference theoretically. Very low stiffness values failed to translate the particles, while high-stiffness cantilevers moved the particles during imaging. The desired cantilever stiffness is that which enables both dynamic imaging (typically higher than 1 N/m) and particle manipulation. Commercial AFM cantilevers can have stiffness values up to about 200 N/m.

Figure 4.10 displays the forces during pushing and pulling manipulations. Note that a quasistatic assumption is made for small stage velocities in relation to the resonant frequencies of the system. In this figure, and in what follows, the superscripts (T) and (S) denote the tip–particle and surface–particle interactions, and subscripts (F), (N), and (A) denote friction, normal, and adhesion, respectively.

Pietrement's contact mechanics model [61] provides an analytical equation that relates the adhesive forces (F_a) and contact radii (a) with the normal forces (F_n) at a given interface. Given the known parameters listed in Table 4.2, F_a and a can be accurately calculated at the particle–tip and particle–substrate interfaces from F_n.

Friction at the nanoscale is dominated by the shear strength (τ) at the contact area of the interface ($A = \pi a^2$) [58]:

$$F_f = \tau A.$$

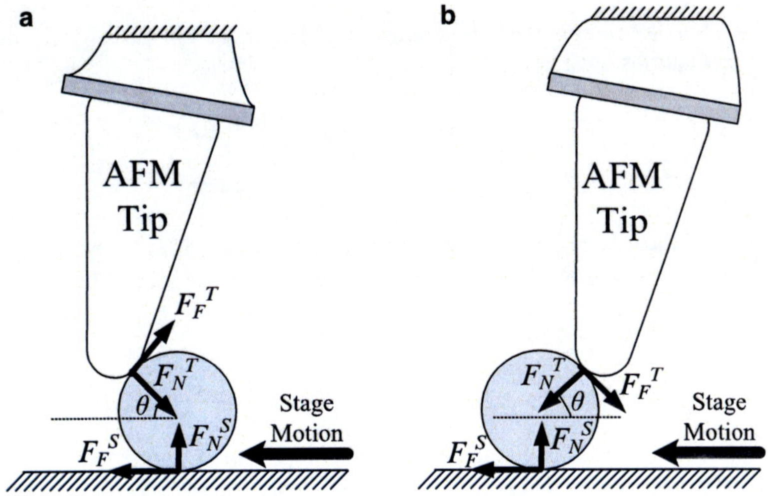

Fig. 4.10 Forces encountered in the tip–particle–surface interfaces for (**a**) pushing–sliding, and (**b**) pulling–sliding cases

Table 4.2 Parameters used for nanoparticle manipulation analysis. Particle radius (R_p), Young's moduli (E), Poisson's ratio (ν), and interatomic distance (z_0) values are known a priori; effective surface energy (γ), interfacial sheer strength (τ), and tip radius (R_t) values are calibrated experimentally

R_p	50 nm	R_t	130 nm
E_t	170 GPa	E_p	78 GPa
E_s	15 GPa	ν_t	0.17
ν_p	0.44	ν_s	0.5
$\gamma_{t,s}$	0.304 N/m	$z_{0t,s}$	0.3 nm
$\tau_{t,s}$	0.326 GPa		

The normal force (F_n^s) at the particle–substrate interface is directly related to the cantilever deflection in the normal direction (δ_n) by $F_n^s = k_n \delta_n$, where k_n is the cantilever normal stiffness. Note that the effective surface energy (γ) and interfacial shear strength (τ) values for both interfaces are equal. This is due to the fact that PLL-like polymers cover the tip as soon as any tip–substrate contact occurs [62]. Consequently, both interfaces consist of Au–PLL contact, assuming that nanoparticles do not roll on the surface [58] and hence do not become coated with PLL. The tip radius (R_t) is calibrated, using the convolution effect of the particles in AFM images, by geometric relations between the actual and apparent particle sizes.

For successful pushing-sliding of the particle on the surface, the following inequalities must hold:

$$F_f^{s\max} \leq \cos\theta F_n^t + \sin\theta F_f^s \tag{4.21}$$

$$F_f^{t\max} \geq \cos\theta F_f^s + \sin\theta F_n^s. \tag{4.22}$$

Since there is no way to determine simultaneous values for normal and frictional forces, the following worst-case equations will be used instead:

$$F_f^{s\max} \leq \cos\theta F_n^t \tag{4.23}$$

$$F_f^{t\max} \geq \cos\theta F_f^{s\max} + \sin\theta F_n^s. \tag{4.24}$$

Here, (4.23) means that the horizontal component of the tip–particle forces should be larger than the friction on the surface. This condition can easily be satisfied by the lateral deformation of the cantilever. That is, if the second inequality holds, tip–particle contact will not be lost, and as the cantilever base is moved, the necessary horizontal force will be achieved to move the particle (since otherwise, the cantilever will break). Therefore, the necessary condition in pushing is (4.24), which means that the tip does not slide over the particle (tip–particle friction is high enough to keep the tip on the particle).

Using the above-mentioned contact mechanics equations, a simulation of the net force ($F_f^{t\max} - \cos\theta F_f^{s\max} - \sin\theta F_n^s$) according to equation (4.24) for different contact angles (θ) reveals that there is a lower limit of cantilever stiffness to push–slide a particle as seen in Fig. 4.11. This result shows that for a realistic contact angle of about 45 degrees, about 3 N/m normal stiffness is required for pushing the particle.

Similarly, for a successful pulling–sliding manipulation, the following inequalities should hold:

$$F_f^s \leq F_f^t \sin\theta - F_n^t \cos\theta \tag{4.25}$$

$$F_n^s \leq F_n^t \sin\theta + F_f^t \cos\theta. \tag{4.26}$$

Note that the first inequality is the same as in the microscale case (see equation (4.1)) except that the contact angle θ is no longer constant and not equal to the tip half-cone angle due to the tip contact with the particle in comparison to the side contact in the microscale. Also, the microparticle analysis assumes that the tip–sample separation is constant (i.e., the cantilever has a high stiffness, and its deflections are negligible), while at the nanoscale, cantilever deflections are an important part of the analysis.

The normal forces for the pulling case are negative and cannot be larger than the adhesive force value in the negative direction. The cantilever bends down to generate the pulling force, which pulls the particle forward and upward. Since the cantilever needs to bend downward to generate the necessary pulling force, the

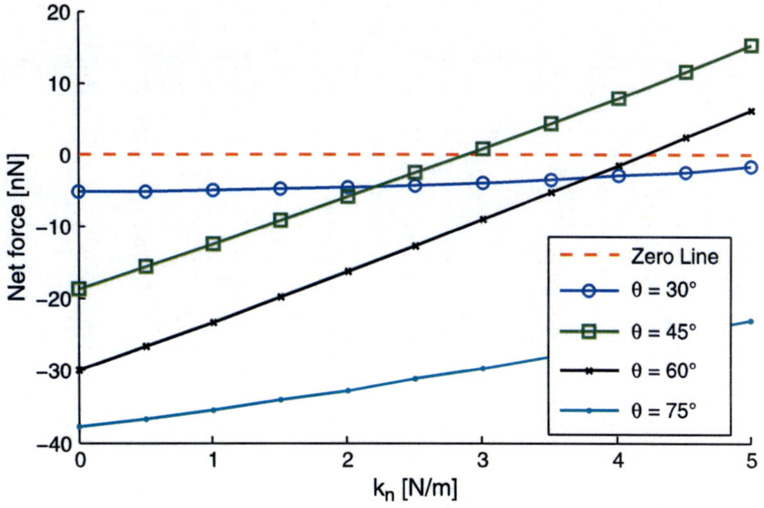

Fig. 4.11 The net force acting on the particle for different contact angles during pushing–sliding

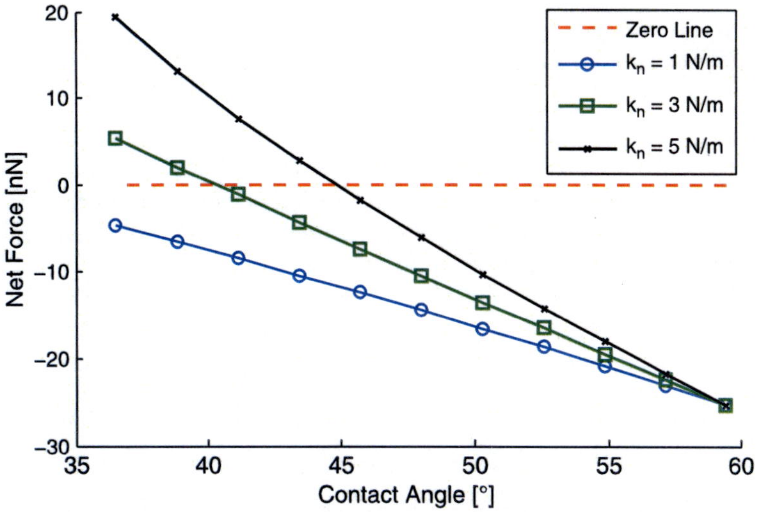

Fig. 4.12 The net force acting on the particle for different cantilever normal stiffness values during pulling–sliding

initial tip–substrate separation (δ_0) or similarly the initial tip–particle contact angle (θ_0) needs to be higher. Assuming an initial separation of $\delta_0 = 75$ nm ($\theta_0 \approx 60°$) for pulling purposes, Fig. 4.12 displays the results of our theoretical analysis.

Note that this figure should be read from right to left, since the tip contacts the particle with higher angles and bends down due to adhesive forces and initiates

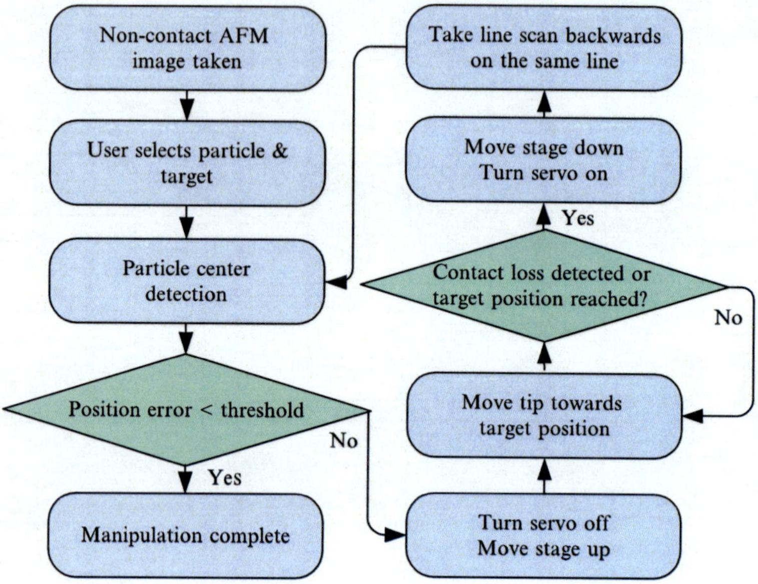

Fig. 4.13 A flowchart description of the automated nanoparticle-manipulation algorithm. Reprinted with permission. © 2009 IEEE

pulling when the net force ($F_f^t \sin\theta - F_n^t \cos\theta - F_f^s$) becomes positive. The second inequality (4.26) is found to be always satisfied. An obvious lower limit on the cantilever normal stiffness is again visible in this simulation, suggesting that a normal stiffness of about 2 N/m is required to pull a gold nanoparticle with the AFM tip and given parameter set that were used.

These results are in line with our preliminary experimental findings. The normal stiffness value of our AFM cantilever is 3.2 N/m, which is suitable for both pushing and pulling manipulations according to these simulation results.

4.3.5 Automated Nanoparticle Manipulation Scheme

The general algorithm for the autonomous manipulation of nanoparticles is outlined in a flowchart in Fig. 4.13. The procedure starts with an initial 10 μm $\times$ 10 μm noncontact image of the sample. On this image, the operator clicks on a particle and a target position. The program takes a fast low-resolution 1 μm $\times$ 1 μm image around the particle and finds the highest point of this image as an initial guess for the particle center. Then, the particle-center-detection algorithm described in Algorithm 2 is used on successive line scans in two orthogonal directions over the estimated center position to converge to the actual particle center position.

The second algorithm is an altered version of the watershed algorithm [63] widely used in image processing. Its main principle is to use the edges of a particle

Algorithm 2 Particle-Center Detection from Linear Topography Data

1: Low-pass filter and take derivative of topography data.
2: Take the points of absolute derivative values larger than a threshold as initial guesses.
3: Iterate guess points according to their corresponding derivative values until convergence (i.e., $x_k = x_{k-1} + K_x dz/dx$, $y_k = y_{k-1} + K_y dz/dy$, and $K_x, K_y \in \Re^+$).
4: Remove one from among the guesses closer to each other than a threshold.
5: Take weighted averages of data points around the remaining guesses.

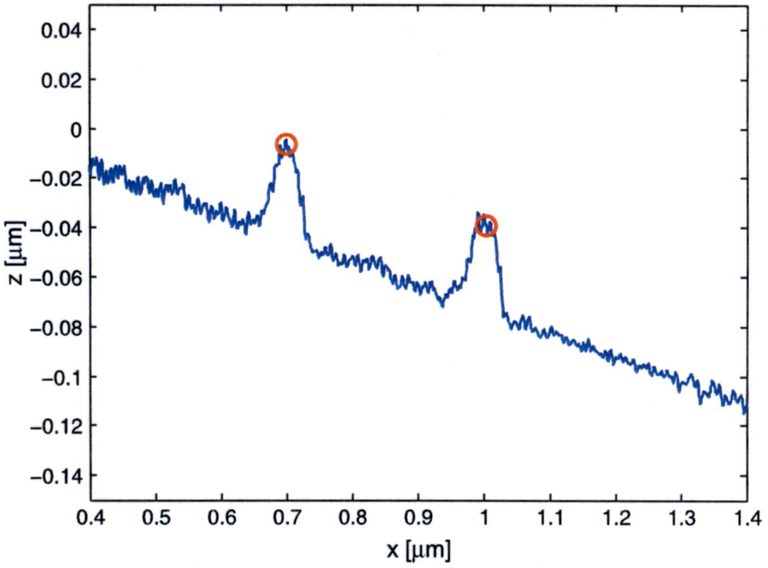

Fig. 4.14 Sample AFM line scan data of two close particles; *circles* indicate the centers of the particles detected using the center-detection algorithm. From all detected centers, the one closest to the previous estimate is taken as the actual particle to be manipulated, and others are discarded. Reprinted with permission. © 2009 IEEE

as initial guesses and move these estimates according to the derivative information to converge on local maxima. This operation detects all of the particles that appear in the same line-scan data as seen in Fig. 4.14.

During tip travel, contact loss can be detected using the algorithm described in Algorithm 3. As seen in Fig. 4.15, the V_{A-B} signal is almost zero, with minimal deviation when the tip and the particle are not in contact. However, for tip–particle contact during manipulation, the V_{A-B} signal starts to oscillate with a higher magnitude around a nonzero offset. The contact-loss-detection algorithm uses this fact as its basis. It continuously checks for a nonzero signal during manipulation, using the mean and the standard deviation values of the normal deflection signal. The overall algorithm runs in a loop until the final particle-positioning error drops below a threshold value, which is defined here as 100 nm.

Algorithm 3 Contact Loss Detection from Normal Deflection Signal

1: Wait for contact.
2: Take the latest N cantilever normal deflection signal data points; calculate their mean and standard deviation.
3: If mean and the standard deviation values are both close to 0, contact has been lost.

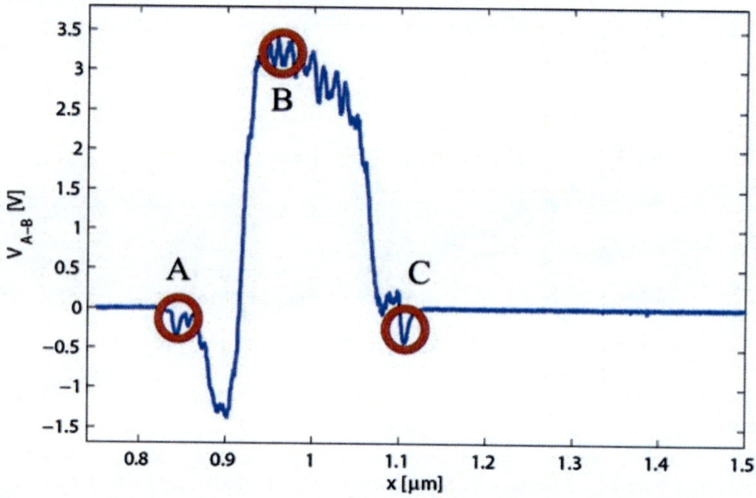

Fig. 4.15 Sample AFM normal force data during a particle-pushing operation: point A is where the tip and nanoparticle snap into contact, point B is where tip is jumping over the particle, and point C is where the contact loss occurs. Reprinted with permission. © 2009 IEEE

4.3.6 *Experimental Results*

Manipulation experiments are conducted using the procedure described in Sect. 4.3.5. Figures 4.16 and 4.17 show sample 10 μm $\times$ 10 μm AFM images demonstrating results of the automated nanoparticle-manipulation method.

During preliminary experiments, manipulated particles are sometimes pulled instead of being pushed. During the manipulations that are achieved by pushing, the final tip positions are behind the particles. The V_{A-B} signal, which is the signal that reflects the normal deflection of the cantilever, has a positive offset during pushing, which means simply that the cantilever is pushed up by the particle.

On the other hand, during the manipulations that are achieved by pulling, the final tip positions are in front of the particles that are being manipulated. The V_{A-B} signal has a negative offset in these cases, which means simply that the cantilever is pulled down by the particle. Figure 4.18 shows sample V_{A-B} data for pushing and pulling manipulations.

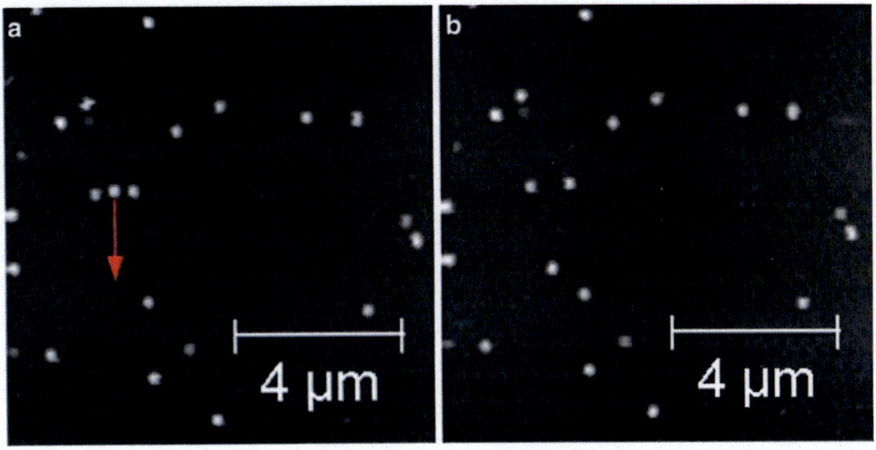

Fig. 4.16 (**a**) Before and (**b**) after manipulation of a 100-nm-diameter gold nanoparticle indicated by the *arrow*. It is possible to manipulate a particle without disturbing the close neighbor particles. Reprinted with permission. © 2009 IEEE

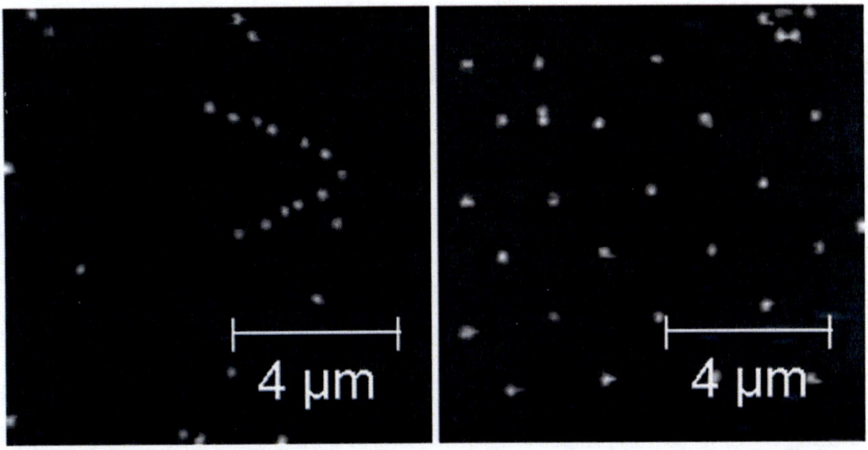

Fig. 4.17 Two resulting sample patterns after a sequence of automated single-particle manipulations. Reprinted with permission. © 2009 IEEE

Pulling is a more stable manipulation technique than pushing. During pushing, the nanoparticle can spin off the tip, or the tip can jump over the particle. For pulling, it is easier to define the trajectory and target position of the tip to decrease the final particle positioning error.

We also investigated the overall success rate of the system. Manipulation was attempted on 50 different nanoparticles, and the outcomes were grouped into three categories. Success was achieved in 86% of all manipulation trials. The particles in

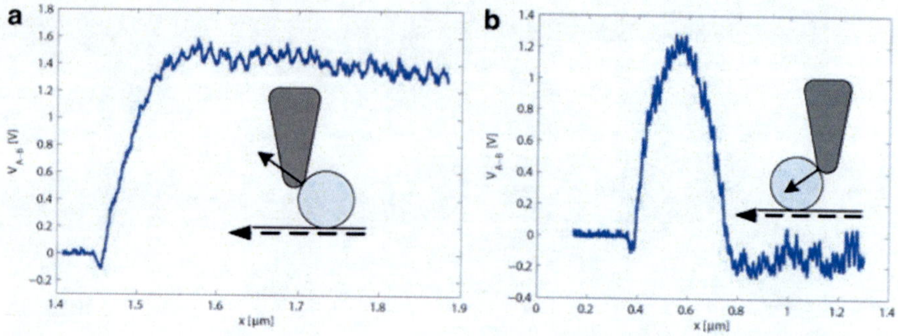

Fig. 4.18 V_{A-B} signal during (**a**) pushing and (**b**) pulling. Sample is moving left to manipulate the particle to the right. For pushing, particle exerts a force on the tip directed upward in the normal direction; for pulling, particle exerts a force on the tip directed downward in the normal direction. Reprinted with permission. © 2009 IEEE

this category were positioned with a final error lower than the positioning-error threshold of 100 nm. In 6% of the manipulation trials, the particles could not be moved at all, and in 8% of the trials, nanoparticles got stuck on the tip, which resulted in the loss of the nanoparticle and contamination of the tip. This further caused distortions in the successive images due to tip convolution. Before experimentation is continued, the tip can be mechanically cleaned after such trials by scanning a clean area in contact mode.

The particles that could not be moved at all were observed to be smaller than the manipulated ones. Particles of height 90–120 nm can be manipulated easily, whereas particles whose height is between 50 and 70 nm cannot be moved at all. We believe that the normal stiffness of the cantilever used in the experiments is not great enough to move these smaller particles.

Besides the overall success rate, the performance of successful manipulations was also evaluated by calculating the speed and the final positioning error of the manipulations with different angles (the direction of the manipulation trajectory in the horizontal plane) and different distances. The final position error was defined as the actual distance between the target position and the particle center position after the last run of manipulation. Speed was defined as the ratio between the distance of the manipulation and the time elapsed during the manipulation. Since there is a constant amount of time required for finishing certain tasks before the actual manipulation of a particle begins, the average speed is expected to increase as manipulation distance increases.

Figure 4.19 shows the variation of final positioning error and speed for different pushing angles. As can be seen from the plots, final positioning error and speed have no correlation with the pushing angle. Figure 4.20 shows the variation of final positioning error and speed for different manipulation distances. The speed increases with the distance due to the overhead operations but positioning error has no correlation with manipulation distance.

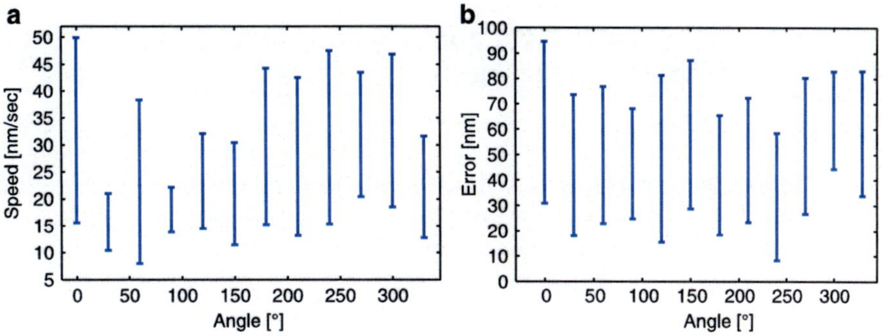

Fig. 4.19 Average manipulation speed (**a**) and final positioning error (**b**) results for 12 different manipulation angles using five data points for each angle. Reprinted with permission. © 2009 IEEE

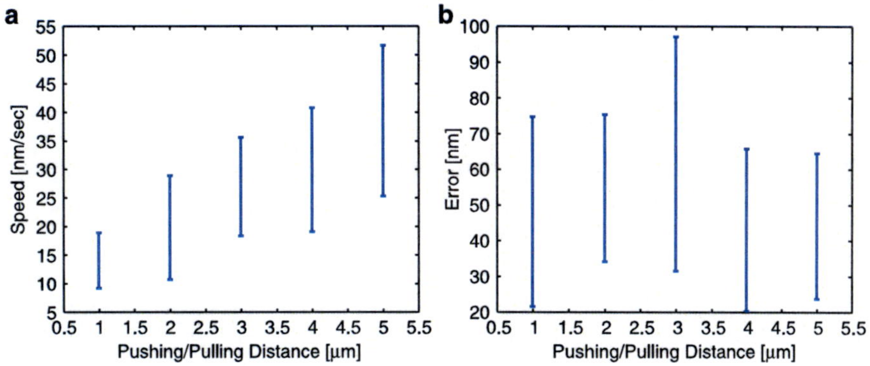

Fig. 4.20 Average manipulation speed (**a**) and final positioning error (**b**) results for five different manipulation distances using 12 data points for each distance. Reprinted with permission. © 2009 IEEE

The absence of correlation between the pushing angle and the performance parameters shows the reliability of the manipulation procedure for all angles. The speed increase in longer-distance manipulations is due to initial overhead operations (making a small low-resolution image around the particle and locating the center) taking a smaller percentage of the total time. Contact-loss detection does not cause additional speed loss, which demonstrates that the manipulation procedure does not need to divide a long manipulation into smaller steps. This is a direct outcome of detecting the contact loss between the tip and the particle in comparison to blind manipulation, where the program manipulates the particle and then determines whether it has moved.

4.4 2-D Arrangement and Assembly of Multiple Micro/Nanoparticles

Since a single probe is used for pushing/pulling particles in their respective reference frames, generating nonlinear trajectories is inconvenient and should be avoided for simplicity. The constraint in using linear trajectories is that one particle cannot exist between the initial and final positions of another particle.

At this point, the first reasonable approach is to use distance-based planning, so that the nearest particle is pushed/pulled to each reference. However, in most initial and many final configurations, two of which are shown in Figs. 4.21 and 4.23, such a planner would fail, since the placed particles would block others.

To overcome this problem, a metric that defines the problem of blockage caused by each particle was developed. The metric simply counts the number of blockages that each particle and reference position causes for every possible trajectory. The particle to be pushed/pulled is the particle that cuts the greatest number of potential trajectories, and the reference position to which it will be pushed/pulled is the one that cuts the minimum number of potential trajectories. This approach will take the most "problematic" particle to the least-interfering position. The algorithm for pattern-formation planning is summarized in Algorithm 4.

The fact that the pattern-formation-planning algorithm minimizes the number of cuts on all possible trajectories can be interpreted as an optimization in terms of this blockage metric. Moving the particle that blocks the greatest number of trajectories to the target position that blocks the fewest trajectories maximizes the number of one-trajectory manipulations. The success of the manipulation (and assembly) of multiple particles with linear trajectories depends on this blockage metric, and

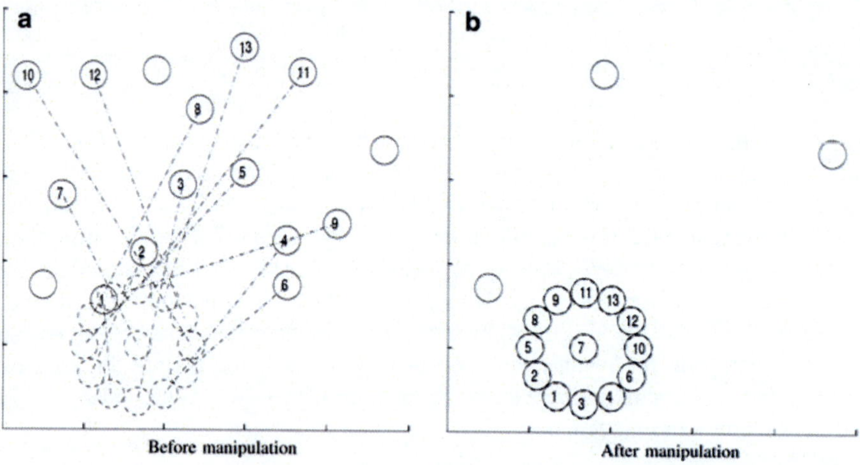

Fig. 4.21 Simulated assembly operation with a complicated target pattern using Algorithm 2. (Numbers depict the order of manipulation.) Reprinted with permission. © 2007 IEEE

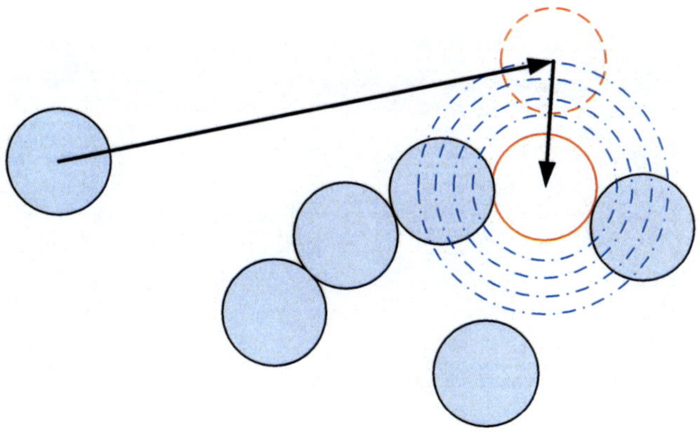

Fig. 4.22 The search algorithm for assembly. A suitable "first stop" is determined by searching in radially increasing circles around the final target position. Reprinted with permission. © 2007 IEEE

therefore, the algorithm yields a suboptimal solution to the problem. It is to be complemented by a second, "assembly," algorithm. With this approach, it becomes unnecessary to implement trajectories that are more complex than simple linear ones, which in turn decreases the total manipulation time and distance traveled. Assuming that particles do not move by themselves, this planning could be realized offline, before beginning any manipulation, reducing the online processing burden.

To realize an assembly of particles, there should by definition be some attachment between them. This attachment can be in terms of chemical bonding, a bonding material that acts as a glue, or simply adhesive forces (primarily van der Waals). It has been experimentally verified that once the particles are brought into contact, they have a rather strong and stable adhesion that keeps them together. This makes it difficult to unmake a pattern once it has been made, and for this reason, it was easier to choose new particles for continued experimentation.

The manipulation of a particle typically ends once it has been pushed/pulled to its target position with an error less than a specified threshold. However, even with the above scheme to minimize obstructions, it is still not always possible to pull all particles into an assembly, since other particles can become unmovable obstacles to the linear trajectory connecting the initial and final positions of the particle to be manipulated. This becomes a problem especially if the number of particles is not greater than the number of target positions and is generally more of an issue for later manipulations.

The solution to this problem, without losing the simplicity of linear trajectories, has been to divide the manipulation for the inaccessible target positions into two submanipulations in linear trajectories. Another algorithm, which searches for a second target position that is accessible from the initial position and from which the final position of the particle is also accessible, is included.

The second algorithm is activated once there is a target position that is inaccessible for every particle. It searches in circles with the center at the target position,

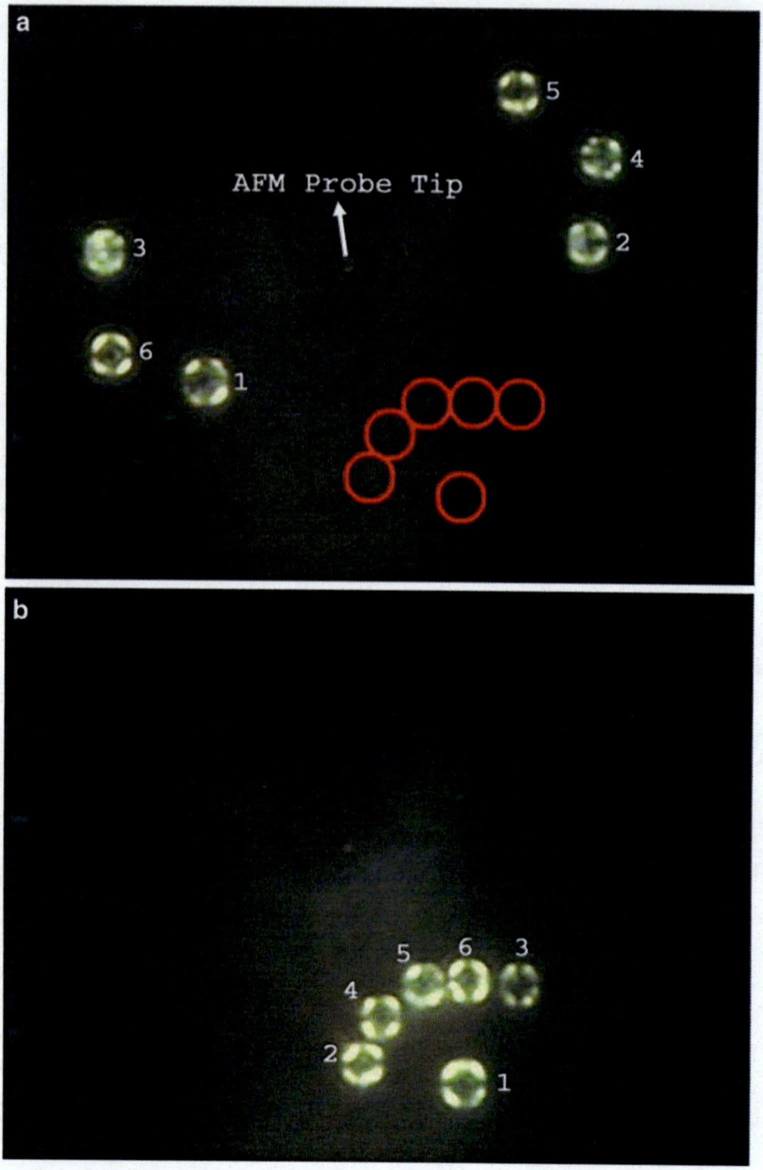

Fig. 4.23 Optical microscope top-view images of (**a**) before and (**b**) after autonomous manipulation is used to assemble six 4.5-μm-diameter PS particles. (Numbers depict the order of manipulation.) Reprinted with permission. © 2007 IEEE)

increasing the radius until a suitable position is found. The entire search can be done offline before manipulation is begun, so that it does not add any computational burden during assembly. This algorithm is depicted in Fig. 4.22.

Algorithm 4 Pattern-Formation Planning

1: Take a frame.
2: Detect all particles.
3: Generate all possible linear trajectories between each particle and target position.
4: Determine number of cuts on the generated trajectories by the particle and target positions.
5: **for** $n = 1 \rightarrow$ number of targets **do**
6: Determine the particle to be manipulated and the target position to which it will be moved.
7: Update number of cuts.
8: **end for**

Another problem might arise if a particle should be pulled to a position such that the tip is located between the current particle and another previously manipulated particle. This can occur if a closely packed target assembly (e.g., a 3×3 square consisting of nine particles) is attempted. However, even in a closely packed target assembly, this is a small probability, since even if the AFM tip lands between two particles, there is a good possibility that the two particles will snap into contact at the point where they are closest to each other once the tip is moved away. Therefore, in this work, the probability of errors due to this possibility is assumed to be negligible.

Figure 4.23 shows an experimental result of the assembly procedure. As seen in this figure, microparticles are pulled into their target positions utilizing the pattern-formation and assembly algorithms. The average error and standard deviation in the demonstrated manipulation in Fig. 4.23 are 0.64 and 0.44 μm, respectively. A movie of the complete assembly operation can be seen in [60].

Similarly, we applied our task planner for the automation of nanoparticle manipulations. The addition of a task planner for multiple nanoparticle manipulations has an additional benefit at the nanoscale, since it allows us to remove the overhead of taking intermediate AFM images between each manipulation operation. Figure 4.24 demonstrates the result of a proof-of-concept multiple-particle manipulation experiment. Six nanoparticles are autonomously positioned in a pentagonal arrangement. Average manipulation time is about 1 min per particle.

4.5 Conclusion

In this work, automated 2-D manipulation of micro- and nanoparticles with a single AFM probe is demonstrated based on visual and force feedback, respectively. Smooth linear trajectories that pass through the centers of the manipulated particles are generated for the motion of the AFM tip relative to the substrate for speed and ease of operation.

For microparticle manipulation, PS particles of diameter 4.5 μm are positioned based on visual information by an optical microscope with an average accuracy of less than 0.64 μm. A globally stable iterative discrete-sliding-mode observer estimates the parameters of the transformation between the image and world

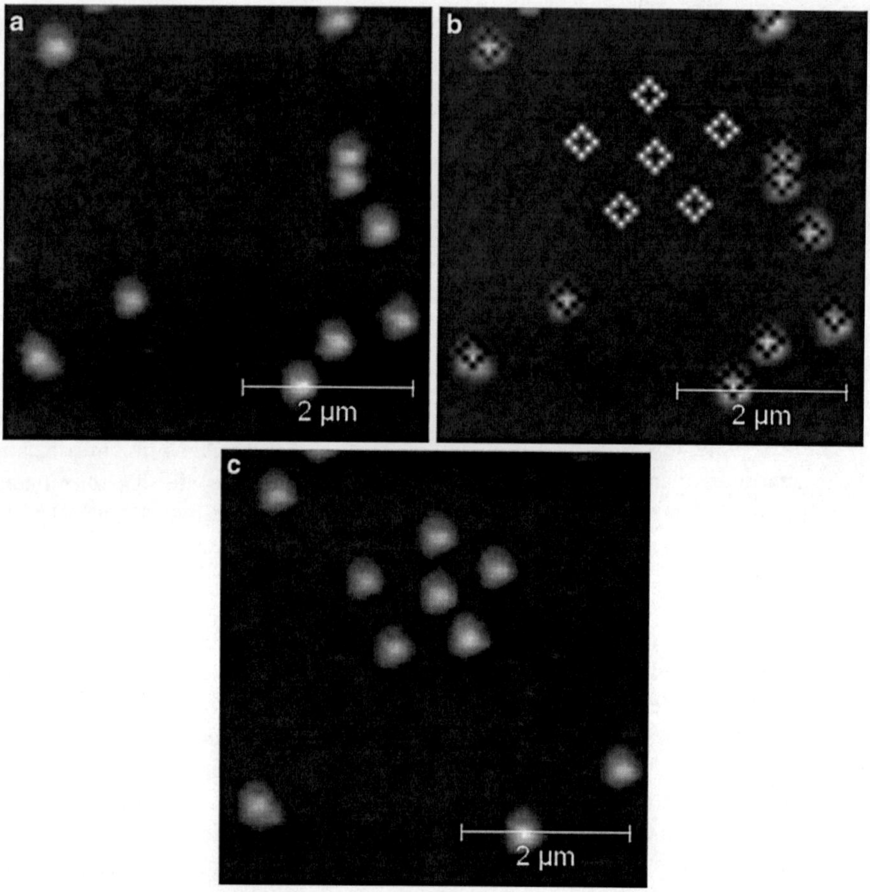

Fig. 4.24 Automated manipulation of six nanoparticles to form a pattern that cannot be formed with a distance-based planner. An initial image is taken in (**a**). All particles are detected using a watershed algorithm, and the target positions are given to the program in (**b**). The task planner decides which particle will be manipulated to which target position. A final image after the experiment is shown in (**c**)

coordinates. This estimation is used not only for conversion of reference positions of the piezoelectric stage to world coordinates, but for an extrapolation of particle positions according to motions of the stage between two frames fed back from the camera.

For nanoparticle manipulation, 100-nm-diameter gold particles are positioned based on force feedback from the AFM system with an average accuracy of less than the particle diameter. Robust particle-center detection and contact-loss detection algorithms are developed to overcome speed and reliability issues of AFM-based nanomanipulation. Unlike "blind" manipulation techniques, manipulation distances are not artificially divided into parts to increase the reliability. The performance of the designed manipulation system is statistically investigated.

Moreover, multiple particles are successfully positioned with a task planner that chooses the order of manipulation based on a metric that defines the blockage of particles and target positions on the generated linear trajectories. The task planner effectively minimizes the number of obstacles for greater efficiency and yields a fully automated pattern-formation and assembly procedure that eliminates the necessity to take AFM images between individual nanoparticle manipulation attempts.

Designing and implementing a fast and reliable technique for multiple-particle manipulation will increase the use of AFM for micro/nano-manufacturing applications where AFM would be inexpensive in comparison to most of the techniques and machines that are currently used for micro/nanofabrication. We believe that forming micro/nanofabrication masks and templates for plasmonic, optoelectronic, or MEMS/NEMS devices and contact printing methods for maskless micro/nanofabrication techniques would be possible using such procedures. Manipulating nanoparticles into predefined positions could also potentially be used for gluing or soldering at the nanoscale.

As can be seen from this automated 2-D micro/nano-manipulation study, there are many difficulties in AFM control, such as accurate modeling of the plant at the micro/nanoscale, limited force and visual feedback, and significant system noise, drift, and other time-varying disturbances such as vibrations, electrical noise, and temperature and humidity changes. Additionally, AFM control systems have many other broad challenges beyond manipulation. First, AFM systems are typically slow due to stability issues and hardware limitations. Therefore, new high-speed and high-precision AFM control methods with relevant hardware need to be developed. Next, for improving speed and adding multiple functions, some have proposed AFM systems with multiple AFM probes (up to 1,000 probes to date). These multiprobe systems will require parallel and distributed control, which brings up new control challenges. Finally, full automation of AFM systems is necessary for reliable and high-throughput nanotechnology applications such as data storage using AFM probes. For example, online system identification methods should be developed to tune the AFM control parameters automatically. All of these AFM control methods will enable new AFM-based applications in such areas as micro/nanoscale manufacturing, biological or inorganic micro/nanomaterials characterization, video rate imaging, micro/nano-device prototyping, and high-density data storage.

References

1. A. A. G. Requicha, "Nanorobots, nems, and nanoassembly," *Proc. IEEE*, vol. 91, no. 11, pp. 1922–1933, Nov. 2003.
2. B. Chui and L. Kissner, "Nanorobots for mars eva repair," in *Proc. Int. Conference on Environmental Systems (ICES)*, July 2000.
3. P. Pillai, J. Campbell, G. Kedia, S. Moudgal, and K. Sheth, "A 3D fax machine based on claytronics," in *Proc. IEEE/RSJ International Conference on Intelligent Robots and Systems*, Oct. 9–15, 2006, pp. 4728–4735.

4. A. Menciassi, A. Eisinberg, I. Izzo, and P. Dario, "From "macro" to "micro" manipulation: models and experiments," *IEEE/ASME Trans. Mechatronics*, vol. 9, no. 2, pp. 311–320, 2004.
5. R. Fearing, "Survey of sticking effects for micro parts handling," in *Proc. IEEE/RSJ International Conference on Intelligent Robots and Systems 95. "Human Robot Interaction and Cooperative Robots"*, vol. 2, 1995, pp. 212–217.
6. M. Sitti, "Microscale and nanoscale robotics systems [grand challenges of robotics]," *IEEE Robotics Automation Magazine*, vol. 14, no. 1, pp. 53–60, Mar. 2007.
7. S. Wu, J. Serbin, and M. Gu, "Two-photon polymerisation for three-dimensional microfabrication," *Journal of Photochemistry and Photobiology A: Chemistry*, vol. 181, no. 1, pp. 1–11, 2006.
8. C. N. LaFratta, J. T. Fourkas, T. Baldacchini, and R. A. Farrer, "Multiphoton fabrication," *Angew. Chem. Int. Ed.*, vol. 46, no. 33, pp. 6238–6258, 2007.
9. S. Marua and J. T. Fourkas, "Recent progress in multiphoton microfabrication," *Laser & Photonics Reviews*, vol. 2, no. 1-2, pp. 100–111, 2008.
10. J. J. Gorman and N. G. Dagalakis, "Probe-based micro-scale manipulation and assembly using force feedback," in *Proc. International Conference on Robotics and Remote Systems for Hazardous Environments*, 2006.
11. A. Requicha, S. Meltzer, R. Resch, D. Lewis, B. Koel, and M. Thompson, "Layered nanoassembly of three-dimensional structures," in *Proc. ICRA Robotics and Automation IEEE International Conference on*, S. Meltzer, Ed., vol. 4, 2001, pp. 3408–3411 vol.4.
12. M. Sitti, "Atomic force microscope probe based controlled pushing for nanotribological characterization," *IEEE/ASME Trans. Mechatronics*, vol. 9, no. 2, pp. 343–349, 2004.
13. A. Burkle, F. Schmoeckel, H. Worn, B. P. Amavasai, F. Caparrelli, and J. R. Travis, "A versatile vision system for micromanipulation tasks," in *Proc. Int. Conference on Multisensor Fusion and Integration for Intelligent Systems*, 2001.
14. Y. Ahn, T. Ono, and M. Esashi, "Si multiprobes integrated with lateral actuators for independent scanning probe applications," *J. Micromech.Microeng*, vol. 15, pp. 1224–1229, 2005.
15. F. Dionnet, D. Haliyo, and S. Regnier, "Autonomous micromanipulation using a new strategy of accurate release by rolling," in *Proc. IEEE International Conference on Robotics and Automation ICRA '04*, D. Haliyo, Ed., vol. 5, 2004, pp. 5019–5024 Vol.5.
16. T. Kasaya, H. T. Miyazaki, S. Saito, K. Koyano, T. Yamaura, and T. Sato, "Image-based autonomous micromanipulation system for arrangement of spheres in a scanning electron microscope," *Review of Scientific Instruments*, vol. 75, no. 6, pp. 2033–2042, 2004.
17. S. Saito, T. Motokado, K. J. Obata, and K. Takahashi, "Capillary force with a concave probe-tip for micromanipulation," *Applied Physics Letters*, vol. 87, no. 23, p. 234103, 2005.
18. K. Castelino, S. Satyanarayana, and M. Sitti, "Manufacturing of two and three-dimensional micro/nanostructures by integrating optical tweezers with chemical assembly," *Robotica*, vol. 23, no. 4, pp. 435–439, 2005.
19. C. Gosse and V. Croquette, "Magnetic tweezers: Micromanipulation and force measurement at the molecular level," *Biophysical Journal*, vol. 82, pp. 3314–3329, June 2002.
20. L. Zheng, S. Li, P. Burke, and J. Brody, "Towards single molecule manipulation with dielectrophoresis using nanoelectrodes," in *Proc. Third IEEE Conference on Nanotechnology IEEE-NANO 2003*, S. Li, Ed., vol. 1, 2003, pp. 437–440 vol. 2.
21. A. Subramanian, B. Vikramaditya, B. Nelson, D. Bell, and L. Dong, "Dielectrophoretic micro/nanoassembly with microtweezers and nanoelectrodes," in *Proc. th International Conference on Advanced Robotics ICAR '05*, B. Vikramaditya, Ed., 2005, pp. 208–215.
22. K.-F. Bohringer, K. Goldberg, M. Cohn, R. Howe, and A. Pisano, "Parallel microassembly with electrostatic force fields," in *Proc. IEEE International Conference on Robotics and Automation*, K. Goldberg, Ed., vol. 2, 1998, pp. 1204–1211 vol. 2.
23. C. Ropp, R. Probst, Z. Cummins, R. Kumar, A. J. Berglund, S. R. Raghavan, E. Waks, and B. Shapiro, "Manipulating quantum dots to nanometer precision by control of flow," *Nature Photonics*, 2010 (submitted).

24. A. J. Berglund, K. McHale, and H. Mabuchi, "Feedback localization of freely diffusing fluorescent particles near the optical shot-noise limit," *Opt. Lett.*, vol. 32, pp. 145–147, 2007.
25. C. Onal and M. Sitti, "Visual servoing-based autonomous 2-D manipulation of microparticles using a nanoprobe," *IEEE Trans. Control Syst. Technol.*, vol. 15, no. 5, pp. 842–852, 2007.
26. V. Utkin, "Variable structure systems with sliding modes," *IEEE Transactions on Automatic Control*, vol. 22, no. 2, pp. 212–222, 1977.
27. C. Onal and A. Sabanovic, "Plant behaviour dictation using a sliding mode model reference controller," in *Proc. 9th IEEE International Workshop on Advanced Motion Control*, A. Sabanovic, Ed., 2006, pp. 243–248.
28. A. Sabanovic, L. M. Fridman, and S. Spurgeon, Eds., *Variable structure systems: from principles to implementation.* IEE Control Series, vol. 66, The Institute of Electrical Engineers, London, 2004.
29. B. Bhushan, *Handbook of Micro/Nano Tribology, 2nd ed.* CRC Press, 1999.
30. S. Saito, H. T. Miyazaki, T. Sato, and K. Takahashi, "Kinematics of mechanical and adhesional micromanipulation under a scanning electron microscope," *Journal of Applied Physics*, vol. 92, no. 9, pp. 5140–5149, 2002.
31. G. V. Dedkov, "Friction on the nanoscale: new physical mechanisms," *Materials Letters*, vol. 38, pp. 360–366, 1999.
32. J. A. Hurtado and K.-S. Kim, "Scale effects in friction of single-asperity contacts. I. from concurrent slip to single-dislocation-assisted slip," *Proceedings: Mathematical, Physical and Engineering Sciences*, vol. 455, no. 1989, pp. 3363–3384, 1999.
33. K. L. Johnson, "The contribution of micro/nano-tribology to the interpretation of dry friction," *Proc. Instn. Mech. Engrs.*, vol. 214 Part C, pp. 11–22, 2000.
34. R. O. Duda and P. E. Hart, "Use of the Hough transformation to detect lines and curves in pictures," *Commun. ACM*, vol. 15, no. 1, pp. 11–15, 1972.
35. D. H. Ballard, "Generalizing the Hough transform to detect arbitrary shapes," *Pattern Recognition*, vol. 13, no. 2, pp. 111–122, 1981.
36. L. Xu, E. Oja, and P. Kultanen, "A new curve detection method: randomized Hough transform (RHT)," *Pattern Recogn. Lett.*, vol. 11, no. 5, pp. 331–338, 1990.
37. A. A. Rad, K. Faez, and N. Qaragozlou, "Fast circle detection using gradient pair vectors," in *Proc. VIIth Digital Image Computing: Techniques and Applications*, O. S. Sun C. and, Talbot H. and and A. T., Eds., 2003.
38. B. Yilmaz, M. Unel, and A. Sabanovic, "Rigid and affine motion estimation in vision using sliding mode observers," in *Proc. IEEE International Symposium on Industrial Electronics ISIE 2005*, M. Unel, Ed., vol. 1, 2005, pp. 31–36.
39. G. Li, N. Xi, M. Yu, and W. K. Fung, "3D nanomanipulation using atomic force microscopy," in *Proc. IEEE International Conference on Robotics and Automation ICRA '03*, vol. 3, 2003, pp. 3642–3647.
40. C.-K. Liu, S. Lee, L.-P. Sung, and T. Nguyen, "Load-displacement relations for nanoindentation of viscoelastic materials," *Journal of Applied Physics*, vol. 100, no. 3, p. 033503, 2006.
41. X. Tian, N. Jiao, L. Liu, Y. Wang, N. Xi, W. Li, and Z. Dong, "An AFM based nanomanipulation system with 3D nano forces feedback," in *Proc. International Conference on Intelligent Mechatronics and Automation*, 2004, pp. 18–22.
42. R. W. Stark, F. J. Rubio-Sierra, S. Thalhammer, and W. M. Heckl, "Combined nanomanipulation by atomic force microscopy and UV-laser ablation for chromosomal dissection." *Eur Biophys J*, vol. 32, no. 1, pp. 33–39, Mar 2003.
43. J. H. Lü, "Nanomanipulation of extended single-DNA molecules on modified mica surfaces using the atomic force microscopy." *Colloids Surf B Biointerfaces*, vol. 39, no. 4, pp. 177–180, Dec 2004.
44. T. Junno, K. Deppert, L. Montelius, and L. Samuelson, "Controlled manipulation of nanoparticles with an atomic force microscope," *Applied Physics Letters*, vol. 66, no. 26, pp. 3627–3629, 1995.

45. D. M. Schaefer, R. Reifenberger, A. Patil, and R. P. Andres, "Fabrication of two-dimensional arrays of nanometer-size clusters with the atomic force microscope," *Applied Physics Letters*, vol. 66, no. 8, pp. 1012–1014, 1995.
46. C. Baur, B. C. Gazen, B. Koel, T. R. Ramachandran, A. A. G. Requicha, and L. Zini, "Robotic nanomanipulation with a scanning probe microscope in a networked computing environment," vol. 15, no. 4. AVS, 1997, pp. 1577–1580.
47. T. R. Ramachandran, C. Baur, A. Bugacov, A. Madhukar, B. E. Koel, A. Requicha, and C. Gazen, "Direct and controlled manipulation of nanometer-sized particles using the non-contact atomic force microscope," *Nanotechnology*, vol. 9, no. 3, pp. 237–245, 1998.
48. M. Martin, L. Roschier, P. Hakonen, U. Parts, M. Paalanen, B. Schleicher, and E. I. Kauppinen, "Manipulation of Ag nanoparticles utilizing noncontact atomic force microscopy," *Applied Physics Letters*, vol. 73, no. 11, pp. 1505–1507, 1998.
49. S. Hsieh, S. Meltzer, C. Wang, A. Requicha, M. Thompson, and B. Koel, "Imaging and manipulation of gold nanorods with an atomic force microscope," *Journal of Physical Chemistry B*, vol. 106, no. 2, pp. 231–234, 2002.
50. B. Mokaberi and A. Requicha, "Towards automatic nanomanipulation: drift compensation in scanning probe microscopes," in *Proc. IEEE International Conference on Robotics and Automation ICRA '04*, vol. 1, 2004, pp. 416–421 Vol.1.
51. G. Li, N. Xi, H. Chen, C. Pomeroy, and M. Prokos, "Videolized" atomic force microscopy for interactive nanomanipulation and nanoassembly," *IEEE Trans. Nanotechnol.*, vol. 4, no. 5, pp. 605–615, 2005.
52. B. Mokaberi and A. A. G. Requicha, "Drift compensation for automatic nanomanipulation with scanning probe microscopes," *Automation Science and Engineering, IEEE Transactions on*, no. 3, pp. 199–207, July 2006.
53. B. Mokaberi, J. Yun, M. Wang, and A. Requicha, "Automated nanomanipulation with atomic force microscopes," in *Proc. IEEE International Conference on Robotics and Automation*, 2007, pp. 1406–1412.
54. C. D. Onal, B. Sumer, and M. Sitti, "Cross-talk compensation in atomic force microscopy," *Review of Scientific Instruments*, vol. 79, no. 10, p. 103706, 2008.
55. F. Krohs, C. Onal, M. Sitti, and S. Fatikow, "Towards automated nanoassembly with the atomic force microscope: A versatile drift compensation procedure," *Journal of Dynamic Systems, Measurement, and Control*, vol. 131, no. 6, p. 061106, 2009.
56. J. E. Sader, J. W. M. Chon, and P. Mulvaney, "Calibration of rectangular atomic force microscope cantilevers," *Review of Scientific Instruments*, vol. 70, no. 10, pp. 3967–3969, 1999.
57. C. D. Onal, O. Ozcan, and M. Sitti, "Automated 2-D nanoparticle manipulation with an atomic force microscope," in *Proc. IEEE International Conference on Robotics and Automation ICRA '09*, 2009, pp. 1814–1819.
58. B. Sumer and M. Sitti, "Rolling and spinning friction characterization of fine particles using lateral force microscopy based contact pushing," *J. Adh. Scie. Techn.,*, 2008.
59. M. Sitti and H. Hashimoto, "Controlled pushing of nanoparticles: modeling and experiments," *IEEE/ASME Trans. Mechatronics*, vol. 5, no. 2, pp. 199–211, 2000.
60. C. Onal, 2007. [Online]. Available: http://www.andrew.cmu.edu/user/cdonal/uassembly.avi
61. O. Pitrement and M. Troyon, "General equations describing elastic indentation depth and normal contact stiffness versus load." *J Colloid Interface Sci*, vol. 226, no. 1, pp. 166–171, Jun 2000.
62. I. M. Nnebe, R. D. Tilton, and J. W. Schneider, "Direct force measurement of the stability of poly(ethylene glycol) polyethylenimine graft films," *Journal of Colloid and Interface Science*, vol. 276, pp. 306–316, 2004.
63. R. C. Gonzalez and R. E. Woods, *Digital Image Processing*, 2nd ed. Prentice Hall, 2002.

Chapter 5
Atomic Force Microscopy: Principles and Systems Viewpoint Enabled Methods

Srinivasa Salapaka and Murti Salapaka

5.1 Introduction

In 1959, Richard Feynman [12] offered an impressive vision of engineered devices at the nanoscale, where he asserted that there are no fundamentally limiting reasons that would disallow manipulation of matter at the nanoscale. He offered an elaborate hierarchy of contraptions that would facilitate altering matter at the nanoscale from a macroscale environment.

A significant milestone in this direction, the invention of the atomic force microscope (AFM) that realized an elegant and simple means of interrogating as well as manipulating matter at the nanoscale, was first reported in [6]. Here, a cantilever beam with a sharp tip at one end (see Fig. 5.1) was shown to be an effective means of sensing interatomic forces between the atoms on the tip and the atoms on the sample being interrogated. The AFM borrows a number of operating principles from its predecessor, the scanning tunneling microscope (STM). Similar to STM operation, the positioning of the sample with respect to the main probe (the cantilever-tip in the case of the AFM) is provided by using piezoelectric actuation, where the actuating material deforms when an external voltage is applied. A remarkable and enabling discovery by the inventors of STM [7] is that it is possible to deform piezoelectric material with Angstrom (an Angstrom is roughly the dimension of an atom) precision reliably by applying voltages that can be easily generated. In a typical AFM setup (see Fig. 5.1), the sample is positioned using piezoelectric scanners that provide motion of the cantilever tip with respect to all three directions; the two lateral x and y directions as well as the vertical z

S. Salapaka (✉)
University of Illinois, Urbana Champaign, Champaign, IL 61820-5711, USA
e-mail: salapaka@illinois.edu

M. Salapaka
University of Minnesota, Minneapolis, MN 55455, USA
e-mail: murtis@umn.edu

J.J. Gorman and B. Shapiro (eds.), *Feedback Control of MEMS to Atoms*,
DOI 10.1007/978-1-4419-5832-7_5, © Springer Science+Business Media, LLC 2012

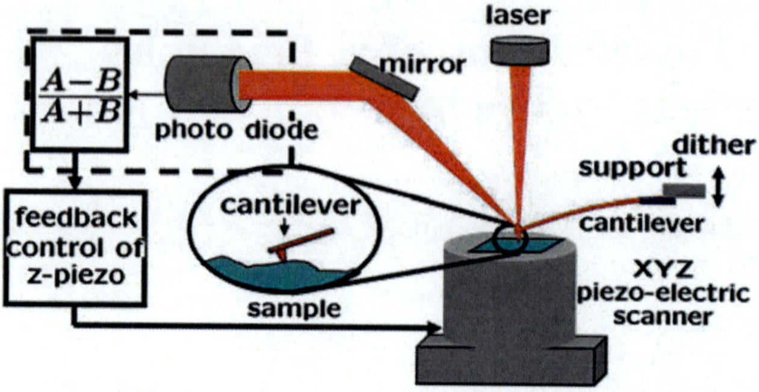

Fig. 5.1 An atomic force microscope schematic illustrates a cantilever with a tip that probes a sample. The support of the cantilever can be actuated by a dither piezo. The cantilever deflection is sensed by a laser beam that reflects from the cantilever surface into a split photodiode. The sample can be positioned with sub-nanometer accuracy using piezoactuated material

direction. The first AFM reported in [5] used an STM to sense the motion of the cantilever. However, using an STM to sense the cantilever motion proved to be too cumbersome. The laser beam-bounce method, where a laser beam incident on the cantilever reflects into a split photodiode, is the preferred means of sensing the cantilever deflection. The beam-bounce method leverages the large optical path that effectively amplifies a small motion of the cantilever into a relatively large motion of the laser at the photodiode. Photodiodes provide bandwidths in MHz regime with high accuracy, and the associated shot noise is dominated by other noise sources that limit the AFM operation.

At the heart of the AFM is a force sensor; a micro-cantilever that has a sharp tip at its free end. The atoms on the tip of the cantilever and the atoms on the surface of the sample exert a force on each other. These forces are typically in the piconewton range and qualitatively have the characteristics of an attractive nature for large separations and sharp repulsive nature at short ranges. A good qualitative model for the interatomic forces is the Lennard–Jones potential, where the force F between the atoms is given by $F(r) = -A/r^7 + B/r^{13}$, where r represents the separation between the atoms (see Fig. 5.2). For the cantilever flexure to have a large enough deflection (above the dominant noise sources) due to the interatomic forces the cantilevers need to be soft. Also, the first resonant frequency of the cantilever has to be away from the frequency content in the 0–2 kHz range of the disturbances of the ambient environment, which include building vibrations. Thus, a high first resonant frequency is needed for the cantilever. A small stiffness (in the range 0.06–100 N/m) and a large resonant frequency implies small mass for the cantilevers and therefore the cantilevers employed in AFM have dimensions at the micrometer scale (a length, width, and thickness on the order of 100, 10 and 5 μm, respectively).

A prevalent use of AFMs is in unraveling the force profile between various materials, where a quantitative description of the force variation as a function of the separation between the material is reconstructed. Such a description is obtained by moving the sample relative to the tip in the vertical z direction while

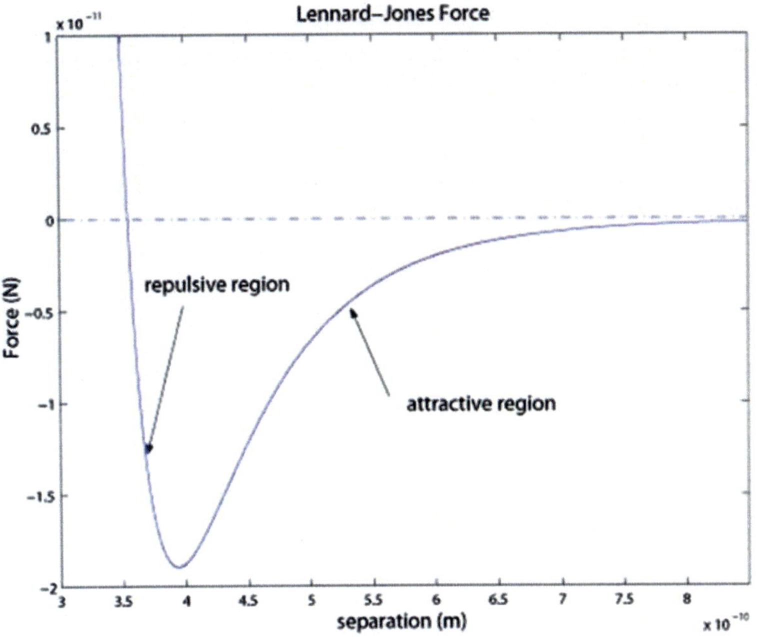

Fig. 5.2 The qualitative nature of forces between tip atoms and atoms on the sample is shown. These forces are characterized by weak attractive forces for large tip-sample separations and strong repulsive forces for small tip-sample separations

keeping the lateral position fixed. Often the cantilever tip is suitably modified to characterize forces between different materials of interest. For example, in a number of studies the cantilever tip is modified biologically with a molecule of interest and the interaction forces between this molecule and the molecules on the sample are interrogated with respect to the separation between them to obtain the force profile (often such force separation relationships differ even qualitatively from the Lennard–Jones force characteristics).

AFMs are mainly used to obtain the topography of the sample with nanometer or sub-nanometer resolution. The primary means of obtaining the topography of the sample is to move (scan) the sample laterally with respect to the tip. As the sample is scanned, at each lateral coordinate (x,y) the tip feels the interatomic force $h(x,y)$ that results in a deflection of the tip. Assuming that the material properties are the same along the lateral directions, the changes in the forces felt by the cantilever tip are due to the tip atoms moving closer or farther from the sample atoms due to the topography change in the sample. Thus the forces felt by the cantilever tip, as the sample is scanned, do not vary linearly with respect to the sample topography since the dependence of the tip-sample interaction force on the separation is qualitatively described by the nonlinear Lennard–Jones potential. Also, since there is no quantitative description of the nonlinear dependence, the problem of extracting the sample-surface topography from the forces felt by the cantilever becomes challenging. The inventors of AFM employed feedback (the same principle was employed for STM) to overcome these challenges, which also provided a significant enabling method for interrogation at the nanoscale.

Feedback plays a pivotal role in the operability of the AFM, particularly for estimating the topography of the sample. In methods based on the *force-balance principle*, the sample is moved vertically in the z direction by a controller (see Fig. 5.1) to maintain a constant deflection of the cantilever (which is equivalent to maintaining a constant force on the cantilever) while the sample is scanned laterally in the x and y directions. The vertical movement of piezo scanner is considered an estimate of the topography of the sample with the reasoning that the sample is appropriately moved to negate the variations of the topography to regulate a constant force on the cantilever tip. This methodology proves remarkably effective and elegantly overcomes challenges in estimating the topography without the knowledge of the nonlinear force profile that exists between the tip and the sample atoms.

The demands on AFM technology are considerable, many of which primarily stem from the need to interrogate large samples with resolution at the nanometer scale. The AFM technology enables interrogation of the sample at the atomic scale with unparalleled ease. However, it interrogates the sample a single location at a time which makes it impractical for high-throughput applications. This shortcoming can be addressed partly through parallel deployment of multiple cantilevers. Another strong motivation for high-speed scanning requirements arise from studies that investigate nanoscale *dynamics* of samples. For example, a significant effort in AFM research is the study of the dynamics of bio-molecules, where the motion of molecular motors is typically examined [1]. These needs cannot be met by introducing parallelism. Thus, the need to improve the interrogation bandwidth remains central to AFM instrumentation research. Such an improvement in bandwidth implies related challenges on the positioning systems and on cantilever related technologies.

Another important objective of future AFM technologies is to provide a measure of the fidelity of the data for the sample that is being generated. Current commercially available AFMs provide scant or no information on the interpretation accuracy of the data.

As will be seen in this chapter, control systems viewpoints provide an effective means of addressing these challenges. In this chapter, after presenting the basic operational principles of AFM, we present research related to nanopositioning followed by research on the cantilever dynamics in the presence of sample forces.

5.2 Operational Principles

As described earlier, the AFM (see Fig. 5.1) uses a micro-cantilever with a sharp tip at one end to probe the sample being interrogated. The other end is fixed to a support, which can be oscillated by using a *dither piezo*. The laser beam that bounces off the cantilever surface is collected by a quadrant photodiode that provides a measure of the cantilever's vertical deflection as well as any torsional twist of the cantilever. The controller is typically realized through analog components or digital components such as DSP and FPGA boards. The controller uses the photodiode signal to regulate the relative separation of the tip and the sample in the vertical z direction and also provides the capability to control the lateral x and y motions. In earlier AFMs, feedback was used for the z direction but no sensors were used for the lateral positioning and therefore the lateral positioning was achieved in open-loop. Lateral closed-loop operation was first realized in [9].

5.2.1 Noise Sources

The main noise sources are the laser, the photodiode, structural vibrations of the surroundings such as building sways and floor vibrations, acoustic sources, and thermal noise sources. Considering the cantilever as a linear time-invariant system that processes various input forces to yield the cantilever deflection as the output, some of the noise sources act at the input while others affect the measurement of the output of the cantilever system. The thermal noise, acoustic, and building noise sources appear at the input of the cantilever system and thus these inputs get processed by the cantilever system before these effects are perceived at the measurement. These form the main contributors to the process noise input to the cantilever system. Since cantilever behavior is well approximated by a one-mode model for most studies, the dynamics are described by

$$\ddot{p}+\frac{\omega_0}{Q}\dot{p}+\omega_0^2 p=f, \tag{5.1}$$

where p is the deflection of the cantilever, ω_0 is the first resonant frequency, Q is the quality factor of the cantilever that characterizes the damping, and f is the net external force action on the cantilever system. The associated transfer function is

$$G=\frac{1}{s^2+\frac{\omega_0}{Q}s+\omega_0^2}. \tag{5.2}$$

The laser noise and photodiode noise are sources that affect the measurement of the cantilever deflection and therefore appear as measurement noise ϑ. The measurement is given by

$$y=p+\vartheta. \tag{5.3}$$

One simple and elegant method of assessing the force resolution of the micro-cantilever is to obtain the power spectral density of the photodiode output when the laser beam bounces off the cantilever surface into the photodiode without any sample present. Assuming that appropriate vibration isolation is in place, the main

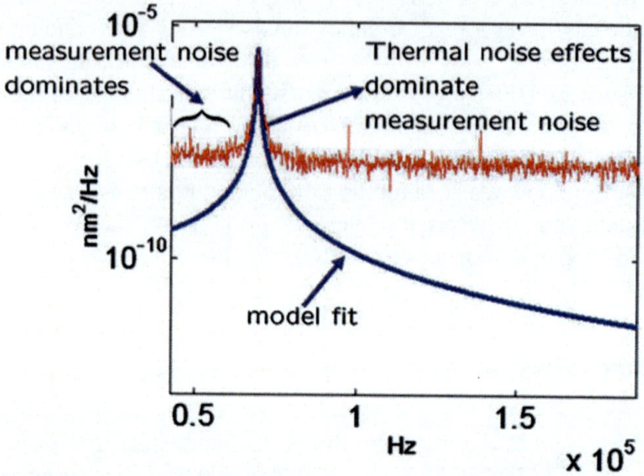

Fig. 5.3 The power spectral density of the cantilever deflection when the cantilever is forced by thermal noise is shown. The response of the cantilever due to thermal noise peaks above the noise floor only near the first resonant frequency of the cantilever demonstrating that the cantilever can sense forces as small as the thermal Langevin force near the resonant frequency. Away from the resonant frequency of the cantilever, measurement noise dominates

process noise source is the thermal noise. The thermal noise is a white noise forcing term at the input to the cantilever system caused by the interaction of the cantilever at temperature T in thermal equilibrium with its environment (see [22] for a detailed multi-mode analysis of thermal noise in cantilevers). Considering the thermal forcing as the signal of interest, it is interesting to assess if this signal can be deciphered from the measurement of the cantilever deflection. The power spectral density (psd) plot of the cantilever deflection (when the sample is absent) is shown in Fig. 5.3. A sharp peak is evident in the psd plot of thermal response. This peak is at the first resonant frequency ω_0 of the cantilever. Indeed, this also confirms that the effect of the thermal signal can indeed be deciphered from the photodiode measurement since thermal noise is dominant over other noise at frequencies near the cantilever resonant frequency ω_0. The observed cantilever deflection in Fig. 5.3 is predominantly a response to thermal noise. The other noise sources only additively corrupt the deflection measurement and are dominant only at frequencies away from the resonant frequency of the cantilever.

5.2.2 *Traditional Modes of Imaging*

5.2.2.1 Contact Mode Imaging

In contact mode imaging, the cantilever is not excited externally and the dither piezo is not used. The cantilever tip deflects due to interatomic forces.

In lift-mode contact-mode imaging, there is no feedback in the vertical z direction and the deflection of the cantilever is considered as an estimate of the sample topography. This method is suitable for imaging relatively flat samples over small areas since any incline or high aspect ratio features will lead to the cantilever tip either losing contact with the sample or crashing into the sample. It also has the disadvantage of not providing a quantitative estimate of the topography. However, it has the advantage of high-bandwidth operation since no feedback is required that obviates the need for high-speed positioning of the sample with respect to the tip.

In the constant-force contact-mode imaging scheme, a constant deflection of the cantilever is regulated by the controller that positions the sample with respect to the tip. The controller effort is considered as an estimate of the topography of the sample. This method provides a highly reliable estimate of the topography particularly when the frequency content of the sample topography is below the bandwidth of the z direction positioning system. It also allows for easy interpretation of data. Contact mode imaging has the drawback of imposing large lateral and vertical forces on the sample that precludes its use when imaging soft samples.

5.2.2.2 Dynamic Modes

In the dynamic modes of imaging, the cantilever is forced externally near the resonant frequency of the cantilever and information about the sample is obtained by monitoring how the cantilever's nominal motion (for example, in air) is altered under the influence of the sample. Typically, the information of the sample is modulated in a frequency range near the resonant frequency of the cantilever, where, as described before, the cantilever probe is thermally limited and thus has better signal to noise ratio.

The traditional dynamic modes of imaging can be classified into two classes: amplitude-modulation and frequency-modulation dynamic modes. In the amplitude-modulation scheme, the dither piezo drives the cantilever support sinusoidally near the resonant frequency of the cantilever that sets the cantilever tip into a periodic motion (see Fig. 5.4). The amplitude of the first harmonic of the drive frequency in the measured cantilever deflection is monitored to obtain the topography of the sample. In the constant-amplitude mode the amplitude is regulated at a set-point amplitude by the controller that alters the tip-sample separation by moving the sample with respect to the tip. Similar to the constant-force contact-mode operation, the control signal forms an estimate of the sample topography. In another mode (sometimes referred to as the error-mode), the feedback is rendered ineffective and the error between the measured amplitude and the set-point amplitude is considered as the imaging signal. In the frequency modulation mode (see Fig. 5.5), the cantilever is forced in a manner that maintains the drive at a phase of $\pi/2$ with respect to the cantilever oscillation, so that the cantilever is always maintained at the resonant condition (note that for a second order under-damped system, at resonance the output of the system is at a phase lag of 90° with respect to the input). The resonant frequency of the cantilever-sample system is different from

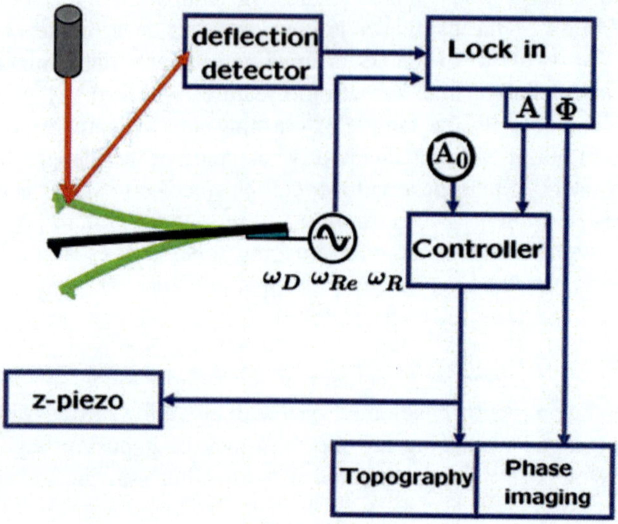

Fig. 5.4 A schematic describing the amplitude modulation AFM is shown. In this mode the cantilever is forced at a frequency ω_D that is near the first resonant frequency ω_0 of the cantilever. The amplitude and the phase of the first harmonic of the cantilever deflection are found by lock in methods. The controller moves the z piezo vertically to maintain a constant setpoint A_0 and the control signal forms an estimate of the sample topography

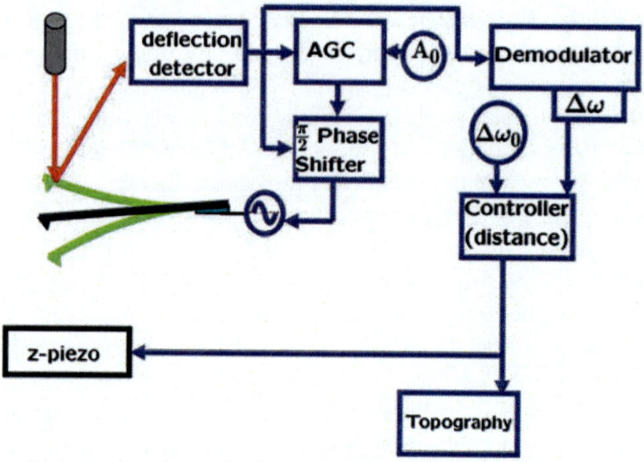

Fig. 5.5 The figure describes the frequency modulation AFM scheme where the cantilever is driven by the cantilever deflection that is phase shifted by 90°. This scheme maintains the cantilever forced at the equivalent resonant frequency of the cantilever. The automatic gain controller alters the magnitude of the driving force to maintain a set amplitude A_0. The resonant frequency shift $\Delta\omega$ is measured and the controller positions the sample vertically to maintain a reference frequency shift $\Delta\omega_0$. The control signal forms an estimate of the sample topography

the resonant frequency of the cantilever alone as the sample introduces a periodic force every oscillation cycle on the cantilever tip. The changed resonant frequency of the cantilever-sample system is called the equivalent resonant frequency of the cantilever. The measured equivalent resonant frequency is compared to a set-point frequency and the controller positions the sample with respect to the tip to regulate the set-point frequency. As in previously discussed approaches, the control signal provides the estimate of the sample topography.

It is amply evident that feedback control forms an essential part of the traditional imaging modes. The feedback controllers employed are primarily of proportional, derivative, and integral (PID) controllers. However, as will be seen later in the chapter, modern control theory has a lot to offer to AFM instrumentation.

5.3 Control Design for Nanopositioning Systems

In both static and dynamic modes of operation, the lateral motion required for scanning is obtained by moving the sample. The positioning system, which provides this motion, plays a vital role in AFM. A significant aim of positioning-system design in AFM is to preserve the high vertical resolution provided by the cantilever sensor. High-resolution, high-bandwidth, and reliable positioning are the main performance criteria that most nanoscientific studies and applications seek from the positioning systems. One of the main challenges with current systems is achieving high tracking bandwidth. Even though the cantilever sensors have large resonant frequencies on the order of 100 kHz, the AFMs are severely limited by the vertical and lateral positioning systems that have bandwidths only on the order of 1 kHz. Another challenge that is typically neglected in typical positioning system design, is reliability. Reliability in terms of repeatability of experiments is essential for validation of the underlying studies. The delicate nature of AFM experiments, the diverse operating conditions, and lack of tools for obtaining accurate models for AFM emphasize the importance of reliable positioning. The challenge, in this context, arises in developing methods for characterizing, evaluating, and designing positioning systems for reliability. Typical nanopositioning systems comprise a flexure stage that provides frictionless motion through elastic deformation, an actuator typically made from piezoelectric material that provides the required force to deform the flexure stage, and a sensing system along with the control system. The main obstacles in the design of robust broadband nanopositioning systems stem from flexure-stage dynamics that limit the bandwidth of the positioning stage, nonlinear effects of piezoelectric actuation such as hysteresis and creep that are difficult to model, and sensor noise issues that can potentially hamper the tracking resolution of the device.

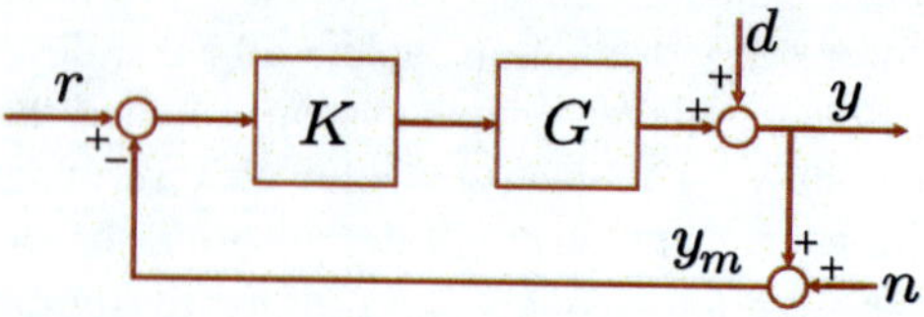

Fig. 5.6 A block diagram schematic for a typical nanopositioning system

5.3.1 *Performance Criteria and Limitations*

One of the important contributions of control systems theory to the design of nanopositioning systems is the quantification of performance objectives and fundamental design limitations. Figure 5.6 shows a block-diagram schematic of a typical nanopositioning system. The transfer function G represents the *scanner*, which comprises the actuator, the flexure stage, and the sensor. It is the transfer function from the voltage input u applied to the actuator to the flexure-stage displacement y. The signals r, d, n, and y_m, respectively, represent the reference trajectory for tracking, the *mechanical noise*, or the effects of dynamics that are not incorporated in the model G, the sensor noise, and the noisy measurement, while the transfer function K represents the feedback law. The main objective for the design of the controller K is to make the *tracking error* small, that is to make the difference $r-y$ between the desired and actual motions small.

The performance criteria are quantified by characterizing the tracking error. For a given controller K, the tracking error for the system in Fig. 5.6 is given by

$$e = r - y = S(r-d) + Tn, \tag{5.4}$$

where the *sensitivity transfer function* $S = (1+GK)^{-1}$ and the *complementary sensitivity transfer function* $T = 1 - S = (1+GK)^{-1}GK$. Thus, high resolution can be achieved by designing the feedback law K such that S and T are small in ranges where the frequency contents of r and n, respectively, are large. The resolution of the closed-loop positioning system is determined by the term Tn, whereby low values of T over a larger range of frequencies guarantee better resolution. More specifically, the resolution of a positioning system is determined by the standard deviation σ of the position signal when reference signal is identically zero, where

$$\sigma^2 = \int_0^\infty |T(j\omega)|^2 P_n(\omega)\,\mathrm{d}\omega, \tag{5.5}$$

and $P_n(\omega)$ denotes the power spectral density of the noise signal n. Thus the smaller the bandwidth of T, which is characterized by the roll-off frequency ω_T, the smaller the standard deviation σ, and hence the better the resolution of the closed-loop device. The tracking bandwidth is determined by the bandwidth ω_{BW} of the sensitivity transfer function. The reliability criterion translates to robustness of positioning systems to modeling uncertainties and operating conditions. That are insensitive to diverse operating conditions give repeatable measurements, and are hence reliable. The peak value $\|S\|_\infty$ of the magnitude of the sensitivity function serves as a good measure to characterize the robustness of the positioning system. Thus the performance specifications translate to control design objectives of achieving high values of ω_{BW} for high tracking bandwidth, high roll-off rates of T, smaller values of ω_T for better positioning resolution, and low values of $\|S\|_\infty$ for better robustness to modeling uncertainties.

The challenges in achieving the above design objectives mainly originate from hardware and fundamental algebraic limitations on the design of the feedback law K. Good control design trades off one objective for another in the needed frequency ranges. For instance, the simple algebraic constraint of $S(j\omega)+T(j\omega)\equiv 1$ clearly implies that S and T cannot be made small simultaneously in all frequencies, which reflects the conflict between the bandwidth and the resolution objectives. Besides, for scanner G with phase margin less than $90°$, which is true for most practical systems, the bandwidth ω_{BW} cannot be larger than ω_T [31]. This limitation prevents the feedback control to achieve noise attenuation over the target reference frequency range. Another fundamental limitation that imposes a trade-off between bandwidth, resolution, and robustness requirements can be explained in terms of the Bode integral law [8, 13], which imposes the following constraint on any stable system G with the relative degree of the transfer function $K(s)G(s)$ greater than or equal to two,

$$\int_0^\infty \log|S(j\omega)|\,\mathrm{d}\omega = 0. \tag{5.6}$$

For positioning systems, the condition on the relative degree is typically satisfied. This is so since T needs a sufficiently fast roll-off rate at high frequencies for noise attenuation (and therefore better resolution). The open-loop transfer function $K(s)G(s)$ is designed such that it has relative degree greater than or equal to 2. Also when the control is discrete and the system is analog, then the relative degree condition is inherently satisfied [18]. The limitations from this algebraic law can be explained in terms of the *waterbed effect* [11] – since the area under the graph of $\log|S(j\omega)|$ over the entire frequency range is zero, $S(j\omega)$ made small at a frequency range has to be compensated by making it large in some other frequency ranges. One direct consequence of this law is that $S(j\omega)$ cannot be made less than 1 over all frequencies. Therefore $\|S\|_\infty$, the measure for robustness is at least 1. Moreover, for positioning systems with real non-minimum phase zeros and design objectives that demand high roll-off rates on T, the following stricter fundamental algebraic limitation can be derived [13],

$$\int_0^\infty \log|S(j\omega)|W(z,\omega)\,\mathrm{d}\omega = 0, \tag{5.7}$$

$W(z,\omega)=2z/(z^2+\omega^2)$ for a real positive zero z. Typical scanner systems have non-collocated actuators and sensors that are separated by flexure stages. The transfer function models of such systems generally exhibit non-minimum phase zeros. In this

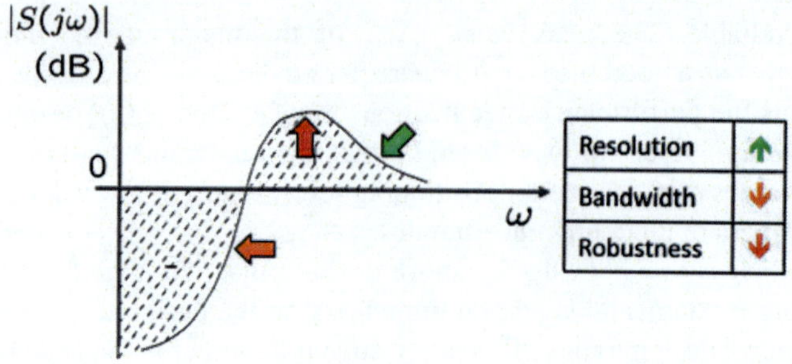

Fig. 5.7 Trade-offs due to the finite-waterbed effect. The Bode integral laws manifest themselves as waterbed effects, where decreasing the magnitude of the sensitivity function at a certain frequency range results in its increase in some other frequency range. For instance, making the sensitivity function small (near to 1) at high frequencies to ensure a high roll-off rate of the complementary sensitivity function (since $T = 1 - S$) for better resolution results in lower robustness to modeling uncertainties (due to higher values of the peak ($\|S\|_\infty$)) and lower values of tracking bandwidth (since ω_{BW} decreases). Similar trade-offs where one performance objective is sacrificed at the cost of others can be analyzed by studying the finite-waterbed effect (© IOP 2009), reprinted with permission

case, it can be shown that the integral of $\log|S(j\omega)|$ over a *finite* frequency range can be bounded from below, thus manifesting a waterbed effect over a finite frequency range. Thus the simultaneous requirements of low $|S(j\omega)|$ over a large frequency for a high tracking bandwidth, high order roll-off rates of T at high frequencies for high resolution, and small peaks of the $S(j\omega)$ compete against each other. For instance, small $|S(j\omega)|$ over a specified bandwidth might not leave out enough frequency range to be compatible with the integral bound in the *finite-waterbed effect* even with $S(j\omega)$ at the allowed peak value for the remaining frequencies (see Fig. 5.7).

Besides these algebraic limitations, further constraints come in the form of hardware constraints. For instance, high-order controllers require larger computation times by digital signal processor (DSP), which limit sampling rates, and therefore the tracking bandwidth. Another important limitation on control design arises from saturation limits on the actuation signals imposed by the hardware.

5.3.2 *Design for Resolution, Bandwidth, and Robustness*

The algebraic and practical limitations on the control design severely restrict the space of achievable performance specifications. The model-free based designs (such as proportional-integral-derivative (PID) designs) that are typically used in the nanopositioning industry, as well as designs based on loop-shaping of the open-loop transfer functions, further restrict the achievable range of specifications due to their inherent structural limitations. These techniques are inadequate to achieve simultaneously the multiple objectives of resolution, bandwidth, and robustness under the design challenges and fundamental limitations described above. The robust optimal control theory provides an apt framework for control design for nanopositioning systems [15, 24, 29]. In this framework, it is possible to determine if a set of design specifications are feasible, and when feasible the control law K is obtained by posing and solving an optimization problem. The main advantage of using this optimization framework is that it incorporates performance objectives directly into its cost function. This eliminates the tedious task of tuning gains (in trial-and-hit manner) as in the PID designs, where even the exhaustively tuned gains may fail to yield acceptable performance. The robust control optimization problems are of the form

$$\min_K \|\Phi(K)\|, \tag{5.8}$$

where Φ is a matrix transfer function whose elements are in terms of closed-loop transfer functions in (5.4) and $\|(\cdot)\|$ represents a metric on transfer functions. The design specifications are interpreted in terms of closed-loop signals z (such as tracking error e in (5.4)) and these set Φ as the transfer function from external variables w (such as reference signal r and noise n) to signals z. Demonstration of this framework and discussion of its advantages are presented through experimental results in the following section.

The fundamental limitations on control design presented in the previous section can be effectively used for designing better nanopositioning systems in addition to designing trade-offs between performance objectives in control design. These limitations are on the control design for a *given* scanner stage G. The study of the limitations can be used to design new scanners which result in a larger space of achievable performance specifications. This requires integration of control design into the device design, which is rarely done in the nanopositioning industry. For instance, most of the nanopositioning systems designs focus on single degree of freedom axis modules and multi-axis systems are realized by stacking individual units together. One of the main reasons for the popularity of these serial-kinematic mechanisms is that in contrast to single-mass systems, having a separate mass for each axis avoids the coupling between different axial motions. However, these multi-mass nanopositioning systems are heavier and therefore provide lower tracking bandwidth than parallel-kinematic (single-mass) mechanisms. By delegating the decoupling of axial-motions to the control design, single-mass positioning systems can be made that achieve significantly higher bandwidths for similar resolution as guaranteed by multiple-mass devices.

5.3.3 *Experimental Demonstration of Optimal Control Framework*

In Fig. 5.8, a parallel-kinematic xyz nanopositioning scanner stage and a schematic of its kinematic model are shown. The detailed design, kinematics, and dynamics analysis are described in [10]. The mechanical component of the stage is fabricated

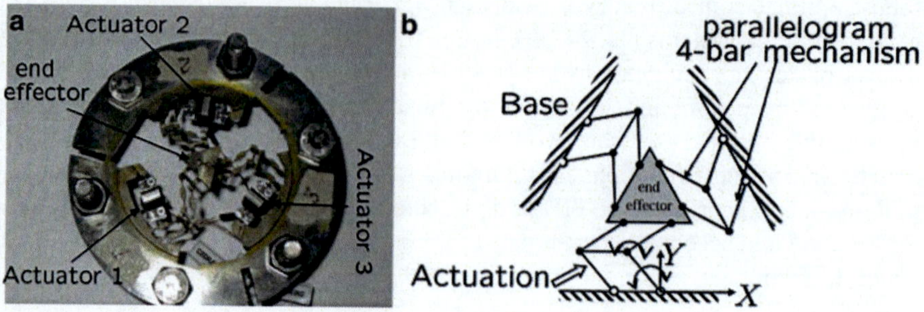

Fig. 5.8 (**a**) A prototype of a parallel-kinematic *xyz* nanopositioning system and (**b**) its schematic. In this design, three independent kinematic chains, which are piezoactuated, connect the base and the end effector. Each kinematic chain is composed of two parallelogram four-bar mechanisms, which makes the connector always parallel to the base. Together the three kinematic chains restrict all rotational degrees-of-freedom at the table, leaving it with three translations to satisfy the constraints imposed by the three kinematic chains (© ASME 2008), reprinted with permission

as a monolithic structure using electro-discharge machining (EDM). The triangular end effector at the center undergoes translation in the x, y, and z directions when the kinematic chains are actuated. A set of piezoelectric actuators were chosen for this stage (APA35XS by CEDRAT, free stroke 55 μm, blocking force 27 N, and maximum driving voltage 150 V) to actuate each of these kinematic chains at the first four-bar mechanism, that is connected to the base. The position sensing system consists of three capacitance displacement sensors which have a measuring range of ±50 μm and sub-nanometer resolution (0.3 nm up to 1 KHz bandwidth). Time-domain identification techniques resulted in a 13th order 3×3 multi-input multi-output (MIMO) transfer function from the input voltages to piezoelectric actuators to the sensor outputs. The six dominant modes are in the range 150–900 Hz, all of which are lightly damped with damping factors less than 0.01 (see [10] for details). Several models were identified at different operating points to characterize the modeling uncertainty. The uncertainties were dominant at low frequencies (<5 Hz). This uncertainty is primarily due to nonlinear effects of piezoelectric actuation such as hysteresis and creep.

The device design of this scanner system is such that the motion along the x, y and z directions are strongly coupled. Hence the off-diagonal terms in the MIMO transfer function are not small. For this system, it is extremely difficult and impractical to design controllers based on tuning-based or open-loop shaping techniques. The strongly coupled MIMO system makes optimal-control based design almost a necessity. The primary objective of the control design is to achieve a stage with high precision positioning and high bandwidth tracking capability that is robust to uncertainties in the operating conditions. Further, these objectives are to be met under the hardware limitation that requires the control signal (actuator input) to be within the range −1–7.5 V. The signal z in the optimal-control framework was chosen as $[W_P e \; W_T y \; W_u u]'$, where W_P, W_T and W_u are transfer functions that

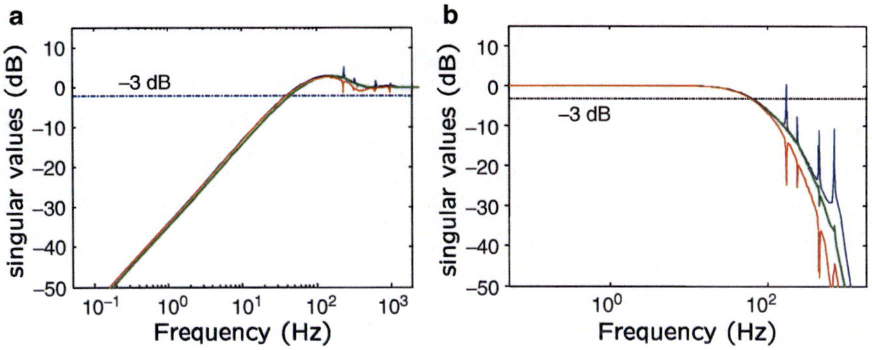

Fig. 5.9 Demonstration of the optimal-control design on a positioning system. (**a**) Singular-value plots of the sensitivity function (*solid*) from models at different operating points. The crossover frequency is 32 Hz for all the plots emphasizing the robustness in the closed-loop design. (**b**) A comparison of the corresponding complementary sensitivity transfer functions (crossover frequency is at 59 Hz) (© ASME 2008), reprinted with permission

reflect the relative frequency weighting of the performance objectives. The resulting transfer function Φ from $w = [r\ n]'$ to z for the formulation in (5.8) is given by $[W_P S\ W_T T\ W_u KS]$. We used the $\mathscr{H}_\infty$ norm to determine the control design. The transfer function W_P is chosen to have high gain at low frequency and low gain at high frequency, which forces the solution to the optimal control problem to have low input-to-error gain and achieves a prescribed bandwidth (the weight W_P in [10] was designed for input-to-error gain of 0.01% at low frequencies and a bandwidth of 50 Hz). The high-frequency noise attenuation is imposed by designing the weight W_T for the complimentary sensitivity transfer function T. W_T is chosen to have high gains at high frequencies so as to make T small at high frequencies (W_T was designed to ensure a high-frequency gain of 120 dB and a roll off rate of a 40 dB slope for T). The weighting function W_u is chosen to be a constant, so that the input to the actuator does not cross its saturation limits.

Figure 5.9 depicts experimental results when the controller obtained from the design described above is implemented on the positioning system. The maximum peak $\|S\|_\infty$ in the singular-value plot of the sensitivity transfer function is less than 1.85, which indicates good robustness to the modeling uncertainties and operating conditions. The singular-value plots of the sensitivity transfer function and the closed-loop transfer function have crossover frequencies at 32 and 59 Hz, respectively. The closed-loop system demonstrates a good tracking performance, as shown in Fig. 5.10, where the stage tracked the triangular reference signals well for frequencies in the range 5–40 Hz.

Figure 5.11a, b demonstrates the practical elimination of hysteresis and justification of the linear model. When an actuation voltage is applied to an actuator gradually from 0 to 1.65 V in the open-loop configuration, a maximum output hysteresis of about 1.5 μm (15%) and a maximum input hysteresis of about 0.23 V (14%) is observed. However, the feedback control design effectively compensates

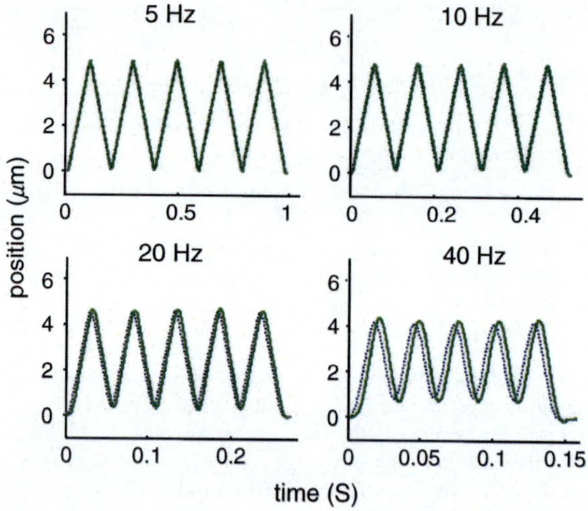

Fig. 5.10 Demonstration of tracking performance. The closed-loop system demonstrates good tracking performance (*solid*), where the stage tracked the triangular reference signals (*dashed*) with frequencies in the range 5–40 Hz. The reference and stage trajectories are practically indistinguishable at low frequencies (5 and 10 Hz) (© ASME 2008), reprinted with permission

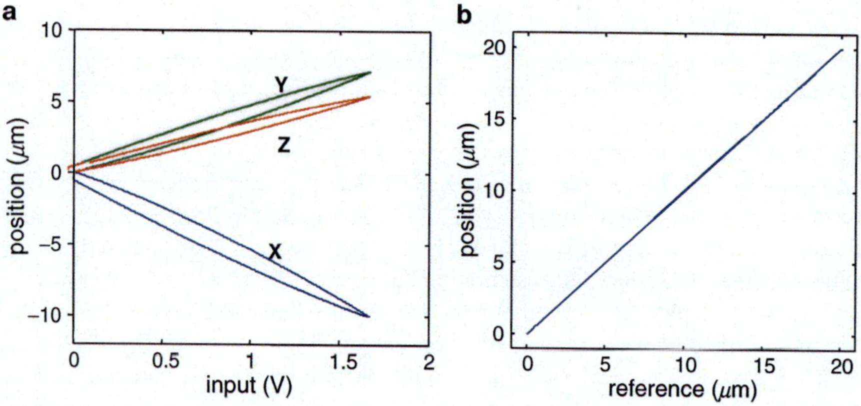

Fig. 5.11 Demonstration of practical elimination of nonlinear effects of piezoelectric actuation. Hysteresis observed in open-loop configuration (**a**) is eliminated in the closed-loop configuration (**b**) (© ASME 2008), reprinted with permission

this nonlinear effect as shown in Fig. 5.11b. A low-frequency (<1 Hz) high-amplitude (20 μm) triangular reference signal is given for the closed loop to track. The resulting experiments show maximum output and input hysteresis of about 20 nm, only 0.1% of the overall input and output range. Also, as evident from these experimental results, the closed-loop system gives a linear input–output characteristic for the entire traversal range, which validates the use of the linear

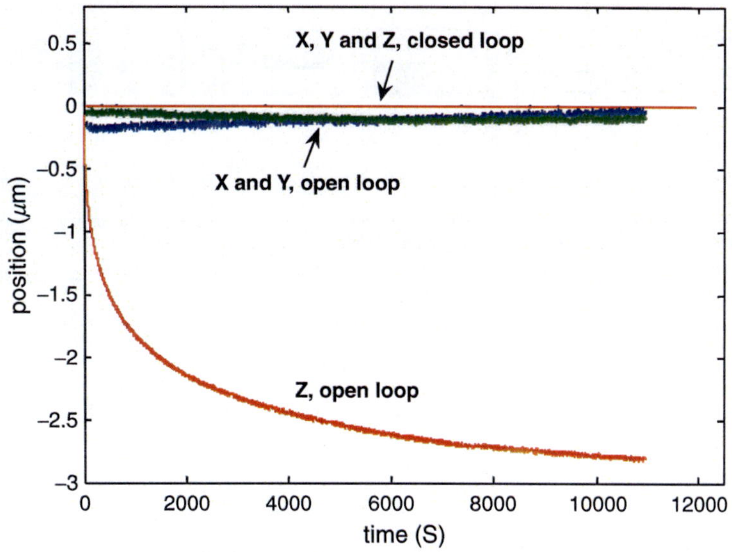

Fig. 5.12 Demonstration of practical elimination of creep effects of piezoelectric actuation by the feedback design (© ASME 2008), reprinted with permission

nominal model G for our control design. Figure 5.12 shows the creep effect of piezoelectric actuation and its elimination by the feedback design. A reference signal is designed to move the stage to the origin in open-loop and closed-loop configurations. The open-loop experiments demonstrate creep in all the actuators, especially prominent in the z direction since similar forcing on the actuators result in motions that are primarily along the z-axis due to a symmetry in the design of the positioning system. The experimental results with the closed-loop system, however, show no creep along any direction, which demonstrates that the control design practically removes this nonlinear effect.

5.3.4 Design for Ultrahigh Resolution

For a nanopositioning scanner, the positioning resolution is the smallest motion that can be differentiated from the noise in the system (signal values greater than three times the standard deviation of noise gives a 99.7% confidence in the signal). A widely held belief is that any feedback-based design strategy can only deteriorate resolution of the device since sensor (electronic) noise is fed back into the system. This argument, however, assumes no modeling uncertainty (which is always present). In fact, in the presence of significant modeling uncertainties, feedback strategies can achieve a better resolution by making better trade-offs between the effects of modeling uncertainties and the sensor noise. For instance,

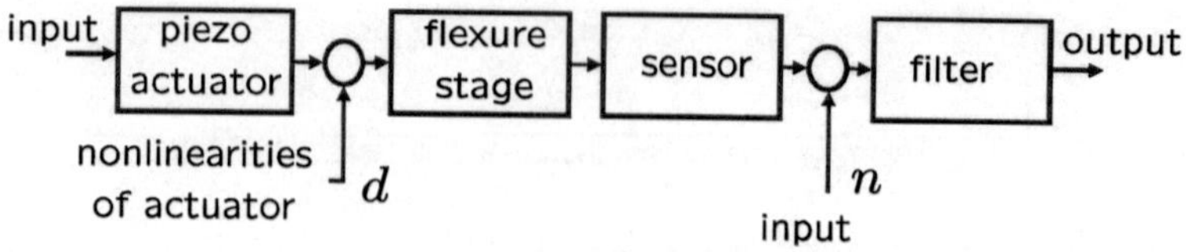

Fig. 5.13 A schematic of a single-axis open-loop system: The piezo actuator drives the flexure stage and its movement is measured by a sensor. The main challenges to achieving high resolution stem from modeling uncertainties d that arise from the nonlinear effects of piezoelectric actuation and the sensor noise n (© AIP 2007), reprinted with permission

in the context of Fig. 5.13, the resolution of a nanopositioning system is determined mainly by the mechanical and sensor noises d and n, respectively. The mechanical noise, which represents modeling uncertainties, mainly consists of the slowly varying drift and creep, which are therefore prominent in slow scans, and the inertial lag at high frequencies that is prominent in high speed scans. These nonlinear effects of drift, creep, and hysteresis are sensitive to changes in operating conditions such as ambient temperature, residual polarization in piezoelectric actuators, and the operating point – that is, the reference value on the nonlinear input–output (input voltage vs stage displacement) graph about which stage motions are calibrated. Therefore, including their precise behavior in device models is practically impossible and hence they are treated as noise. Feedback-based schemes can achieve effective compensation for the creep, drift, hysteresis, and inertial lag problems without requiring precise models. They compensate for the mechanical noise but at the cost of feeding back relatively smaller electronic noise. Hence, designing the feedback law to limit the effect of additional electronic noise from the sensor is critical. In view of the fundamental limitations on the control design presented in Sect. 5.3.1, the feedback design has to achieve the right trade-off between the resolution and the bandwidth of the resulting closed-loop positioning system. While trying to reduce the overall error e in (5.4), the error due to a large range of frequencies in the reference signal can be made small only at the expense of increased errors due to noise at those frequencies. Since the electronic noise has components in practically all the frequencies, obtaining high resolution requires high rejection bandwidth of T implying a very low rejection bandwidth of S. Fortunately, for applications such as raster scanning, the range of appreciable frequency components in reference signals is small (some scans even have only one frequency), thus making it possible to achieve high resolution. For high resolution in low-frequency scans (where r has only low-frequency components), this design can be made to exploit the resolution-bandwidth trade-off by making T have a very small *pass* bandwidth which forces S to have a very small *rejection* bandwidth, which results in a high-resolution output. Making S small at low frequencies over a small bandwidth ensures that the sensor noise is dominant over modeling uncertainties and therefore determines the resolution. This design results in low resolution since T has a very low bandwidth,

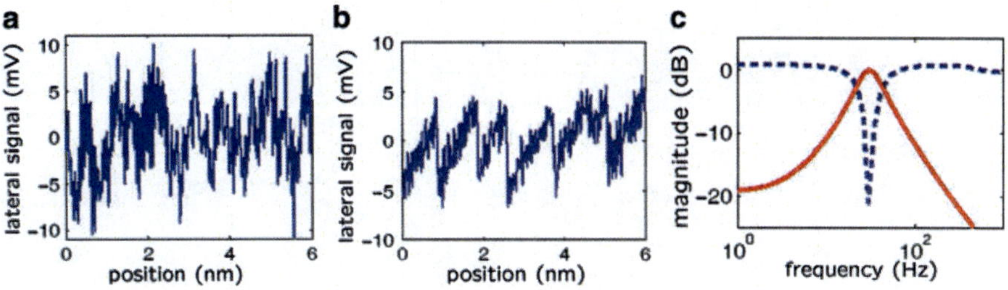

Fig. 5.14 Demonstration of sub-nanometer positioning resolution: In order to prove the high resolution capability, we used mica as the calibration sample. The friction image of mica has been established to show lattice structure (which repeats every 5.2 Å) through stick–slip dynamics that manifests as a sawtooth wave structure. The atomic-scale stick–slip is not observed in (**a**) when a high-bandwidth control design, which does not exploit the trade-off between the bandwidth and resolution, is implemented. The stick–slip is detected in (**b**) when the proposed band-focused control design in (**c**) is implemented on the nanopositioning system. With this design, lateral positioning resolution is very high (sub-nanometer) which enables detection of the atomic-scale stick–slip dynamics (© AIP 2007), reprinted with permission

which results in low value of the integral in (5.5) that determines the standard deviation of the effective noise in the closed-loop system.

The noise-management scheme described above is verified by an experiment performed on the nano-scanning stage of a commercial AFM, Molecular Force Probe-3D (MFP-3D) from Asylum Research Inc., Santa Barbara. In order to prove the high resolution capability, Mica was imaged over a scan range of 6–8 nm at slow scan rates (0.02 Hz). Under these conditions, the friction signal from mica is expected to show a triangular waveform representing the stick–slip phenomenon [4, 33] with a periodicity of 5.2 Å, which can be detected only if the positioning system has a sub-nanometer positioning resolution. The experimental results are shown in Fig. 5.14. The stick–slip phenomenon could not be observed when common (high-bandwidth or open-loop) feedback designs were used as shown in Fig. 5.14a. When a very low bandwidth (2.85 Hz) controller was designed for the scanning axis with the complementary sensitivity T as a lowpass filter (see Fig. 5.14c), sub-nanometer positioning resolution was obtained. This enabled capturing the lattice averaged atomic scale stick–slip phenomenon in the friction signal in Fig. 5.14b (see [30] for details).

This trade-off between the designs of the sensitivity and complementary sensitivity transfer functions can be extended for large scans. This approach, in fact, results in making tracking precision practically independent of the scanning frequency. For large scans, where the amplitude of the reference r is large, the tracking error e in (5.4) is determined by the Sr component which dominates over the Tn term. By making the sensitivity transfer function small over a small bandwidth around the scanning-frequency, the Sr term can be made extremely small which results in small tracking error e. In view of the finite waterbed-effect resulting from the limitation (5.6), since, the constraint on S is only over a small frequency range, a

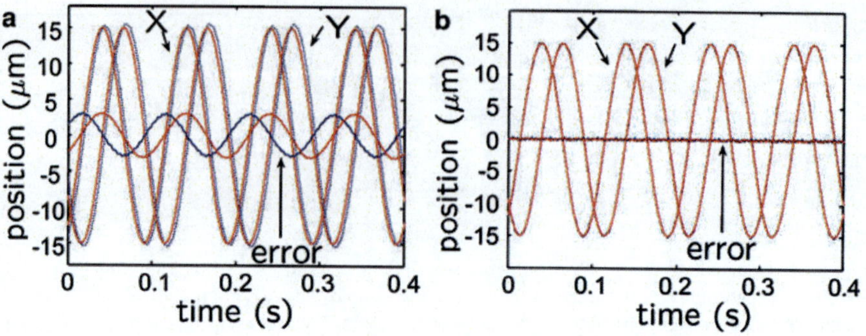

Fig. 5.15 Demonstration of better tracking with band-focused designs. Comparison of tracking of a 10 Hz sinusoidal reference signal (*dotted*) using a high-bandwidth controller in (**a**) and band-focused design in (**b**). The results show about 20 times reduction of error amplitudes by the band-focused design when compared to the high-bandwidth control design (© ASME 2008), reprinted with permission

feedback law can be obtained that makes S very small in this frequency range. This design was implemented on the positioning system shown in Fig. 5.8. The control design was obtained by implementing the optimal-control scheme described in Sect. 5.3.3, where the weighting functions W_P and W_T were chosen, respectively, as low-bandwidth band-reject and band-pass filters in the scanning frequency range. The comparison of experimental results from this design with the high-bandwidth robust controller designed in Sect. 5.3.3 are presented in Fig. 5.15. The reference $r = [r_x\, r_y]'$ is such that the positioning system tracks a circular motion with a 15 μm radius running at 10 Hz (the maximum velocities in x and y directions is about 0.95 mm/s). It is evident that the tracking error is significantly reduced. The maximum error decreased from 3 μm from the high-bandwidth controller to 150 nm for the band-focused controller.

5.3.5 *Analysis and Design of 2DOF Control for Nanopositioning Systems*

In this section, we show that the feasible space of performance specifications, which are constrained by the limitations described above in feedback-only configurations, can be extended by using a 2DOF design scheme. In contrast to the feedback-only scheme described in Sect. 5.3.1, where the controller acts only on the difference between the reference r and the position-measurement y_m, in the 2DOF scheme, the controller acts independently on them (see Fig. 5.16).

In the 1DOF scheme, the robustness to modeling uncertainties as well as resolution of the device are determined only by the feedback part of the controller, that is, the transfer function from d to y that characterizes robustness to modeling uncertainties is still determined by the sensitivity function $S = (1 + GK_{\text{fb}})^{-1}$, and

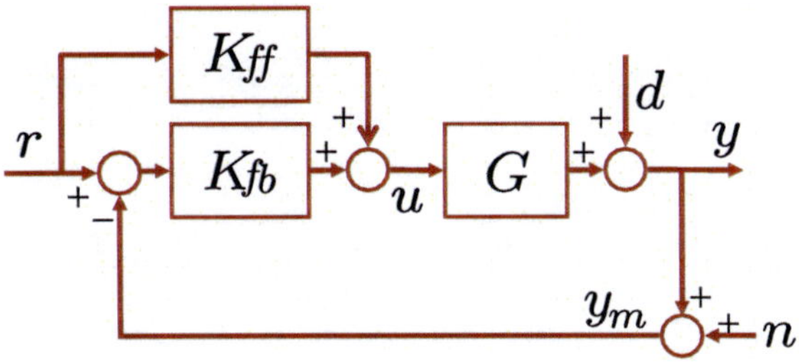

Fig. 5.16 2DOF control architecture: The feedforward–feedback scheme where the actuation signal is $u = K_{\rm ff}r + K_{\rm fb}(r - y_m)$

the transfer function from n to y that characterizes resolution is still determined by the complementary sensitivity function $T = (1+GK_{\rm fb})^{-1}GK_{\rm fb}$. The main difference and advantage in 2DOF control design compared to the feedback-only design stems from the fact that the transfer functions from r to y and from n to y are different and can be designed independently. This difference gives greater independence in designing for better trade-offs between different performance objectives. In this setting, the relevant closed-loop signals are given by

$$y = T_{\rm yr}r - Tn + Sd, \quad e = S_{\rm er}r + Tn - Sd,$$
$$u = S(K_{\rm ff} + K_{\rm fb})r - SK_{\rm fb}n - SK_{\rm fb}d, \tag{5.9}$$

where $T_{\rm yr}$ and $S_{\rm er}$ denote the transfer function from r to y and from r to e, respectively, that is, $S_{\rm er} = S(1 - GK_{\rm ff})$ and $T_{\rm yr} = SG(K_{\rm ff} + K_{\rm fb})$. The control objectives translate to small roll-off frequency as well as high roll-off rates for T to have good resolution, long range of frequencies for which $S_{\rm er}$ is small to achieve large bandwidth and low (near 1) values of the peak in the magnitude plot of $S(j\omega)$ for robustness to modeling uncertainties. Even though the 2DOF control design has greater flexibility than the feedback-only design, the main challenges to design still arise from implementation and algebraic (albeit less severe) limitations. The constraints on hardware implementation in terms of sampling frequencies as well as saturation limits of actuation signals limit the scope of this design. Similarly, the algebraic limitations constrain the control design in this setting too; for instance, the constraint $S(j\omega) + T(j\omega) \equiv 1$ has the same ramifications on the trade-off between the resolution and robustness to modeling uncertainties as in the feedback-only design.

A control synthesis scheme based on the optimal-control framework is discussed in [15]. In this scheme, both the feedforward and the feedback control laws are solved in an optimal control setting. Following the same rationale as in the 1DOF design, the regulated output z was chosen as $[W_Se\; W_Ty\; W_uu]'$ for the optimal control problem, and Φ, the transfer function from weighted $w = [r\; n]'$ to z is given by

$$\begin{bmatrix} z_s \\ z_t \\ z_u \end{bmatrix} = \underbrace{\begin{bmatrix} W_sS_{\rm er}W_r & -W_sSW_n \\ W_tT_{\rm yr}W_r & -W_tTW_n \\ W_uS(K_{\rm ff}+K_{\rm fb})W_r & -W_uSK_{\rm fb}W_n \end{bmatrix}}_{=\Phi(K)} \begin{bmatrix} r \\ n \end{bmatrix}, \tag{5.10}$$

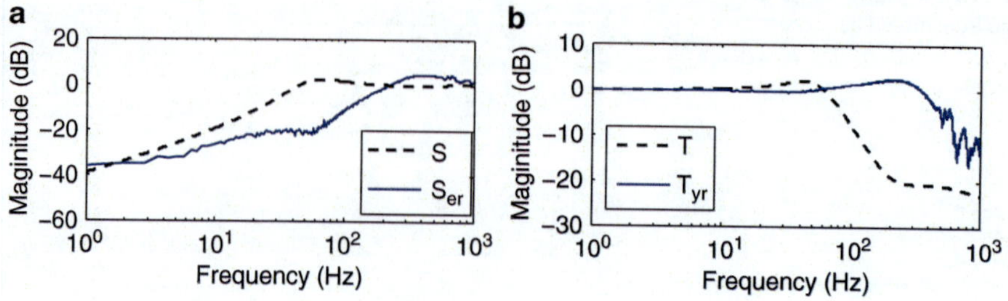

Fig. 5.17 Magnitude of $S_{\mathrm{er}}(s)$ in **(a)** and $T_{\mathrm{yr}}(s)$ (*solid*) in **(b)** obtained from experiments and compared to S and T(*dashed*). The 2DOF design achieves over 290% improvement in the bandwidth over the optimal 1DOF design for the same resolution and robustness specifications (© IOP 2009), reprinted with permission

where the weights W_r and W_n, which reflect the frequency content of the reference and the noise signals, provide greater independence in specifying trade-offs to the optimization problem. Accordingly, the $\mathscr{H}_\infty$ optimal control problem that we pose is $\min_K \|\Phi(K)\|_\infty$. The minimization of z_s reflects the tracking-bandwidth requirement. If we design the weight function $W_s(j\omega)$ to be large in a frequency range of $[0\ \omega_{\mathrm{BW}}]$ and ensure that z_s is small over the entire frequency range (through the above optimization problem), then the tracking-error e will be small in the frequency range $[0\ \omega_{\mathrm{BW}}]$; that is the closed-loop positioning device has a bandwidth of ω_{BW}. Alternatively, note that the transfer function from r to z_s is $W_s S_{\mathrm{er}} W_r$. The optimization problem along with our choice of W_s and W_r ensures that the transfer function S_{er} is small in the frequency range $[0\ \omega_{\mathrm{BW}}]$. Similarly the transfer function from n to z_t is the weighted complementary sensitivity function $W_t T W_n$, whose minimization ensures better resolution as it forces low control gains at high frequencies, and the transfer function from r to z_u is $W_u S(K_{\mathrm{ff}} + K_{\mathrm{fb}})W_r$, which measures the control effort. Its minimization reflects imposing the limitation that the control signals be within the saturation limits. Figure 5.17a shows the experimentally obtained transfer function from reference to error, i.e. $S_{\mathrm{er}}(s)$ which represents the tracking performance ($\omega_{\mathrm{BW}} = 148.2$ Hz) and the transfer function from reference to output, i.e. $T_{\mathrm{yr}}(s)$ is shown in Fig. 5.17b. There was an improvement of 290% in bandwidth for the same values of the resolution and robustness when compared to the feedback-only design. Similar improvement in other performance objectives (resolution and robustness) can be obtained by appropriately designed weight functions.

An analysis of the resulting 2DOF closed-loop system demarcates the roles of the feedforward and the feedback control. Since $S_{\mathrm{er}} = S(1 - GK_{\mathrm{ff}})$, and S cannot be made small over the entire bandwidth range (in order to allow for noise attenuation), the feedforward control is "active" in making S_{er} small beyond the frequency where $|S|$ is not small (say greater than $1/\sqrt{2}$). Also since $S = (1 + GK_{\mathrm{fb}})^{-1}$ is completely determined by K_{fb}, the feedback component is dominant in frequencies where S is small. Therefore the main contribution of the feedforward component is in the frequency range where S is no longer small. However, this frequency range is limited. Typically nanopositioning systems have very low gains beyond their flexure resonance frequencies. Therefore very high control inputs are needed to make the positioning systems practical beyond their flexure resonances. The saturation limits on control signals form the main constraints on attaining bandwidths beyond flexure resonances. Thus the feedforward components provide performance *enhancements* over feedback-only designs in the frequency range from the corner frequency of S to the flexure resonant frequency. This separation of roles becomes evident from the optimal feedback law, where the resulting design is such that $K_{\mathrm{ff}} \approx 0$ when K_{fb} is large, even when the optimization cost function in the problem formulation does not discriminate between feedforward and feedback components.

Another important benefit of the 2DOF design is that it does yield performance specifications that are impossible for 1DOF designs. For instance, the experimental results in Fig. 5.17 show that the tracking bandwidth ω_{BW} of the closed-loop device can be made larger than the roll off frequency ω_T, which determines the resolution. The corner frequency ω_{BW} can never be made larger than ω_T in feedback-only (1DOF) designs, which suffer from a stricter trade-off between the resolution and the bandwidth. The 2DOF control based on a $\mathscr{H}_\infty$ controller has the bandwidth ω_{BW} of 148.2 Hz and ω_T of 60.1 Hz while the 1DOF optimal-controller yielded bandwidth of 49.4 Hz and the same roll-off frequency.

In this section, we have presented the demands, challenges, and control designs for nanopositioning devices. It has been shown that they fall naturally into the modern control theory paradigm. This theory provides an appropriate approach to quantify, incorporate, and achieve both the performance and robustness objectives. Using this approach, we have obtained significant improvements in bandwidth and repeatable sub-nanometer resolution for these devices. Even though the experimental results presented here have focused on bandwidth enhancements in existing positioning systems, the control theoretic framework allows for extreme specifications on resolution or robustness by making appropriate trade-offs. For instance the extreme specification of sub-nanometer resolution is achieved by a large sacrifice of tracking bandwidth in Sect. 5.3.4. Similar trade-offs can be imposed to achieve extreme specifications on robustness. The control design here is focussed on improving performance of single nanopositioning systems. There is a large scope for control systems theory in extending these design methods to an array of positioning systems specially developed for high-throughput applications.

5.4 Dynamic Mode Operation in AFM

In the dynamic mode operation of AFM, the cantilever is forced externally by a known signal. Typically the forcing is induced by moving the support of the cantilever using a dither piezo. Traditionally the support is forced sinusoidally at or near the first resonant frequency of the cantilever. However, recently, methods that force the cantilever with a superposition of sinusoids are being proposed.

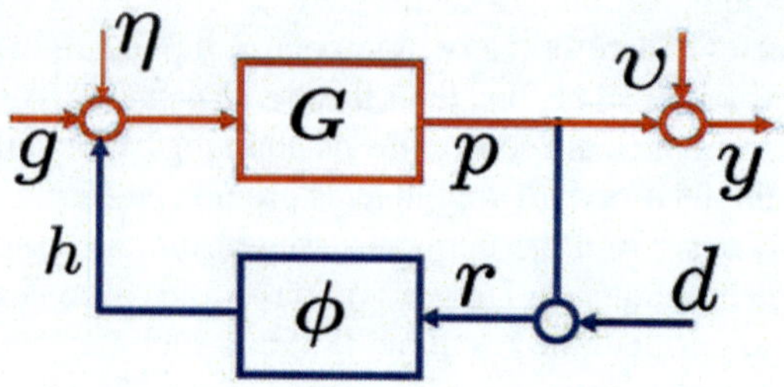

Fig. 5.18 The figure shows a feedback interconnection of a linear time invariant system G that models the cantilever with a memoryless non-linearity ϕ. Here g, η, v, and d model dither forcing, measurement noise, and sample topography, respectively

For the dynamic mode, the cantilever dynamics in (5.1) can be rewritten as

$$\ddot{p} + \frac{\omega_0}{Q}\dot{p} + \omega_0^2 p = \eta + g + h, \tag{5.11}$$

where the total force at the cantilever tip is the sum of the thermal noise force η, the external dither forcing g, and the force h on the tip due to the interaction between the tip atoms and the atoms on the sample. Note that the tip-sample interaction forces depend on the tip-deflection and possibly the tip velocity. Assuming a static dependence of the tip-sample interaction force on the tip-sample separation, r a system description of the dynamic mode AFM is provided by the feedback interconnection of the cantilever transfer function and the static nonlinearity ϕ as shown in Fig. 5.18. In this figure, the sample topography (or the initial offset between the sample and the cantilever) appears as d and the tip-sample separation r is modeled as the sum of the tip-deflection p and the tip-sample offset or the topography d . This viewpoint, where the cantilever tip-sample dynamics is posed as a feedback interconnection of a linear filter (representing cantilever dynamics) and static nonlinearity (representing tip-sample interaction), enables significant analysis and design in AFM. Such a viewpoint was first introduced in [27, 28].

For most practical implementations of control, it is necessary to identify the transfer function G from input–output experiments instead of deriving it from (5.1), since various latencies and dynamics of the AFM system components are not captured by (5.1). One means of obtaining such a description is to obtain the frequency response of the cantilever with the forcing at the dither as the input and the photodiode voltage signal as the output. It is to be remarked that in Fig. 5.18 all the forces η, h, and g see the same transfer function G. However, η is a distributed force, g is realized by forcing the base of the cantilever, whereas h appears at the tip end of the cantilever. The validity of the assumption that all these force inputs appear at the same point in the feedback interconnection and that they see the same transfer function G needs to be verified for the particular application and the dynamic method being analyzed.

In many applications, the identification of the function ϕ for a fixed lateral coordinate on the sample is the goal. In such cases, the sample position with respect to the cantilever is changed quasi-statically using the vertical piezo and the tip-deflection p is measured. That is, d is altered in a quasi-static manner by using the z piezo. Typically the dither forcing is of the form $g = A_f \sin \omega_0 t$, where ω_0 is the first resonant frequency of the cantilever. The cantilever deflection p, when the sample's influence is negligible, is sinusoidal with phase $\angle G(j\omega_0)$ and amplitude $|G(j\omega_0)|A_f$. Under the sample's influence, for a given offset d, the cantilever deflection is assumed to be periodic with period $2\pi/\omega_0$. This assumption implies a periodic force h on the cantilever tip. The sample force h on the cantilever tip can be reconstructed by realizing that the measured tip-deflection is

$$y = G(\eta + h + g) + \vartheta. \tag{5.12}$$

Therefore, reconstruction of h requires inversion of the cantilever transfer function G and appropriate methods to address measurement noise ϑ and process noise η. Once h is estimated, a parametric model of ϕ can be assumed and the parameters of ϕ identified (see [28]), or, as introduced in the preliminary work [16], it is possible to obtain the force–separation relationship from the reconstructed h and deflection p data without assuming any model of the relationship.

When the topography image is being sought, $d = d(x,y)$ models the topography, where (x,y) represent the lateral coordinates of sample position being imaged. The lateral position of the sample is changed typically by a piezoactuated positioning system. Therefore, the sample-topography estimation problem is translated to reconstructing the sample profile d as a function of (x,y) from the measured deflection signal y.

The dynamical system depicted in Fig. 5.18 can lead to complex behavior (see [3, 14, 23]). An analysis of the amplitude and phase of the first harmonic of the cantilever oscillations when forced sinusoidally is provided in [25,26]. The study of complex behavior is significant in its own right. However, for imaging and imaging related analysis, it is evident that simplifying assumptions need to be made that render the models tractable while capturing the dynamics relevant to imaging.

One such simplifying assumption is to break the feedback interconnection of Fig. 5.18 and consider the tip sample force $h(t)$ to be a train of impulses that model the cantilever tapping the sample as it oscillates. This assumption is also well motivated as the cantilever oscillation amplitude is typically in the 25 nm regime and the tip-sample interaction forces are effective for separations less than 5–6 nm. Thus the cantilever tip spends only a small fraction of its trajectory under the sample's influence and therefore can be approximated as an impulse every oscillation cycle. Thus a reasonable approximation for h is given by

$$h(t) = \sum_n a_n \delta(t - nT), \tag{5.13}$$

where δ denotes an impulse function, T is the time period of the dither forcing and a_n models the strength of the impulsive force during the nth cycle. The problem then

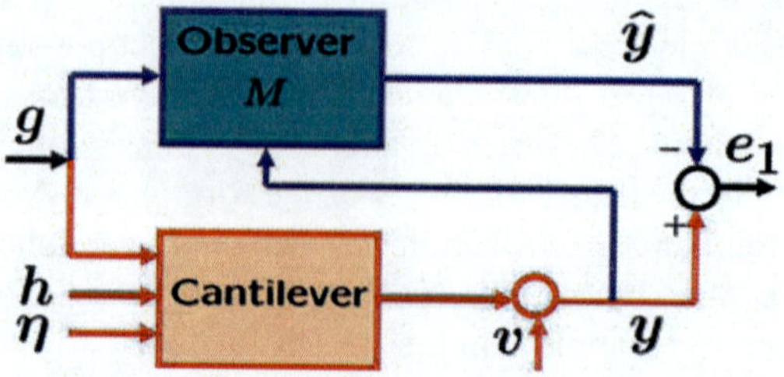

Fig. 5.19 Figure shows that an observer can be implemented on the cantilever model parameters. The observer mainly comprises a model of the cantilever and its role is to estimate the cantilever dynamics. The observer is provided with the dither forcing and the measured cantilever deflection. The mismatch in the cantilever deflection and the observer-based estimated signal is primarily due to the sample force ϕ, thermal noise η, and measurement noise ϑ

reduces to estimating the information source a_n. Such a modeling simplification has resulted in the high speed sample detection scheme described next, the transient force atomic force microscopy (TfAFM), which forms the first instance of the use of observers for AFM applications.

5.4.1 Observer-Based Dynamic Mode Methods

For h given by (5.13), the problem of estimating a_n that represent the magnitude of the sample interaction strength, becomes closely related to the problem of estimating the initial condition reset of the cantilever system state (represented by the transfer function $G(s)$) caused by the impulsive force during the nth cycle. The impulsive force contribution during the nth cycle changes the initial condition of the cantilever state and the initial condition change response appears at the measured output y. For many applications, the transfer function $G(s)$ has a large quality factor that translates to an impulse response that has a long duration. Thus, unless impulsive force inputs are separated by long intervals, the output suffers from the interference of the effects of two initial condition resets. Such an effect leads to inter-symbol interference, where the sample is considered the information source to be deciphered. An effective way of shortening the channel $G(s)$ response time is to implement an observer as shown in Fig. 5.19. The impulse response of the system viewed from the input h to the output e_1 in Fig. 5.19 can be considerably shortened (effectively from Q oscillation cycles to four cycles, where Q is the quality factor of the cantilever). Thus the channel $G(s)$ is effectively equalized using communications terminology and the source symbol sequence a_n can be deciphered considerably faster. This effectively translates to faster detection of the sample features. The details of this method can be found in [20, 21] (see Fig. 5.20).

Another challenge in dynamic mode AFM methods is the need for a measure of the fidelity of the data that is being interpreted as the image of the sample.

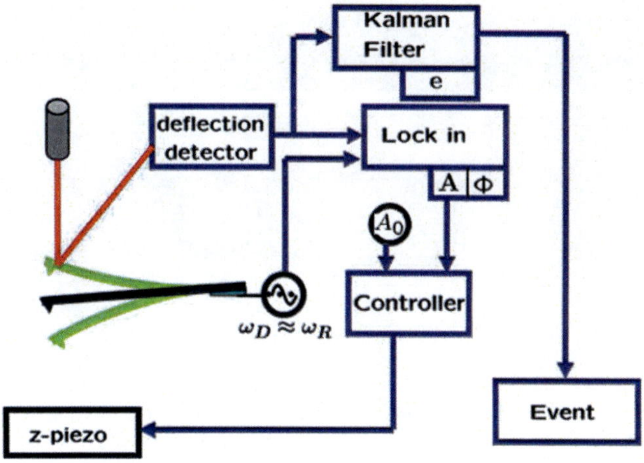

Fig. 5.20 The figure shows a schematic of transient force atomic force microscopy where in addition to the Kalman observer, a closed-loop system regulates a setpoint amplitude A_0 primarily to maintain the cantilever tip engaged with the sample surface. Every time the cantilever impacts a surface feature that is modeled as an impulsive force, an event is registered

For example, in amplitude-modulation AFM, it is often difficult to estimate in real time whether the cantilever tip is interacting with the sample or not. This issue is particularly acute for high-speed AFM applications and for samples with high aspect ratio features where the cantilever "parachutes" [2, 32] off into a valley in the sample's topography. In the absence of the sample, in steady state, after the initial condition mismatch between the cantilever state and the observer state has died out, the error e_1 is close to zero. The remnant mismatch in steady state is due to photodiode noise and the thermal noise that affect the measured cantilever deflection without affecting the observer circuit. In the presence of the sample, the cantilever behaves as a modified cantilever with changed quality factor and changed resonant frequency (see Fig. 5.21a). If the sample force is given by $h(p,\dot{p})$, the estimate of the equivalent cantilever frequency and the quality factor are given by [17]

$$\omega_0'^2 = \omega_0^2 + \frac{2}{a}\frac{1}{2\pi}\int_0^{2\pi} h(a\cos\psi, -a\omega\sin\psi)\cos\psi \,\mathrm{d}\psi \tag{5.14}$$

and

$$\frac{\omega_0'}{Q'} = \frac{\omega_0}{Q} + \left(\frac{1}{a\omega}\frac{1}{\pi}\int_0^{2\pi} h(a\cos\psi, -a\omega\sin\psi)\sin\psi \,\mathrm{d}\psi\right), \tag{5.15}$$

respectively, where a is the steady-state amplitude of the cantilever in the presence of the sample. Thus the cantilever behavior under the sample presence is given by

$$\ddot{p} + \frac{\omega_0'}{Q'}\dot{p} + \omega_0'^2 p = g(t) + \eta(t). \tag{5.16}$$

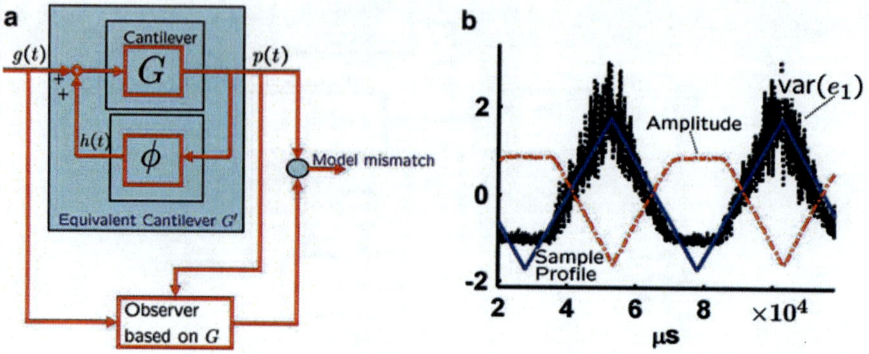

Fig. 5.21 (**a**) The cantilever-sample combination behaves as an equivalent cantilever with changed resonant frequency and changed damping factor. This combined system can be compared to an observer based on the parameters of the cantilever alone that provides a measure of the samples influence on the cantilever trajectory. (**b**) shows the experimentally obtained model mismatch signal (in black), which is the rms value of the signal e_1 when the sample first approaches the cantilever tip and then is retracted (shown by the blue triangular signal). The amplitude signal is shown in red dash-dot format

Note that the observer still behaves according to nominal cantilever parameters, thus the difference between the observer and the cantilever is due to the change in the model parameters, which do not decay with time. Thus the persistent error between the observer and the cantilever dynamics can be used to provide a signal that indicates the presence of the sample. This is unlike, and in contrast to, the change due to initial condition mismatch used for transiently detecting the sample's presence in [21]. Figure 5.21a shows the implementation architecture used, where the cantilever-sample system is viewed as an equivalent cantilever with modified quality factor and stiffness. The measured output and the observer estimated output are compared. Figure 5.21b shows the root mean square of the estimation error e_1 as the sample-cantilever separation is first decreased in the approach phase and then the separation is increased in the retract phase. As the sample approaches the cantilever, the effect of the nonlinearity ϕ on the cantilever is increased and thus the equivalent cantilever dynamics deviates more from the nominal cantilever model. The observer is based on the parameters of the nominal cantilever model and thus there is increased mismatch between the observer estimated cantilever deflection and the measured cantilever deflection. The trend of increased rms value of the signal e_1 is evident from the data shown in Fig. 5.21b. Note that this signal has a near linear relationship with the tip-sample separation. This experimentally observed linear behavior may be exploited in future research. In related work [19], real-time estimation of equivalent cantilever parameters using modified recursive least squares method is presented.

Figure 5.22 shows the AFM operated in a amplitude modulation AFM mode where the control signal (the vertical z signal to the piezo scanner) is used as an estimate of the sample height with the probe-loss architecture implemented. It is

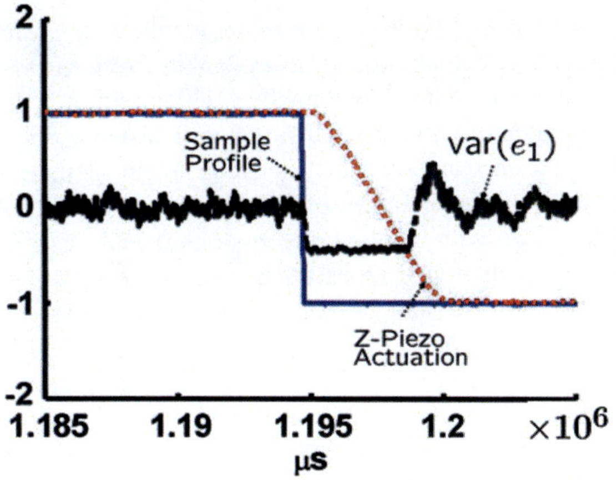

Fig. 5.22 (Experimental data) The blue solid line depicts the sample topography, red dotted line shows the conventional signal used for imaging; the control signal and the black dotted points represent the model-mismatch (probe-loss) signal. It is evident that the probe-loss affected region of the sample can be identified in real-time

evident that as soon as a valley in the sample topography is encountered the probe-loss signal goes low. The controller actuates the z positioner to move the sample closer to the cantilever tip. Eventually the cantilever again engages with the sample and the probe-loss signal goes high. The probe-loss is detected within a couple of oscillations of the dither forcing. This remarkable speed is possible because the measurement noise in an AFM setup is quite small allowing for the observer gain to be high. It is also evident that the probe-loss signal can be used as an effective real-time means of estimating which parts of the image can be trusted and which cannot be trusted. The related observer-based framework thus provides for quantitative measurement of the image fidelity.

5.5 Conclusions

This chapter shows the impact of system-theoretic tools on nanotechnology, especially in scanning probe microscopy. Feedback formed an important and integral part of the scanning probes right in their original designs. The various applications presented in this article confirm that system-theoretic tools will continue to play a fundamental role in realizing major objectives of nanotechnology – in achieving practical control, manipulation, and investigation of matter at atomic scales.

This chapter focused on control systems theoretic modeling, analysis, and synthesis of new modes of operations that significantly expand the range of performance specifications and capabilities of scanning probe microscopes. A systems perspective was presented that enabled a study of fundamental limitations on the performance of these devices. For instance, this framework made it possible to study inherent trade-offs between resolution, tracking-bandwidth, and reliability specifications on positioning systems. In addition to determining fundamental limitations, this framework led to a better understanding of existing technology and allowed us to exceed some technological hurdles that were previously thought to be fundamentally limiting. The system theoretic viewpoint leads to a new generation of techniques that can potentially enable probing material at the sub-angstrom level at significantly higher bandwidths. For instance, TfAFM, which achieves enormous detection bandwidth that is independent of the quality factor of the probe, and, therefore, independent of the resolution, or, ThNcAFM, that has made it possible to image with resolution as high as 0.25 Å in ambient conditions. The orders-of-magnitude improvements achieved in areas such as precision positioning, sample imaging, and sample detection rates emphasize the potential of systems tools in nanotechnology.

Many nanoscientific applications and studies, especially scanning probe microscopy, require analytical and design tools that facilitate parallelization, integration of multi-range multi-resolution actuators and sensors, real-time estimation of material features and properties from dynamic measurements, detection of artifacts, and validation of the interpreted data. These requirements can be viewed as new challenges for the areas of system identification, distributed control, robustness analysis, nonlinear estimation, and system verification and validation. In addition to extending existing methodologies in systems theory, new tools need to be developed that will specifically address the above challenges.

References

1. T. Ando, N. Kodera, E. Takai, D. Maruyama, K. Saito, and A. Toda. A high-speed atomic force microscope for studying bilological macromolecules. *Proceedings of National Academy of Science* **98**(22), 034,106 (2001)
2. T. Ando, T. Uchihashi, N. Kodera, A. Miyagi, R. Nakakita, H. Yamashita, and M. Sakashita. High-speed atomic force microscopy for studying the dynamic behavior of protein molecules at work. *Japanese Journal of Applied Physics Part 1 Regular Papers Short Notes And Review Papers* **45**(3B), 1897 (2006)
3. M. Ashhab, M.V. Salapaka, M. Dahleh, and I. Mezic. Melnikov-based dynamical analysis of microcantilevers in scanning probe microscopy. *Nonlinear Dynamics* (November 1999)
4. B. Bhushan. *Handbook of micro/nano tribology*, second edn. CRC Press (1999)
5. G. Binnig, C. Gerber, E. Stoll, T.R. Albrecht, and C.F. Quate. Atomic resolution with atomic force microscopy. *Europhys. Lett.* **3**, 1281 (1987)
6. G. Binnig, C.F. Quate, and C. Gerber. Atomic force microscope. *Physical Review Letters* **56**(9), 930–933 (1986). DOI 10.1103/PhysRevLett.56.930
7. G. Binnig and H. Rohrer. Scanning tunnelling microscopy. *Helv. Phys. Acta* **55**, 726 (1982)
8. H. Bode. *Network analysis and feedback amplifier design*. New York, Van Nostrand Reinhold (1945)
9. A. Daniele, T. Nakata, L. Giarre, M.V. Salapaka, and M. Dahleh. Robust identification and control of scanning probe microscope scanner. In: 2ND IFAC Symposium on Robust Control Design. Budapest, Hungary (1997)
10. J. Dong, S. Salapaka, and P. Ferreira. Robust control of a parallel kinematics nano-positioner. *Journal of Dynamic Systems, Measurement, and Control* **130**, 041,007(1—15) (2008)
11. J.C. Doyle, B.A. Francis, and A.R. Tannenbaum. *Feedback control theory.* New York, MacMillan (1992) URL citeseer.ist.psu.edu/doyle90feedback.html

12. R. Feynman. There's plenty of room at the bottom – an invitation to enter a new field of physics. *A talk given on December 29, 1959 at the annual meeting of APS at Caltech*
13. J.S. Freudenberg and D.P. Looze. Right half-plane poles and zeros and design tradeoffs in feedback systems. *IEEE Transactions on Automatic Control* **30**(6), 555—565 (1985)
14. S. Hu and A. Raman. Chaos in atomic force microscopy. *Physics Review Letters* pp. 036,107:1—3 (2006)
15. C. Lee and S. Salapaka. Robust broadband nanopositioning: fundamental tradeoffs, analysis and design in two degree of freedom control framework. *Nanotechnology* **20**, 035,501 (2009)
16. D. Materassi, M. Salapaka, M. Basso. Identification of interaction potentials in dynamic mode atomic force microscopy. Decision and control, *2006 45th IEEE Conference*, pp. 3702—3705 (2006). DOI 10.1109/CDC.2006.377702
17. Y.A. Mitropolskii and N. Van Dao. *Applied asymptotic methods in non-linear oscillations*. The Netherlands, Kluwer Academic Publishers (1997)
18. C. Mohtadi. Bode's integral theorem for discrete-time systems. Control theory and applications, *IEE Proceedings D* **137**, 57—66 (1990)
19. A. Pranav and M.V. Salapaka. Real time estimation of equivalent cantilever parameters in tapping mode atomic force micrscopy. *Applied Physics Letters* (2009)
20. D. Sahoo and M.V. Salapaka. Observer based imaging metods for atomic force microscopy. *Proceedings of the IEEE conference on Decision and Control*, p. Accepted for publication (2005)
21. D.R. Sahoo, A. Sebastian, and M.V. Salapaka. Transient-signal-based sample-detection in atomic force microscopy. *Applied Physics Letters* **83**(26), 5521 (2003)
22. M.V. Salapaka, H.S. Bergh, J. Lai, A. Majumdar, and E. McFarland. Multi-mode noise analysis of cantilevers for scanning probe microscopy. *Journal of Applied Physics* **81**(6), 2480—2487 (1997)
23. S. Salapaka, M. Dahleh, and I. Mezic. On the dynamics of a harmonic oscillator undergoing impacts with a vibrating platform. *Nonlinear Dynamics* **24**, 333—358 (2001)
24. S. Salapaka, A. Sebastian, J.P. Cleveland, and M.V. Salapaka. High bandwidth nano-positioner: A robust control approach. *Review of Scientific Instruments* **73**, 3232—3241 (2002)
25. A. Sebastian, A. Gannepalli, and M. Salapaka. A review of the systems approach to the analysis of dynamic-mode atomic force microscopy. Control Systems Technology, *IEEE Transactions*, **15**(5), 952—959 (2007) DOI 10.1109/TCST.2007.902959
26. A. Sebastian and M. Salapaka. Amplitude phase dynamics and fixed points in tapping-mode atomic force microscopy. *Proceedings of the American Control Conference*, Boston, Massachusetts pp. 2499—2504 (2004)
27. A. Sebastian, M.V. Salapaka, D. Chen, and J.P. Cleveland. Harmonic balance based analysis for tapping-mode AFM. *Proceedings of the American Control Conference*, San Diego (June 1999)
28. A. Sebastian, M.V. Salapaka, D. Chen, and J.P. Cleveland harmonic and power balance tools for tapping-mode atomic force microscope. *Journal of Applied Physics* **89 (11)**, 6473—6480 (June 2001)
29. A. Sebastian and S. Salapaka. Design methodologies for robust nano-positioning. *IEEE Transactions on Control Systems Technology* **13**(6), 868–876 (2005)
30. A. Shegaonkar and S. Salapaka. Making high resolution positioning independent of scan rates: A feedback approach. *Applied Physics Letters* **91**, 203,513 (2007)
31. S. Skogestad and I. Postlethwaite. *Multivariable feedback control, analysis and design*, 2nd edn. John Wiley and Sons (2005)
32. T. Sulchek, R. Hsieh, J. Adams, G. Yaralioglu, S. Minne, C. Quate, J. Cleveland, A. Atalar, and D. Adderton. High-speed tapping mode imaging with active q control for atomic force microscopy. *Applied Physics Letters* **76**(11), 1473 (2000)
33. M.D. Yi Sang and M. Grant. Thermal effects on atomic friction. *Phys. Rev. Lett.* **87**(17), 174,301 (2001)

Chapter 6
Feedback Control of Optically Trapped Particles

Jason J. Gorman, Arvind Balijepalli, and Thomas W. LeBrun

6.1 Introduction

Optical trapping is a family of approaches for localizing and manipulating atoms, molecules, and nano- and microscale particles using optical forces (see [1–4] for an introduction). Among these approaches, the gradient force optical trap, or *optical tweezers*, discovered by Ashkin et al. [5] is the most widely used due to its simplicity in implementation, applicability across both the micro- and nanoscales, and passive trapping stability. Ashkin's gradient force optical trap is the subject of this chapter and from herein will be referred to simply as an optical trap. In its simplest form, an optical trap is formed by directing collimated light from a laser into the back aperture of a microscope objective with high numerical aperture (NA), which is typically located on an inverted microscope. The microscope objective focuses the light to a diffraction-limited spot inside the sample of interest and then the light diverges after passing the focal plane, as shown in Fig. 6.1a. An optical trap is created near the focal spot and imparts optical forces on micro- and nanoscale particles within the vicinity of the trap. The shape of the beam results in the creation of two force components, a gradient force that pushes the particle towards the focus of the beam and a scattering component that directs the particle along the direction of propagation of the beam. A stable optical trap is formed when the gradient force is greater than the scattering force along all three Cartesian axes. Once in the trap, the particle will remain there until it escapes due to its own Brownian motion [6, 7] or is pushed out by an external force. Figure 6.1b shows multiple microscale particles held in optical traps.

The basic principles behind the generation of forces in an optical trap are well understood, although the development of accurate physical descriptions remains an

J.J. Gorman (✉) • A. Balijepalli • T.W. LeBrun
National Institute of Standards and Technology, 100 Bureau Drive,
Gaithersburg, MD 20899, USA
e-mail: gorman@nist.gov; arvind@nist.gov; lebrun@nist.gov

J.J. Gorman and B. Shapiro (eds.), *Feedback Control of MEMS to Atoms*,
DOI 10.1007/978-1-4419-5832-7_6,

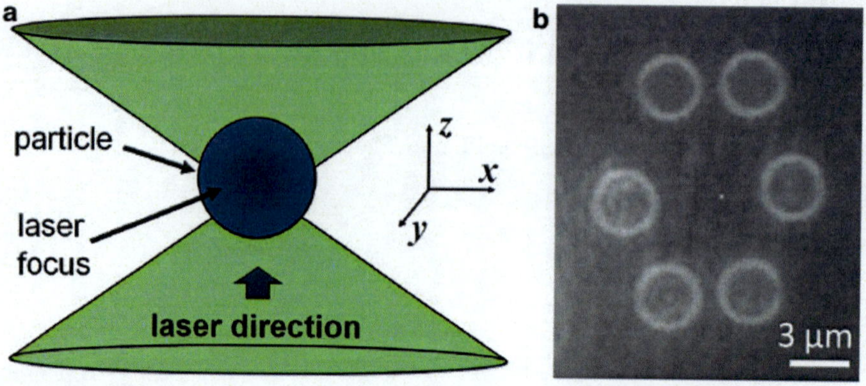

Fig. 6.1 Optically trapped particles. (**a**) Schematic of an optical trap formed at the focus of a laser beam with a particle trapped at its center. (**b**) Simultaneous trapping of six microparticles using the time-sharing approach (optical microscope image). See [4] for a description of the time-sharing approach

open area of research (e.g., see [8–10]). For particles with critical dimensions larger than the wavelength of light used for trapping ($d \gg \lambda$, referred to as regime 1), these phenomena are generally described from the perspective of ray optics. The particle acts as a lens, thereby refracting the light and changing the direction of the momentum of the photons. Due to the conservation of momentum, the change in momentum of light causes an equal but opposite force on the particle. The sum of the forces from the rays of light pushes the particle towards the center of the trap as a result of the particle shape and the sharply focused laser. When the particle is smaller than the wavelength of light ($d \ll \lambda$, referred to as regime 2), a different perspective is used to explain the optical forces. The light's electric field imparts an oscillating electric dipole moment on the particle. This dipole moment is attracted to the point of highest intensity gradient, which is at the laser focus, resulting in the particle moving to the center of the trap. These two regimes provide a convenient way to think about the trapping physics but do not address the situation when the particle is approximately the same size as the wavelength of light. In reality, there is a continuum between these regimes but unified theories that describe the trapping phenomena across these regimes for all materials (e.g., generalize Lorenz–Mie theory [11, 12]) are complex and require extensive numerical calculations. Regardless of the regime, both descriptions above support the fact that in a stable trap, particles are drawn to the center of the trap and are held there until larger external forces push them out.

Particles ranging from tens of nanometers to tens of micrometers in diameter can be trapped (e.g., see [9]), although successful trapping across this range is heavily dependent on the shape and material of the particle. Spherical particles trap reliably due to their symmetry and ability to refract light across a wide entrance angle for the incoming light. As a result, spherical particles are most commonly used, although many other shapes have been trapped, as discussed shortly. The particle material also has a large impact on the trapping forces, which can be easily explained using the two regime descriptions discussed above. In regime 1, the particle must be able to refract light and, therefore, is typically made of a dielectric material with the appropriate index of refraction, such as glass or polystyrene. Metals, which reflect and absorb light, do not trap well in this regime because they make poor lenses and they experience significant heating. Looking at regime 2, we find that the opposite is true. When particles are smaller than the wavelength of light they must be capable of generating a strong electric dipole moment. The strength of the dipole moment is directly related to the material's polarizability; the more polarizable the material, the larger the force. Since the polarizability of metals is significantly larger than for dielectrics, metal nanoparticles yield stronger trapping forces than their dielectric counterparts [13, 14]. In general, metal nanoparticles only trap when they are below several hundred nanometers in diameter, whereas dielectric particles trap from the microscale to the nanoscale, but not as effectively as metals at the nanoscale.

The effective restoring forces that push the particle toward the center of the trap can be conceptualized as a three-dimensional nonlinear spring. The magnitude of the forces is directly related to the intensity of the trapping light. The trapping forces can range from fractions of a piconewton to hundreds of piconewtons with the corresponding laser power ranging from a few milliwatts to hundreds of milliwatts. Photodamage of cells and macromolecules is one limiting factor in the laser intensity used [15, 16], and as a result limits the maximum forces that can be generated when manipulating biological components. Rohrbach [9] has shown that the trapping forces follow a nonlinear trend for dielectric particles as their diameter is reduced. Starting in regime 1, as the particle diameter decreases the trapping forces increase. When the wavelength of the trapping light and the particle diameter are equal, the trapping forces peak. As the diameter is reduced further, moving into regime 2, the trapping forces get smaller and approach zero for diminishing size. Similar behavior has also been demonstrated for gold nanoparticles, where the trapping forces decreased with decreasing particle diameter [17].

Once the particle is trapped it can be manipulated in all three Cartesian coordinates by moving the trap. Various actuators are used to move the trap within the sample to achieve a desired position or dynamic motion. In some cases, manipulation of multiple particles is necessary. A single laser can be used to create multiple traps to localize the position of multiple particles in different locations using a time-sharing approach. This is achieved by scanning the laser rapidly between several different locations where traps are to be formed [4]. By scanning the beam much faster than the response time of an individual particle, the intermittent trapping forces are still capable of maintaining the desired positions of the particles. An example of time-shared trapping is shown in Fig. 6.1b, where six microparticles are trapped simultaneously and rotated in a circle using one laser. Trap scanning can also be used to create traps of arbitrary shape by scanning the beam rapidly in a pattern to localize nonspherical particles. For example, line traps have been shown to be effective at trapping nanowires lengthwise in the image plane [18].

The applications of optical trapping are quite broad. In addition to simply manipulating a particle, a trapped particle can be used to impart forces on another object or be used to measure small forces applied to the trapped particle by some other system of interest. A single trap or multiple traps can be used to pull or push objects, which can then be used for experimental mechanical measurements with exceptional precision due to the nanoscale positioning capabilities of an optical trap. More than any other research area, biophysics has adopted optical trapping as a primary research tool for measurement and manipulation and it has been instrumental in many important discoveries with respect to the mechanics of molecules and cells over the last two decades. Optical trapping has been used to investigate the nonlinear stiffness exhibited by DNA [19] and to measure the velocity of RNA polymerase steps during the transcription of DNA, as well as the pulling force of the RNA polymerase [20]. In these cases, one end of the DNA molecule was attached to a microsphere that was trapped and the other end was stuck to a glass slide either directly or through an additional molecule, such that the trapped microsphere could apply a tensile load to the DNA. Similar experimental approaches have been used to observe the folding and unfolding of single proteins [21] and to measure the pulling force and velocity of molecular motors, including myosin (critical in muscle contraction) [22] and kinesin (used in many cell functions) [23]. Optical trapping has also made a large impact on cell-level studies. For example, traps have been used to stretch red blood cells to measure their elasticity [24,25], which is an excellent indicator of cell health, and to perform automated cell sorting for large assays that require high-quality cells [26].

Numerous measurement applications at the micro- and nanoscales beyond biophysics have also made use of optical trapping. Trapped particles have been used as local probes of fluid flow velocity [27] and viscosity [28] within microfluidic channels, providing *in situ* measurements that are otherwise difficult to achieve. The force characteristics of an optically trapped particle have also been used as a probe for measuring the topography of structures within an aqueous sample with minimal contact forces (<5 pN). Similar to scanning probe microscopy, the trapped particle is scanned over a structure and the displacement of the particle is recorded to produce a representation of the structure topography [29–31]. Optical traps have also aided exploration into the fundamental physics of mesoscopic systems. For example, an optical trap was used to demonstrate that the instantaneous velocity of a Brownian particle can be measured [32], which was long believed to not be possible.

Another important application domain for optical trapping is the assembly of particles into more complex and functional micro- and nanostructures. Early work in this area focused on microstructures. For example, multiple optical traps were used to assemble linkages composed of cells and polystyrene microparticles that bond due to protein molecules that provide biospecific adhesion [33]. Linear chains of microparticles have also been assembled using a combination of optical trapping and photopolymerization to bond the particles together [34]. The resulting structures have been used to control particle flow in microfluidic channels and could be used as a probe for manipulating other microstructures. Trapping has also been used to automatically assemble a sixteen-piece microscale puzzle [35], which demonstrated that this approach is viable for low-throughput manufacturing. More recently, there has been a focus on assembling structures with nanoparticles. Nanowires have been trapped and rotated in the plane and then assembled into free-floating linkages using laser-based cutting and bonding techniques [36]. It has also been shown that multi-nanowire structures can be assembled on a surface using an adhesive layer [37]. Recently, gold nanoparticles have been assembled into ordered patterns of forty or more using optical trapping with position deviations from the desired locations on the order of ± 100 nm [38]. Assembling nanoparticles and nanowires into nanodevices using optical trapping can be more time and cost efficient than many top–down and bottom–up fabrication approaches when prototyping devices. However, it is likely that it will not be appropriate for high-volume production due to its low throughput.

It is clear from the applications discussed above that users of optical trapping typically require a high level of control over both force and position, whether it be a measurement, manipulation, or assembly task. Furthermore, due to the optical forces, the interaction forces between the trapped object and the trapping medium, and the object inertia, the dynamics of a trapped particle are quite complex. As a result, the application of feedback control to optical trapping is an obvious approach to improving the system performance, whether it be for force or position control. This chapter focuses on the use of feedback control in optical trapping and is intended as an introduction to both the technical and practical issues. An overview of the existing research in this area is provided in the following section along with a discussion of the various control objectives encountered. This is followed by a discussion of the sensing and actuation technologies typically used in optical trapping and the tradeoffs in their performance with respect to closed-loop control. The dynamics of an optically trapped particle are then explored, with an emphasis on the control challenges and tractable models that can be used for control design. A control approach that uses the optical trap position as the control input (scan control) to suppress Brownian motion is then presented. Finally, the outstanding challenges in this field from our perspective are discussed.

6.2 Overview of Feedback Control in Optical Trapping

As with so many other fields, feedback control has been integrated with optical trapping in order to improve system performance and enable new capabilities. The earliest example of feedback control is the stabilization of levitated microparticles with optical forces, as demonstrated by Ashkin and Dziedzic [39]. However, in this case, the trapping forces for optical levitation are largely along the laser's direction of propagation rather than in all three Cartesian directions, as found in the gradient force optical trap. Feedback control was first applied to a three-dimensional gradient force optical trap by Finer et al. [22] to maintain a constant position for a trapped particle while a motor protein (myosin) pulls on the particle. The added stability of the particle provided by feedback control yielded force and position measurements that had the best sensitivity and resolution to date. Since then, single molecule biophysical experiments have been a driving force for control innovations in optical

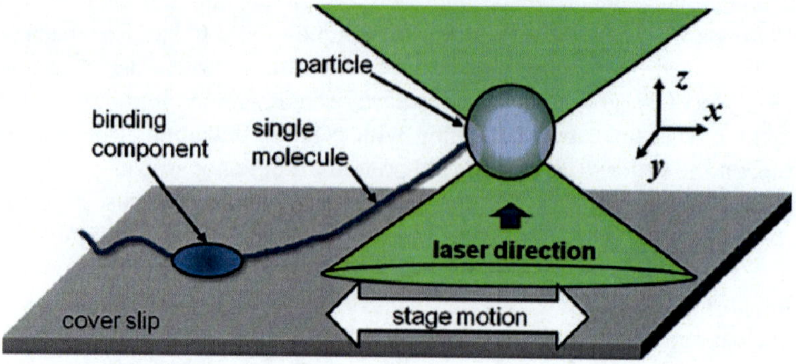

Fig. 6.2 Schematic of a common configuration for single molecule measurements using an optical trap. A single molecule is attached at one end to a trapped microparticle and attached to a binding component at another location along its length. Synchronized motion of the trap and coverslip is used to measure the mechanical properties of the single molecule, the binding component, or both

trapping and the use of feedback control has played a major role in a number of important studies [19, 20, 23, 40–44].

The majority of these single molecule studies can be viewed as variations on the trapping configuration depicted in Fig. 6.2. A microparticle is held in an optical trap close to the sample coverslip. A single molecule is attached to the microparticle at one of its ends and is attached to a binding component at another location along its length. The binding component is adhered to the coverslip through a selective chemical bond. In some cases, the binding component is a motor protein that can move along the single molecule or move on the slide, which in both cases will impart a force on the microparticle. Specific examples of this configuration include an actin molecule and myosin molecule [22]; a DNA molecule and RNA polymerase [19]; and a kinesin molecule and a microtubule [23], for the single molecule and binding component, respectively. The particle motion in an optical trap can be measured accurately using one of the methods described in more detail later in this chapter. For a properly calibrated optical trap, the trapping force can then be accurately measured as a function of the displacement of the particle, under the assumption that the trap acts like a nonlinear spring. Additionally, the coverslip, and therefore the binding component, can be moved relative to the trap using a precise motion stage. As a result, this configuration can be used to measure the motion and force characteristics of the single molecule or binding component through coordinated motion between the trap and stage. This coordination is where feedback control plays a critical role. Since the range of the particle in the trap where linear force measurements can be made is small (e.g., ∼250 nm for a 1 μm particle), precise control over the position of the trap is needed to maintain the experiment in this range, which has been an enabling factor in the biophysical experiments discussed above.

Depending on the measurement of interest, the experimental configuration shown in Fig. 6.2 is typically implemented either isometrically (constant particle position) or isotonically (constant particle force). Isometric measurements are used to

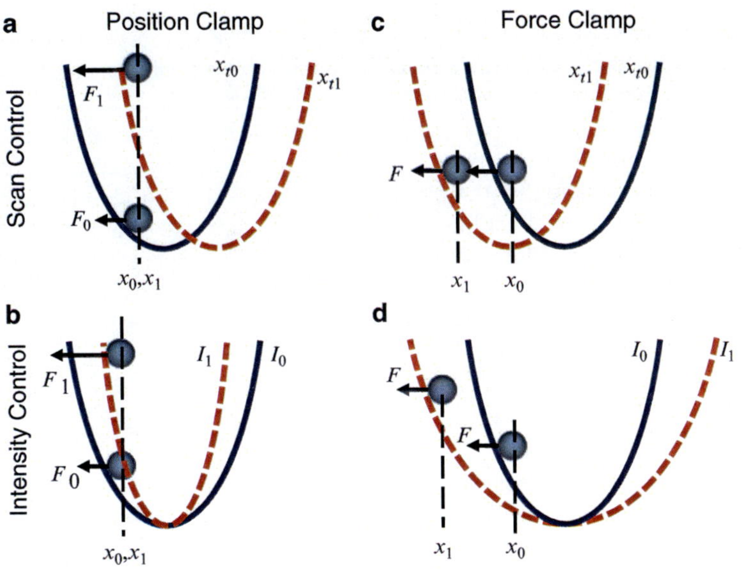

Fig. 6.3 A visual description of position and force clamps implemented with either scan control or intensity control. The optical trap is represented as a potential well and in each case is shown in two states, before (0) and after (1) control action. (**a**) Position clamp using scan control – the trap position is adjusted to keep the particle stationary. (**b**) Position clamp using intensity control – the laser intensity is adjusted to keep the particle stationary. (**c**) Force clamp using scan control – the trap position is adjusted to keep the force on the particle constant. (**d**) Force clamp using intensity control – the laser intensity is adjusted to keep the force on the particle constant. Adapted from Visscher and Block [43]

measure the mechanical properties of the single molecule or binding component (e.g., stiffness or pulling force of motor protein), whereas isotonic measurements are useful for tracking their motion over longer ranges (e.g., motor protein step size). In both cases, they are implemented using feedback control, either with a position clamp (isometric) [19,20,22,40–42] or a force clamp (isotonic) [23,43,44]. Feedback control for either clamping methods can be realized using one of the two control inputs to the system: the position of the trap or the intensity of trapping laser. As a result, there are four distinct control modes that are defined by whether position or force clamping is desired and whether the control input is the laser intensity or trap position, as depicted in Fig. 6.3.

The trap is drawn in Fig. 6.3 as a potential well and the trapped particle is shown in two states for each control scenario, before (state 0) and after (state 1) control action. A force, F, is applied to the particle through the single molecule, which pulls the particle away from the center of the trap. The trapping force that counteracts the external force F is proportional to the slope of the potential well at the position of the particle. When using a position clamp, the position, x, is held constant either by scanning the trap position, x_t, to the right (Fig. 6.3a) or by increasing the laser intensity, I (Fig. 6.3b), to oppose the force. Alternatively, a force clamp can keep the reaction force constant either by scanning the trap to track the particle position (Fig. 6.3c) or by reducing the laser intensity as the particle is pulled away from the center of the trap (Fig. 6.3d). Note that the particle shown in Fig. 6.3d is at two different heights for the two states. In order to maintain a constant force on the particle, the location of the particle in the well shifts to maintain the same slope for the two states. Each of the control modes has its advantages depending on the biophysical experiment of interest, which are discussed in [43]. These modes provide the control architecture but the control law that determines the laser intensity or trap position at a given time instant must still be defined. In general, the position and force clamps demonstrated in [19, 20, 22, 23, 40–44] have been implemented using a proportional–integral–derivative (PID) control law, or some subset of this (e.g., P, PI), with static position or force setpoints.

Closed-loop position and force clamps have been the enabling factor in all of the biophysical measurements discussed above. However, as would be expected, the emphasis in this research has been on the biophysics and not on the optimization of the controller design. As a result, the controllers up to this point were implemented with minimal control design effort and without analysis of the tracking performance or suppression of the Brownian motion of the trapped particle. Furthermore, the bandwidth of these control systems was relatively low given the known dynamics of the particle. In most cases, the bandwidth is below 2 kHz due to the use of averaging filters. Finally, the results discussed above were almost exclusively focused on particles that are tethered to the coverslip by a molecule. The molecule localizes the particle and reduces its Brownian motion, thereby making it easier to control. Since there are many applications for optical trapping outside of biophysics that do not use a tether, there is a need for control systems for freely suspended particles that go beyond the force and position clamp control modes.

Within the past decade, control system researchers have begun to investigate general control design approaches for optically trapped particles based on the previous work on force and position clamps [45–52]. This research has been motivated by the large impact of the biophysical measurements described above and other broad applications for precise optical traps, including nanoscale assembly and material characterization. One major thrust has been the suppression of Brownian motion to improve the position localization of the particle. In general, this research has focused on the control of free particles rather than those that are tethered because the control approach can then be adapted for any specific application. Wulff et al. [45] investigated the performance of a proportional–integral (PI) controller that is tuned to achieve Bode's ideal first-order loop gain and found that it could suppress Brownian motion by 20% below 100 Hz for a 1 μm polystyrene particle, but it amplified the motion above this frequency. The limited bandwidth was partially due to the fast scanning mirror used to move the trap, which required an observer to estimate its scan angle and has slower dynamics than other commonly used scanners. An adaptive control approach was also demonstrated by this group, which can automatically identify the system parameters for different particle sizes, particle materials, and laser power and then tune the controller appropriately [46]. High-bandwidth Brownian motion suppression was first demonstrated by Wallin et al. [47], where a linear proportional controller was used to increase the trap stiffness by a factor of 13 over a wide frequency range (DC to ~5 kHz). A proportional predictive controller was also investigated to extend the bandwidth of the controller by compensating for loop delays with an infinite impulse response (IIR) filter, resulting in an even greater trap stiffness and frequency range for motion suppression [48]. Most recently, we have presented an analysis of a general PID controller for Brownian motion suppression, including expressions for H_2 and H_∞ norms for the closed-loop system, and have experimentally demonstrated a reduction in the RMS motion of a 1 μm particle by approximately 40% compared to a static trap [49].

All of the controllers discussed above have been linear, although the stiffness of the trap is known to be nonlinear. The trapping force can be approximated by a linear stiffness near the center of the trap but the stiffness decreases as the particle moves away from the center and asymptotically approaches zero. From an energy perspective, the potential well formed by the trap has a finite depth and the particle can escape the potential well if its energy is large enough, which could either be due to the particle's thermal energy or an external force. As a result, there has been interest in developing nonlinear controllers that can guarantee that the particle remains in the trap, thereby enhancing manipulation capabilities. Such controllers have been developed [50–52] but have yet to be demonstrated experimentally. However, stability analysis and simulation results indicate that these nonlinear approaches can render the particle's motion globally asymptotically stable, such that the particle will remain trapped indefinitely barring any physical limitations of the experiment. These approaches warrant further study and need to be implemented in practice.

These advances in controller design have brought clear improvements in system performance, including enhanced disturbance rejection and wider operational bandwidth, compared to those used in previous force and position clamp research. However, control researchers have only begun to scratch the surface with optical trapping. The achievements discussed above only address particle localization or the zero setpoint regulation problem. Nonzero setpoint regulation, as found in position clamps, and tracking or following, as found in force clamps, have not been analyzed from the control perspective. Additionally, the majority of this work uses scan control instead of intensity control and as a result does not address the control of the particle position along the optical axis. One exception is our recent work on using intensity control to keep nanoparticles in the trap for much longer durations than found with a passive trap [53]. Finally, it is clear that there is still room for improvement in both the closed-loop bandwidth and resolution for optical traps. Increasing the bandwidth will enable added suppression of Brownian motion at higher frequencies, thereby improving the overall localization of the trapped particle. Increasing the resolution will improve the manipulation precision for particles during the assembly of heterogeneous structures, particularly at the nanoscale. Further, there are typically tradeoffs between bandwidth and resolution that must be understood through analysis of the closed-loop system. One way of addressing some of these issues is through the selection of the best trap actuators and particle sensing methods for a given application.

6.3 Actuation and Sensing

The performance of the sensors and actuators within a servo control loop has a direct impact on the performance of the closed-loop system. Although in theory the controller can be designed to compensate for some shortcomings in actuator performance (e.g., dynamic inversion, pole/zero cancellation), in practice this often results in high control effort, which can have negative effects on the actuator. Therefore, it is important that the actuators and sensors are selected based on known performance goals. Optical trapping has many options when it comes to sensing and actuation with a wide range of tradeoffs. This section discusses some of the best options for feedback control and highlights the issues that affect closed-loop operation. A broader overview of these technologies with respect to optical trapping can be found in [3, 43].

As described previously, an optically trapped particle has two distinct control inputs: the laser intensity and the trap position. Therefore, there are two classes of actuators of interest: laser intensity modulators and trap scanners. As shown in Fig. 6.4, both actuators can be used simultaneously or separately depending on the application. Obviously, the particle position sensing system is important because it provides feedback for the controller but it is also used to perform measurements with the particle, such as indirectly measuring the external forces applied to the particle. These three components, laser trap scanner, intensity modulator, and position sensing system, are now examined.

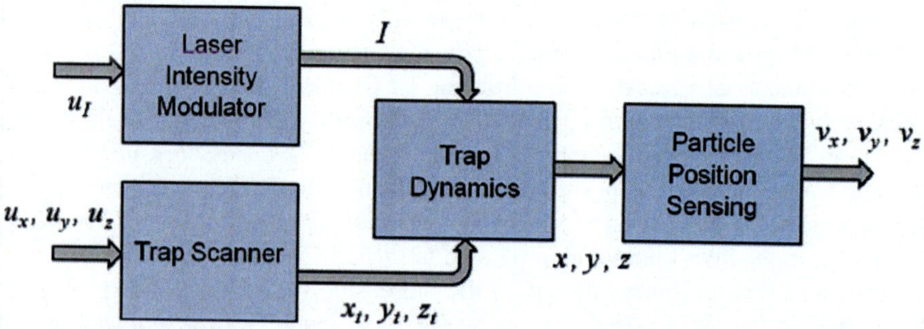

Fig. 6.4 Block diagram showing the relationships between the trap actuation, trap dynamics, and particle position sensing

6.3.1 Trap Scanners

In order to achieve trap scanning in the x and y directions (in-plane scanning), the laser beam must rotate about the back aperture of the microscope objective to avoid clipping the beam. This is achieved by locating the laser scanner, which alters the angle of the beam, at a plane conjugate to the back aperture (i.e., an object at the back aperture can be imaged at the plane of the scanner location) [54]. Rotating the beam about the back aperture also ensures that the beam does not get clipped at the aperture before entering the microscope objective. Spatial light modulators are widely used for laser scanning in optical trapping, particularly for holographic optical tweezers [55] and the generalized phase contrast method [56]. However, spatial light modulators have modest update rates that are not appropriate for closed-loop control and, therefore, are not covered further in this chapter. The most prevalent approach for scanning the beam for feedback control, thereby translating the trap, is with an acousto-optic deflector (AOD). AODs use acoustic waves to scan a laser at high rates. A radio frequency (RF) acoustic wave is generated inside a crystal using a piezoelectric transducer. This creates a periodic structure inside the crystal with areas of increased and decreased index of refraction that is analogous to a diffraction grating. This grating causes the incoming beam to be split into two beams, a diffracted and undiffracted beam, when the incoming beam is properly aligned to the Bragg angle, θ_{Bragg} (see Fig. 6.5a). Changing the frequency of the sound wave alters the period of the grating, resulting in an angular deflection of the diffracted beam (see [57, 58] for more details). The xy trap position can then be controlled through fine control of the input frequency. One major benefit of using AODs is that the scan bandwidth is typically on the order of 1 MHz, which is achieved with tunable RF frequency sources. Furthermore, AODs are widely available for different laser wavelengths and beam sizes. However, their main drawback is that they exhibit an inherent time delay caused by the time of flight of the sound wave between the transducer and the far side of the deflected beam. The time delay ranges from 1 μs to 30 μs depending on the size of the crystal and the beam diameter. Although this may seem small, within a closed-loop system this level of time delay will reduce the effective bandwidth of the system to tens of kilohertz and will set limits on the system stability. Most of the results reported to date have not been affected by this issue because their controller bandwidth was the limiting factor. However, recent results on Brownian motion suppression [47–49] have shown that the time delay sets a lower bound on the particle RMS motion that can be achieved with feedback. An additional drawback of AODs is that the sound wave causes wavefront distortion in the beam, causing nonideal behavior in the trap (e.g., the intensity profile may deviate from the expected Gaussian profile).

Another important scanner option is the electro-optic deflector (EOD) [58], which has seen limited use in optical trapping [59] but is well suited for high-bandwidth control. An electro-optic crystal undergoes a change in its index of refraction when a high voltage (0 V to 1,000 V, depending on the material) is applied

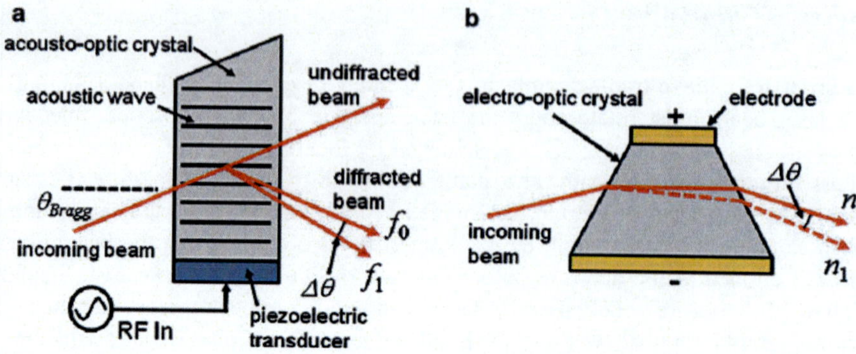

Fig. 6.5 Operating principles for acousto-optic and electro-optic deflectors. (**a**) Acousto-optic deflector (AOD). A laser beam enters the acousto-optic crystal at the Bragg angle, θ_{Bragg}. Half of the beam is unaffected and passes through. The other half is diffracted by the RF acoustic wave generated by the piezoelectric transducer. The exit angle of the diffracted beam is controlled by changing the acoustic wave frequency (exit beam shown at frequencies f_0 and f_1). (**b**) Electro-optic deflector (EOD). The index of refraction of an electro-optic prism is controlled by applying a voltage across the prism. As the index of refraction is changed, the beam's exit angle changes (shown at two different indices of refraction, n_0 and n_1)

across the crystal. As shown in Fig. 6.5b, when the crystal is shaped as a prism, the change in the index of refraction causes a change in the exit angle of the beam from the prism. In practice, the prism design is more complex than shown but the basic principal remains the same. Similar to the AOD, EODs have high scanning bandwidth but they do not suffer from a time delay like the AOD. As a result, EODs offer the fastest scanning response of any commercially available actuator. The bandwidth is only limited by the voltage amplifier rather than the electro-optic effect and scan rates on the order of hundreds of kilohertz are possible. The main disadvantages of EODs are that they have the smallest scan range among the most common scanner technologies and they can generate electromagnetic interference (EMI) due to the high drive voltage that can introduce noise into surrounding instrumentation.

Due to the high scanning bandwidth of AODs and EODs, they are the two best options for closed-loop operation of optical traps. However, scanning mirrors have also seen significant use in optical trapping and are, therefore, worth mentioning here. Scanning mirrors are electromechanical actuators that provide control over the tip and tilt of an attached mirror. The actuation mechanism for the rotational degrees of freedom can be achieved with a piezoelectric actuator, a voicecoil, or a galvanometer; each with its own advantages and disadvantages. The main benefits of using a scanning mirror for optical trapping are that they are easy to align and use, they reflect almost all of the beam power into the microscope objective, so very little power is wasted, and they have a negligible effect on the wavefront of the reflected beam. However, they almost universally have low scanning bandwidth

Table 6.1 Critical performance parameters for trap scanners (values based on commercially available systems)

Actuator type	Range (mrad)	Resolution (μrad)	Bandwidth (MHz)	Time delay (μs)
AOD	4–56	0.01–6	0.1–4	1–30
EOD	0.6–5	0.01–1	0.2–0.25	Negligible
Scanning mirror	1–350	0.02–1	0.0002–0.002	Negligible

due to the mechanical motion involved, with most systems scanning below 1 kHz. Additionally, they typically have more complex dynamics than AODs and EODs, including mechanical resonances, hysteresis, and nonminimum phase behavior. As a result, scanning mirrors are in general not well suited to fast feedback control for optical traps. However, they can be of use in low-bandwidth applications where their simplicity of use and minimal impact on beam quality are needed.

The most important performance parameters for trap scanners with respect to feedback control are scanning range, resolution, bandwidth, and the time delay if present. These parameters are summarized in Table 6.1 for AODs, EODs, and scanning mirrors based on commercially available systems known to have been used for optical trapping. Although AODs typically have the highest bandwidth of the three, the time delay in AODs limits their effective bandwidth within a feedback loop. As a result, EODs offer the best closed-loop performance if the small scan range is acceptable. Otherwise, AODs with short time delays are preferred for control. Compression mode, or longitudinal mode, AODs typically have the shortest time delays because the acoustic wave is propagated along a crystal axis that has a higher acoustic velocity, whereas shear mode AODs have longer time delays but larger scan ranges. Proper selection of the beam scanner can greatly simplify the control design for a given application.

Scanning the trap along the optical axis (z-axis) is much more difficult than in the x- and y-axes for feedback control. One approach is to use a piezoelectric scanner to translate either the sample or the microscope objective, thereby moving the optical trap relative to the sample. This approach typically results in low scanning bandwidth ($<200\,\text{Hz}$) due to the mass of the objective or sample holder, which is generally inadequate for fast feedback control. Additionally, the complex dynamics of the scanner and its mechanical interactions with the coverslip on the sample make it a challenge to control. Surprisingly, few other options have been implemented. One intriguing possibility is to use AODs to control the height of the trap. Two AODs can be used in series to control the focus of a beam [60], which would in turn change the height of the trap, providing scanning bandwidth similar to that obtained with AODs for in-plane motion. Another possibility is to use a MEMS membrane mirror to control the beam focus [61], which could achieve a scanning bandwidth on the order of 5 kHz. The current lack of scanning options along the optical axis is a major limiting factor in obtaining precision three-dimensional control of particle position.

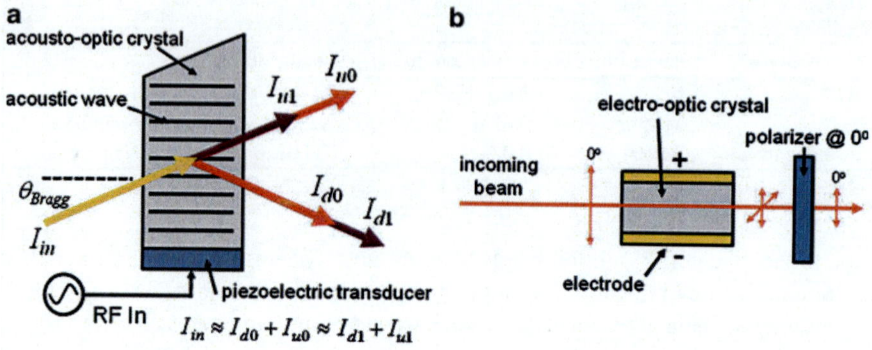

Fig. 6.6 Operating principles for acousto-optic and electro-optic modulators. (**a**) Acousto-optic modulator (AOM). The diffraction effect caused by the RF acoustic wave splits the incoming beam into two beams. The RF input power is used to tune the intensity ratio between these beams. (**b**) Electro-optic modulator (EOM). The incoming beam enters an electro-optic crystal that alters its polarization as a function of the control voltage. The beam then passes through a polarizer set at 0°. The exiting intensity is reduced by shifting the laser polarization in the crystal away from 0°

6.3.2 Laser Intensity Modulators

The stiffness of the trap is directly proportional to the laser power. Therefore, particle motion within a trap can be controlled by adjusting the laser intensity accordingly. The acousto-optic modulator (AOM) is the most common actuator used to control laser intensity and has been used previously in optical trapping to implement position clamps [19, 20]. An AOM operates similarly to an AOD in that an acoustic wave inside a crystal is used to create a diffraction grating that alters the incoming beam's direction (see Fig. 6.6a). However, instead of adjusting the frequency of the acoustic wave, its amplitude is controlled, which changes the amount of light diffracted by the wave. The sum of the optical intensities for both the diffracted and undiffracted beams remains relatively constant as the wave amplitude is changed. AOMs generally have time delays shorter than most AODs ($<5\,\mu$s). Therefore, the effective bandwidth in AOMs is typically higher in comparison to AODs, with most systems achieving at least 100 kHz. Similar to AODs, AOMs can cause wavefront distortion in the trapping beam due to the diffraction grating, which can result in undesirable deviations from the ideal beam profile.

An electro-optic modulator (EOM) can also be used to control laser intensity. An EOM is effectively a tunable waveplate, as shown in Fig. 6.6b. The beam entering the electro-optic crystal is linearly polarized at 0°. The polarization of the beam exiting the crystal is controlled by the voltage applied across the crystal crosssection. After exiting the crystal, the beam passes through a linear polarizer set at 0°. If the beam polarization is rotated away from 0° due to an applied voltage on the crystal, the intensity of the beam exiting the polarizer will be reduced accordingly. Similar to EODs, the bandwidth of EOMs is limited by the achievable slew rate for

Table 6.2 Critical performance parameters for laser intensity modulators (values based on commercially available systems)

Actuator type	Bandwidth (MHz)	Extinction ratio	Time delay (μs)
AOM	1–30	300:1–2,000:1	0.5–5
EOM	0.25–100	200:1–1,000:1	Negligible
Modulated laser diode	0.35–20	>1,000 : 1	Negligible

high-voltage amplifiers and is typically above 200 kHz. Unlike AOMs, EOMs have a negligible effect on the beam wavefront and can transmit up to 95% of the incoming intensity.

One other option for intensity control worth mentioning is intensity modulation of a diode laser. This requires the use of a diode laser for trapping, which has become prevalent over the last decade. It is common to achieve an intensity modulation bandwidth in the tens of megahertz using this approach, making it the fastest actuation mechanism discussed here. The main drawback is that the beam characteristics can change when going from high to low intensity, including the phase and wavelength. However, this may have only a marginal effect on controlling the trapped particle. The critical performance parameters for the three intensity modulators discussed here are shown in Table 6.2. All three approaches are capable of achieving bandwidth greater than 100 kHz, which has been found to be sufficient in previous research. Hence, one advantage of intensity control over scan control is that it is straightforward to achieve high actuation rates with any of the modulator options, which is not the case for scanners. Additionally, intensity control can be used to localize a particle along the z-axis, which is difficult with the available trap scanner options, although it cannot control the position of the particle arbitrarily.

6.3.3 Particle Position Sensing

A number of different methods have been developed to measure the position of an optically trapped particle with resolution approaching 1 nm, both for in-plane and out-of-plane motion. However, only a few of these methods are appropriate for closed-loop control. These methods are described here along with some of the shortcomings of other methods that are less suitable for feedback. By far, the most commonly used position sensing method has been back-focal-plane detection [4, 43, 62, 63], which can measure the in-plane or x and y motion of the trapped particle. As shown in Fig. 6.7a, the back-focal-plane detection method uses a second laser to measure displacement by focusing the beam with the trapping objective and measuring the motion of the light scattered by the trapped particle. The advantage of using an independent laser for position measurement is that it provides a fixed reference, which yields absolute position measurements. The detection laser intensity is much lower than the trapping laser to minimize the trapping forces generated by this second beam. An idealized depiction of the relationship between the particle displacement and the motion of the light for dielectric particles is shown

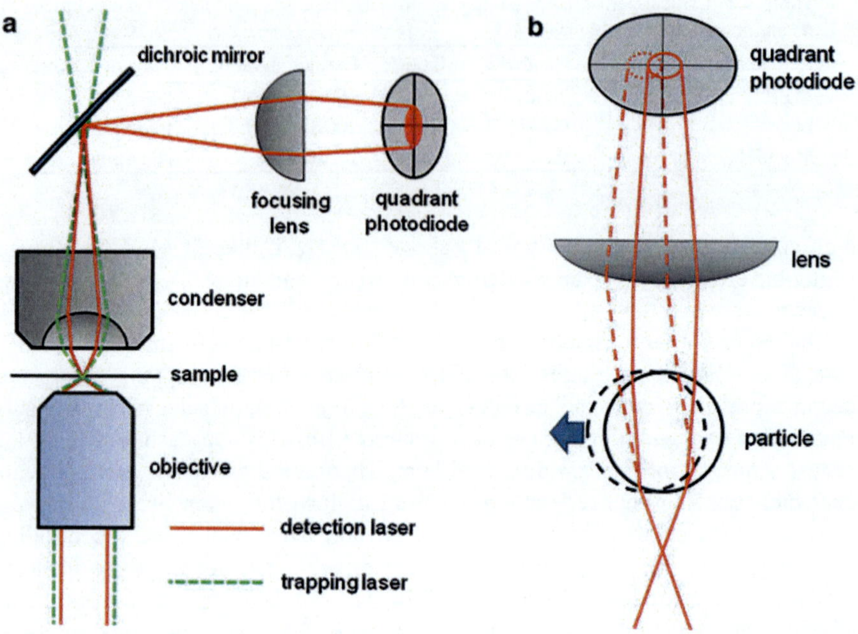

Fig. 6.7 Description of the back-focal-plane detection method. (**a**) Both the trapping laser and detection laser pass through the sample and are then separated with a dichroic mirror. The detection laser is directed onto a quadrant photodiode positioned near a conjugate plane of the back-focal-plane of the condenser. (**b**) Lateral motion of the particle results in scanning of the detection laser, which can be measured on the quadrant photodiode

in Fig. 6.7b. The detection laser is focused just below the trapped particle such that the particle collimates the light passing through it. In-plane motion of the particle results in angular scanning of the beam. The detection laser exiting the microscope condenser is then directed onto a quadrant photodiode by a dichroic mirror. The detection laser is generally a different wavelength than the trapping laser so that the two can be separated optically. The quadrant photodiode measures the motion of the beam, which can then be related to the particle motion through calibration (e.g., see [44]). This approach is referred to as back-focal-plane detection because the quadrant photodiode is placed at a plane conjugate to the back-focal-plane of the condenser. However, this exact placement is not required and the position of the quadrant photodiode along the optical axis is usually optimized empirically to maximize the displacement signal. Achieving a resolution and bandwidth of 1 nm and 100 kHz, respectively, is typical.

Several improvements on this method have been developed that could be beneficial for closed-loop control. One is the use of a differential back-focal-plane measurement in which two detection beams are used, one to track the particle of interest and the other to measure the motion of the microscope [64]. This approach has been used to counteract instrumentation drift and has yielded displacement resolution below 0.4 nm due to the cancellation of common mode disturbances.

Another approach has been developed to improve the sensing bandwidth, where the forward-scattered beam is segmented into two halves and each half is directed onto a separate fast photodiode rather than a quadrant photodiode [32]. Taking the difference between the two photodiode signals provides the same information for a single axis as for the quadrant photodiode but with a higher bandwidth, which was 75 MHz in this case. Control experiments have yet to require bandwidths in this range but it may be needed as experiments progress towards air or vacuum trapping environments where there is significantly less damping from the trapping medium surrounding the particle.

There are also variations on back-focal-plane detection pre-dating the use of two separate lasers, one for trapping and one for sensing, that are less appropriate for feedback control. Microscope illumination can be used to measure particle motion by imaging the particle onto a quadrant photodiode [22, 40, 41]. However, the light intensity is significantly lower than that from a laser and, therefore, the shot noise is much higher for a given bandwidth. The increased noise results in worse position resolution and as a result this method is not the best option for control. Another option is to use the trapping laser for both trapping and measuring the position of the particle by directing the forward-scattered light of the trapping laser onto a quadrant photodiode [4, 62]. This provides a measure of the particle's position with respect to the center of the trap, which can be useful for some experiments. However, as pointed out in [43], decoupling the position sensing from the trap scanning is necessary to provide particle position with respect to an inertial reference frame for feedback control. As a result, this method is typically not used for control.

Other methods for in-plane particle position measurement include laser interferometry based on differential interference contrast [65, 66] and image processing using digitized video of the particle [67, 68]. The former method can yield excellent resolution ($\sim$0.3 nm resolution over 100 kHz in [65]) but can only measure motion in one in-plane direction making it impractical for most applications. The latter method currently suffers from strict bandwidth limitations. Compensation for instrument drift using feedback from image processing has been demonstrated with a video sampling rate of 25 Hz [69]. However, this bandwidth is not suitable for the suppression of Brownian motion or fast trajectory following. Due to recent innovations in the design of high-speed digital cameras, images of trapped particles can be captured in excess of 2.5 kHz [70] but real-time image processing of the particle position remains a challenge. A conservative estimate for the current limit on the sampling rate of an image processing-based controller is 500 Hz but this will likely increase as parallel computation and field programmable gate arrays (FPGAs) are applied to this problem.

Particle sensing along the optical axis (z-axis) has rarely been used for feedback in a gradient force optical trap. However, there is clearly a need to move from the two-dimensional control described in [19, 20, 22, 23, 40–44] to complete three-dimensional control for nanoassembly and lower noise biophysical experiments. Two laser-based methods have been demonstrated for optical trapping that can be used for this purpose. The first method assumes that the detection laser passes through the particle and the refraction of the exiting light is modulated by the axial motion of the particle [29, 71, 72]. As shown in Fig. 6.8, the particle motion causes

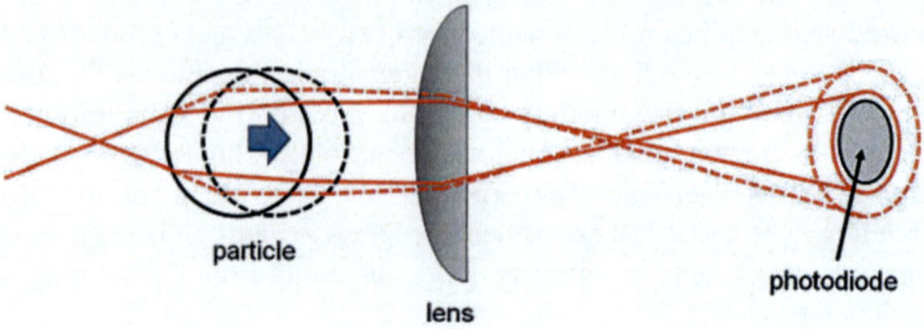

Fig. 6.8 Schematic of the optical detection mechanism for motion along the optical axis. The particle sits just upstream from the focus of the trapping and detection lasers. Light passing through the particle is refracted by the particle and is then passed through a lens (condenser and focusing optics). The exiting beam is focused to a point and then diverges. The diverging beam is positioned on a photodiode such that half of the intensity is on the photodiode for the nominal position. As the particle moves along the optical axis, the angle of refraction for the light passing through particle changes, thereby changing the intensity on the photodiode

more or less intensity to hit a photodiode located near a conjugate plane of the back-focal-plane of the condenser. The best position sensitivity is typically achieved when the photodiode is overfilled such that only half of the laser intensity is detected by the photodiode. The second method assumes that some of the detection laser is refracted by the particle while the rest is reflected, which is typically true for particles smaller than the wavelength of the detection laser [73, 74]. Interference between the refracted and reflected light results in a position-dependent intensity at the back-focal-plane of the condenser, which can be measured with a photodiode. Although these two methods leverage different physical phenomena it is difficult to separate them, particularly for particles below 1 μm. As with the back-focal-plane detection method described above, an independent detection laser is recommended for both of these methods to decouple the sensing and actuation. In-plane and out-of-plane sensing have also been combined for *xyz* measurement of the particle [72–74].

One of the major benefits of the laser-based position sensing methods described above is that the sensing bandwidth is only limited by the photodiode response and the dynamics of the readout electronics. In general, 1 MHz can be achieved with modest effort, which is more than suitable for the control applications proposed to date. However, sensor noise sets the ultimate limit on resolution and as a result, there is generally a direct trade-off between sensor resolution and bandwidth since wider bandwidth introduces more noise. Designing the positioning sensing system to have only the required bandwidth will maximize resolution and reduce the constraints on the controller design. The main sources of noise in these laser-based methods are intensity fluctuations and pointing instabilities in the detection laser, shot noise and Johnson noise in the photodiode and respective electronics, and mechanical vibrations due to acoustic or base excitations. Readers interested in an overview of noise considerations should refer to Gittes and Schmidt [75].

6.4 Dynamic Behavior of Optically Trapped Particles

The dynamics of an optically trapped particle are unique in that the optical forces have a limited operational range around the center of the trap. In this section, the finite trapping lifetime of particles is discussed to highlight some of the challenges in controlling trapped particles. This is followed by the development of an empirical dynamic model of a trapped particle that is suitable for control design.

6.4.1 Finite Lifetime in an Optical Trap

One of the most distinctive characteristics of the dynamics of an optically trapped particle is that the particle has a finite lifetime within the trap, meaning that the particle can escape from the potential well. This can be seen experimentally when trapping particles with low laser intensity, where the low trapping forces can only keep the particle in the trap for a short time. This phenomenon can be explained by noting that the potential well formed by the trap has finite depth since there is a limit on the trap width and depth set by the optics and laser power. If the particle reaches the edge of the well due to Brownian motion it can escape the trap or go back down into the well. Since Brownian motion is stochastic it is necessary to discuss the escape time, or lifetime, in a probabilistic sense, typically as a mean value. In this section, two methods for investigating trapping lifetime are discussed to highlight the challenges in controlling trapped particles.

An optically trapped particle can be considered as a mass–spring–damper system with a softening nonlinear spring. Assuming that the trapped particle is an ideal sphere, the inertial and damping components in the dynamics are relatively straightforward and are a function of the particle size and the material properties of the particle and trapping medium. However, the trapping forces are more difficult to model. Both experimental (e.g., see [41]) and theoretical results (e.g., see [76]) have shown that the qualitative behavior of the conservative optical trapping forces can be approximated with reasonable accuracy using a Gaussian potential well model. In reality, the optical forces along each axis are coupled to the motion along all three axes. However, for the purpose of discussing trapping lifetime, a one-dimensional Gaussian potential is considered here. The potential energy of the particle, V, can be written as $V = -\alpha e^{-\mu x^2}$, where x is the particle position with respect to the center of the trap. Ignoring the parameters α and μ for the time being, one can see that the potential energy resulting from optical forces approaches a constant value as x gets larger, as shown in Fig. 6.9a. The trapping force resulting from this model is found by taking the negative derivative of V with respect to x such that $F_t = -\partial V/\partial x = -2\mu\alpha x e^{-\mu x^2}$. As shown in Fig. 6.9b, the optical force is zero at the trap center, it increases while moving away from the center until it peaks, and then decreases until it reaches zero again. Intuitively, when the particle moves past the point of peak force the force pulling the particle back to the center is decreasing

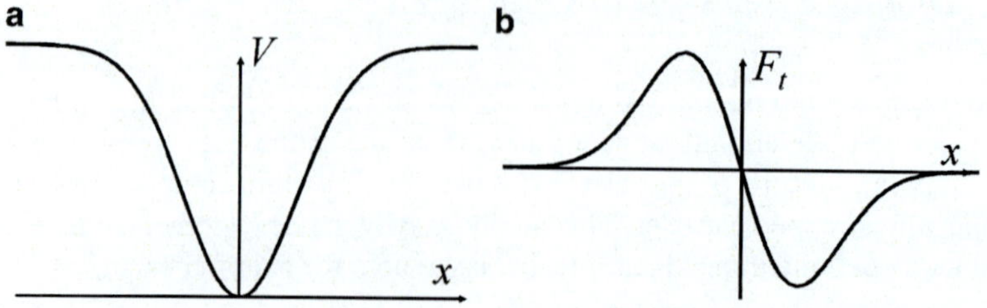

Fig. 6.9 The potential energy and force curve for the optical forces on a trapped particle. (**a**) The potential energy of a particle along one axis. (**b**) The resulting optical force as a function of displacement from the trap center

and, therefore, the particle can escape if Brownian motion drives the particle further away from the trap.

The Fokker–Planck equation [77] can be used to estimate the lifetime of a trapped particle. The time evolution of the probability density function (PDF), $W(x,t)$, for a particle under the influence of an optical trapping force is given by the Fokker–Planck equation shown in (6.1), which can be solved numerically. In this case, the optical trapping force, $F(x)$, is modeled using a Gaussian potential. The other parameters in (6.1) are the diffusion constant, $D = k_B T/\gamma$, Boltzmann's constant, k_B, the temperature of the fluid, T, and Stokes' constant, $\gamma = 6\pi\eta a$, where a is the particle radius and η is the fluid viscosity.

$$\frac{\partial}{\partial t}W(x,t) = -\frac{\partial}{\partial x}D\left(\frac{1}{k_B T}[W(x,t)F(x)] - \frac{\partial}{\partial x}W(x,t)\right). \tag{6.1}$$

The initial conditions for (6.1) are defined by a normalized PDF, $\varphi(x)$, (6.2), which describes the initial distribution of the particles in the trap. The function $\varphi(x)$ is selected so that the particle ensemble has filled the well at $t = 0$. The boundaries of the trap are defined over the closed interval $[-x_R, x_R]$. This absorbing boundary condition is shown in (6.3).

$$W(x,0) = \varphi(x), \tag{6.2}$$

$$W(-x_R,t) = W(x_R,t) = 0. \tag{6.3}$$

The Fokker–Planck equation has been solved numerically using Mathematica. The solver first converts the partial differential equation (PDE) into a system of ordinary differential equations (ODEs) using a technique called the method of lines. It then solves this system of ODEs using either an implicit backward difference formula numerical integration technique or the Adams multi-step method, depending on the stiffness of the equations. The numerical solution of the PDE yields the PDF, $W(x,t)$, which can then be integrated over the density in the well to extract trapping

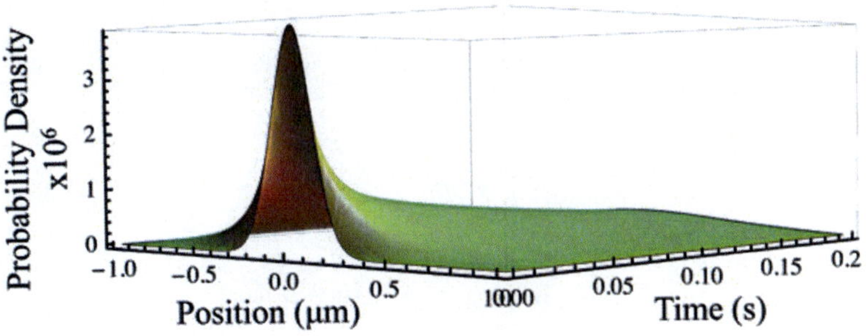

Fig. 6.10 A plot of the probability density function (PDF) as a function of time for a 100 nm gold nanoparticle in a weak optical trap. The flattening of the PDF as time progresses shows that the particle is likely to leave the trap within the time of this simulation

lifetime. This approach can also be used to determine the stability of a controller by augmenting the analyzed dynamic system with the controller and then solving the equations for an array of control parameters.

A representative PDF for a 100 nm gold nanoparticle in a weak optical trap is shown in Fig. 6.10. The power of the trapping beam is 1.5 mW and the corresponding depth of the trapping potential is calculated to be $1.14 k_B T$. It can be seen in the figure that initially, the particles are located close to the center of the well, as indicated by the large peak at the center at $t = 0$ s. However, as the particles diffuse, they quickly escape the trap and the central peak, which is proportional to the number of particles inside the trap, decays exponentially. All of the relevant information pertaining to the behavior of the particle in the trap, such as the trap lifetime and average power absorbed by the particle, can be obtained by operating on this PDF. Trapping lifetime is directly related to the strength of the trapping laser intensity and the size of the particle. A microscale particle trapped with tens of milliwatts of power can in theory stay in the trap for weeks or months on average whereas a nanoparticle trapped with a few milliwatts of power may only stay in the trap for a few seconds.

A more physically accurate approach for simulating the stochastic nature of optically trapped particles is to combine a trapping force model obtained using generalized Lorentz–Mie theory (GLMT) [11,78] with Brownian dynamics simulations [79]. GLMT, which is applicable to a wide range of particle sizes and materials, has been used to calculate the optical force on the particle as a function of position in the x, y, and z directions for a 350 nm particle trapped with a laser beam with a wavelength, $\lambda = 1064$ nm. The force field is calculated over a closed interval of $\pm 3\,\mu$m along the x- and y-axes and $(-2, +6)\,\mu$m along the z-axis, with a uniform grid spacing along all axes of 75 nm. Figure 6.11 shows 100 simulated trajectories of 350 nm glass nanoparticles under the influence of a GLMT-calculated force field with a beam power of 5 mW. The total trapping force along the z-axis generated by the GLMT model is overlaid on top of the nanoparticle trajectories in Fig. 6.11. The thick black line in the figure represents the total axial trapping force, including

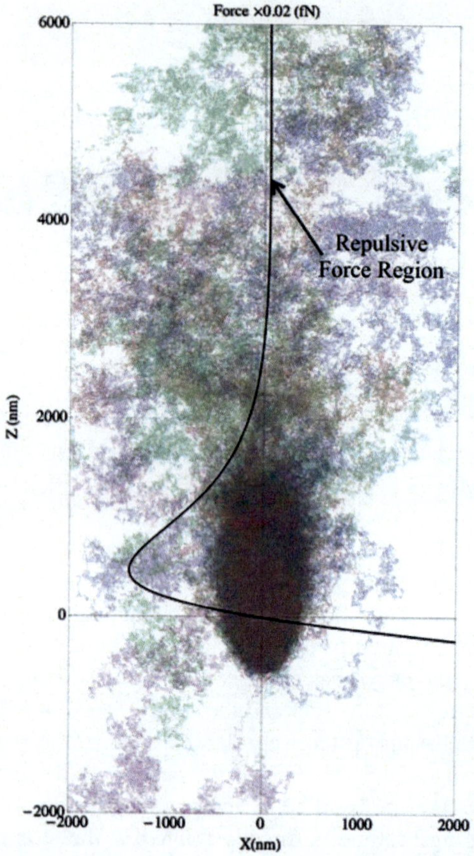

Fig. 6.11 Simulated motion of nanoparticles in an optical trap in the *xz* plane. 100 trajectories are shown with most of them exiting along the *z*-axis. The densely populated region around (0, 0) represents the stable part of the trap. The black curve is the force profile of the trap along the optical axis. Note that the force becomes repulsive approximately 3 μm above the center of the trap

gradient and scattering components (directed along the positive *z*-axis). From the figure, it can be seen that the trapping force is attractive on either side of $z = 0$ and the magnitude of the attractive force is much larger on the negative side of the origin. On the other hand, when $z > 0$, the trapping force reaches a maximum at approximately 500 nm and then decreases, eventually becoming repulsive at approximately 3.0 μm. The combination of weak gradient forces and axially directed scattering forces results in particles escaping preferentially along the positive *z*-axis as seen from the figure. The preferential exit along the optical axis is not surprising since the axial stiffness is always found to be lower than the lateral stiffness and the disruptive scattering component of the trapping force has a strong longitudinal component. These simulations show that a vast majority of particles exit along the optical axis, confirming that this is the primary issue in particle trapping lifetime for a single beam optical trap. This behavior has also been observed qualitatively in trapping experiments in our laboratory.

These two numerical examples provide insight into the behavior of optically trapped particles that can influence controller design. Maybe even more important, both the Fokker–Planck equation and Brownian dynamics simulations have the ability to be used as control design tools through iterative solution for different control parameters and functions, providing a method for controller optimization. These numerical approaches have largely not been explored but may provide a viable solution for stochastic control systems in which the dynamic behavior is beyond the capabilities of standard optimal control methods.

6.4.2 An Empirical Model for Control Design

In this section, an empirical closed-form dynamic model for an optically trapped particle is derived for the purposes of control design. This model represents the in-plane dynamics (x and y directions) and does not describe motion along the optical axis (z direction). Following this derivation, the model is linearized for the control design discussed in the next section.

The equations of motion are derived using the energy method. The position of the particle in the plane is represented by the generalized coordinates x and y, and the inputs to the system for scan control are the trap coordinates, x_t and y_t, where all variables are defined with respect to the sensing coordinate system, which is an inertial reference frame (see Fig. 6.12). The kinetic energy, T, and potential energy, V, for the particle, and the generalized forces applied to the particle, Q_x and Q_y, can be written as:

$$T = \frac{1}{2}m\left(\dot{x}^2 + \dot{y}^2\right), \tag{6.4}$$

$$V = \frac{1}{2}\alpha\left(1 - e^{-\mu r_t^2}\right), \tag{6.5}$$

$$Q_x = -\beta\dot{x} + \gamma\Gamma_x(t), \tag{6.6}$$

$$Q_y = -\beta\dot{y} + \gamma\Gamma_y(t), \tag{6.7}$$

where m is the mass of the particle, the parameter α is proportional to the laser power, and μ is a function of the beam waist of the trap and the radius of the trapped particle. The potential energy of the trapped particle is described by a two-dimensional Gaussian potential, where r_t is the distance between the center of the particle and the center of the trap (see Fig. 6.12), and can be written as:

$$r_t = \left((x - x_t)^2 + (y - y_t)^2\right)^{1/2}. \tag{6.8}$$

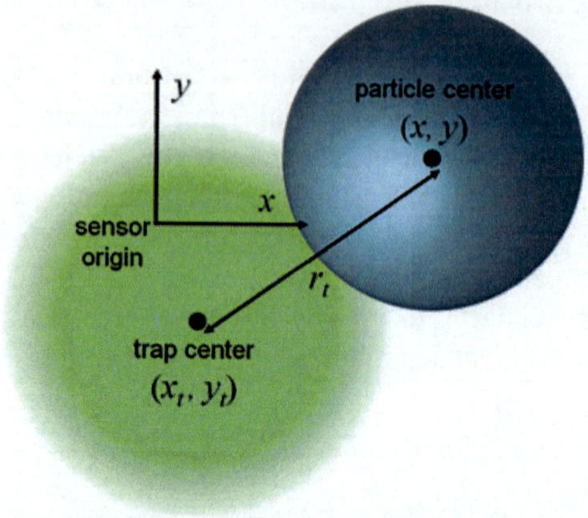

Fig. 6.12 The trapping coordinate system

In (6.6) and (6.7), β is the damping coefficient, which for a spherical particle is defined by Stokes' Law as $\beta = 6\pi\eta r_{\mathrm{p}}$. The parameter η is the viscosity of the trapping medium, which is a function of the absolute temperature of the trapping medium, T, and r_{p} is the radius of the particle. The second term in both (6.6) and (6.7) is the Langevin force, which is a stochastic force that results in Brownian motion. The parameter $\gamma = (2\beta k_{\mathrm{B}}T)^{1/2}$, and $\Gamma_x(t)$ and $\Gamma_y(t)$ are Gaussian white noise processes [6, 7], and the parameter k_{B} is the Boltzmann constant.

Applying Lagrange's equations [80] using (6.4)–(6.8) results in the following equations of motion:

$$m\ddot{x} + \beta\dot{x} + \alpha\mu(x - x_{\mathrm{t}})e^{-\mu r_{\mathrm{t}}^2} = \gamma\Gamma_x(t), \tag{6.9}$$

$$m\ddot{y} + \beta\dot{y} + \alpha\mu(y - y_{\mathrm{t}})e^{-\mu r_{\mathrm{t}}^2} = \gamma\Gamma_y(t). \tag{6.10}$$

It can be seen in (6.9) and (6.10) that the motion along the x- and y-axes is coupled through the nonlinear stiffness of the trap. This nonlinear model can be simplified by: (1) noting that the inertial components are small compared to the damping and stiffness terms, and (2) the trapping force can be linearized about the center of the trap for small displacements. Therefore, the order of the system can be reduced by setting $m = 0$ and linearized about the origin using a Taylor series approximation, yielding the final reduced-order linearized equation of motion.

$$\dot{x} = -\frac{\alpha\mu}{\beta}(x - x_{\mathrm{t}}) + \frac{\gamma}{\beta}\Gamma_x(t). \tag{6.11}$$

The equation of motion for the y-axis is the same as (6.11) except for the substitution of y for x and y_{t} for x_{t}. This linearized dynamic model is valid within a range about the center of the trap that is dependent on the beam spot size and particle diameter (e.g., ±250 nm for a 1 μm particle). This model has been assumed in the majority of previous work on controlled optical trapping (e.g., see [41,45–48]) and will be used throughout the rest of the chapter.

The product $\alpha\mu$ in (6.9)–(6.11) is the linear trap stiffness such that $k = \alpha\mu$. When implementing scan control, the control inputs are the trap position coordinates x_t and y_t. Therefore, the control inputs are not affine in the nonlinear dynamics, (6.9) and (6.10), but are affine in the linearized model. When implementing intensity control, the laser intensity control input is introduced through the parameter α, such that $\alpha = \alpha_0 I(t)$, where $I(t)$ is the laser intensity. One can see that the laser intensity directly affects the trap stiffness since $k = \mu\alpha_0 I(t)$.

Before closing this section, simple models of the sensor and actuator for scanning control are described. In the following section, scan control using an acousto-optic deflector (AOD) is presented. The AOD and drive electronics have first-order dynamics and a time-delay, but these dynamics do not affect the closed-loop system in the frequency range of interest in this discussion. Therefore, they are modeled as a linear gain, and can be written in the Laplace domain as:

$$x_t(s) = k_{\mathrm{AOD}_x} u_x(s), \tag{6.12}$$

where k_{AOD_x} is the gain. The output signal from the quadrant photodiode used in back-focal-plane detection has been shown to be nonlinear [44, 62], and it approximately matches the derivative of a Gaussian potential. Therefore, the sensor function can be written as:

$$v_x = k_x x \mathrm{e}^{-\varepsilon_x x^2}, \tag{6.13}$$

where k_x is the sensitivity of the sensor, and ε_x determines the zero-slope point for the error function. Calibration data for back-focal-plane detection is shown in Fig. 6.13 along with a least-square fit, showing a close fit between the data and model. In the control design discussed in the next section, the sensor response is linearized for the purpose of control design, such that $v_x = k_x x$.

6.5 Brownian Motion Suppression Using Feedback Control

In this section, an experimental instrument for optical trapping is described and a method for selecting the gains for a PID controller is presented for optimal suppression of the Brownian motion of the particle in the trap. Experimental results for this control system are then discussed.

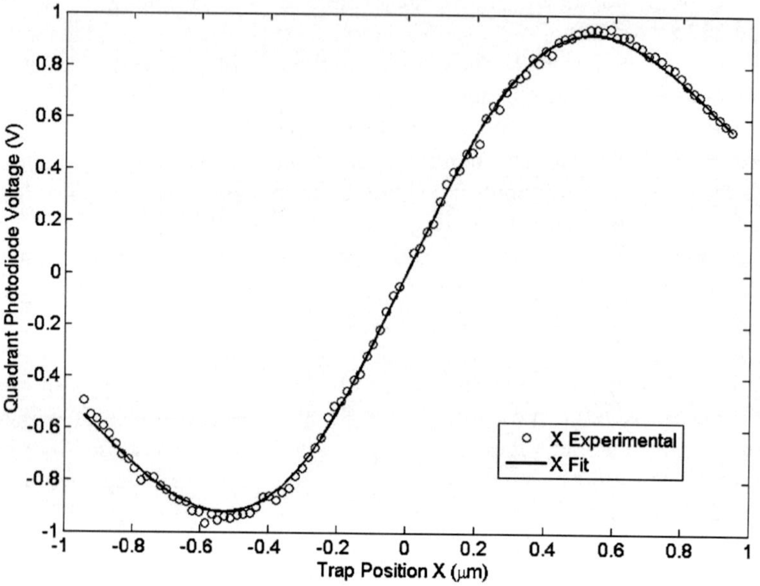

Fig. 6.13 Quadrant photodiode output voltage as a function of particle displacement for back-focal-plane detection (experimental data and fit curves for (6.13))

6.5.1 Instrument Design

A diagram of the optical layout of the optical trapping instrument is shown in Fig. 6.14. The optical layout utilizes two lasers: a $Nd:YVO_4$ laser with a 1064 nm wavelength for trapping and a diode laser with a 640 nm wavelength for detecting the trapped particle's position. The collimated trapping laser passes through an XY acousto-optic deflector (AOD), where it is scanned along two axes. A telescope then expands the beam to overfill the back aperture of the microscope objective. The telescope also ensures that the exit aperture of the AOD is at a plane conjugate to the back aperture of the microscope, so that the beam rotates about the back aperture rather than translate. The trapping beam is combined with the detection beam through a dichroic mirror. Both beams are directed into the microscope objective using another dichroic mirror and the beams then pass through the sample. The power of the detection laser at the focal plane is less than 1 mW, whereas the trapping laser power is typically between 30 mW and 150 mW. Therefore, the detection laser only has a negligible effect on the trapping dynamics. The microscope objective and dichroic mirror are mounted on a commercial inverted microscope.

Back-focal-plane detection is used to measure the position of the trapped particle, as described in Sect. 6.3.3. After exiting the sample, the detection beam is reflected by a dichroic mirror and passed through a lens, which focuses the beam onto a quadrant photodiode, where the deflection of the beam due to particle motion is measured. The quadrant photodiode in the back-focal-plane detection system has a

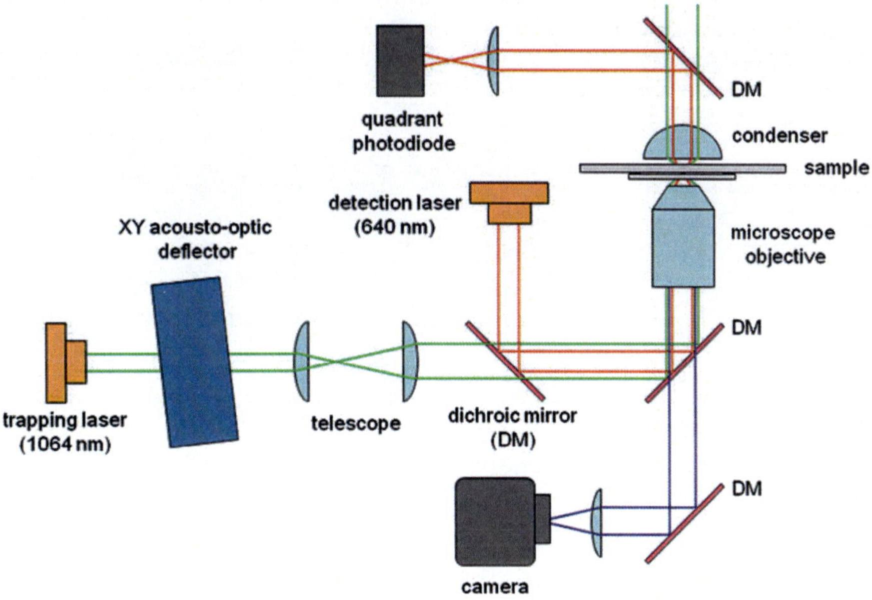

Fig. 6.14 A simplified diagram of the optical trapping instrument

bandwidth of 100 kHz. The sample is composed of a glass slide with a coverslip attached using double-sided tape. The chamber formed by this coupling is filled with a dilute solution of 1 µm diameter silica spheres in deionized water.

The position of the optical trap is controlled by the AOD. As is typical of acousto-optic crystals, the AOD introduces a time delay due to the time of travel of the acoustic wave in the crystal (approximately 27 µs in this case). Its scanning bandwidth is 30 kHz. As a result, the AOD is the limiting component in terms of loop rate. Although not considered within the following analysis, this limitation will be discussed with respect to the experimental results.

6.5.2 *Design of a Scan Controller*

As described previously, there are many different control objectives that can be pursued with optical trapping, including position tracking, force control, and particle localization. In this section, a linear controller design is presented for the suppression of Brownian motion. In (6.9) and (6.10), the Gaussian white noise processes $\Gamma_x(t)$ and $\Gamma_y(t)$ act as disturbances to the particle motion, resulting in Brownian motion. A PID controller that provides disturbance rejection of the white noise is presented here. The reduced-order linearized system is utilized in order to simplify the closed-loop analysis. Equation (6.11) can be written in the Laplace domain as:

$$x(s) = \frac{a}{s+a} x_t(s) + \frac{b}{s+a} \Gamma_x(s), \tag{6.14}$$

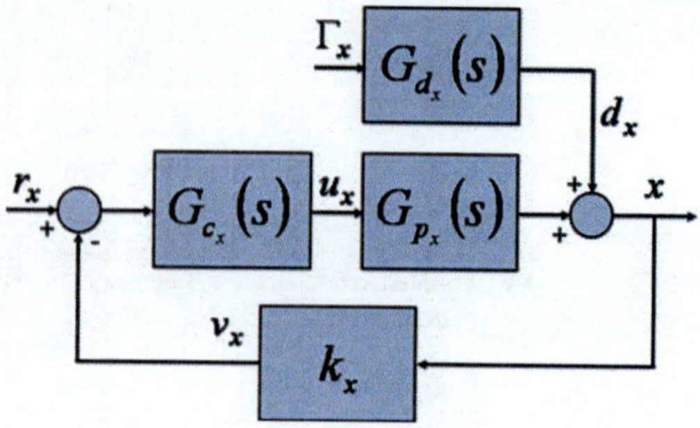

Fig. 6.15 Block diagram of the linearized closed-loop system

where $a = k/\beta$ and $b = \gamma/\beta$. Substituting (6.12) into (6.14) results in the following equation:

$$x(s) = G_{p_x}(s)u_x(s) + G_{d_x}(s)\Gamma_x(s), \tag{6.15}$$

where

$$G_{p_x}(s) = \frac{ak_{\mathrm{AOD}_x}}{s+a}, \tag{6.16}$$

$$G_{d_x}(s) = \frac{b}{s+a}. \tag{6.17}$$

The linearized closed-loop system can be drawn as shown in Fig. 6.15, where $G_{c_x}(s)$ is the controller transfer function, $r_x(s)$ is the command input, and the input to the controller is the control error $e_x(s)$, which can written as:

$$e_x(s) = r_x(s) - v_x(s). \tag{6.18}$$

However, in this case we are not interested in the tracking problem and only look at the disturbance rejection properties of the controller. Therefore, r_x is set equal to zero and the closed-loop transfer function can then be written as:

$$x(s) = G_{\Gamma_x x}(s)\Gamma_x(s), \tag{6.19}$$

where

$$G_{\Gamma_x x}(s) = \frac{G_{d_x}(s)}{1 + k_x G_{p_x}(s) G_{c_x}(s)}. \tag{6.20}$$

The standard PID form [81] is used for the controller,

$$G_{c_x}(s) = \frac{K_{\mathrm{d}}s^2 + K_{\mathrm{p}}s + K_i}{s}, \tag{6.21}$$

where K_{p} is the proportional gain, K_{i} is the integral gain, and K_{d} is the derivative gain.

Equation (6.19) defines the closed-loop response of the particle motion to the thermal noise that causes Brownian motion. Given (6.21) as the controller, the transfer function $G_{\Gamma_x x}(s)$ can be written as:

$$G_{\Gamma_x x}(s) = \frac{bs}{(1+\bar{a}K_\mathrm{d})s^2 + (a+\bar{a}K_\mathrm{p})s + \bar{a}K_\mathrm{i}}, \tag{6.22}$$

where $\bar{a} = ak_{\mathrm{AOD}_x}k_x$. The analysis for motion along the y-axis is identical to that described here.

A number of different methods can be used to tune the PID controller gains to maximize performance based on the transfer function, (6.22) (e.g., see [82]). Here, the control gains are selected based on the H_∞ and H_2 norms for (6.22), which can be written in closed form for this system. The H_∞ norm of $G_{\Gamma_x x}$, $|G_{\Gamma_x x}|_\infty$, represents the maximum amplitude of $|G_{\Gamma_x x}|$, and can be viewed as the largest standard deviation of the particle displacement at any given frequency. The expression can be shown to be:

$$||G_{\Gamma_x x}||_\infty = \frac{b}{a+\bar{a}K_\mathrm{p}}, \tag{6.23}$$

where the maximum is located at $\omega = \sqrt{\bar{a}K_\mathrm{i}/(1+\bar{a}K_\mathrm{d})}$ (see [49] for details on this derivation of this expression and those that follow). Therefore, the H_∞ norm of the particle motion can only be reduced by increasing K_p, but its location in frequency is determined by both K_i and K_d.

The H_2 norm of $G_{\Gamma_x x}$ is the root mean square, or standard deviation, of the particle displacement. The resulting expression is:

$$||G_{\Gamma_x x}||_2 = \frac{b}{\sqrt{2(a+\bar{a}K_\mathrm{p})(1+\bar{a}K_\mathrm{d})}}. \tag{6.24}$$

It is clear that both K_p and K_d can be used to reduce the standard deviation, while K_i has no effect.

The PID control system was implemented on the trapping instrument described above using an analog PID controller with 100 kHz bandwidth. The x- and y-axes were controlled simultaneously using the same controller gain values. The power spectrum of the motion of the particle in the trap was fit to the reduced-order model (6.11) using least-squares parameter estimation, where the cutoff frequency is $\omega_\mathrm{c} = k/\beta$ (see [4]). In this case, $\omega_\mathrm{c} = 1{,}046.0\,\mathrm{rad/s}$ for 30 mW laser intensity, so $k = 9.56\times10^{-6}\,\mathrm{N/m}$. Using Stokes' law, $\beta = 9.14\times10^{-9}\,\mathrm{Ns/m}$ where $\eta = 1.0\times10^{-3}\,\mathrm{Ns/m^2}$ and $r_\mathrm{p} = 0.485\,\mu\mathrm{m}$. The gains for the AOD and associated electronics and the back-focal-detection have been measured to be $k_{\mathrm{AOD}_x} = 0.1885\,\mu\mathrm{m/V}$ and $k_x = 2.8781\,\mathrm{V}/\mu\mathrm{m}$, respectively.

The amplitude of Brownian motion (as measured by the quadrant photodiode) when using proportional control is shown in Fig. 6.16 for various K_p values. As expected, an increase in K_p decreases the H_∞ and H_2 norms, where the largest

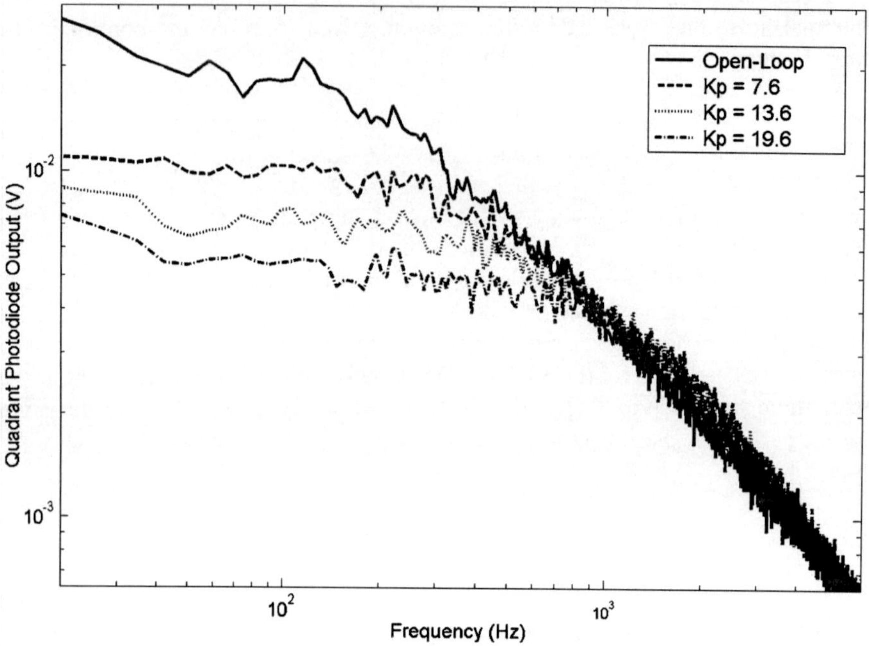

Fig. 6.16 Power spectrum of the Brownian motion of the particle when using proportional control

K_p value reduces the H_∞ norm by 73% and the H_2 norm by 44% compared to the uncontrolled particle (i.e., no feedback).

Results for a proportional–derivative (PD) controller are shown in Fig. 6.17, in comparison with the open-loop case and a proportional controller with the same K_p value ($K_p = 7.6$). Although the PD controller reduces the amplitude out to approximately 2 kHz compared to the proportional controller, the time delay in the acousto-optic deflector electronics causes a spike in the amplitude around 4 kHz. Even so, the PD controller reduces $||G_{\Gamma_x x}||_2$ by 13.1% compared to the proportional controller alone. Finally, results for a proportional–integral (PI) controller are shown in Fig. 6.18 in comparison to the open-loop case and proportional controller with an equivalent K_p value. As predicted by the analysis in the previous section, the integrator has no effect on the H_∞ and H_2 norms. However, the integrator is effective in shifting the particle's energy from one frequency band to another. This would be useful if the particle experiences additional disturbances that were at frequencies below 200 Hz in this example.

Complete PID controllers were also tested but the results did not improve compared to the cases listed here since the PD controller reduces the H_∞ and H_2

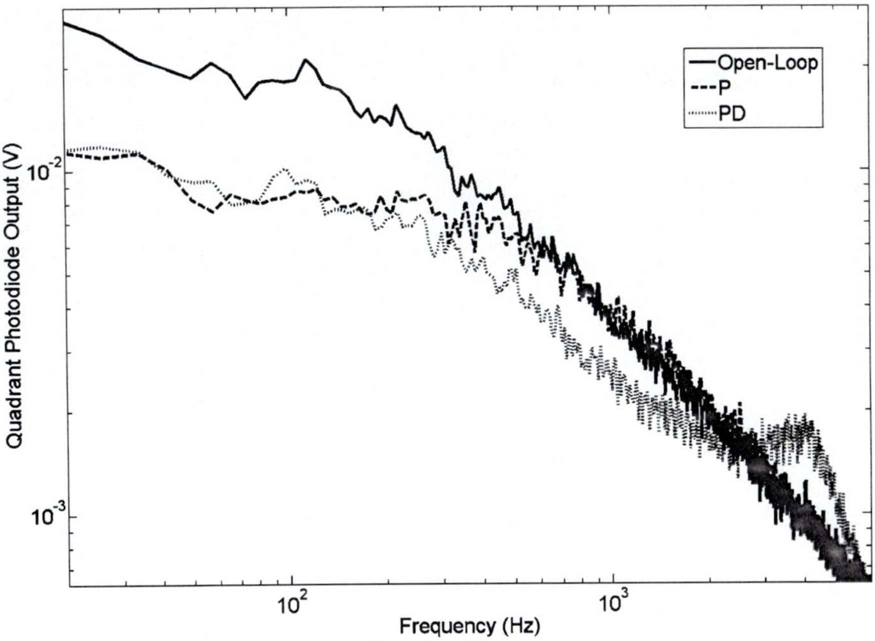

Fig. 6.17 Power spectrum for the Brownian motion of the particle when using a PD controller

norms just as much as the PID controller, as shown in (6.23) and (6.24). However, the addition of the derivative term clearly does provide dissipation of the thermal energy and has a clear benefit over proportional control. Minimizing the time delay in the AODs will increase the effectiveness of the derivative control term. Additional experimental results for the PID controller are presented in [49] and proportional control results are shown in [47].

The PID controller is effective in reducing the Brownian motion of the particle even though the control design was based on a linearized model of the particle dynamics. This is due to the laser intensity used in the experiments, which is moderately high, and the size of the particle, which is approximately 1 μm in diameter. Due to these two features, the particle explores a small region in the trap where the dynamics are essentially linear. When the gains are increased, the particle explores a wider region where the forces a more nonlinear and eventually the system becomes unstable due to the use of the approximate model. When using smaller particles and lower laser intensity the particle will be more likely to explore the outer part of the trap and, therefore, the dynamics will be inherently more nonlinear. Under these conditions, it is unclear how effective the PID controller will be in suppressing Brownian motion.

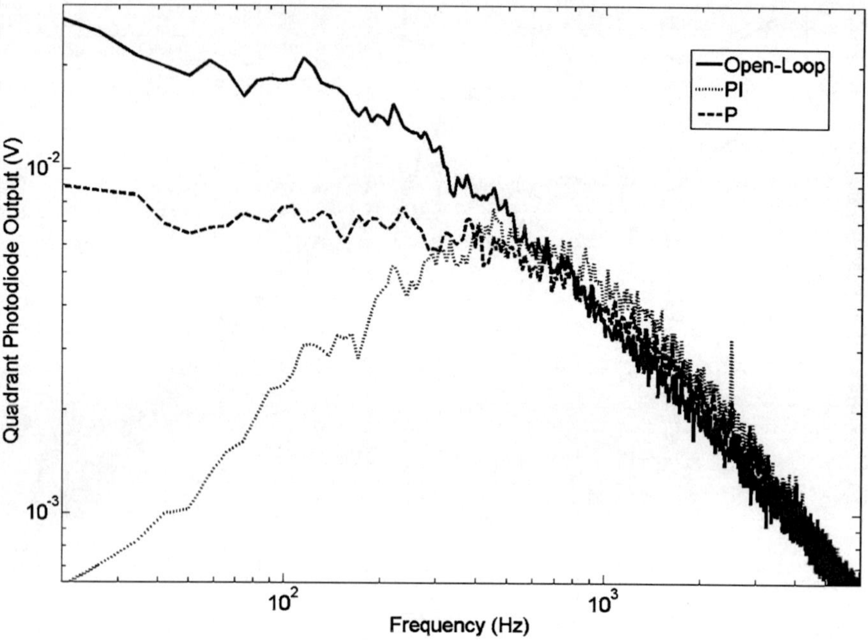

Fig. 6.18 Power spectrum for the Brownian motion of the particle when using a PI controller

6.6 Research Challenges and Future Directions

This chapter has presented an overview of optical trapping and results on the control of optically trapped particles. As with so many servo control problems, feedback control has been shown to provide significant performance benefits in optical trapping and has the potential to enable numerous new capabilities. In many ways, the integration of control systems with optical trapping is in its infancy, with only the most basic analysis and implementations pursued to date. Before closing, we end with a discussion of some exciting technical challenges and opportunities that could have a large impact on the field if solved.

6.6.1 *Three-Dimensional Position Control*

All of the implementations of feedback control in gradient force optical trapping reported to date have been two dimensional. Achieving three-dimensional position control will open many new applications. However, adding control along the z-axis requires a suitable actuator. As described in Sect. 6.3.1, there are only a few scanning actuator options for the z-axis and none of them have been demonstrated for high-bandwidth controlled optical trapping. Although intensity control can alter the trapping dynamics along the z-axis, it can only be used to improve the localization of the particle about the center of the trap. It cannot change the vertical position of the trap, which is desirable in manipulation applications. Some combination of both scanning and intensity control that exploits the coupling between these actuation approaches is needed for three-dimensional position control.

6.6.2 Extending Trapping Lifetime

In Sect. 6.4.1, it was shown that a trapped particle will exit the trap in finite time due to Brownian motion and the fact that the trap is a finite-depth potential well. The scanning control approach discussed in Sect. 6.5.1 can reduce the Brownian motion of the particle but it also narrows the effective width of the trap while maintaining the same potential depth. As a result, the particle's time in the trap, or lifetime, is shortened. Nonlinear control methods that can simultaneously reduce particle Brownian motion and increase trapping lifetime must be developed. Intensity control appears to be the best method for increasing trapping lifetime because it can control the potential depth of the trap, which increases the trapping forces along all three axes [53].

6.6.3 Controlled Trapping of Nanoparticles

Almost all of the work on controlled optical trapping has been with particles that are 1 μm in diameter or larger. However, nanoparticle research could strongly benefit from the manipulation capabilities offered by optical trapping. It is known that the trapping lifetime for nanoparticles is fairly short in comparison to the microparticles because the trapping forces decrease strongly with the diameter of the particle. Feedback control could be used to localize nanoparticles if the forces can be modeled and adequate position sensitivity can be achieved with laser-based sensing methods (i.e., the sensitivity also decreases with particle diameter). Reliable nanoparticle trapping could be used to print catalysts for nanowire growth and to observe the absorption of nanoparticles into living cells for toxicological studies, among many other applications.

6.6.4 Trapping in Air and Vacuum

Optical trapping of particles has been almost exclusively pursued in liquids because (1) the biophysical experiments of interest must be performed in liquid, (2) the viscous damping provide by the liquid makes the trap more stable, and (3) the liquid trapping medium facilitates the introduction of particles into the trap.

However, there is interest in trapping in both air and vacuum for nanoassembly applications and for fundamental studies in quantum mechanics [32]. The biggest challenge in trapping in air or vacuum is that the lack of damping, which radically changes the particle dynamics, turning the system into a nonlinear lightly damped oscillator. Depending on the trap strength, the oscillations can be large enough to immediately eject the particle from the trap. Advances in trap modeling, high-bandwidth oscillation control systems, and sample handling must be pursued to make trapping in air and vacuum more reliable.

6.6.5 Nonlinear Stochastic Control Theory

A particle in an optical trap is a nonlinear stochastic system with fairly complex dynamics. Most of the work to date has simplified the problem by either linearizing the system or by excluding the stochastic terms within the stability analysis. A complete stability analysis of the nonlinear stochastic system is needed, in which the stability of the probability distribution of the trapped particle is investigated and the performance of trapping lifetime and position variance is determined as a function of the controller parameters. It is unclear whether the existing control theory can be applied to this problem or whether there is a need for the development of new theory.

The challenges described here are important for optical trapping but also from a general control systems perspective because they are found in many other micro- and nanoscale applications (e.g., nonlinear stochastic control, Brownian motion, and finite escape time). Upon solving them, the resulting innovations must be fed back into the optical trapping applications that have motivated the use of feedback control in the first place. It is expected that the resulting new control approaches will enable exciting new applications of optical trapping in the areas of nanomanufacturing and biophysics, among others.

References

1. A. Ashkin. History of optical trapping and manipulation of small-neutral particle, atoms, and molecules. *IEEE J. Sel. Top. Quantum Electron.*, 6:841–856, 2000.
2. D.C. Grier. A revolution in optical manipulation. *Nature*, 424:810–816, 2003.
3. K.C. Neuman and S.M. Block. Optical trapping. *Rev. Sci. Instrum.*, 75:2787–2809, 2004.
4. K. Visscher, S.P. Gross, and S.M. Block. Construction of multiple-beam optical traps with nanometer-resolution position sensing. *IEEE J. Sel. Top. Quantum Electron.*, 2:1066–1076, 1996.
5. A. Ashkin, J.M. Dziedzic, J.E. Bjorkhom, and S. Chu. Observation of a single-beam gradient force optical trap for dielectric particles. *Opt. Lett.*, 11:288–290, 1986.
6. D.T. Gillespie. The mathematics of Brownian motion and Johnson noise. *Am. J. Phys.*, 64: 225–240, 1995.

7. R.M. Mazo. *Brownian motion: Fluctuations, dynamics and application*, New York, Oxford, 2002.
8. A. Ashkin. Forces of a single-beam gradient laser trap on a dielectric sphere in the ray optics regime. *Biophys. J.*, 61:569–582, 1992.
9. A. Rohrbach. Stiffness of optical traps: Quantitative agreement between experiment and electromagnetic theory. Phys. Rev. Lett., 95:168102, 2005.
10. A.A.R. Neves et al. Electromagnetic forces for an arbitrary optical trapping of a spherical dielectric. *Opt. Express*, 14:13101–13106, 2006.
11. G. Gouesbet, B. Maheu, and G. Grehan. Light-scattering from a sphere arbitrarily located in a Gaussian beam, using a Bromwich formulation. *J. Opt. Soc. Am. A*, 5:1427–1443, 1988.
12. J.A. Lock. Calculation of the radiation trapping force for laser tweezers by use of generalized Lorenz–Mie Theory. I. Localized model description of an on-axis tightly focused laser beam with spherical aberration. *Appl. Opt.*, 43:2532–2544, 2004.
13. K. Svoboda and S.M. Block. Optical trapping of metallic Rayleigh particles. *Opt. Lett.*, 19: 930–932, 1994.
14. Y. Seol, A.E. Carpenter, and T.T. Perkins. Gold nanoparticles: enhanced optical trapping and sensitivity coupled with significant heating. *Opt. Lett.*, 31:2429–2431, 2006.
15. Y. Liu, D.K. Cheng, G.J. Sonek, M.W. Berns, C.F. Chapman, and B.J. Tromberg. Evidence for localized cell heating induced by infrared optical tweezers. *Biophys. J.*, 68:2137–2144, 1995.
16. K.C. Neuman, E.H. Chadd, G.F. Liou, K. Bergman, and S.M. Block. Characterization of photodamage to *Escherichia coli* in optical traps. *Biophys. J.*, 77:2856–2863, 1999.
17. P.M. Hansen, V.K. Bhatia, N. Harrit, and L. Oddershede. Expanding the optical trapping range of gold nanoparticles. *Nano Lett.*, 5:1937–1942, 2005.
18. A. Balijepalli, T.W. LeBrun, and S.K. Gupta. A flexible system framework for a nanoassembly cell using optical tweezers. *Proceedings of the ASME IDETC/CIE*, Philadelphia, PA, 2006, DETC2006–99563.
19. M.D. Wang, H. Yin, R. Landick, J. Gelles, and S.M. Block. Stretching DNA with optical tweezers. *Biophys. J.*, 72:1335–1346, 1997.
20. M.D. Wang, M.J. Schnitzer, H. Yin, R. Landick, J. Gelles, and S.M. Block. Force and velocity measured for single molecules of RNA polymerase. *Science*, 282:902–907, 1998.
21. C. Cecconi, E.A. Shank, C. Bustamante, and S. Marqusee. Direct observation of the three-state folding of a single protein molecule. *Science*, 309:2057–2060, 2005.
22. J.T. Finer, R.M. Simmons, and J.A. Spudich. Single myosin molecule mechanics: piconewton forces and nanometer steps. *Nature*, 368:113–119, 1994.
23. K. Visscher, M.J. Schnitzer, and S.M. Block. Single kinesin molecules studied with a molecular force clamp. *Nature*, 400:184–189, 1999.
24. J. Sleep, D. Wilson, R. Simmons, and W. Gratzer. Elasticity of the red cell membrane and its relation to hemolytic disorders: an optical tweezers study. *Biophys. J.*, 77:3085–3095, 1999.
25. M. Dao, C.T. Lim, and S. Suresh. Mechanics of the human red blood cell deformed by optical tweezers. *J. Mech. Phys. Solids*, 51:2259–2280, 2003.
26. M.M. Wang et al. Microfluidic sorting of mammalian cells by optical force switching. *Nat. Biotechnol.*, 23:83–87, 2005.
27. B.A. Nemet and M. Cronin-Golomb. Microscopic flow measurements with optically trapped probes. *Opt. Lett.*, 27:1357–1359, 2002.
28. B.A. Nemet, Y. Shabtai, and M. Cronin-Golomb. Imaging microscopic viscosity with confocal scanning optical tweezers. *Opt. Lett.*, 27:264–266, 2002.
29. L.P. Ghislain and W.W. Webb. Scanning-force microscope based on an optical trap. *Opt. Lett.*, 18:1678–1680, 1993.
30. M.E.J. Friese, A.G. Truscott, H. Rubinsztein, and N.R. Heckenberg. Three-dimensional imaging with optical tweezers. *Appl. Opt.*, 38:6597–6603, 1999.
31. A. Rohrbach, C. Tischer, D. Neumayer, E.-L. Florin, and E.H.K. Stelzer. Trapping and tracking a local probe with a photonic force microscope. *Rev. Sci. Instrum.*, 75:2197–2210, 2004.
32. T. Li, S. Kheifets, D. Medellin, and M.G. Raizen. Measurement of the instantaneous velocity of a Brownian particle. *Science*, 328:1673–1675, 2010.

33. R.E. Holmlin, M. Schiavoni, C.Y. Chen, S.P. Smith, M.G. Prentiss, and G.M. Whitesides. Light-driven microfabrication: assembly of multicomponent, three-dimensional structures by using optical tweezers. *Angew. Chem. Int. Ed.*, 39:3503–3506, 2000.
34. A. Terray, J. Oakey, and D.W.M. Marr. Fabrication of linear colloidal structures for microfluidic applications. *Appl. Phys. Lett.*, 81:1555–1557, 2002.
35. P.J. Rodrigo, L. Kelemen, C.A. Alonzo, I.R. Perch-Nielsen, J.S. Dam, P. Ormos, and J. Glückstad. 2D optical manipulation and assembly of shape-complementary planar microstructures. *Opt. Express*, 15:9009–9014, 2007.
36. R. Agarwal, K. Ladavac, Y. Roichman, G. Yu, C.M. Lieber, and D.G. Lieber. Manipulation and assembly of nanowires with holographic optical traps. *Opt. Express*, 13:8906–8912, 2005.
37. P.J. Pauzauskie, A. Radenovic, E. Trepagnier, H. Shroff, P. Yang, and J. Liphardt. Optical trapping and integration of semiconductor nanowire assemblies in water. *Nat. Mater.*, 5: 97–101, 2006.
38. M.J. Guffey and N.F. Scherer. All-optical patterning of Au nanoparticles on surfaces using optical traps. *Nano Lett.*, 10:4302–4308, 2010.
39. A. Ashkin and J.M. Dziedzic. Feedback stabilization of optically levitated particles. *Appl. Phys. Lett.*, 30:202–204, 1977.
40. J.E. Molloy, J.E. Burns, J. Kendrick-Jones, R.T. Tregear, and D.C.S. White. Movement and force produced by a single myosin head. *Nature*. 378:209–212, 1995.
41. R.M. Simmons, J.T. Finer, S. Chu, and J.A. Spudich. Quantitative measurements of force and displacement using an optical trap. *Biophys. J.*, 70:1813–1822, 1996.
42. W.H. Guilford, D.E. Dupuis, G. Kennedy, J. Wu, J.B. Patlak, and D.M. Warshaw. Smooth muscle and skeletal muscle myosins produce similar unitary forces and displacements in the laser trap. *Biophys. J.*, 72:1006–1021, 1997.
43. K. Visscher and S.M. Block. Versatile optical traps with feedback control. *Methods Enzymol.*, 298:460–489, 1998.
44. M.J. Lang, C.L. Asbury, J.W. Shaevitz, and S.M. Block. An automated two-dimensional optical force clamp for single molecule studies. *Biophys. J.*, 83:491–501, 2002.
45. K.D. Wulff, D.G. Cole, and R.L. Clark. Servo control of an optical trap. *Appl. Opt.*, 46: 4923–4931, 2007.
46. K.D. Wulff, D.G. Cole, and R.L. Clark. Adaptive disturbance rejection in an optical trap. *Appl. Opt.*, 47:3585–3589, 2008.
47. A.E. Wallin, H. Ojala, E. Hæggström, and R. Tuma. Stiffer optical tweezers through real-time feedback control. *Appl. Phys. Lett.*, 92:224104, 2008.
48. H. Ojala, A. Korsbäck, A.E. Wallin, and E. Hæggström. Optical position clamping with predictive control. *Appl. Phys. Lett.*, 95:181104, 2009.
49. J.J. Gorman, A. Balijepalli, and T.W. LeBrun. Control of optically trapped particles for Brownian motion suppression. *IEEE Trans. Control Syst. Technol., in press*, 2011.
50. A. Ranaweera, B. Bamieh, and A.R. Teel. Nonlinear stabilization of a spherical particle trapped in an optical tweezer. *IEEE Conference on Decision and Control*, Maui, HI, 2003, 3431–3436.
51. A. Ranaweera and B. Bamieh. Modeling, identification, and control of a spherical particle trapped in an optical tweezer. *Int. J. Robust Nonlinear Control*, 15:747–768, 2005.
52. C. Aguilar-Ibañez, M.S. Suarez-Castanon, and L.I. Rosas-Soriano. A simple control scheme for the manipulation of a particle by means of optical tweezers. *Int. J. Robust Nonlinear Control*, 21:328–337, 2011.
53. A.K. Balijepalli. *Modeling and experimental techniques to demonstrate nanomanipulation with optical tweezers*. Ph.D. Thesis, University of Maryland, 2011.
54. E Fallman and O Axner. Design for fully steerable dual-trap optical tweezers. *Appl. Opt.*, 36:2107–2113, 1997.
55. E.R. Dufresne, G.C. Spalding, M.T. Dearing, S.A. Sheets, and D.G. Grier. Computer-generated holographic optical tweezer arrays. *Rev. Sci. Instrum.*, 72:1810–1816, 2001.
56. P.J. Rodrigo, V.R. Daria, and J. Glückstad. Real-time three-dimensional optical micromanipulation of multiple particles and living cells. *Opt. Lett.*, 29:2270–2272, 2004.

57. A.P. Goutzoulis and D.R. Pape. *Design and fabrication of acousto-optic devices*, New York, Marcel Dekker, 1994.
58. M. Gottlieb, C.L.M. Ireland, and J.M. Ley. *Electro-optic and acousto-optic scanning and deflection*, New York, Marcel Dekker, 1983.
59. M.T. Valentine, N.R. Guydosh, B. Gutiérrez-Medina, A.N. Fehr, J.O. Andreasson, and S.M. Block. Precision steering of an optical trap by electro-optic deflection. *Opt. Lett.*, 33:599–601, 2008.
60. N. Kaplan, A. Friedman, and N. Davidson. Acousto-optic lens with very fast focus scanning. *Opt. Lett.*, 26:1078–1080, 2001.
61. V.X.D. Yang et al. Doppler optical coherence tomography with a micro-electro-mechanical membrane mirror for high-speed dynamic focus tracking. *Opt. Lett.*, 31:1262–1264, 2006.
62. F. Gittes and C.F. Schmidt. Interference model for back-focal-plane displacement detection in optical tweezers. *Opt. Lett.*, 23:7–9, 1998.
63. M.W. Allersma, F. Gittes, M.J. deCastro, R.J. Stewart, and C.F. Schmidt. Two-dimensional tracking for ncd motility by back focal plane interferometry. *Biophys. J.*, 74:1074–1085, 1998.
64. L. Nugent-Glandorf and T.T. Perkins. Measuring 0.1 nm motion in 1 ms in an optical microscope with differential back-focal-plane detection. *Opt. Lett.*, 29:2611–2613, 2004.
65. W. Denk and W.W. Webb. Optical measurement of picometer displacements of transparent microscopic objects. *Appl. Opt.*, 29:2382–2391, 1990.
66. K. Svoboda, C.F. Schmidt, B.J. Schnapp, and S.M. Block. Direct observation of kinesin stepping by optical trapping interferometry. *Nature*, 365:721–727, 1993.
67. J.C. Crocker and D.G. Grier. Methods of digital video microscopy for colloidal studies. *J. Colloid Interface Sci.*, 179:298–310, 1996.
68. M.K. Cheezum, W.F. Walker, and W.H. Guilford. Quantitative comparison of algorithms for tracking single fluorescent particles. *Biophys. J.*, 81:2378–2388, 2001.
69. M. Capitanio, R. Cicchi, and F.S. Pavone. Position control and optical manipulation for nanotechnology applications. *Eur. Phys.* J. B, 46:1–8, 2005.
70. O. Otto, C. Gutsche, F. Kremer, and U.F. Keyser. Optical tweezers with 2.5 kHz bandwidth video detection for single-colloid electrophoresis. *Rev. Sci. Instrum.*, 79:023710, 2008.
71. L.P. Ghislain, N.A. Switz, and W.W. Webb. Measurement of small forces using an optical trap. *Rev. Sci. Instrum.*, 65:2762–2768, 1994.
72. I.M. Peters, B.G. de Grooth, J.M. Schins, C.G. Figdor, and J. Greve. Three dimensional single-particle tracking with nanometer resolution. *Rev. Sci. Instrum.*, 69:2762–2766, 1998.
73. A. Pralle, M. Prummer, E.-L. Florin, E.H.K. Stelzer, and J.K.H. Hörber. Three-dimensional high-resolution particle tracking for optical tweezers by forward scattered light. *Microsc. Res. Tech.*, 44:378–386, 1999.
74. A. Rohrbach and E.H.K. Stelzer. Three-dimensional position detection of optically trapped dielectric particles. *J. Appl. Phys.*, 91:5474–5488, 2002.
75. F. Gittes and C.F. Schmidt. Signals and noise in micromechanical measurements. *Methods in Cell Biol.*, 55:129–156, 1998.
76. A. Rohrbach and E.H.K. Stelzer. Trapping forces, force constants, and potential depths for dielectric spheres in the presence of spherical aberrations. *Appl. Opt.*, 41:2494–2507, 2002.
77. H. Risken. *The fokker-planck equation: Methods of solution and applications*. New York, Springer, 1996.
78. Y.K. Nahmias and D.J. Odde. Analysis of radiation forces in laser trapping and laser-guided direct writing applications. *IEEE J. Quantum Electron.*, 38:131–141, 2002.
79. A. Balijepalli, T.W. Lebrun, and S.K. Gupta. Stochastic simulations with graphics hardware: Characterization of accuracy and performance. *J. Comput. Inf. Sci. Eng.*, 10: 011010, 2010.
80. J.H. Ginsberg, *Advanced engineering dynamics*, 2nd edition, New York, NY, Cambridge University Press, 1995.
81. B.J. Kuo, *Automatic Control Systems*, 7th edition, Englewood Cliffs, NJ, Prentice-Hall, 1995.
82. K.J. Åström and T. Hägglund, *Advanced PID control*, Research Triangle Park, NC, ISA, 2005.

Chapter 7
Position Control of MEMS

Michael S.-C. Lu

7.1 Introduction

The earliest development of micromachined sensors and actuators, often referred as the field of MEMS [47], can be traced back to the 1960s, when Westinghouse Research Laboratories developed the resonant gate transistor [61]. Using technologies originally developed for the semiconductor industry, miniaturized transducers can be fabricated from silicon and other materials. Key devices for commercial applications include ink-jet heads [77], inertial sensors [13], pressure sensors [14], fingerprint sensors [76], microphones [68], and micro-mirrors for projection displays [79], among others.

Combining micromachined actuators and sensors to form feedback control systems brings together two unique features that give rise to interesting research problems and exciting applications. The integration of actuators, sensors, and control circuitry enables operation of a device array for either enhancing system performance or accomplishing a complicated task. One example is the MEMS-based AFM (Atomic Force Microscopy) system, which uses multiple closed-loop controlled cantilevers to increase scanning throughput of surface imaging [6]. Another example is a large-scale router for optical communication, which can be built by using micro-mirrors with controlled angles for deflecting light signals to the receiving ports [18]. The second feature of MEMS control is improved response time and control accuracy due to increased actuator bandwidths. Related applications include read/write-head positioning in magnetic hard disk drives for improving access time [21].

In many cases, it is advantageous to choose feedback control over open-loop operation even though the latter is cheaper and simpler to implement. The object to

M.S.-C. Lu (✉)
Department of Electrical Engineering, National Tsing Hua University, Hsinchu,
Taiwan, Republic of China
e-mail: sclu@ee.nthu.edu.tw

J.J. Gorman and B. Shapiro (eds.), *Feedback Control of MEMS to Atoms*,
DOI 10.1007/978-1-4419-5832-7_7, © Springer Science+Business Media, LLC 2012

be controlled in a feedback control system is called the plant. Feedback is desirable primarily for sensitivity reduction in the presence of plant uncertainties and external disturbances. For example, the spring constant of a microactuator varies around a nominal value due to fabrication tolerances, and consequently the situation can lead to positioning error. The error can be reduced by using an independent motion sensor and a large steady-state loop gain in a feedback system, which is assumed stable. Closed-loop control is also desirable for system stabilization when the plant is inherently unstable, such as the parallel-plate electrostatic actuator beyond the pull-in point [56] and the tunneling-based accelerometer [54].

Two types of control systems exist, and which is used depends on the specific control problems being addressed. Systems designed to maintain an output signal at a fixed set point while rejecting unknown disturbances are called regulators. This operation is often considered to be feedback linearization around an operating point, and it has been applied to extend the dynamic range of MEMS accelerometers [22]. In addition to regulating systems, there are tracking or servo systems where the output signal (e.g., position, velocity, acceleration) is designed to track a reference signal, such as positioning a read/write head in a magnetic hard disk drive.

Control design methods and stability analysis for linear time-invariant control systems based on conventional frequency-domain and time-domain approaches can be readily found in textbooks [15, 30]. For many MEMS actuators (e.g., electrostatic, thermal) and sensors that are inherently nonlinear, stability analysis requires advanced knowledge in nonlinear control systems [67, 74]. For control problems that deal with multiple inputs and multiple outputs (MIMO), such as three-axis motion control of a MEMS actuator, the theory of multivariable control systems can be applied [57, 73]. It should be noted that some MIMO systems can be decoupled into subsystems with a single input and single output (SISO) such that SISO design methods can still be applied. Moreover, methodologies of robust control [57, 72, 84] can be considered when designers need to achieve robust performance under plant uncertainties and external disturbances. The level of design difficulty increases when more attributes (e.g., MIMO, nonlinearity, and robustness) of a control system need to be considered [24, 27].

Linear time-invariant controller designs are usually preferred, on the premise that desired specifications are satisfied, for convenient implementation using either analog circuits or digital signal processors [63]. Nonlinear and time-varying controllers [3, 48] are more versatile for improving system performance, but the analysis of stability and performance is also more complex than for linear time-invariant design. In addition, the implementation of a nonlinear or time-varying controller usually requires a digital signal processor due to the complexity of the control laws.

Before resorting to a more sophisticated methodology, it is always helpful to gain some design insights by exploring a linear/linearized control system design using the classical frequency-domain approach. The block diagram of a conventional linear control system in Fig. 7.1a is used to explain the important design issues. A typical system has to deal with added external disturbances, sensor noise, and plant uncertainties while maintaining robust performance. In the classical loop-shaping

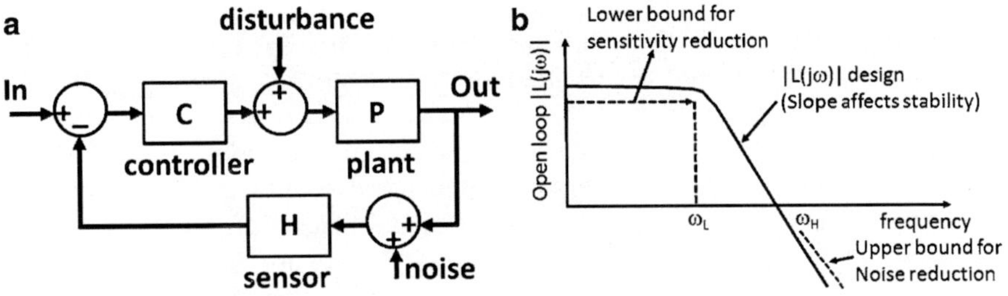

Fig. 7.1 (**a**) Block diagram of a conventional closed-loop control system. (**b**) Classical loop-shaping on the frequency domain

method shown in Fig. 7.1b, the open-loop transfer function, which is the product of all the functional blocks inside the loop, is desired to have a large loop gain $|L(j\omega)|$ at low frequencies for sensitivity reduction and a large steady-state gain for reducing the steady-state error. At high frequencies, it is desired that the loop gain roll off quickly to avoid noise amplification and excitation of unmodeled high-frequency poles. In mechatronic systems, these poles are usually the higher-order mechanical modes of the actuator, and their underdamped response leads to a significant reduction in phase margin when excited. The control bandwidth is usually designed around the actuator's primary mode of vibration. The loop gain should roll off properly at intermediate frequencies to produce a proper phase margin. For stable minimum-phase systems, Bode's gain-phase relationship [10] states that a large gain reduction within a frequency range is accompanied by a large phase reduction, which consequently leads to a reduced phase margin. Design trade-offs often have to be made when requirements on system stability, sensitivity reduction, noise attenuation, and output performance cannot be simultaneously met.

The output produced by sensor noise is desired to be small in order to enhance control resolution. The noise-induced output represented by the Laplace transform is given by

$$Y_N(s) = L(s)/[1 + L(s)] \cdot N(s), \tag{7.1}$$

which indicates that the output is close to the sensor noise at the low and intermediate frequencies, where the open-loop gain is large. Designers thus should avoid using a larger loop bandwidth than necessary in order to minimize induced noises. At high frequencies the loop gain becomes small, thus driving down the noise-induced output. Both thermomechanical and electronic noise exist in micromechanical sensors. False signals produced by the thermomechanical noise [26] are due to air-particle impingement on mechanical structures above absolute zero. Discussion of electronic noise can be found in textbooks on analog integrated circuit design ([32, 43]).

The selection of the MEMS fabrication platform is an important issue in the control of a device array. As the array size grows significantly, the number of input/output connections becomes unacceptable, with MEMS devices and sense/control circuits being placed on separate chips. For I/O reduction, an integrated circuit (IC) process, such as the complementary metal oxide semiconductor (CMOS) process, can be considered as the platform for monolithic integration. MEMS processing can take place before, after, or in between the normal CMOS fabrication steps [5]. Either analog or digital circuitry can be used for implementing control laws. Designers have to be aware of the chip area occupied by circuits, especially the passive elements for implementing low-frequency poles or zeros.

The rest of this chapter is organized as follows. Control of electrostatic microactuators is presented in the next section, with the focus on stabilizing parallel-plate-type microactuators beyond the pull-in instability point. In Sect. 7.3, the sigma-delta modulator for discrete-time feedback linearization of inertial sensors is presented, followed by the control of ultrasensitive tunneling-based accelerometer. Control of an array of thermally driven MEMS actuators for fast surface scanning is presented in Sect. 7.4. Section 7.5 presents MEMS control for data storage applications, including the media actuator control for a MEMS-based data storage device and the dual-stage control scheme for conventional magnetic hard disk drives. Finally, discussion and conclusions are given in Sect. 7.6.

7.2 Control of Electrostatic Microactuators

Electrostatic actuation is widely used in MEMS because of its low power consumption and ease of fabrication. Electrostatic microactuators have been fabricated for a variety of applications, such as torsional micro-mirrors for optical projection displays [79], tunable capacitors for wireless communication [85], and read/write head actuators for hard disk drives [21, 41]. The parallel-plate actuator (PPA), as shown in Fig. 7.2a, is capable of producing a larger force than the electrostatic comb drive ([75], Fig. 7.2b), since the actuator displacement is in parallel with respect to the major electric field lines. The drawback of a PPA is that the maximum displacement by open-loop operation is limited to one-third of the initial gap, the so-called pull-in limit [62].

Comb-electrode actuators provide another type of electrostatic actuation based on fringing electric fields. It usually requires a high drive voltage but is a stable plant

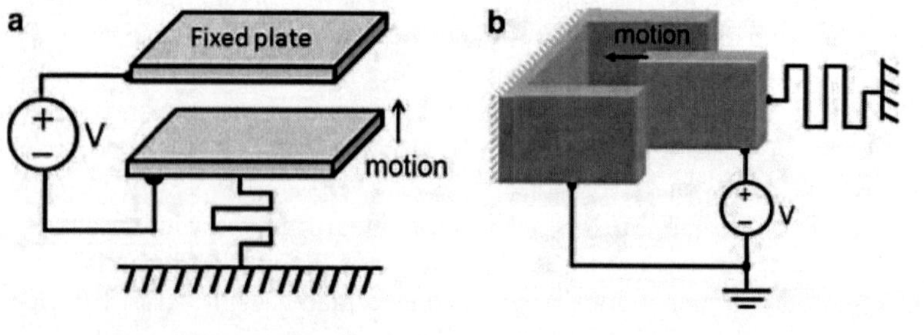

Fig. 7.2 (**a**) Schematic of the parallel-plate actuator. (**b**) Schematic of the comb-electrode actuator

capable of producing a large displacement without pull-in. A comb-drive actuator is a nonlinear plant with the produced force proportional to the applied voltage squared. Reference [17] adopted a state-variable feedback approach [15], in which a Kalman filter was used to estimate position and velocity to control a comb-drive actuator. Note that the pull-in phenomenon can appear in a comb-drive actuator when the structure is not completely symmetric due to manufacturing imperfections. Destabilization caused by pull-in can be prevented by applying feedback control based on the sensed signals [12, 42].

Extending the travel range of a PPA beyond the pull-in point is a challenging MEMS control problem. The PPA is a nonlinear plant and becomes unstable beyond the pull-in. Both the voltage-controlled and the charge-controlled approaches have been demonstrated for extending the PPA travel range. In terms of implementation, it is more straightforward to get the desired electrostatic force by adjusting the voltage rather than controlling the accumulated charges between parallel plates, since it is not easy to estimate the amount of charges of a MEMS capacitor due to the associated parasitic capacitances and leakage paths. However, in terms of expanding the stable range of operation, the theoretical pull-in range can be extended in charge control when certain conditions are met. Examples from both approaches will be presented.

7.2.1 Modeling of a Parallel-Plate Actuator

The conventional PPA in Fig. 7.2a is represented by a lumped mass, spring, and damper. Squeeze-film damping arises from the vertical plate motion, which creates a pressure in the thin film of air between the plates [9, 80]. The electrostatic force derived by the energy method [81] is given by

$$F_\mathrm{e} = \frac{\varepsilon_0 A}{2(g-z)^2} V^2, \tag{7.2}$$

where ε_o is the permittivity of air, A is the plate area, V is the applied voltage, g is the gap between plates, and z is the plate displacement. The force is inherently nonlinear, with its magnitude proportional to the driving voltage squared, and inversely proportional to the instantaneous gap squared. The dynamic response of a PPA can be approximated by

$$m\ddot{z} + b\dot{z} + kz = \frac{\varepsilon_0 A}{2\,(g-z)^2} V^2, \tag{7.3}$$

where m is the mass, b is the damping coefficient, and k is the spring constant. By linearizing around an operating point (Z_o, V_o), the above equation becomes

$$m\Delta\ddot{z} + b\Delta\dot{z} + (k + k_{e,v})\Delta z = \frac{2kZ_\mathrm{o}}{V_\mathrm{o}} \Delta v, \tag{7.4}$$

where $k_{e,v}$ is a negative spring constant induced by the electrostatic force gradient $\partial F_e/\partial z$. Its value is $-2\alpha/(1-\alpha)\cdot k$, where α is the normalized displacement, Z_o/g. Therefore, the electrical spring constant completely negates the mechanical spring constant when Z_o is one-third of the gap. The static pull-in voltage is

$$V_{pi} = \sqrt{8kg^3/(27\varepsilon_0 A)}, \tag{7.5}$$

which is useful for estimating the maximum driving voltage of a PPA design. Based on the linearized model in (7.4), one unstable pole is produced when a PPA travels beyond the pull-in limit and the pole magnitude varies with respect to the displacement. The controller design has to ensure a stable closed-loop system with a satisfactory time-domain response.

7.2.2 The Voltage-Control Approach

The linearized PPA model has an unstable pole after the pull-in point. From the classical control standpoint, the unstable pole affects the shaping of the sensitivity function $S(s) = 1/(1+L(s))$ and the complementary sensitivity function $T(s) = L(s)/(1+L(s))$. The unstable pole needs to be pulled into the left half s-plane for stabilization by extra loop gain, which can otherwise be used for sensitivity reduction; in other words, the unstable pole causes the reduction of feedback benefit. In addition, in order to prevent peaking in $|T(j\omega)|$ leading to a reduced phase margin, the loop gain should roll off slowly such that the crossover frequency increases. The crossover frequency is suggested to be at least twice the unstable pole frequency [25]. The increased bandwidth for controlling an unstable plant leads to increased power consumption and noise amplification.

Figure 7.3a shows a released PPA fabricated in a conventional CMOS process for read/write head positioning in a MEMS-based data storage device [56]. The PPA has two capacitive sensing plates and an actuation plate. An external electrode (the storage media) is placed on top of the actuator to form the initial gap and the actuated and sensing capacitances as shown in Fig. 7.3b. The capacitively sensed signal is amplified, demodulated, and low-pass filtered, followed by subtraction of the sensor's d.c. offset before being processed by the controller, whose output is applied to the actuation plate. The control system diagram depicted in Fig. 7.3c has two functional blocks to be designed [38]: the controller $C(s)$ inside the loop is intended to achieve robust stability in the presence of the unstable pole, plant uncertainty, and disturbances; the pre-filter $F(s)$ shapes the input command for the feedback loop to track and follow. Selection of this two-degree-of-freedom control system is crucial, since it decouples the design processes for system stability and output tracking. A satisfactory response cannot be obtained simply by increasing the phase margin of the feedback loop, especially when the plant is unstable. The plant consists of

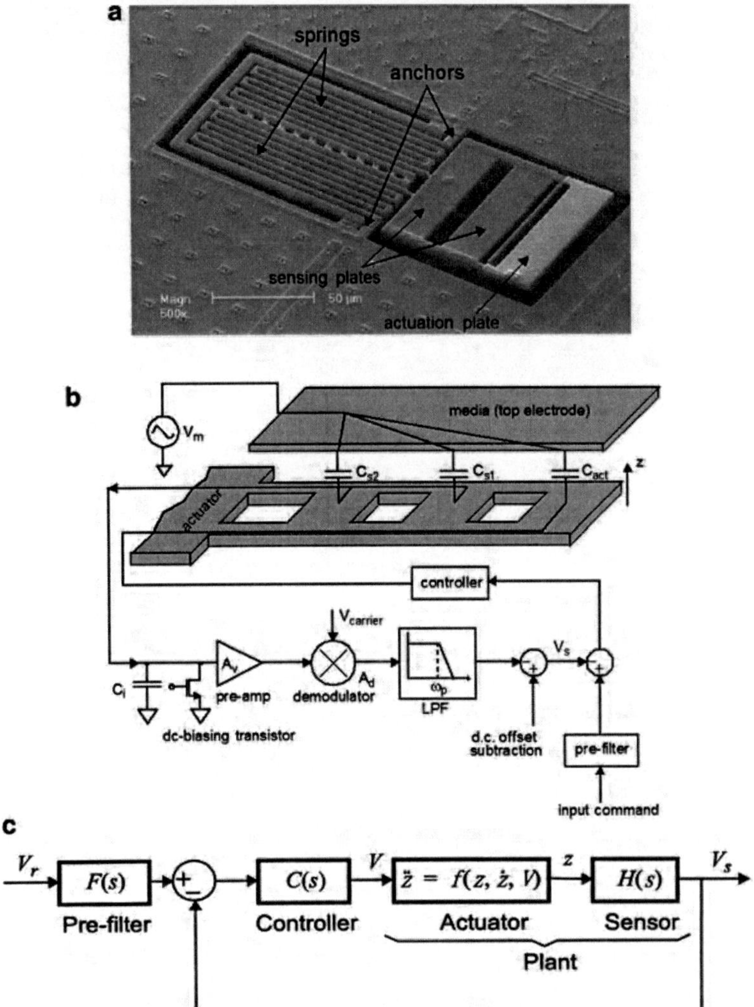

Fig. 7.3 (**a**) Micrograph of the released parallel-plate microactuator. (**b**) Schematic representation of feedback control system around the parallel-plate microactuator. (**c**) Block diagram of the control system (Lu and Fedder [56] © IEEE)

the PPA represented by a nonlinear differential equation and the capacitive position sensor represented by $H(s)$. The sensor at the displacement $z = Z_o$ is represented by a gain $G(Z_o)$ and a pole at $\omega = \omega_p$.

The PPA has a resonant frequency at 11.9 kHz. Controller design is realized through linearization of the nonlinear plant. The pre-filter shapes the input command such that the actuator moves quasistatically along the motion trajectory with a rise time less than 3 ms. Stability is analyzed using the gain and phase margins in the frequency domain for a proportional-gain controller ($C(s) = k_p$). The effects of initial conditions and higher-order terms of electrostatic force omitted during linearization are formulated as a disturbance-rejection problem [55].

The open-loop transfer function at the operating point (Z_o, V_o) in the unstable regime ($Z_o/g \geq 1/3$) is expressed as

$$L(s) = \frac{2k_p\omega_n^2 G(Z_o)Z_o}{V_o}\frac{1}{(1+s/\omega_p)(s^2+2\xi_e\omega_e s-\omega_e^2)}, \tag{7.6}$$

where

$$\omega_e = \sqrt{\frac{3\alpha-1}{1-\alpha}}\omega_n,\ \xi_e = \sqrt{\frac{1-\alpha}{3\alpha-1}}\xi, \tag{7.7}$$

ω_n and ξ are the natural frequency and damping ratio of the PPA, and α is Z_o/g. Following (7.6), the phase margin ϕ_m and the crossover frequency $\omega = \omega_\phi$ are related by

$$\phi_m = \tan^{-1}\left(\frac{\dfrac{2\xi_e\omega_e\omega_\phi}{\omega_\phi^2+\omega_e^2} - \dfrac{\omega_\phi}{\omega_p}}{1+\dfrac{2\xi_e\omega_e\omega_\phi}{\omega_\phi^2+\omega_e^2}\cdot\dfrac{\omega_\phi}{\omega_p}}\right). \tag{7.8}$$

Two gain margins are produced on the Bode plots as the phase $\angle L(j\omega)$ for $Z_o/g \geq 1/3$ increases from $-180°$ at $\omega = 0$ and then decreases to $-270°$ at $\omega = \infty$. The key for design is to find the maximum phase margins at all operating points by differentiating $\angle L(j\omega)$ with respect to ω at $\omega = \omega_\phi$. The resultant crossover frequency with maximum phase margin is the solution of the following quadratic equation:

$$(\omega_p + 2\xi_e\omega_e)\,\omega_\phi^4 + (2\xi_e\omega_p^2\omega_e + 2\omega_p\omega_e^2 + 4\xi_e^2\omega_p\omega_e^2 - 2\xi_e\omega_e^3)\cdot\omega_\phi^2$$
$$-(2\xi_e\omega_p^2\omega_e^3 - \omega_p\omega_e^4) = 0. \tag{7.9}$$

The crossover frequencies are solved based on the values of ω_e, and ξ_e at different displacements. Then the associated maximum phase margin, gain margins, and controller k_p can be obtained accordingly. The phase margin and controller gain with respect to Z_o/g are plotted in Fig. 7.4a, b. A linear time-varying controller is required to preserve the maximum phase margin at different displacements. A linear time-invariant controller, $k_p = 16$, is implemented with satisfactory phase margins. Figure 7.4c shows the controller output waveforms corresponding to a series of increasing input commands. The driving voltage decreases after the pull-in voltage (10.4 V) to reduce the stored charge on the actuated plate for stabilization beyond the pull-in. Figure 7.4d shows the steady-state plate displacements, as calculated based on the measured sensor outputs, with respect to input command values. The maximum displacement achieved by the design is 2 µm, equivalent to 60% of the gap of 3.3 µm. The failure of the control loop beyond 60% of the gap is mainly

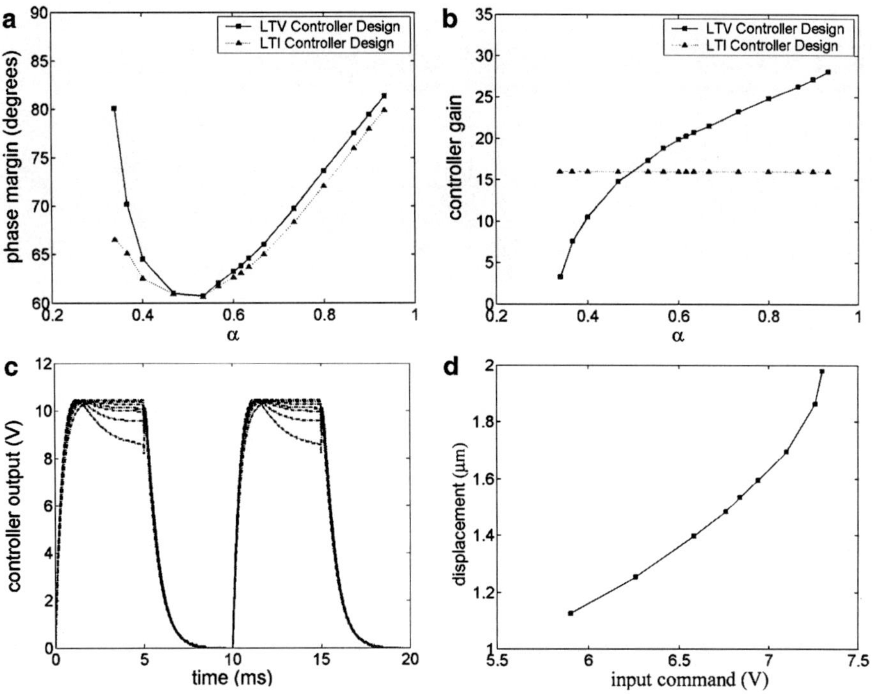

Fig. 7.4 (**a, b**) Phase margin and the corresponding controller gain with respect to the normalized position (**c**) Controller output waveforms for different input commands. (**d**) Maximum plate displacements with respect to input commend values (Lu and Fedder [56] © IEEE)

attributed to the loss of phase margin as a result of the reduced squeeze-film damping coefficient. Reduced damping is caused mainly by the plate tilt. In addition, the actuator dynamics have a lower damping coefficient when moving away from rather than close to the opposing electrode. Integral control can be included to eliminate the steady-state error found with the simple proportional-gain controller. A very low-frequency zero has to be included in the controller to compensate for the 90° phase lag of an integrator.

From the controller design standpoint, use of a linear time-invariant controller facilitates the implementation but is less versatile than a nonlinear or adaptive controller. Figure 7.5a shows a PPA fabricated on a silicon-on-insulator (SOI) wafer with dedicated comb electrodes for sensing and actuation [66]. The capacitive sensing interface is implemented by a commercial chip. The control law as shown in Fig. 7.5b is implemented on a computer and consists of three parts: (1) a nonlinear inversion model to cancel the actuator nonlinearities; (2) a linear controller with an integrator and two zeros to cancel the underdamped poles of the actuator for enhancing closed-loop damping and a fast pole to make the controller proper and attenuate noise; and (3) an on-line adaptive estimator for determining the gap distance, which varies due to fabrication imperfections and as a result, affects the

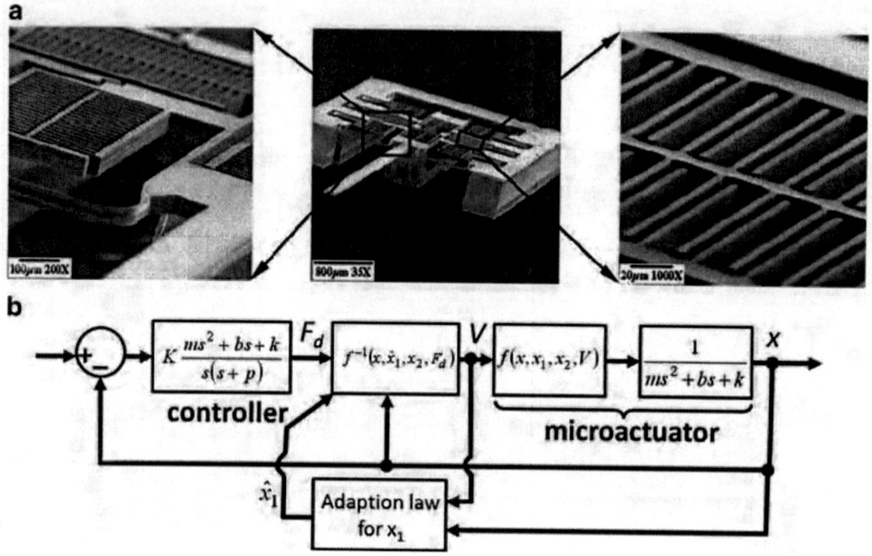

Fig. 7.5 (**a**) Micrograph of the PPA fabricated on an SOI wafer. (**b**) Block diagram of closed-loop control (Piyabongkarn et al. [66] © IEEE)

servo performance. The extended travel range is up to 80% of the gap; however, the tracking bandwidth is relatively low at 20 Hz, as limited by sensor noise. This phenomenon is expected, because the capacitive sensing circuit is not integrated, so that electronic noise is amplified owing to the large parasitic capacitance seen at the sensing nodes.

In addition to the position servo of a PPA in the translational directions, the optical communication industry is interested in angular control of electrostatic micro-mirrors in order to make large-scale optical switches. The light-deflecting mirror has to be light and rigid for quick response and vibration immunity. A flat and highly reflective surface is required for low optical loss. Many devices are made of either single-crystal silicon or polysilicon with a thin reflective gold coating. The steady-state angular noise must be low to maintain the transmitted optical power stability. The mirror can use a gimbaled design to nearly decouple the angular motions in the x and y axes. Large motions are desired to accommodate numerous receiving ports in a large-scale optical switch. Curvature induced by thermal mismatch of materials affects the optical power stability under different temperatures. This type of plant uncertainty cannot be distinguished and dealt with by feedback for the following systems that use external optical sensors. A conventional design usually has four electrodes in the four quadrants beneath a mirror, making it a four-input–two-output multivariable and nonlinear control problem.

On-chip angle detection for mirror control remains a challenging issue. Capacitive sensing is more favorable than the piezoresistive and piezoelectric mechanisms, since it requires no additional film deposition and is nearly temperature insensitive.

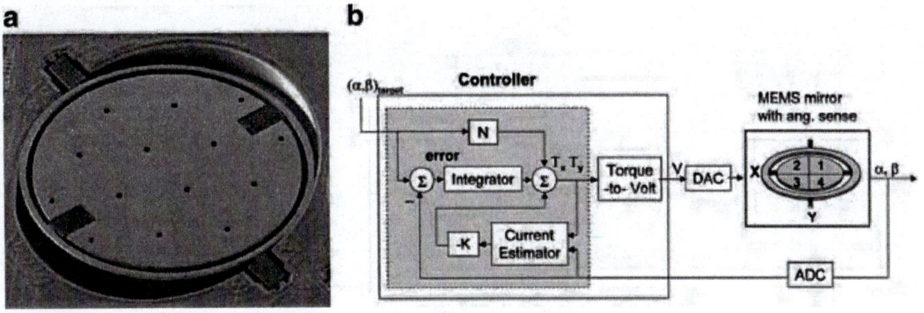

Fig. 7.6 (**a**) Micrograph of the gimbaled two-axis micro-mirror fabricated on an SOI wafer. (**b**) Control block diagram (Chu et al. [18] © IEEE)

However, the implementation is rather difficult because the sensing and actuating electrodes have to share a limited area, leading to a large driving voltage and a degraded sensitivity. In addition, it is quite complicated to resolve the exact angles based on a limited number of sensing electrodes. An external position-sensitive detector (PSD) is usually used instead. After signal conditioning and A/D conversion, a variety of control algorithms can be implemented in a digital signal processor. For example, nonlinearities of the parallel-plate force can be partially canceled by adding an inverse nonlinear function. In the trajectory-planning sense, a digital signal processor can store a lookup table containing the detected angles and the associated voltages as the control algorithm. Robust stability and performance analysis can become complicated, since heuristic rules and intelligence are included in the algorithm.

Figure 7.6a shows a gimbaled two-axis micro-mirror fabricated on an SOI wafer [18]. The digital control system is shown in Fig. 7.6b. The applied torque is calculated by a state-variable controller, with the angular position and velocity calculated by a state estimator based on the sensed signal from an external PSD. The conversion of torques to the applied voltages is completed through a lookup table, in which at most two driving voltages are nonzero in the simplified actuation scheme. The voltage values beyond the snapdown angle are obtained by numerical simulation in the static sense, and revised after the dynamic servo test. An integrator is included for removing the steady-state error. A feedforward term is added from outside the feedback loop to compensate for the low-frequency pole introduced by the integrator. A feed-forward controller [31] is similar to the aforementioned pre-filter in providing an extra degree of freedom to cancel the undesired dynamics in the feedback loop. The servo achieves up to 80% of the angular range in a few milliseconds. For reducing the loss of phase margin due to the unstable plant and signal delay, the sampling frequency of the digital controller is almost two orders of magnitude higher than the closed-loop bandwidth.

Knowing that the voltage across the plates must decrease as the angle increases, the controller can be designed in a static trajectory-planning sense according to the measured angle. As shown in Fig. 7.7a, the controller can first supply a constant

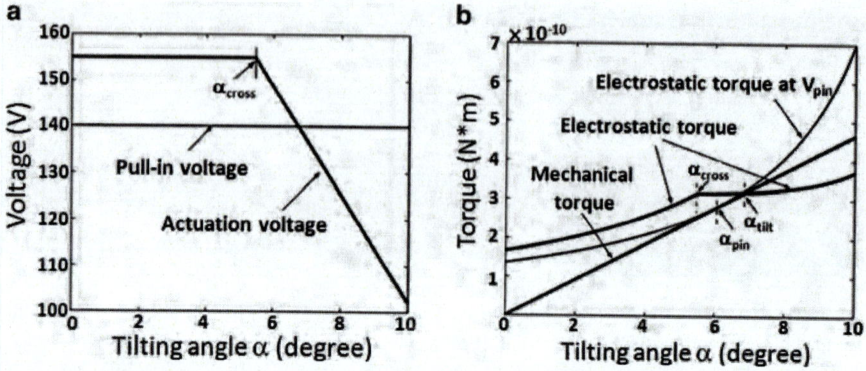

Fig. 7.7 Feedback control algorithm: (a) Applied voltage versus tilting angle. (b) Applied torque versus tilting angle (Chen et al. 2004 © IEEE)

Fig. 7.8 Phase portrait of sliding-mode control containing two states

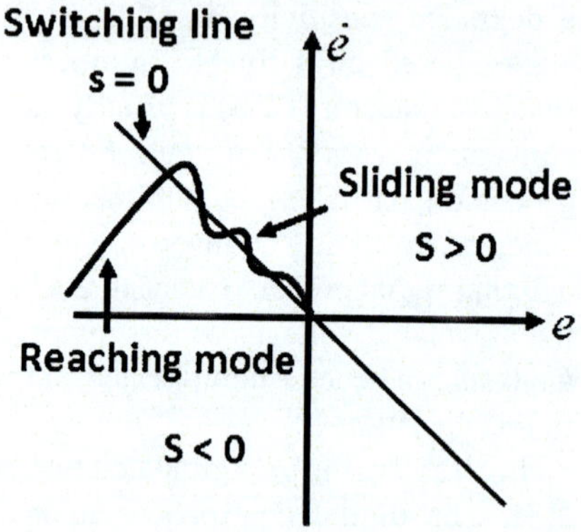

voltage slightly larger than the pull-in value at small angles, and then linearly reduce the voltage when approaching the desired angle [16]. This control law leads to one intersection angle between the mechanical and electrostatic torque curves in Fig. 7.7b. This angle would be a stable equilibrium, since the electrostatic torque is larger than the mechanical torque for angles below the intersection, and is smaller for angles beyond. This static approach does not consider the closed-loop dynamics involving the moment of inertia and damping coefficient, and hence a large overshoot can occur. This nonlinear control law provides the advantage of convenient implementation by analog circuits, which enhances scalability in building large-scale optical switches.

Sliding-mode control [78] is another nonlinear control method that can be used for pull-in stabilization of a micro-mirror [83]. The basic concept can be explained using the phase portrait in Fig. 7.8, where the error and its derivative are the two states for state feedback and a switching line forms the control algorithm. The method has two operating modes: the reaching mode and the sliding mode. The control law is altered whenever the states cross the switching line.

Hence, sliding-mode control is a form of variable structure control (VSC). The motion in the neighborhood of the switching line is guided repeatedly toward and along the line until the states arrives at the origin, where the steady-state error becomes zero. The switching action implies that the driving voltage of the PPA is continuously switched between two values, with the average force being determined by the switching frequency. The equation of the switching line is used to construct the Lyapunov function for stability analysis. A switching surface is needed when there are more than two states. The method can provide control robustness because the simple switching actions, which need not be precise, are less sensitive to parameter variations that enter into the control channel. The method can achieve a shortened settling time and a small overshoot for highly underdamped cases. Chattering around the steady state is a known issue resulting from the switching action. The chattering can produce wide-bandwidth noise, and the sensor noise is also amplified by instantaneous high gain during switching. The amount of steady-state error is related to the hardware resolution of analog-to-digital (A/D) and digital-to-analog (D/A) converters and the induced noise. This work introduces an integrator, which is essentially a low-pass filter, to partially alleviate the noise issue and improve the steady-state error.

7.2.3 *The Charge-Control Approach*

The pull-in range can be extended in charge control when certain conditions are met. Assume that a PPA has an initial capacitance C_0 with a constant capacitance C_{p0} connected in parallel. Then the electrical spring constant for charge control is expressed as [71]

$$k_{e,q} = -\frac{2C_{p0}\alpha}{C_0 + C_{p0}(1-\alpha)} \cdot k. \tag{7.10}$$

For $C_{p0} < C_0/2$, charge pull-in will not occur, because the negative $k_{e,q}$ value does not completely negate the mechanical spring constant k. When C_{p0} is significantly larger than C_0, this expression reduces to the form of the voltage-controlled electric spring, and charge pull-in occurs at one-third of the gap.

The switch-capacitor (SC) circuit is a natural choice for implementing charge control, for it is known to be capable of transferring a fixed amount of charge to a capacitor. Figure 7.9 shows the SC control circuit design, where the charge $q = C_s(V_s - V_{icm})$ is first sampled onto the input capacitor C_s, and then transferred to the actuated capacitor C at phase ϕ_2 [71]. The alternating reset and charge-transferring processes mean that the actuator motion depends on an average force, and the amplifier needs to settle quickly and accurately during charge transfer to reduce the switching effect. Issues such as charge leakage and injection through switches should be considered as well.

Design of the SC circuit with a PPA in the feedback path is similar to designing other SC circuits. The main difference is that there are additional dynamics in the

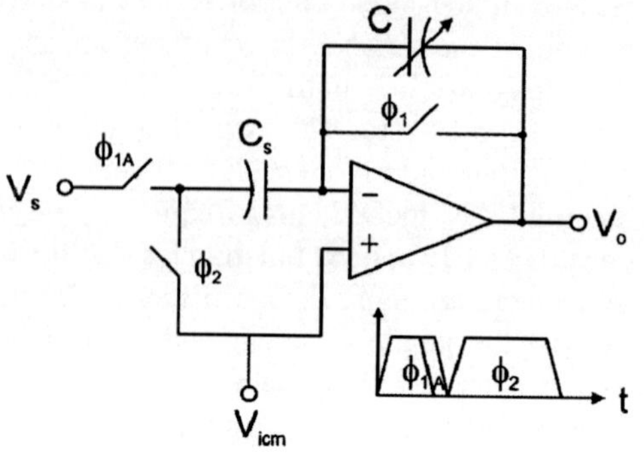

Fig. 7.9 The switched-capacitor circuit for charge control of a PPA represented by the capacitor C (Seeger and Boser [71] © IEEE)

feedback factor, which is expressed by the actuated capacitance $C(s)$ and the other fixed capacitances seen at the negative node of the op-amp. The closed-loop stability is determined from the open-loop gain, which is the product of the op-amp's open-loop gain and the feedback factor. For stability analysis, $C(s)$ is expressed in the small-signal sense by

$$C(s) = \frac{\delta q}{\delta v} = \frac{ms^2 + bs + k + k_{e,q}}{ms^2 + bs + k + k_{e,v}}. \tag{7.11}$$

It has one pole in the right half-plane for deflections greater than voltage pull-in, and one right-half-plane zero for deflections greater than charge pull-in. The Nyquist criterion can be used for stability analysis of this non-minimum phase system.

The effect of fixed capacitances seen at the negative node of the op-amp, including C_S, the capacitance from plate to ground, and the op-amp's input capacitance, are suppressed with a large op-amp gain such that the charge pull-in point depends only on the ratio C_{p0}/C_0. However, these capacitances still reduce the closed-loop bandwidth during charge transfer and thus increase the settling time. Determination of the op-amp's gain-bandwidth is also crucial for stabilizing PPAs with different quality factors. The required gain-bandwidth is proportional to the product of quality factor and natural frequency of the actuator. Lead compensation can be added to alleviate the phase loss due to insufficient bandwidth. This charge-control scheme extends the travel range to 83% of the gap for mirror structures. The failure is due to tip-in instability, not charge pull-in, since charges redistribute on the surface when a mirror is tilted. This phenomenon is both unobservable and uncontrollable, because there is no sensor to detect the charge redistribution. Since no angular position sensor is used, this charge control scheme is operated in the open-loop sense but implemented under the feedback configuration of an SC circuit.

Another charge-control approach uses a current source and a current sink to put a fixed amount of charge on a PPA. Since the capacitor value of a MEMS is usually small, on the order of pF or less, the implementation thus requires low-leakage transistors to maintain a fixed amount of charge. A good current charging/discharging source is required to provide a high-output resistance and a faster time constant than that of the actuator. It should also ensure a constant current

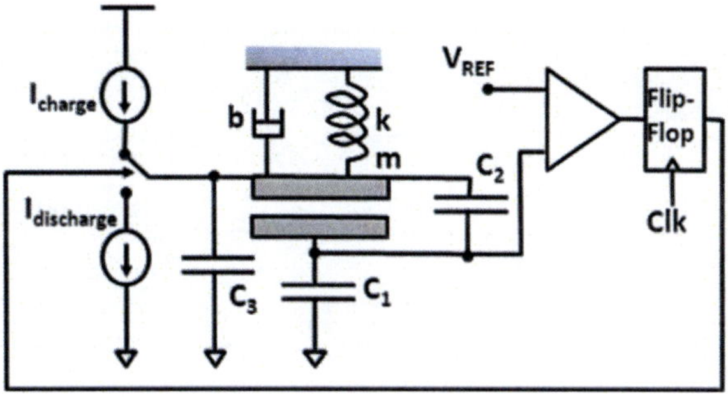

Fig. 7.10 Closed-loop charge control using a current source and a current sink. A series capacitor is used for detection of charges on the actuator (Nadal-Guardia et al. [60] © IEEE)

under a high-output voltage swing on the actuated capacitor. Figure 7.10 shows the closed-loop configuration that uses a charging source and a discharging source to add and remove charge on the actuator [60]. A sensing capacitor in series detects the voltage drop on the actuator. The associated sensed voltage is compared with a reference voltage, and either the source or sink is activated to maintain a fixed amount of charge. Ringing around the final position and position drift due to the leakage current are the major issues with this approach.

7.3 MEMS Control for Inertia Sensors

MEMS inertial sensors have been widely used in automotive, industrial, and consumer applications such as airbag deployment, vibration detection [29], and motion-controlled video games. Many commercial inertial sensors are based on capacitive detection due to its high sensitivity and low-temperature dependency. Since the impedance of a MEMS capacitor is high, monolithic integration in a CMOS process is advantageous in minimizing parasitic capacitance and enhancing the signal-to-noise ratio.

To enhance sensitivity, the sensing capacitors are usually operated based on the parallel-plate motion, which produces a nonlinear capacitive sensitivity with respect to the electrode displacement. A closed-loop accelerometer performs feedback linearization to extend the operating range and sensitivity reduction with respect to parameter variations. As shown in Fig. 7.11, in the presence of an input force, the proof mass moves causing a sense signal, and then a driving voltage is produced by the controller and converted to an electrostatic force that pulls the proof mass in the direction opposite to its original deflection. The Brownian motion due to particle impingement in air is reduced by feedback as a disturbance-rejection problem. Feedback does not reduce the electronic noise in the sensing interface, which in turn limits the minimum detectable signal. In situations in which low-pressure operation is required to reduce thermomechanical (Brownian) noise, closed-loop control can increase the system damping to yield a reasonable settling time. Note that force balancing is not desirable for low-cost and low-power applications in which the

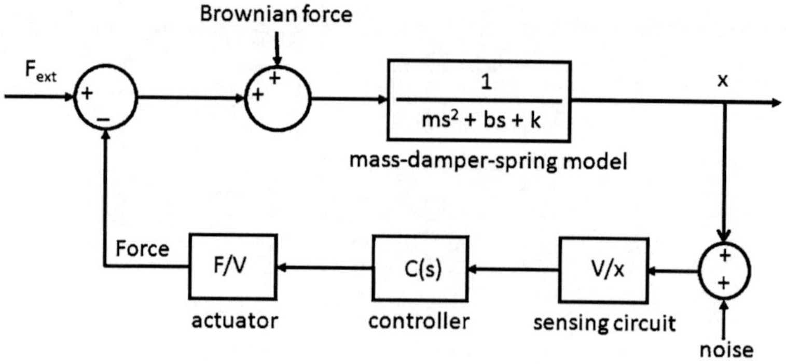

Fig. 7.11 Closed-loop feedback configuration for a capacitive MEMS accelerometer

added complexity, cost, and power consumption are of concern. The tunneling-based accelerometer presented in Sect. 7.3.2 is a special case that needs feedback stabilization because it is inherently open-loop unstable.

7.3.1 Sigma-Delta Control for Capacitive Accelerometers

Both continuous- and discrete-time controllers can be considered for feedback implementation. The latter commonly adopt sigma-delta ($\Sigma\Delta$) modulation, which is a well-established technique for A/D signal conversion, to perform feedback linearization and provide a high-resolution digital output of the acceleration signal. From the perspective of A/D conversion, a $\Sigma\Delta$ modulator operates on the basis of oversampling such that it has more relaxed requirements on the analog components than does a Nyquist-rate converter [43]. With advances in digital IC technology, $\Sigma\Delta$ modulation has become popular for low-frequency applications such as speech processing, where a high oversampling ratio can efficiently reduce quantization noise and thus increase bit resolution. The method is also suitable for accelerometers since the associated signal bandwidth is only in the kilohertz range.

The conventional $\Sigma\Delta$ modulator in Fig. 7.12a has a feedback configuration containing a digital loop filter $H(z)$ along with A/D and D/A blocks. For analysis, the A/D and D/A are replaced by a unity gain and added quantization noise in Fig. 7.12b. The in-band noise decreases as the filter's order increases as shown in Fig. 7.12c, leading to an increased number of resolution bits. The reduction of quantization noise is the same as in the disturbance-rejection problem that requires a high loop gain at low frequencies; hence integrators are often selected as the loop filter $H(z)$. Note that closed-loop stability is affected as the system order increases. Since an accelerometer is intrinsically a second-order micromechanical filter, it can assume the role of a loop filter to form a second-order micromechanical $\Sigma\Delta$ modulator. The A/D and D/A blocks in a $\Sigma\Delta$ modulator do not necessarily require

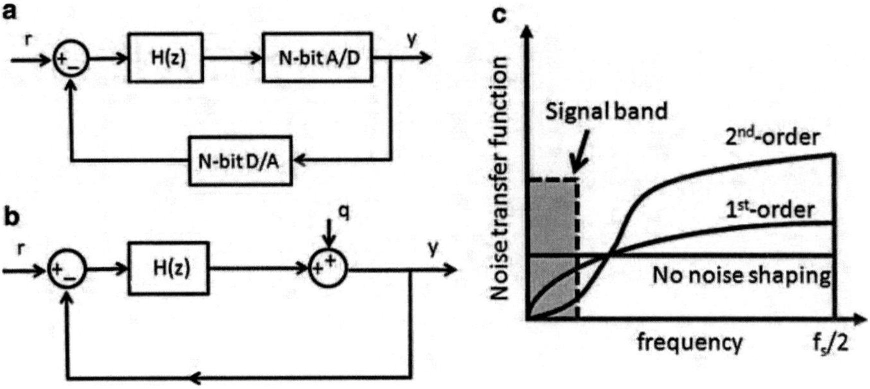

Fig. 7.12 (**a**) Block diagram of a ∑Δ modulator. (**b**) Its linear model by discrete z transformation. (**c**) Noise shaping by modulators of different orders

multibit resolution to improve the quantization noise; instead, a one-bit quantizer, which is relatively easy to design, is often used in ∑Δ accelerometers.

The quantization noise is modeled as a white noise that has a mean-square value given by

$$q_{\text{rms}}^2 = \Delta^2/12, \tag{7.12}$$

where Δ is the step size of the quantizer. The noise transfer function (NTF) represented by z transform is the ratio of the noise-induced output over the quantization noise given by:

$$\text{NTF}(z) = \frac{n(z)}{q(z)} = \frac{1}{1+H(z)}, \tag{7.13}$$

where $z = e^{j2\pi fT}(T = 1/f_s$. f_s is the sampling frequency). The Mth-order ∑Δ modulator with M integrators results in,

$$\text{NTF}(z) = (1 - z^{-1})^M, \tag{7.14}$$

which has the characteristic of a high-pass filter for suppressing in-band noise power spectral density as the filter order increases. The closed-loop system of a ∑Δ modulator will be followed by low-pass and decimation stages to remove the enhanced out-of-band noise observed in Fig. 7.12c.

The oversampling ratio in a ∑Δ modulator is defined as

$$\text{OSR} = \frac{f_s}{2f_0}, \tag{7.15}$$

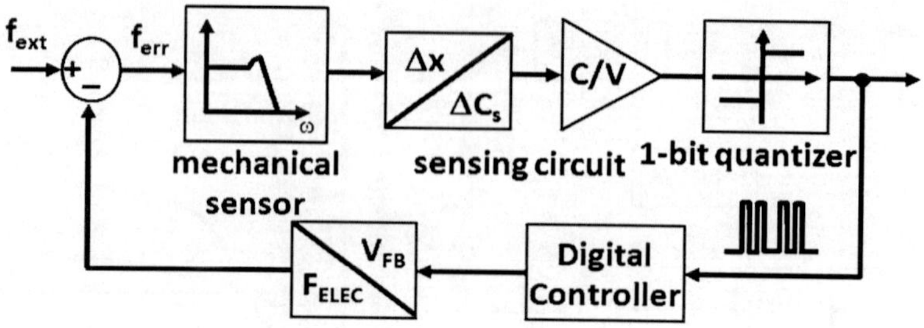

Fig. 7.13 Block diagram of a $\sum\Delta$ capacitive inertial sensor

where f_0 is the signal bandwidth. If one integrator ($M = 1$) and one-bit quantization are used, the in-band noise power is given by [69]

$$n_{\mathrm{rms}}^2 = \frac{\pi^2 q_{\mathrm{rms}}^2}{3(\mathrm{OSR})^3}. \tag{7.16}$$

The noise is reduced by 9 dB, or 1.5 bits are obtained as the oversampling ratio doubles. For $M = 2$, the in-band noise power is

$$n_{\mathrm{rms}}^2 = \frac{\pi^4 q_{\mathrm{rms}}^2}{5(OSR)^5}. \tag{7.17}$$

Doubling the oversampling ratio gives about 2.5 extra bits, better than the first-order modulator.

The benefits of $\sum\Delta$ control extend to the micromechanical system, providing superior sensing linearity and large dynamic range using simple electronics. The loop filter in the $\sum\Delta$ loop is replaced by the second-order accelerometer to shape the high-frequency quantization noise. Capacitive sensing and electrostatic force-balancing are mostly adopted in the literature [2,22,49,51,65]. As shown in Fig. 7.13, the proof mass displacement is converted by a capacitive sensing circuit and further amplified before being quantized by a fast-sampling comparator. The electrostatic actuator converts the bitstream from the comparator into force vectors to counterbalance inertial forces with amplitudes determined by the pulse density. Information on the input acceleration is obtained by digital filtering of the pulse stream afterward. A periodic motion of the proof mass is produced under zero acceleration. Increasing the sampling frequency effectively reduces this movement, since it is inversely proportional to f_s^2. By maintaining a small deflection, nonlinearities in the capacitive interface and mechanical springs are minimized. Raising the feedback pulse value extends the step size of the quantizer and the full-scale range.

It is desired that the added quantization noise in the $\sum\Delta$ modulator be negligible compared to the electronic noise of the sensing interface, which typically sets the resolution for open-loop operation. Theoretically, it is possible to reduce quantization noise by choosing a sufficiently large oversampling ratio. Sub-μg resolution with negligible quantization noise has been achieved for a bulk-micromachined second-order $\sum\Delta$ accelerometer [49]. The effectiveness is doubtful for the less-sensitive surface-micromachined devices. The reason is that the sensed signal has to be amplified by a large gain, which consequently increases the overall noise variance. As a result, the effective quantizer gain decreases, leading to increased quantization noise at the modulator output. A fourth-order $\sum\Delta$ interface uses additional filtering between the sensing front end and the quantizer to reject the out-of-band noise so that the effective quantizer gain can remain high [65].

A $\sum\Delta$ modulator with A/D and D/A elements is inherently nonlinear. Stability analysis is typically based on the linear model in Fig. 7.12b whereby the root-locus, Bode, and Nyquist techniques can be applied. The second-order $\sum\Delta$ can be applied to overdamped mechanical sensors with ensured stability. Lead compensation may be needed for underdamped devices. Note that the output amplitude of the quantizer is fixed such that effective quantizer gain varies with respect to the input signal during stability analysis.

7.3.2 Control of Tunneling Accelerometers

Tunneling-based accelerometers are known to be highly sensitive, with demonstrated displacement resolution approaching $10^{-4}\,\text{Å}/\text{Hz}^{1/2}$ [44]. Electron tunneling is observed when the gap between a conductive sharp tip and an opposing metal electrode is on the order of a few angstroms. Closed-loop control is required to maintain the proof mass in a neutral position for stable operation. The tunneling current is given by

$$I_t \propto V_B \cdot \exp(-\beta\sqrt{\phi} z_0), \tag{7.18}$$

where V_B is the bias across the electrodes, β is a constant with a typical value of about $1.025\,\text{eV}^{-0.5}\,\text{Å}^{-1}$, ϕ is the tunneling barrier height in eV, and z_0 is the gap separation. The linearized model is used for controller design.

Figure 7.14a shows a cross-sectional view of a tunneling accelerometer using electrostatic actuation for force balancing [54]. For operation, the tip and the proof mass are first pulled close into the tunneling range. When the device is accelerated, the proof mass moves relative to the tip and results in a change of tunneling current. The control circuit subsequently adjusts the electrostatic force acting on the proof mass to maintain the operating current. The external acceleration is extracted from the measured control voltage. In general, gap variations less than 1 Å are required to maintain linearity over a large dynamic range. The high displacement sensitivity enables implementation of μg accelerometers with relative ease compared to other types of accelerometers.

The above tunneling accelerometer had to achieve a high resolution of $20\,\text{ng}/\text{Hz}^{1/2}$ over a bandwidth from 5 Hz to 1.5 kHz for the desired underwater acoustic application. The mechanical design, however, requires a low resonant

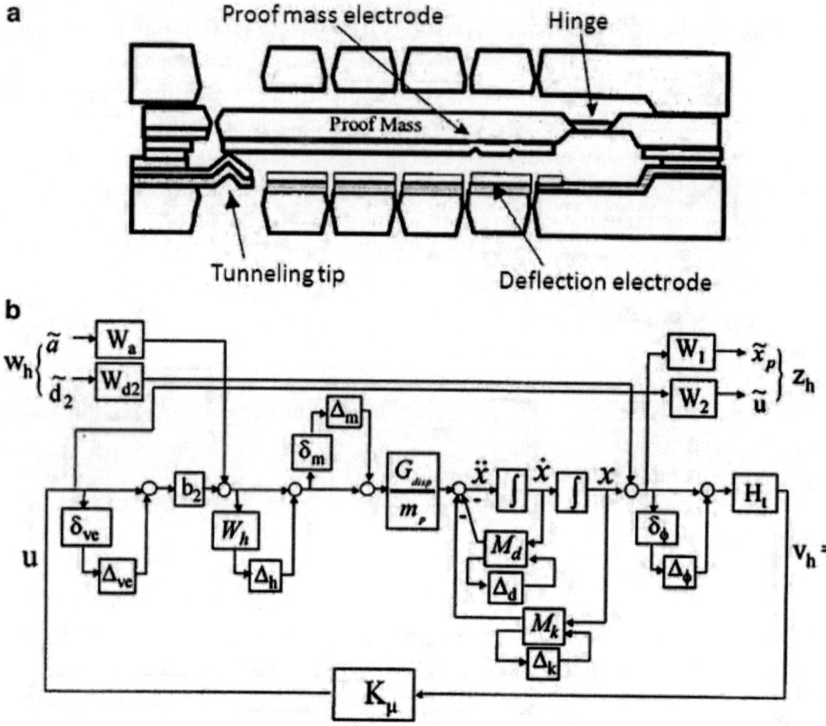

Fig. 7.14 (**a**) Cross-sectional view of the tunneling accelerometer using electrostatic force feedback. (**b**) Mixed μ-synthesis controller design for the tunneling accelerometer (Liu and Kenny [54] © IEEE)

frequency at 100 Hz such that the low acceleration signals are detectable with respect to the flicker noise at low frequencies. In addition, a high quality factor is needed for operation in order to reduce wide-band thermomechanical noise. The feedback controller design faces the challenges of maintaining a tunneling gap and extending the closed-loop bandwidth for a high-Q and low-resonance device. Moreover, stability robustness against plant uncertainties and external disturbances is needed.

There are some robust control design methodologies that can be considered, such as qualitative feedback theory (QFT) [72], H_∞ [57], and mixed μ-synthesis [84]. Mixed μ-synthesis was applied to meet the desired specifications. The uncertainty model in this method includes real and complex blocks, which properly represent variations from physical parameters and unmodeled dynamics, such as variations from feedback voltage, work function, mass, damping coefficient, and spring constant. As shown in Fig. 7.14b, K_μ represents the controller. The selection of the weighting function W_1 is related to the system bandwidth and disturbance rejection. The function W_2 limits the control signal and prevents saturation of the control circuit. The selection and revision of such weighting functions requires the designer's insights in order to harmonize conflicting design specifications. The method typically results in a high-order controller such that order reduction is required before implementation. Integral control is not used to prevent circuit saturation due to integrating the d.c. offset. By feedback linearization, the control above achieved a large dynamic range of 92 dB. Designers can also use simple design methods, such as the proportional-integral-derivative (PID) control, to stabilize a tunneling accelerometer, yet one should be aware of the fact that stability robustness is better considered in the aforementioned approach.

7.4 MEMS Control for Thermally Driven Scanning Cantilevers

Scanning probe microscopy (SPM) has led to many new findings in nanotechnology [28, 33] since its invention in the 1980s [7]. Commercial AFM instruments are rather bulky and have a limited throughput with only one scanning cantilever. The cantilever deflection during surface scanning is detected optically by use of a laser and an optical detector, making it difficult and costly to integrate multiple cantilevers. Research groups have been motivated by the idea of fabricating arrays of scanning cantilevers to increase parallelism, with each cantilever capable of performing closed-loop sensing and actuation. Developed AFM probes have used piezoelectric actuation/piezoresistive detection [46, 58] and thermal actuation/piezoresistive detection [1].

The AFM scanning array can be monolithically integrated into an IC process for miniaturization [6]. The cantilevers in Fig. 7.15 are thermally actuated by a bimorph of silicon and aluminum layers powered by CMOS circuitry. The method for determining the lumped dynamic model of an electrothermal actuator was presented by [11]. The cantilever deflection caused by forces exerted on the tip during a surface scan is detected by a Wheatstone bridge consisting of four diffused piezoresistors.

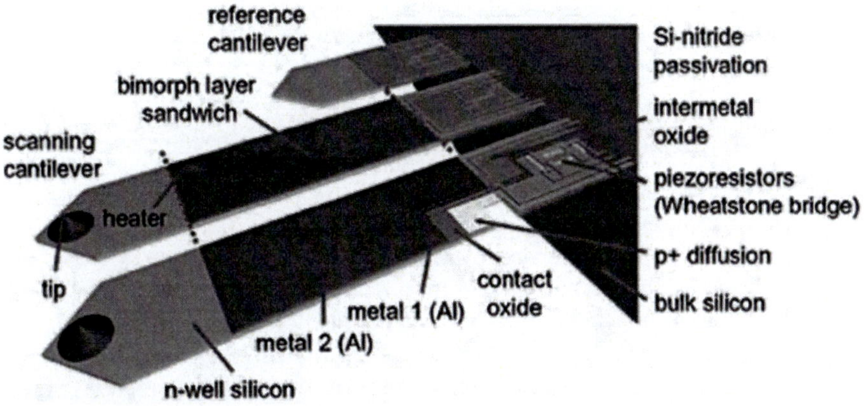

Fig. 7.15 Schematic of the CMOS-integrated scanning cantilevers (Barrettino et al. [6] © IEEE)

The cantilever is inherently a stable plant, and these cantilevers had a resonant frequency of 43 kHz and a quality factor of up to 400. The control bandwidth is limited by the thermal time constant to 3 kHz, which eventually determines the scan rate. Digital PID controllers are used to keep a constant tip–sample contact force for the ten cantilevers.

Note that there are some non-idealities in the control loop: the false force signals produced by thermal crosstalk from the actuator to the sensor are subtracted by digital filters. The offset voltage of the piezoresistive Wheatstone bridge is compensated by the offset value from a reference Wheatstone bridge. An analog square-root circuit in the controller takes care of the quadratic nonlinearity between the control voltage and the dissipated power in the electrothermal actuator. The cantilevers achieve a vertical resolution of better than 1 nm, which is equivalent to a force resolution of better than 1 nN.

Scanning resolution can be enhanced by operating a cantilever at its resonance. The shifts of resonant frequency and quality factor are dependent on the interacting force gradients from the sample and dissipative forces [8]. The oscillation amplitude is maintained by a positive feedback loop that contains specialized circuits such as a phase-locked loop and a variable gain amplifier [19]. The tip–sample spacing is maintained by a negative feedback system. Due to the complexities in design and hardware, it is more challenging to implement a dynamic-mode AFM array in an IC process.

7.5 MEMS Control for Data Storage

The storage density of magnetic hard disk drives (HDD) has doubled every 18 months since they were first invented in the 1950s. The areal density of today's magnetic recording technologies are approaching the limit imposed by the superparamagnetic effect. As the density reaches 10^{12} bits/in^2, the budget for track misregistration is on the order of only a few nanometers. It is therefore natural to consider MEMS technology for improving the read/write head servo bandwidth and performance.

Following the success of scanning cantilevers, the probe-based MEMS data-storage device was proposed in pursuit of a high storage density and a high data rate. Such a MEMS device, as shown schematically in Fig. 7.16a, consists of two chips: the bottom chip has a tip-cantilever array for data read/write, and the top one has large-stroke actuators for moving the storage media. Passive tip-cantilevers are used in IBM's Millipede device to eliminate the complexity of tip servos, since minimal power consumption is desired for portable applications [20]. Without tip servos, variation in tip-cantilever height and thermally induced curl can limit the overall array size. The magnetic medium is actuated with large stroke in the *xy* plane to accommodate read/write action of each cantilever. Access time is reduced, since there is no latency as in an HDD. Control of the media actuator is discussed next.

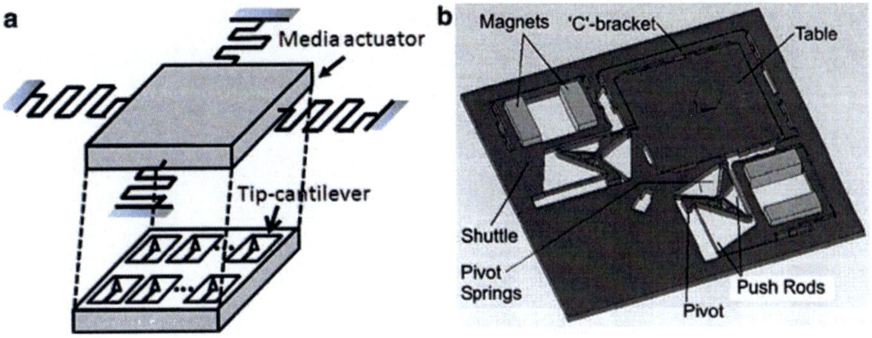

Fig. 7.16 (**a**) Schematic of the MEMS-based data storage device. (**b**) Schematic of the media actuator by IBM (Lantz et al. [50] © IEEE)

By following the infrastructure of a conventional HDD, MEMS can add a microactuator as the secondary actuator to the primary VCM (voice coil motor) actuator for fine positioning of the read/write head. The schemes of dual-stage actuation and control are discussed in Sect. 7.5.2.

7.5.1 Control of Media Actuator for Probe-Based Data Storage

The micro-scanner depicted in Fig. 7.16b consists of a polymer storage medium driven by magnetic actuators in the *x*–*y* plane [50]. The device is inherently a two-input–two-output system capable of providing motion up to $\pm 50\,\mu$m. The resonant frequencies in the *x* and *y* directions are 151 and 137 Hz, respectively, with quality factors of about 6. Thermomechanical read/write on the polymer medium is achieved using tips of the 64×64 AFM cantilevers.

The control system performs two functions: in the seek-and-settle stage, the scanner moves the target track from an arbitrary position close to the read/write probes; then in the track-and-follow stage, the servo system maintains the probe position along the center of the target track as the data read/write proceeds [64]. Thermal position sensors made from doped silicon are used during seek-and-settle. Displacement of the scan table causes a temperature change in the heated sensor and thus a change in electrical resistance. Due to the large sensor noise at low frequencies, a hybrid scheme with dedicated written bits for tracking is used in the track-and-follow stage.

A minimum transient time with nearly no overshoot is the primary requirement in seek-and-settle. The concept of time-optimal control (TOC) [4], or so-called bang-bang control, developed in the 1960s is applied with some modifications. The original method in Fig. 7.17a requires state feedback of displacement and velocity and uses a nonlinear on–off element to drive the plant. Chattering is the main concern of TOC, since the control signal always switches between the maximum and minimum values. The modified version, called proximate TOC (PTOC), as shown

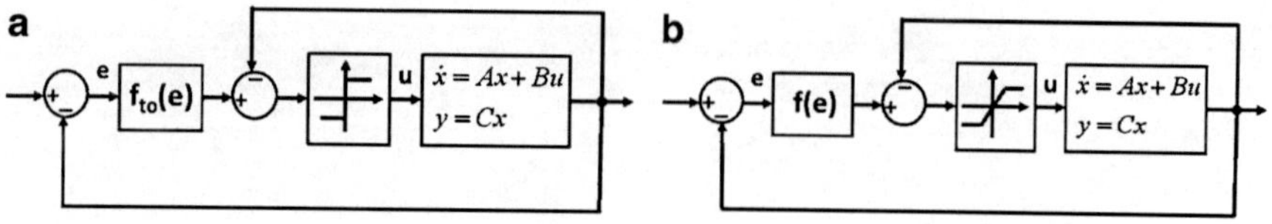

Fig. 7.17 Control systems for minimizing transient time in the seek-and-settle stage: (**a**) Time-optimal control; (**b**) Proximate time-optimal control

in Fig. 7.17b, uses a finite gain saturation element to eliminate chattering at the cost of increased response time [23, 82]. For micro-scanner control, a Kalman state estimator is used to estimate position, velocity, and the dynamics of the thermal sensor. The achieved seek time is 1.6 ms. Higher-order mechanical modes can be excited when one is attempting to further minimize the seek time. A practical solution is to move these modes to higher frequencies.

Next, for track-and-follow, the design challenge is to enhance positioning resolution in the face of noise and external disturbances at a scan rate of a few millimeters per second (equivalent to a desired data rate of 50 kbits/s per probe). The first design approach uses only the position information from the thermal sensors. Two independent closed loops for the x and y directions are designed based on the linear quadratic Gaussian (LQG) regulator [57]. The LQG problem combines the optimal control technique with the design of an observer-based controller. The design procedure reduces to two subproblems: (1) state estimation by Kalman-filter theory; (2) finding the control law that minimizes the cost function represented by the states and the control signal. LQG combined with LTR (loop transfer recovery) guarantees internal stability for minimum-phase plants. Gain shaping of the sensitivity function and the closed-loop transfer function replaces phase compensation to achieve desirable performance and robust properties. Figure 7.18a shows the complete control system, which consists of an LQG regulator, an integrator K_I, and a feedforward term K_{FF}. The feedforward component reduces the effect of higher-order modes using notch filters and enhances the transient speed by canceling the undesired closed-loop pole produced by the integrator.

Since the thermal sensor noise deteriorates at low frequencies, the second design approach adopts a hybrid sensing scheme that combines thermal sensors and medium-derived positional error signals (PES) and takes advantage of their good noise performance at high and low frequencies, respectively. The H_∞ robust control diagram in Fig. 7.18b shows that the controller represented by K has two inputs from the thermal position signal y_{th} and the medium-derived signal y_{PES}. The plant model, noise, and disturbances are represented by G, n, and d. Performance requirements are translated into appropriate weighting functions; for example, the transfer function W_{nL} has a low-pass filter characteristic to force the signal v_1 derived from the thermal sensors to be the preferred signal at high frequencies. The control signal magnitude is limited by the weight W_u to meet the constraint of power consumption. The standard H_∞ control problem is to find a stabilizing and robust controller that minimizes the structured singular value of a transfer function matrix

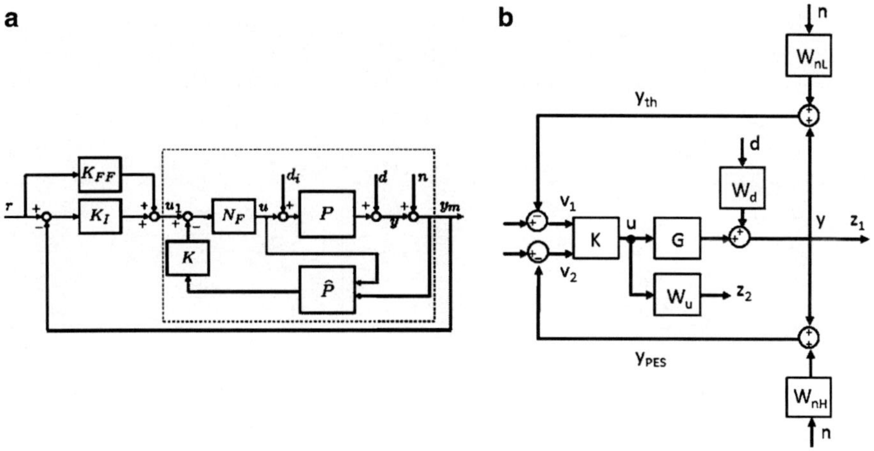

Fig. 7.18 Control systems for track-and-follow: (**a**) based on thermal position sensor and LQG design; (**b**) based on hybrid sensing and H_∞ design (Pantazi et al. [64] © IEEE)

represented by the functional blocks in Fig. 7.18b [57]. The joint controller K has two transfer functions K_1 and K_2; K_1 uses the thermal sensor signal as the input such that it has smaller gain at low frequencies and larger gain at high frequencies than K_2, which uses the PES signal as the input. The drift caused by thermal sensors in the first design is effectively removed.

7.5.2 Dual-Stage Control for Magnetic Hard Disk Drives

The read/write head of a conventional hard disk drive (HDD) is placed at the end of a suspension arm that is actuated by a voice coil motor at the other end. Servo bandwidth is thus limited by the low resonant frequency of the arm. The concept of dual-stage actuation has been proposed to resolve the issue with mainly two different implementations. In the MEMS-based actuation shown in Fig. 7.19a, the microactuator placed at the end of the VCM suspension moves the magnetic head (or slider) relative to the suspension for fine positioning at a larger control bandwidth not limited by the suspension. The second implementation uses piezoelectric actuation as the secondary mechanism to move the suspension for fine positioning. Note that the controller is MIMO for MEMS-based actuation because both the head position relative to the suspension and the suspension motion can be detected [21]. In most piezoelectrically actuated suspensions, relative position sensing is generally not available so that the dual-stage controller is single-input–multiple-output (SIMO).

Electrostatic actuation is preferred for MEMS-based actuation due to its temperature stability. The actuator design is required to be flexible in the in-plane

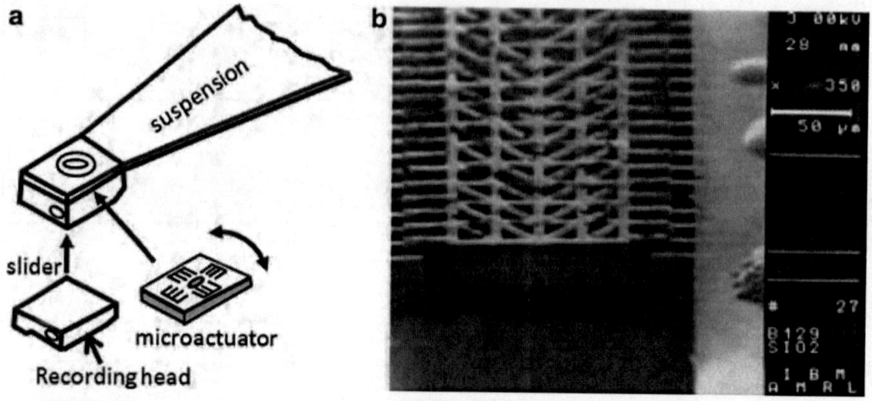

Fig. 7.19 (a) MEMS-based dual-stage actuation using a microactuator placed the end of the VCM suspension to move the magnetic head relative to the suspension. (b) The high-aspect-ratio actuator made by IBM (Hirano et al. [36] © IEEE)

operational mode and very stiff in the other modes for reducing cross-coupled motions. The out-of-plane stiffness must be high enough to withstand the force coming from the slider as the actuator is pressed down to the disk surface. Figure 7.19b shows the electroplated Invar actuator fabricated by IBM with a 40 μm structural height and a 4 μm interelectrode gap [36]. The driving comb electrodes are used simultaneously for capacitive position sensing, with actuation and sensed signals being separated in the frequency domain by modulation [21,41]. With a first resonant frequency at 1.9 kHz, the actuator is suitable for use as a secondary servo actuator.

There are primarily two approaches to dual-stage control design: one is based on classical SISO design methodologies and the other is based on modern MIMO optimal and robust design methodologies [34, 53]. Most of the proposed SISO design methodologies first perform decoupling, followed by multiple SISO loop shaping steps to obtain the overall closed-loop frequency responses. The control architectures include the master–slave method [35,45,52,59] and the PQ method [70].

The master–slave and decoupled sensitivity design approach deals with both MIMO (MEMS-based) and MISO (piezoelectrically actuated) dual-stage schemes [52] as shown in Fig. 7.20a, b, where P_v and P_M represent the VCM and the microactuator, respectively, *PES* is the position error of the head relative to the data track, *RPES* is the microactuator position relative to the tip of the suspension, and *VPES*, as generated by *PES* and *RPES*, is the position error of the suspension tip relative to the data track. The *PES* is generally obtained from encoded information on the magnetic disk. There are three compensators in the MIMO block diagram of Fig. 7.20a, including the VCM loop compensator (C_V), the microactuator loop compensator (C_M), and the microactuator minor loop compensators (C_1 and C_2). The latter is used to damp the microactuator's resonant mode for desired pole placement. The closed-loop sensitivity transfer function from input r to *PES* is the

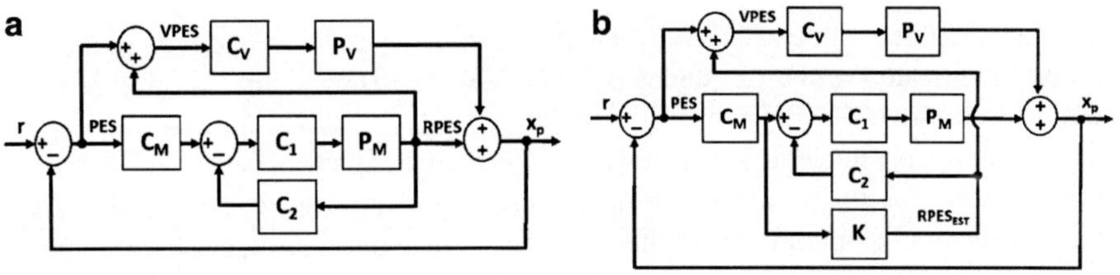

Fig. 7.20 The master-slave and decoupled sensitivity design approach for dual-stage servo: (**a**) MEMS-based MIMO design; and (**b**) piezoelectrically-actuated MISO design (Li and R. Horowitz [52] © IEEE)

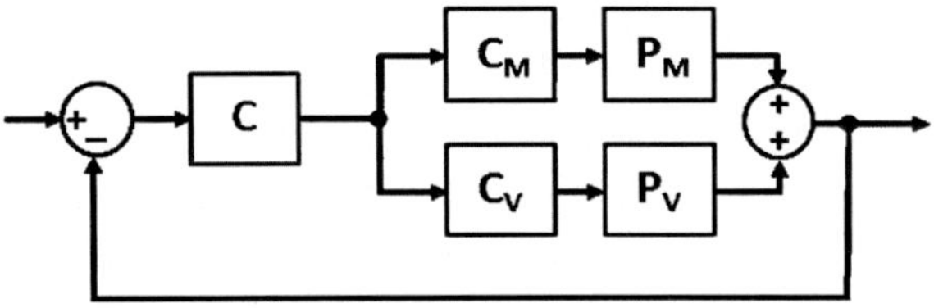

Fig. 7.21 The PQ method using the plants connected in parallel (Schroeck et al. [70] © IEEE)

product of the VCM and microactuator loop sensitivity transfer functions. The dual-stage servo control design is thus decoupled into two independent loop designs. The design procedure can also be applied to the SIMO control of a piezoelectrically actuated suspension where the *RPES* signal is not available and has to be estimated by an observer K as shown in Fig. 7.20b.

The other SISO approach is the PQ method, which connects the high-bandwidth microactuator in parallel with the low-bandwidth VCM as shown in Fig. 7.21. The idea can be intuitively understood as a way to modify the original plant, VCM, for extending servo bandwidth [37]. The added high-bandwidth part becomes effective at high frequencies and almost does not affect the plant behaviors at low frequencies. The design starts with the controllers C_{M} and C_{V} to address the issues of stable zeros and relative output contribution, followed by the design of the controller C.

Cost and reliability of dual-stage servo systems are the key issues to be overcome for commercialization. Other related research includes the use of MEMS strain sensors placed on the suspension arm to detect airflow-induced vibrations for suppressing track misregistration [39].

The control system discussed in this section could be applied to any dual-stage system that requires coarse and fine positioning. There are some issues to be aware of when a MEMS actuator is included. The actuator's lightly damped flexure resonance mode needs to be properly compensated to allow quick settling in the output response [40]; in addition, manufacturing variations result in uncertainties in a microactuator, including those in the spring constant, the resonant frequency, and the required driving voltage. Thus, the controller robustness to uncertainties should be considered for MEMS-based dual-stage servo control design. The contribution of each actuator to the closed-loop sensitivity transfer function attenuation must be properly allocated in the frequency domain. The first-stage coarse motion actuator has a large range but limited bandwidth. Hence its contribution to sensitivity reduction in the presence of stochastic and deterministic disturbances and parametric uncertainties should be primarily in the low-frequency range. The second-stage microactuator can operate in a higher frequency range but it produces a significantly smaller motion. Its contribution to sensitivity reduction should be primarily in this high-frequency range.

7.6 Conclusions

MEMS technology enables system miniaturization, and the resulting large motion bandwidth is desirable for applications that need a fast output response. New control problems arising from MEMS devices have stimulated a colloborative dialogue between the MEMS and control systems communities. Some problems, such as the nonlinear multivariable control of a two-axis electrostatic micro-mirror presented in Sect. 7.2.2, present exciting new challenges that are very good research topics for control theorists and engineers. In particular, robust MEMS control requires more study since it has only been explored to a limited degree so far. Two of the most difficult parts are system implementation and verification, starting from finding a MEMS group with an interest in collaboration on some of the issues we describe below. For MEMS researchers who are interested in MEMS control but not familiar with the theory, learning the material of an undergraduate-level course in control systems would be very helpful in achieving a basic systematic understanding of control objectives and requirements associated with sensor and actuator design.

Some issues related to design and implementation are summarized as follows:

1. *Selection of the actuation and sensing mechanisms*. The criteria are based on the required displacement, driving voltage, sensor resolution, fabrication feasibility, among others. MEMS actuation is often in the micrometer range, and the dynamic range of MEMS sensors is limited, making it a challenge to realize closed-loop control for applications that need large stroke and fine resolution.
2. *Output response time, control resolution, and sensor noise*. Device miniaturization leads to increased actuator bandwidth but also can result in reduced sensor sensitivity, which consequently leads to an increased equivalent noise displacement due to electronic noise. The application of external sensing is one way to avoid compromising sensor and actuator performance. Additionally, as the system bandwidth increases, more sensor noise is induced to the output, which appears as a false signal to reduce the positioning precision. Design trade-offs between the output response time, control resolution, and sensor implementation should be carefully considered.
3. *Sensor readout electronics and their effect on system dynamics*. The additional pole frequencies induced by readout electronics are usually much higher than

those from microactuators, so that they are sometimes overlooked in simulations. Designers should be careful to address the reduced stability margins caused by readout electronics when designing systems with unstable or nonminimum-phase transfer functions, such as the parallel-plate actuator described in Sect. 7.2.

4. *Sensor non-idealities and implementation.* Additional circuit techniques are needed to cancel non-idealities in a sensor, such as the circuit offset due to mismatched transistors and the sensor offset due to mismatched MEMS devices. False signals coupled directly from actuation in an integrated device also need to be removed. Sensor design and implementation becomes challenging when motion detection for more than one axis is required.
5. *Integrated or hybrid implementation.* Integration in an IC process is desirable to reduce I/O complexity in controlling an array of MEMS devices. However, MEMS fabrication does not always match with IC processes. In addition to array implementation, IC integration is preferred for MEMS sensors (e.g., capacitive) to enhance sensor performance.

References

1. T. Akiyama, U. Staufer, N.F. de Rooij et al. Integrated atomic force microscopy array probe with metal-oxide-semiconductor field effect transistor stress sensor, thermal bimorph actuator, and on-chip complementary metal-oxide-semiconductor electronics. *J. Vac. Sci. Technol. B*: 2669–2675, 2000
2. B.V. Amini, R. Abdolvand et al. A 4.5-mW closed-loop $\Sigma\Delta$ micro-gravity CMOS SOI accelerometer. *IEEE J. Solid-State Circuits*: 2983–2991, 2006
3. K.J. Åström and B. Wittenmark. *Adaptive control*. 2nd ed., Dover Publications, 2008
4. M. Athans and P.L. Falb. *Optimal control: an introduction to the theory and its applications*, McGraw-Hill, New York, 1966
5. H. Baltes, O. Brand, A. Hierlemann et al. CMOS MEMS – present and future. *Proc. of IEEE Int. Conf. on Micro Electro Mechanical Systems*: 459–466, 2002
6. D. Barrettino, S. Hafizovic, T. Volden et al. CMOS monolithic mechatronic microsystem for surface imaging and force response studies. *IEEE J. Solid-State Circuits*: 951–959, 2005
7. G. Binnig, C.F. Quate, C. Gerber. Atomic force microscope. *Phys. Rev. Lett.*: 930–933, 1986
8. N. Blanc, J. Brugger, N.F. de Rooij et al. Scanning force microscopy in the dynamic mode using microfabricated capacitive sensors. *J. Vac. Sci. Technol.* B: 901–905, 1996
9. J.J. Blech. On isothermal squeeze films. *J. Lubrication Tech.*: 615–620, 1983
10. H.W. Bode. *Network analysis and feedback amplifier design*, Van Nostrand, Princeton, NJ, 1945
11. B. Borovic, F.L. Lewis, D. Agonafer et al. Method for determining a dynamic state-space model for control of thermal MEMS devices. *IEEE/ASME J. Microelectromech. Syst.*: 961–969, 2005
12. B. Borovic, F.L. Lewis, A.Q. Liu et al. The lateral instability problem in electrostatic comb drive actuators: modeling and feedback control. *J. Micromech. Microeng.*: 1233–1241, 2006
13. B.E. Boser and R.T. Rowe. Surface micromachined accelerometers. *IEEE J. Solid-State Circuits*: 366–375, 1996
14. A.V. Chavan and K.D. Wise. Batched-processed vacuum-sealed capacitive pressure sensors. *IEEE/ASME J. Microelectromech. Syst.*: 580–588, 2001
15. C. T. Chen. *Linear system theory and design*, Oxford University Press, USA, 1998
16. J. Chen, W. Weingartner et al. Tilt-angle stabilization of electrostatically actuated micromechanical mirrors beyond the pull-in point. *IEEE/ASME J. Microelectromech. Syst.*: 988–997, 2004

17. P. Cheung, R. Horowitz, R.T. Howe. Design, fabrication, position sensing, and control of an electrostatically-driven polysilicon microactuator. *IEEE Trans. Magnetics*: 122–128, 1996
18. P. Chu, I. Brener, C. Pu et al. Design and nonlinear servo control of MEMS mirrors and their performance in a large port-count optical switch. *IEEE/ASME J. Microelectromech. Syst.*: 261–273, 2005
19. U. Dürig and H.R. Steinauer et al. Dynamic force microscopy by means of the phase-controlled oscillator method. *J. Appl. Phys.*: 3641–3651, 1997
20. E. Eleftheriou, T. Antonakopoulos et al. Millipede – a MEMS-based scanning-probe data-storage system. *IEEE Trans. Magnetics*: 938–945, 2003
21. L.S. Fan, T. Hirano, J. Hong et al. Electrostatic microactuator and design considerations for HDD applications. *IEEE Trans. Magnetics*: 1000–1005, 1999
22. G. Fedder, R. Howe. Multimode digital control of a suspended polysilicon microstructure. *IEEE/ASME J. Microelectromech. Syst.*:283–297, 1996
23. G.F. Franklin, J.D. Powell, and L.W. Workman. *Digital control of dynamic systems*, Addison Wesley Longman, Inc, 1998
24. R.A. Freeman and P.V. Kokotovic. Robust nonlinear control design: state-space and Lyapunov techniques, Birkhäuser, Boston, 2008
25. J. Freudenberg, D. Looze. *Frequency domain properties of scalar and multivariable feedback systems*. Springer, Berlin, Heidelberg, 1988
26. T.B. Gabrielson. Mechanical–thermal noise in micromachined acoustic and vibration sensors. *IEEE Trans. Elec. Dev.*: 903–909, 1993
27. O. Gasparyan. *Linear and nonlinear multivariable feedback control: a classical approach*, Wiley, 2008
28. J.K. Gimzewski and C. Joachim. Nanoscale science of single molecules using local probes. *Science*: 1683–1688, 1999
29. A. Gola, F. Pasolini, and E. Chiesa et al. A 2.5 rad/s^2 resolution digital output MEMS-based rotational accelerometer for HDD applications. *IEEE Trans. Magnetics*: 915–919, 2003
30. F. Golnaraghi and B.C. Kuo. *Automatic control systems*, 9th ed., Wiley, 2009
31. M. Gopal. *Control systems principals and design*, McGraw-Hill, 2003
32. P.R. Gray, P.J. Hurst, S.H. Lewis et al. *Analysis and design of analog integrated circuits*, 4th ed., Wiley, 2001
33. H.G. Hansma. Surface biology of DNA by atomic force microscopy. *Annu. Rev. Phys. Chem.*: 71–92, 2001
34. D. Hernandez, S. Park, R. Horowitz et al. Dual-stage track-following servo design for hard disk drives. *Proc. American Automatic Control Conference*: 4116–4121, 1999
35. G. Herrmann, C. Edwards, B. Hredzak et al. A novel discrete-time sliding mode technique and its application to a HDD dual-stage track-seek and track-following servo system. *Int. J. Adapt. Control Signal Process*: 344–358, 2008
36. T. Hirano, L.S. Fan, J.Q. Gao et al. MEMS milliactuator for hard-disk-drive tracking servo. *IEEE/ASME J. Microelectromech. Syst.*:149–155, 1998
37. I.M. Horowitz. *Synthesis of feedback systems*, Academic Press, 1965
38. I.M. Horowitz and M. Sidi. Synthesis of feedback systems with large plant ignorance for prescribed time-domain tolerances. *Int. J. Control*: 287–309, 1972
39. R. Horowitz, Y. Li, K. Oldham et al. Dual-stage servo systems and vibration compensation in computer hard disk drives. *Control Eng. Pract.*: 291–305, 2007
40. D.A. Horsley, R. Horowitz, A.P. Pisano. Microfabricated electrostatic actuators for hard disk drives. *IEEE/ASME Trans. Mechatronics*: 175–183, 1998
41. D.A. Horsley, N. Wongkomet, R. Horowitz et al. Precision positioning using a microfabricated electrostatic actuator. *IEEE Trans. Magnetics*: 993–999, 1999
42. A. Izadian, L.A. Hornak, and P. Famouri. Structure rotation and pull-in voltage control of MEMS lateral comb resonators under fault conditions. *IEEE Trans. Control Syst. Technol.*: 51–59, 2009
43. D.A. John and K. Martin. *Analog integrated circuit design*, Wiley, 1997
44. T.W. Kenny, S.B. Waltman, J.K. Reynolds et al. Micromachined silicon tunneling sensor for motion detection. *Appl. Phys. Lett.*: 100–102, 1991

45. Y. Kim and S. Lee. An approach to dual-stage servo design in computer disk drives. *IEEE Trans. Control Syst. Technol.*: 12–20, 2004
46. Y. Kim, H. Nam, S. Cho et al. PZT cantilever array integrated with piezoresistor sensor for high-speed parallel operation of AFM. *Sensors Actuators A*: 122–129, 2003
47. G.T.A. Kovacs. *Micromachined transducers sourcebook*, McGraw-Hill Education, 2000
48. M. Krstic, I. Kanellakopoulos, P.V. Kokotovic. *Nonlinear and adaptive control design*, Wiley-Interscience, 1995
49. H. Külah, J. Chae, N. Yazdi et al. Noise analysis and characterization of a sigma-delta capacitive microaccelerometer. *IEEE J. Solid-State Circuits*: 352–361, 2006
50. M. Lantz, H. Rothuizen, U. Drechsler et al. A vibration resistant nanopositioner for mobile parallel-probe storage applications. *IEEE/ASME J. Microelectromech. Syst.*: 130–139, 2007
51. M. Lemkin and B.E. Boser. A three-axis micromachined accelerometer with a CMOS position-sense interface and digital offset-trim electronics. *IEEE J. Solid-State Circuits*: 456–468, 1999
52. Y. Li and R. Horowitz. Mechatronics of electrostatic microactuators for computer disk drive dual-stage servo systems. *IEEE/ASME Trans. Mechatronics*: 111–121, 2001
53. Y. Li and R. Horowitz. Design and testing of track-following controllers for dual-stage servo systems with pzt actuated suspensions. *Microsystem Technologies*: 194–205, 2002
54. C. Liu and T. Kenny. A high-precision, wide-bandwidth micromachined tunneling accelerometer. *IEEE/ASME J. Microelectromech. Syst.*:425–433, 2001
55. M.S.C. Lu. *Parallel-plate micro servo for probe-based data storage*, Ph.D. thesis, Carnegie Mellon University, Pittsburgh, PA, 2002
56. M.S.C. Lu and G.K. Fedder. Position control of parallel-plate microactuators for probe-based data storage. *IEEE/ASME J. Microelectromech. Syst.*:759–769, 2004
57. J.M. Maciejowski. *Multivariable feedback design*, Addison-Wesley, 1989
58. S.C. Minne, G. Yaralioglu, S.R. Manalis et al. Automated parallel high-speed atomic force microscopy. *Appl. Phys. Lett.*: 2340–2342, 1998
59. K. Mori, T. Munemoto, H. Otsuki et al. A dual-stage magnetic disk drive actuator using a piezoelectric device for a high track density. *IEEE Trans. Magnetics*: 5298–5300, 1991
60. R. Nadal-Guardia, A. Dehé, R. Aigner et al. Current drive methods to extend the range of travel of electrostatic microactuators beyond the voltage pull-in point. *IEEE/ASME J. Microelectromech. Syst.*: 255–263, 2002
61. H.C. Nathanson, W.E. Newell, R.A. Wickstrom et al. The resonant gate transistor. *IEEE Trans. Electronic. Dev.*: 117–133, 1967
62. Y. Nemirovsky and O. Bochobza-Degani. A methodology and model for the pull-in parameters of electrostatic actuators. *IEEE/ASME J. Microelectromech. Syst.*: 601–615, 2001
63. K. Ogata. *Discrete-Time Control Systems*, 2nd ed., Prentice Hall, 1995
64. A. Pantazi, A. Sebastian, G. Cherubini et al. Control of MEMS-based scanning probe data-storage devices. *IEEE Trans. Control Syst. Technol.*: 824–841, 2007
65. V.P. Petkov and B.E. Boser. A fourth-order $\sum\Delta$ interface for micromachined inertia sensors. *IEEE J. Solid-State Circuits*: 1602–1609, 2005
66. D. Piyabongkarn, Y. Sun, R. Rajamani et al, Travel range extension of a MEMS electrostatic microactuator. *IEEE Trans. Control Syst. Technol.*: 138–145, 2005
67. S. Sastry. *Nonlinear systems: Analysis, stability, and control*. Springer, 1999
68. P.R. Scheeper, B. Nordstrand, J.O. Gullov et al. A new measurement microphone based on MEMS technology. *IEEE/ASME J. Microelectromech. Syst.*: 880–891, 2003
69. R. Schreier and G.C. Temes. *Understanding delta-sigma data converters*, IEEE Press, Wiley-Interscience, 2005
70. S.J. Schroeck, W.C. Messner, R.J. McNab. On compensator design for linear time-invariant dual-input single-output systems. *IEEE/ASME Trans. Mechatronics*: 50–57, 2001
71. J.I. Seeger and B.E. Boser. Charge control of parallel-plate, electrostatic actuators and the tip-in instability. *IEEE/ASME J. Microelectromech. Syst.*: 656–671, 2003
72. M.J. Sidi. *Design of robust control systems: from classical to modern practical approaches*. Krieger Publishing Company, 2001

73. S. Skogestad and I. Postlethwaite. *Multivariable feedback control: Analysis and design*, 2nd ed., Wiley-Interscience, 2005
74. J.E. Slotine and W. Li. *Applied nonlinear control*. Prentice Hall, 1991
75. W.C. Tang, T.C.H. Nguyen, M.W. Judy et al. Electrostatic-comb drive of lateral polysilicon resonators. *Sensors Actuators A*: 328–331, 1990
76. M. Tartagni and R. Guerrieri. A fingerprint sensor based on the feedback capacitance sensing scheme. *IEEE J. Solid-State Circuits*: 133–142, 1998
77. F.G. Tseng, C.J. Kim, C.M. Ho. A high-resolution high- frequency monolithic top-shooting microinjector free of satellite drops – Part I: concept, design, and model. *IEEE/ASME J. Microelectromech. Syst.*: 427–436, 2002
78. V.I. Utkin. *Sliding modes in control and optimization*, Springer-Verlag, New York, 1992
79. P.F. Van Kessel, L.J. Hornbeck, R.E. Meier et al. MEMS-based projection display. *Proc. IEEE*: 1687–1704, 1998
80. T. Veijola, H. Kuisma, J. Lahdenperä et al. Equivalent-circuit model of the squeezed gas film in a silicon accelerometer. *Sensors Actuators A*: 239–248, 1995
81. H.H. Woodson and J.R. Melcher. *Electromechanical dynamics, Part I: discrete systems*, Wiley, New York, 1968
82. M.L. Workman. Adaptive proximate time-optimal servomechanisms, Ph.D. thesis, Stanford University, Stanford, CA
83. Y. Zhao, F.E.H. Tay, F.S. Chau et al. Stabilization of dual-axis micromirrors beyond the pull-in point by integral sliding mode control. *J. Micromech. Microeng.*: 1242–1250, 2006
84. K. Zhou and J.C. Doyle. *Essentials of robust control*, Prentice Hall, 1998
85. J. Zou, C. Liu, J. Schutt-Aine et al. Development of a wide-tuning-range two-parallel-plate tunable capacitor for integrated wireless communication systems. *Int. J. RF Microwave Computer-aided Engineering*: 322–329, 2001

Chapter 8
Dissecting Tuned MEMS Vibratory Gyros

Dennis Kim and Robert T. M'Closkey

8.1 Introduction

Vibratory gyros are devices which can detect a change in angle or angular velocity by exploiting the Coriolis force coupling between two degrees of freedom in a mechanical resonator when the equations of motion are considered in a coordinate frame that is fixed with the sensor case. This is a natural coordinate system to choose because deflections of the resonating structure are readily sensed with built-in pick-offs that are fixed with respect to the sensor case. Furthermore, it is often necessary to apply forces to the structure, and this is also easily accomplished with case-fixed electrodes. By monitoring the response of the pick-offs, it is possible to infer the angular velocity, or in certain modes of operation, the change in angle, experienced by the sensor case.

This chapter focuses on the analysis of the Disk Resonator Gyro (DRG) whose development has been sponsored by Boeing. The resonant structure of the DRG is shown in Fig. 8.1 and consists of 16 nested rings (lighter "webbing" in the photo) that are attached to adjacent rings by small "spokes" that bridge the gap between the rings every 45°. The spokes, however, do not create a solid radial structure from the outer ring to the central resonator post but in fact are offset by 22.5° when comparing adjacent rings. This arrangement builds in a high degree of planar elasticity and the modes of interest have in-plane approximately ellipsoidal shapes that are coupled by a coriolis term when the resonator equations of motion are written in a coordinate frame that is fixed at the resonator's center and which rotates with the resonator. The angle subtended by the major axes of the two ellipsoidal modes is nearly 45° in the DRG – the exact angle depends on the details of the mass and stiffness asymmetries that are present in the fabricated resonators. As

---

D. Kim (✉) • R.T. M'Closkey
Mechanical and Aerospace Engineering Department, University of California,
405 Hilgard Avenue, Los Angeles, CA, 90095, USA
e-mail: dongj@seas.ucla.edu; rtm@seas.ucla.edu

J.J. Gorman and B. Shapiro (eds.), *Feedback Control of MEMS to Atoms*,
DOI 10.1007/978-1-4419-5832-7_8,

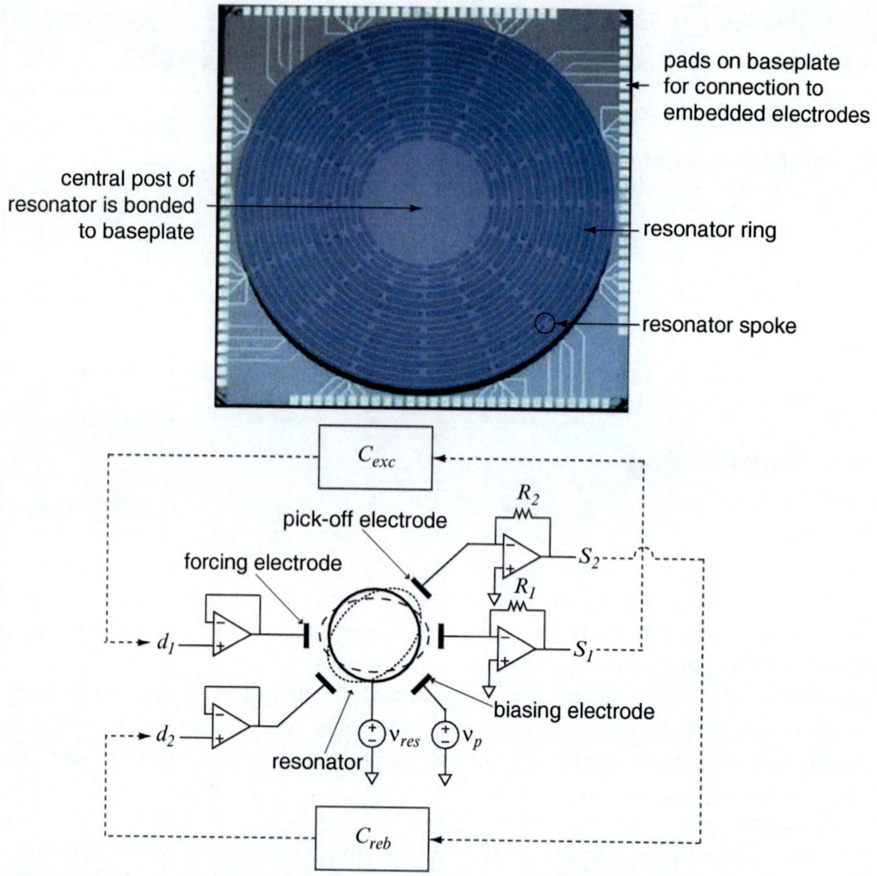

Fig. 8.1 Photo of the 8 mm diameter resonator of the Disk Resonator Gyro (Fig. A1 of [13], copyright/courtesy of Springer). The central resonator post is attached to the electrical base wafer – this arrangement suspends the rings above the baseplate so that they are free to move in the plane. The DRG senses rotation about an axis normal to the plane of the resonator. Electrodes are fixed to the baseplate and embedded between the rings. They can be configured to apply electrostatic forces to the resonator or measure the local deflections of the resonator. The lower schematic shows a simplified version of the electrical interface to the resonator. The resonator is held at a constant potential relative to the electrodes and the transresistance configuration of the pick-off buffers provides a measurement of the local radial velocity of the rings (the s_1 and s_2 voltages). The forcing electrodes apply radial electrostatic forces to the resonator proportional to the d_1 and d_2 voltages. Thus, electrical test data of the DRG consists of four transfer functions from inputs $\{d_1, d_2\}$ to outputs $\{s_1, s_2\}$. The approximately ellipsoidal mode shapes of the two coriolis-coupled modes are shown as dashed lines. The two control loops employed in closed-loop vibratory gyros are also shown (not shown in this schematic is the manner in which an angular rate input induces signals in the closed-loop system – refer to the complete block diagram in Fig. 8.4)

shown in the schematic diagram of Fig. 8.1, the resonator itself is represented by a ring and the electrodes are shown to be distributed around the periphery of the ring. The electrodes are paired in a forcer/pick-off configuration and the pairs are physically oriented at 45° to each other to take advantage of the fact that in a perfectly fabricated resonator, there would be no cross-channel coupling between forcer/pick-off pairs. In the actual DRG, the electrodes are embedded in the gaps between the rings and can be configured to sense an averaged ring displacement or velocity over the electrode area, depending only how the signal conditioning electronics are configured. Other electrodes apply in-plane electrostatic forces to the ring (for resonator forcing), and, finally, a third set of electrodes is reserved for providing constant voltage potentials between the resonator and the electrodes to create electrostatic "springs" which modify the resonator's stiffness matrix. The simplified schematic in Fig. 8.1 shows the roles of the three sets of electrodes. It is clear that a partial differential equation is necessary to describe the dynamics of the resonator due to its distributed mass and stiffness, however, since the two coupled modes of interest are well isolated from neighboring modes, the following two degree-of-freedom model can accurately capture all relevant features of the dynamics,

$$M\ddot{\mathbf{x}} + C\dot{\mathbf{x}} + \alpha S \Omega \dot{\mathbf{x}} + K\mathbf{x} = \mathbf{f}, \tag{8.1}$$

where M, C, and K are real, 2-by-2, positive definite mass, damping, and stiffness matrices, respectively, corresponding to the generalized coordinates $\mathbf{x}$ that rotate with the resonator and the corresponding generalized forces $\mathbf{f}$. The angular velocity of the resonator about the plane normal to the disk is denoted Ω and the coupling strength between these degrees of freedom is denoted α. The matrix S is skew-symmetric,

$$S = \begin{bmatrix} 0 & -1 \\ 1 & 0 \end{bmatrix}.$$

This model makes no assumptions on the structure of the mass or stiffness matrices which are often invoked to be diagonal. While such simplifications are didactically useful for describing the operation of vibratory rate sensors, they are unjustified when considering the dynamics of real resonators.

The generalized coordinates $\mathbf{x}$ may be taken to be the displacements sensed by the two pick-off electrodes embedded between the resonator's rings. These electrodes are sensitive to changes in capacitance between themselves and the resonator and although the electrodes are distributed in the sense that they detect a capacitance change over an arc at several locations on the resonator, the net result is to effectively measure two displacements, denoted x_1 and x_2, which are treated as the generalized coordinates $\mathbf{x} = [x_1, x_2]^{\mathrm{T}}$. Similarly, the forcing electrodes apply electrostatic forces to the resonator, distributed over an arc, at several locations within the resonator that are complimentary to the placement of the pick-off electrodes so that the net electrostatic forces exerted on the modes are represented f_1 and f_2 and combined into $\mathbf{f} = [f_{\mathrm{exc}}, f_{\mathrm{reb}}]^{\mathrm{T}}$. Thus, although the pick-off electrodes and forcing electrodes are not co-located as is assumed when writing the equations of motion in the form (8.1), their placement relative to the two modes of interest permits the treatment of the pick-offs and forcers as being co-located and, thus, (8.1) remains an

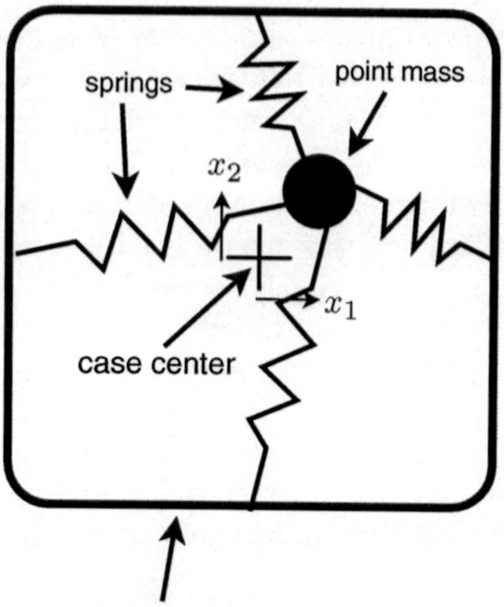

Fig. 8.2 Cartoon of a point mass suspended by springs attached to a case. The equilibrium position of the mass is in the case center and the displacement of the mass relative to this position is given by the variables x_1 and x_2. The case is allowed to rotate about the axis normal to the plane which passes through the case center, however, the $x_1 - x_2$ coordinate axes are fixed to, and rotate with, the case

appropriate model. In terms of actual DRG measurements, the s_1 and s_2 signals in Fig. 8.1 are voltages proportional to the *velocities* $\dot{x}_1$ and $\dot{x}_2$ because the electrodes are connected to transresistance amplifiers, that is the output voltage is proportional to the electrode current, which is proportional to the time-rate-of-change of the capacitance, which is a measure of average ring radial velocity at the electrode.

Immediate analysis of (8.1) would obscure the essential operation of vibratory gyros, so we will initially consider the system in Fig. 8.2, which is often offered as the prototypical example of a vibratory gyro. The model consists of a point mass of mass m suspended in a frame or case that is allowed to rotate in the x_1-x_2 plane about an axis this is located at the center of the frame. The orthogonal x_1 and x_2 degrees of freedom are fixed to the sensor case, that is, they rotate with the case, and we assume that the pick-off arrangement permits the measurement of x_1 and x_2. The springs are chosen so that the restoring force is isotropic for small displacements from the case center (we ignore geometric nonlinearities) and so the effective spring rate for displacements in both the x_1 and x_2 directions is k. Writing the equations of motion for the mass assuming pure rotation about the case center (no translation) yields

$$\begin{bmatrix}\ddot{x}_1\\ \ddot{x}_2\end{bmatrix} + 2\Omega\begin{bmatrix}0 & -1\\ 1 & 0\end{bmatrix}\begin{bmatrix}\dot{x}_1\\ \dot{x}_2\end{bmatrix} + \begin{bmatrix}\omega_{\rm n}^2-\Omega^2 & -\dot{\Omega}\\ \dot{\Omega} & \omega_{\rm n}^2-\Omega^2\end{bmatrix}\begin{bmatrix}x_1\\ x_2\end{bmatrix} = \begin{bmatrix}0\\ 0\end{bmatrix},$$

where Ω is the (time-varying) angular rotation rate of the sensor case (counter-clockwise positive) and $\omega_{\rm n} = \sqrt{k/m}$ is the natural frequency of each degree of freedom. Further simplifications are made by assuming $\Omega \ll \omega_n$ and $\dot{\Omega} \ll \omega_n$ so that the time-varying terms in the stiffness matrix can be dropped,

$$\begin{bmatrix} \ddot{x}_1 \\ \ddot{x}_2 \end{bmatrix} + 2\Omega \begin{bmatrix} 0 & -1 \\ 1 & 0 \end{bmatrix} \begin{bmatrix} \dot{x}_1 \\ \dot{x}_2 \end{bmatrix} + \begin{bmatrix} \omega_n^2 & 0 \\ 0 & \omega_n^2 \end{bmatrix} \begin{bmatrix} x_1 \\ x_2 \end{bmatrix} = \begin{bmatrix} 0 \\ 0 \end{bmatrix}. \tag{8.2}$$

Although this model may be found in almost any reference on vibratory gyros, the amplitude and phase coordinate analysis undertaken here is a novel and expedient way of showing that if the initial conditions are chosen so that the mass oscillates along a straight line that passes through the equilibrium position of the mass (the case center), then, as the frame rotates in the plane, the orientation of the line of oscillation remains inertially fixed. Thus, measurement of the orientation of the line of oscillation *relative* to the case-fixed coordinates yields the orientation of the frame in inertial space, albeit with a polarity change in the measured angle versus inertial angle. Although this mode of operation is not the focus of this chapter it is a worthwhile exercise to understand this property of (8.2). The analysis is accomplished using the following amplitude and phase coordinates [6] for each degree of freedom,

$$x_1(t) = a_1(t)\cos(\omega_n t + \varphi_1(t)) \tag{8.3}$$

$$x_2(t) = a_2(t)\cos(\omega_n t + \varphi_2(t)), \tag{8.4}$$

where a_1 and a_2 are the amplitude functions of x_1 and x_2, respectively, and φ_1 and φ_2 are the phase functions. The objective is to derive first-order differential equations for the dependent variables a_1, a_2, φ_1 and φ_2.

Differentiating (8.3) and (8.4) yields

$$\dot{x}_1 = \dot{a}_1 \cos(\psi_1) - a_1(\omega_n + \dot{\varphi}_1)\sin(\psi_1),$$

$$\dot{x}_2 = \dot{a}_2 \cos(\psi_2) - a_2(\omega_n + \dot{\varphi}_2)\sin(\psi_2),$$

where, for brevity, we define $\psi_k(t) = \omega_n t + \varphi_k(t)$, $k = 1,2$. Following the amplitude and phase coordinate procedure, we set

$$\dot{a}_1 \cos(\psi_1) - a_1 \dot{\varphi}_1 \sin(\psi_1) = 0 \tag{8.5}$$

$$\dot{a}_2 \cos(\psi_2) - a_2 \dot{\varphi}_2 \sin(\psi_2) = 0, \tag{8.6}$$

which yields simplified expressions for $\dot{x}_1$ and $\dot{x}_2$,

$$\begin{aligned} \dot{x}_1 &= -a_1 \omega_n \sin(\psi_1), \\ \dot{x}_2 &= -a_2 \omega_n \sin(\psi_2). \end{aligned} \tag{8.7}$$

Differentiating (8.7) and substituting into (8.2) yields the following differential equations which complement (8.5) and (8.6),

$$-\dot{a}_1\omega_n\sin(\psi_1) - a_1\omega_n\dot{\varphi}_1\cos(\psi_1) + 2\Omega a_2\omega_n\sin(\psi_2) = 0, \tag{8.8}$$

$$-\dot{a}_2\omega_n\sin(\psi_2) - a_2\omega_n\dot{\varphi}_2\cos(\psi_2) - 2\Omega a_1\omega_n\sin(\psi_1) = 0. \tag{8.9}$$

The expressions for $\dot{a}_1$ is obtained by multiplying (8.5) by $-\sin(\psi_1)$ and adding the result to the product of (8.8) with $-\cos(\psi_1)/\omega_n$,

$$\dot{a}_1 - 2\Omega a_2\sin(\psi_1)\sin(\psi_2) = 0. \tag{8.10}$$

Similar manipulation produces differential equations for a_2 and the phases,

$$\dot{a}_2 + 2\Omega a_1\sin(\psi_1)\sin(\psi_2) = 0, \tag{8.11}$$

$$a_1\dot{\varphi}_1 - 2\Omega a_2\cos(\psi_1)\sin(\psi_2) = 0, \tag{8.12}$$

$$a_2\dot{\varphi}_2 + 2\Omega a_1\sin(\psi_1)\cos(\psi_2) = 0. \tag{8.13}$$

Exact analysis of (8.10)– (8.13) is difficult; however, we can definitively conclude

$$a_1\dot{a}_1 + a_2\dot{a}_2 = 0,$$

from manipulation of (8.10) and (8.11). Thus, the norm of the amplitude functions, that is, $a_1^2 + a_2^2$, is constant, which states that the energy in the system is conserved.

An approximate analysis can be carried out by exploiting the fact that the amplitude and phase terms evolve on time scales that are generally much longer than the period of the oscillation associated with the natural frequency ω_n. Thus, averaging analysis [4] yields the following equations, which describe the approximate behavior of the amplitudes and phases,

$$\dot{a}_1 - \Omega a_2\cos(\varphi_2 - \varphi_1) = 0, \tag{8.14}$$

$$\dot{a}_2 + \Omega a_1\cos(\varphi_2 - \varphi_1) = 0, \tag{8.15}$$

$$a_1\dot{\varphi}_1 - \Omega a_2\sin(\varphi_2 - \varphi_1) = 0, \tag{8.16}$$

$$a_2\dot{\varphi}_2 - \Omega a_1\sin(\varphi_2 - \varphi_1) = 0. \tag{8.17}$$

Of considerable interest are situations in which x_1 and x_2 are in-phase for all t, that is $\varphi_1(t) = \varphi_2(t)$. If (8.16) is multiplied by $-a_2$ and summed with the product of (8.17) and a_1, the following differential equation for $\varphi_2 - \varphi_1$ is obtained,

$$a_1a_2(\dot{\varphi}_2 - \dot{\varphi}_1) + \Omega(a_2^2 - a_1^2)\sin(\varphi_2 - \varphi_1) = 0.$$

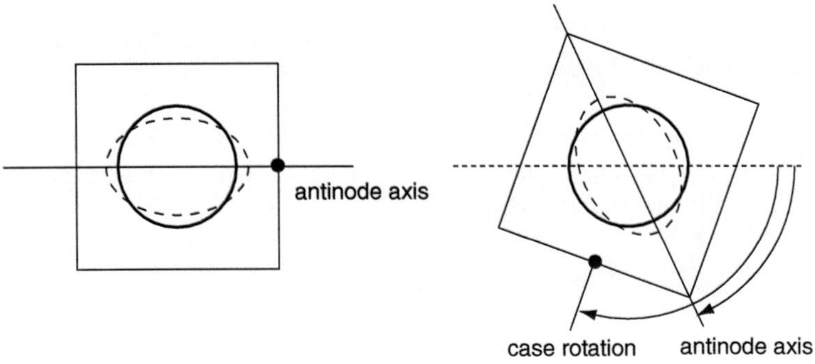

Fig. 8.3 Whole-angle mode of DRG showing the precession of mode antinode when the sensor case is subjected to rotation. In an inertial reference frame, the antinode orientation lags the case orientation by a precisely known factor, hence, the case orientation can be determined, by monitoring with case-fixed pick-offs, the shift in anti-node orientation relative to the sensor case

Note that $\varphi_2(t) - \varphi_1(t) \equiv 0$ is a solution of this differential equation *independent* of the angular velocity Ω of the sensor case. Thus, if the initial phases are chosen to be equal, then they remain equal for all time. In this case, the oscillating point mass traces a straight line through the origin of the $x_1 - x_2$ coordinate frame in Fig. 8.2 and the amplitudes are governed by the coupled equations

$$\dot{a}_1 - \Omega a_2 = 0,$$
$$\dot{a}_2 + \Omega a_1 = 0.$$

The orientation of this line is of interest and the angle it makes relative to the x_1 coordinate axis is $\tan^{-1}(a_2(t)/a_1(t))$. Since

$$\frac{\mathrm{d}}{\mathrm{d}t}\tan^{-1}(a_2(t)/a_1(t)) = \frac{1}{1+(a_2/a_1)^2}\left(\frac{\dot{a}_2}{a_1} - \frac{a_2\dot{a}_1}{a_1^2}\right) = -\Omega(t),$$

the angle of the line that the mass oscillation proscribes relative to the case-fixed coordinates is minus the integral of the angular velocity of the sensor case. In other words, the line that the mass follows remains constant with respect to an inertial reference frame and by monitoring how the orientation of the line changes with respect to the case-fixed $x_1 - x_2$ frame, the angle of rotation of the sensor case can be determined. This is the so-called "whole angle mode" of operation and there is no qualitative difference when considering the response of a "perfect" DRG in which the resonator has no etching errors or damping: the major axis of the ellipsoidal modal response, once excited, will lag the case motion by a precise amount. In other words, the major axis will not remain inertially fixed as in the point mass example, but rather will rotate in same direction as the sensor case, but through a smaller angle as shown in Fig. 8.3. The ratio of the change in major axis angle in an inertial frame relative to a change in case angle in an inertial frame has been experimentally confirmed to be about 0.6 for the DRG.

Advantages of the whole angle mode of operation include the ability to measure angles even when the sensor case experiences very large angular rates of rotation because the "physics" of the device actually does the integration from angular rate into a change in angle. In practice, though, the resonator does exhibit dissipation of energy which must be replaced in order to sustain the oscillation on which the angle sensing mechanism depends, and it is difficult to design a replenishment strategy that does not perturb the orientation of the oscillation. Furthermore, there are challenges associated with the pick-off design because each pick-off must be configured to accept the full dynamic range of amplitudes, and there are also complications introduced by nonlinearities in the structural mechanics when the resonator amplitudes are driven out of their linear regime [2].

It is possible to motivate the *angular rate sensing mode* of operation, as opposed to the whole angle mode, using (8.2). In this case, two "actuators" are introduced that produce forces along the x_1 and x_2 coordinate axes. The x_1 degree of freedom is designated the "excitation" degree of freedom and a control loop is designed whose objective is to maintain a stable harmonic oscillation of the x_1 component of the mass response. In other words, the excitation force, denoted f_{exc}, ensures that $x_1(t) = a\cos(\omega_0 t)$ in the updated equations,

$$\begin{bmatrix} \ddot{x}_1 \\ \ddot{x}_2 \end{bmatrix} + 2\Omega \begin{bmatrix} 0 & -1 \\ 1 & 0 \end{bmatrix} \begin{bmatrix} \dot{x}_1 \\ \dot{x}_2 \end{bmatrix} + \begin{bmatrix} \omega_{\text{n}}^2 & 0 \\ 0 & \omega_{\text{n}}^2 \end{bmatrix} \begin{bmatrix} x_1 \\ x_2 \end{bmatrix} = \frac{1}{m} \begin{bmatrix} f_{\text{exc}} \\ f_{\text{reb}} \end{bmatrix}.$$

There is a second controller, dubbed the "force-to-rebalance" controller, that produces the force f_{reb} along the x_2 degree of freedom. The design objective of this controller is to null, or zero out, the x_2 component of the sensor response. In other words, if we assume $x_2(t) \equiv 0$, then

$$f_{\text{reb}}(t) = 2m\Omega(t)\dot{x}_1(t) = -2ma\omega_0\Omega(t)\sin(\omega_0 t),$$

so the angular rate of the sensor case modulates the "fast" sinusoid $\sin(\omega_0 t)$ which is in-phase with the x_1 velocity component. Thus, a signal proportional to Ω can be recovered by multiplying f_{reb} by a signal proportional to $\dot{x}_1$ and then low-pass filtering the result. The fact that the natural frequency of the mode associated with x_2 degree of freedom is equal to the natural frequency of the mode associated with the x_1 degree of freedom – a situation referred to as *tuned* – is of no consequence at this level of analysis but we will see that it plays a critical role when sensor noise is included in the analysis. An important observation is that achieving tuned modes is an objective of the DRG design: the symmetry of the resonator under discrete rotations of 45° (see Fig. 8.1) can be shown to yield degenerate elliptically shaped modes, although it will be evident later in the chapter that small etch nonuniformities break the resonator symmetry and produce two close, but detuned, modal frequencies – the detuning must be eliminated in order to achieve the best possible gyro performance (tuning the modal frequencies is the subject of Sect. 8.3).

The closed-loop operation of the sensor does have certain advantages over the whole-angle mode of operation. For example, note that the resonator is always maintained in the same dynamic response state relative to the sensor case, that is, x_1 tracks a sinusoid and $x_2 \approx 0$ for all input rates Ω. This simplifies the design of the electronic signal conditioning because the wide dynamic range requirement for both pick-offs in the whole-angle mode is no longer necessary, thus, each pick-off can be optimized for a preset range of signals. The closed-loop operation, however, shifts the dynamic range requirement to f_{reb}, but since f_{reb} is produced by the control electronics and not the mechanics of the resonator, it is much easier to maintain its linearity. Thus, feedback plays an indispensable role in operation of vibratory angular rate sensors. The two control loops are denoted by C_{exc} and C_{reb} in Fig. 8.1 schematic.

The issue that drives closed-loop vibratory gyro performance is the ratio of the angular rate-induced signal in f_{reb} relative to the noise power that is present in this signal. The noise can come from several sources such as mechanical-thermal noise of the resonator [3] and the electronic noise associated with the pick-off signal conditioning electronics. The dominant noise source in the DRG is the electronic noise associated with the pick-off buffers. The noise spectrum measured at s_1 and s_2 in Fig. 8.1 is dominated by the Johnson noise of the feedback resistors in the buffers. The effect of this noise on the angular rate measurement is analyzed in detail in Sect. 8.5. Without the inclusion of noise in the analysis, however, all vibratory gyros perform arbitrarily well so in order to motivate the advantages of tuned, closed-loop vibratory rate sensors over their untuned brethren we embellish (8.2) by assuming that there may be different modal masses and spring rates associated with each degree of freedom,

$$\begin{aligned} m_{11}\ddot{x}_1 + c_{11}\dot{x}_1 - \alpha\Omega\dot{x}_2 + k_{11}x_1 &= f_{\text{exc}}, \\ m_{22}\ddot{x}_2 + c_{22}\dot{x}_2 + \alpha\Omega\dot{x}_1 + k_{22}x_2 &= f_{\text{reb}}, \end{aligned} \tag{8.18}$$

where m_{11} and m_{22} are the modal masses, c_{11} and c_{22} are the modal damping, k_{11} and k_{22} are the modal stiffnesses, and x_1 and x_2 represent the generalized coordinates with corresponding generalized forces f_{exc} and f_{reb}. The subscripts denote, as above, *excitation* and *force-to-rebalance*, since we will continue to assume that there are two control loops that provide for excitation of the x_1 degree of freedom response, and for nulling the x_2 degree of freedom response. This model is still somewhat idealized because cross-coupling terms in the mass, stiffness, and damping matrices are ignored. Furthermore, the point mass model in Fig. 8.2 cannot motivate these equations because this model requires that the modal masses be equal. Thus, (8.18) can be considered to be produced by the analysis of a distributed resonator like the DRG in which all other resonant modes have been ignored. The coriolis term arises from the fact that these equations of motion are still written in coordinates that are fixed to the sensor case and hence rotate with the sensor. The parameter α denotes the coriolis coupling strength between the two modes. The role of the excitation loop, which is closed from x_1 or $\dot{x}_1$ to f_{exc}, is to establish a constant sinusoidal response of x_1. In other words, we assume $x_1(t) = a\cos(\omega_0 t)$, where a denotes the amplitude and ω_0 the frequency. In MEMS sensors, the frequency of the excitation often coincides with the undamped natural frequency because the largest response of the resonator is usually desired given a limit on the forcing magnitude (electrostatic forcing is ubiquitous in MEMS devices and a limit on the potential between the electrode and vibrating structure is typically dictated by the electronics). Thus, we assume in the Introduction that excitation frequency ω_0 is equal to the natural frequency $\sqrt{k_{11}/m_{11}}$. The force-to-rebalance loop is typically a high gain loop whose task is to regulate x_2 to zero. A vibratory gyro is an "AC" sensor in which the angular rate Ω, is modulated, via the device physics, onto a carrier which in this case is the sinusoidal response of $\dot{x}_1$. Thus, the x_2 degree of freedom experiences a disturbance due to the coriolis coupling. Since the measurements in the DRG are proportional to velocities, that is, $\dot{x}_1$ and $\dot{x}_2$, instead of positions, we adopt the perspective that the signals produced by the sensor are proportional to velocity for the remainder of the chapter. The transfer function H is used to represent the sensor dynamics from $[f_{\rm exc}, f_{\rm reb}]^{\rm T}$ to $[\dot{x}_1, \dot{x}_2]^{\rm T}$ when $\Omega = 0$. Subscripts are used to denote the channel of interest, for example, H_{12} is the transfer function from $f_{\rm reb}$ to $\dot{x}_1$, H_{21} is the transfer function from $f_{\rm exc}$ to $\dot{x}_2$, etc. We will assume that the feedback regulation maintains $x_2(t), \equiv 0$ which implies $f_{\rm reb}$ perfectly cancels the "disturbance" from the coriolis-coupled signal, that is,

$$f_{\rm reb}(t) = \alpha\Omega(t)\dot{x}_1(t) = -\alpha\Omega(t)a\omega_0\sin(\omega_0 t).$$

If $f_{\rm reb}$ is demodulated with respect to $\dot{x}_1$ and low-pass filtered, the result is proportional to Ω,

$$\frac{1}{2}\alpha a^2\omega_0^2\Omega(t),$$

where the *scale factor*, denoted $\gamma_{\rm sf}$, is defined as the group of terms multiplying Ω, that is,

$$\gamma_{\rm sf} = \frac{1}{2}\alpha a^2\omega_0^2.$$

Thus, the angular rotation rate is recovered when the demodulated signal is divided by the scale factor. This definition of scale factor is consistent with what is typically termed the "sensitivity" of the gyro. The block diagram for this control scheme is shown in Fig. 8.4, where in the analysis up to this point we have assumed $H_{s_1} = H_{s_2} = 1$, $H_{d_1} = H_{d_2} = 1$, and $n_{s_1} = n_{s_2} = 0$.

The motivation for operating the sensor in a tuned state, that is, the condition $\sqrt{k_{11}/m_{11}} = \sqrt{k_{22}/m_{22}}$ is satisfied, comes from the analysis of the effect of the rebalance loop noise, shown as n_{s_2} in Fig. 8.4, on the estimate of the angular rate. There is also noise associated with the excitation loop pick-off, denoted n_{s_1}, but it will be shown in Sect. 8.4 that the dominant noise source in the estimated angular

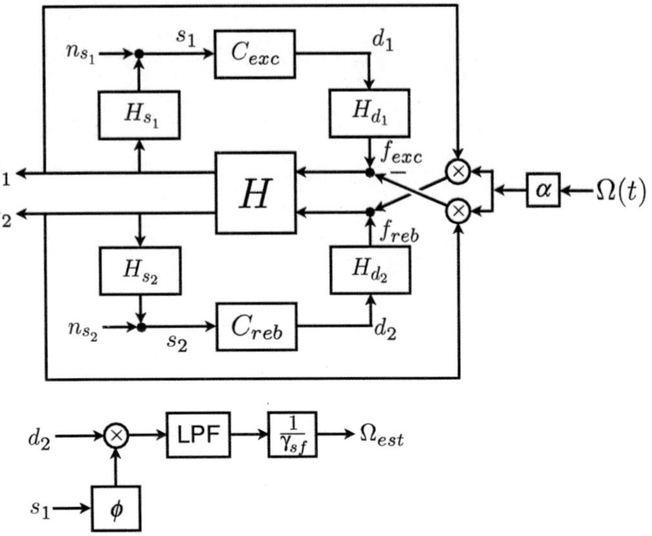

Fig. 8.4 Closed-loop vibratory rate sensor operation. The excitation control loop establishes $x_1(t) = a\cos(\omega_0 t)$ and the force-to-rebalance loop regulates the input to C_{reb}, which is denoted s_2, to zero. The fundamental sensor mechanics are represented by H and the control elements are denoted C_{reb} and C_{exc}. The input and output signal conditioning dynamics are blocks labeled H_{d_1}, H_{d_2}, H_{s_1}, and H_{s_2} and, as shown in the schematic of Fig. 8.1, these blocks capture the combined electromechanical transfer functions of the electrodes and associated electronics. The signals that can be measured and manipulated are the voltages labeled d_1, d_2, s_1 and s_2 (also shown in Fig. 8.1). The estimated angular rate, denoted Ω_{est}, is determined by demodulating d_2 with respect to a phase adjusted copy of s_1 (the phase adjustment is denoted ϕ). "LPF" denotes *L*ow *P*ass *F*ilter

rate of the DRG is associated with the rebalance loop pick-off. If we assume that the rebalance loop has high gain ($\gg 1$) in a neighborhood of the excitation frequency ω_0, then the transfer function from n_{s_2} to f_{reb} is approximately minus the inverse of the $(2,2)$ element of H,

$$-H_{22}^{-1}(s) = -\frac{m_{22}s^2 + c_{22}s + k_{22}}{s}.$$

Thus, the spectrum of the noise-induced component of f_{reb} is

$$P_{\text{reb}}(\omega) = |H_{22}^{-1}(j\omega)|^2 P_{\text{n}_{s_2}}(\omega),$$

where P_{reb} denotes the noise power spectrum of f_{reb}, $P_{\text{n}_{s_2}}$ denotes the power spectrum of the pick-off noise n_{s_2}, and ω is within the frequency band for which the large loop gain assumption holds. The noise spectrum associated with the estimate of Ω, denoted P_Ω, is obtained by dividing the demodulated noise spectrum of P_{reb} by γ_{sf}^2

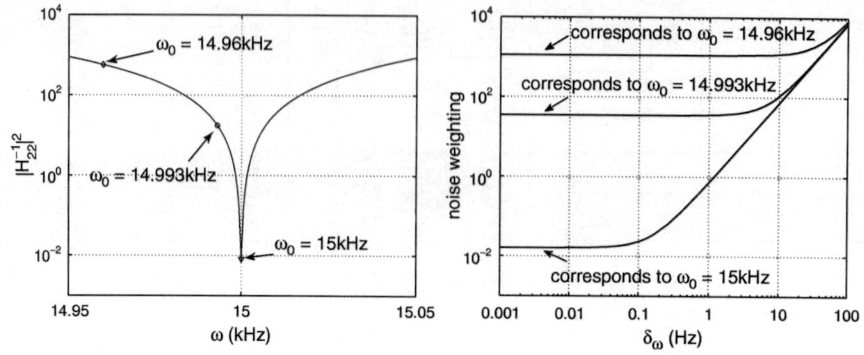

Fig. 8.5 (*Left*) $|H_{22}^{-1}|^2$ when the modal frequency is 15 kHz and $Q = 50$ k (representative of the experimental results for the sensor data in this chapter). Three different excitation frequencies are considered: $\omega_0 = 14.96$ kHz (diamond), $\omega_0 = 14.993$ kHz (circle), $\omega_0 = 15$ kHz (triangle). (*Right*) The "weighting" factor in (8.19) for the noise corrupting the measurement of Ω. It is apparent that the smallest noise intensity is achieved when ω_0 is equal to the modal frequency of the rebalance loop mode. This corresponds to "tuned" sensor dynamics since ω_0 is equal to the modal frequency of the excitation loop mode

$$P_\Omega(\delta_\omega) = \frac{1}{\gamma_{sf}^2} \cdot \frac{1}{2} a^2 \omega_0^2 \left(P_{reb}(\omega_0 + \delta_\omega) + P_{reb}(\omega_0 - \delta_\omega)\right)$$

$$= \frac{1}{\alpha\gamma_{sf}} \Big(\left|H_{22}^{-1}(j(\omega_0 + \delta_\omega))\right|^2 P_{ns_2}(\omega_0 + \delta_\omega)$$

$$+ \left|H_{22}^{-1}(j(\omega_0 - \delta_\omega))\right|^2 P_{ns_2}(\omega_0 - \delta_\omega) \Big),$$

where δ_ω in this case is the frequency associated with the demodulated (baseband) signal. This expression can be simplified when the pick-off noise has a flat spectrum in a neighborhood of ω_0, which we will assume for the remainder of the chapter (this assumption is justified in Sect. 8.5). In other words, if $P_{ns_2}(\omega_0 - \delta_\omega) = P_{ns_2}(\omega_0 + \delta_\omega) = P_{ns_2}(\omega_0)$, then

$$P_\Omega(\delta_\omega) = \frac{1}{\alpha\gamma_{sf}} P_{ns_2}(\omega_0) \left(\left|H_{22}^{-1}(j(\omega_0 + \delta_\omega))\right|^2 + \left|H_{22}^{-1}(j(\omega_0 - \delta_\omega))\right|^2 \right).$$

The noise is minimized at *any* δ_ω when $\omega_0 = \sqrt{k_{22}/m_{22}}$ because the noise "weighting" factor given by

$$\left|H_{22}^{-1}(j(\omega_0 + \delta_\omega))\right|^2 + \left|H_{22}^{-1}(j(\omega_0 - \delta_\omega))\right|^2, \tag{8.19}$$

is minimized by this choice. The weighting is shown in Fig. 8.5 and since ω_0 corresponds to the natural frequency of the excitation channel model, we see that the best signal-to-noise ratio is achieved when the excitation channel and rebalance channel modal frequencies are equal.

The foregoing analysis introduces the basic concepts behind tuned, closed-loop vibratory angular rate sensors. The remainder of this chapter focuses on test results and the analysis of the MEMS Disk Resonator Gyro in Fig. 8.1. Section 8.2 introduces a more realistic sensor model that includes cross-coupling terms and although the cross-coupling can be ignored for the noise analysis it must be included when attempting to understand the source of the angular rate bias and the associated "quadrature" signal. This section also introduces a useful procedure for fitting analytical model parameters to empirical frequency response data. Since a modally tuned resonator is so important, Sect. 8.3 introduces a method for compensating for inevitable fabrication errors which produce modal nondegeneracy. The method relies on electrostatic actuation, which is quite common in MEMS. The model fitting routine from Sect. 8.2 is crucial in guiding the tuning process and can be automated.

Section 8.4 discusses the two basic control loops, the excitation loop and force-to-rebalance loop. Since details of the control architecture have been reported elsewhere [1], only a brief review is given since the primary focus is the closed-loop noise in each channel. Experiments show that the rebalance loop noise dominates the estimated rate noise and hence the effect of the excitation loop noise can be ignored. Section 8.5 provides a detailed comparison of open-loop versus closed-loop operation and revisits the tuned versus detuned sensor and culminaties with the conclusion that open-loop operation can only be practically used when the sensor is highly detuned and, thus, suffers from a vastly degraded signal-to-noise ratio compared to the open-loop tuned sensor. The tuned sensor, however, can be used with feedback to increase its bandwidth and we show how the noise properties of the closed-loop tuned sensor approach those of the open-loop tuned sensor.

Rate sensors of the quality of the DRG can be used for short-term navigation and in this case the rate signal is integrated to estimate a change in angle experienced by the sensor over the interval of integration. The noise that corrupts the rate measurement has a detrimental effect on the estimate of the angle, and Sect. 8.5.3 reveals that the price of increasing the sensor bandwidth beyond the open-loop bandwidth of the coriolis-coupled modes is the creation of *angle white noise* for short integration times. For longer integration times, however, the uncertainty in the estimated angle asymptotically approaches the *angle random walk* characteristic that is associated with the open-loop tuned sensor. Finally, Sect. 8.6 analyzes the rate and quadrature bias terms from the perspective of the cross-coupling terms in the sensor dynamics. It is also shown how perturbation of the phases of two critcal components in the two loops, whether induced by a change in sensor dynamics, signal conditioning electronics, or control filters, can have a detrimental effect on the rate bias by coupling it to the quadrature signal.

8.2 Vibratory Sensor Model

We now return to (8.1), the most general two-degree-of-freedom linear model of the DRG. The fact that we can ignore the other modes in the resonator and focus on a two-degree-of-freedom model is supported by the wideband frequency response in Fig. 8.6 where it is evident that the coriolis coupled modes of interest near 15 kHz have no interaction with other structural modes. This is a consequence of the resonator design and electrode layout (the physical location of the forcer and pick-off electrodes renders many structural modes unobservable and/or uncontrollable). The deep notches are due to parasitic coupling between the forcer and pick-off voltages and can be easily estimated and removed from the frequency response data prior to fitting model parameters.

The frequency response of (8.1), assuming velocity measurements, is denoted H and is given by

$$H(j\omega) = j\omega\left(-\omega^2 M + K + j\omega C\right)^{-1},$$

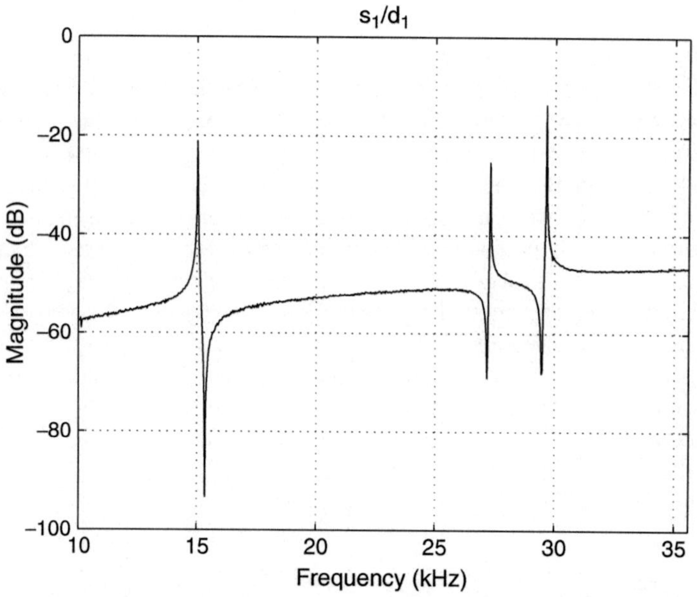

Fig. 8.6 Wideband frequency response of the DRG in Fig. 8.1. The coriolis coupled modes that are exploited for angular rate sensing are a pair near 15 kHz (the two modes cannot be distinguished on this scale). The modes are sufficiently far in frequency from other structural modes so that they can be modeled with the two degree-of-freedom (8.1). Trends in the data that are not supported by this model (like parasitic coupling between d_1, d_2 and s_1, s_2) can be removed prior to estimating model parameters

where ω is the input frequency and where the angular rotation rate Ω is assumed to be zero. From the point of view of the electrostatic forces and capacitive pickoffs, the sensor is a two-input/two-output system and so it is useful to label the channels accordingly,

$$H = \begin{bmatrix} H_{11} & H_{12} \\ H_{21} & H_{22} \end{bmatrix},$$

where H_{11} represents the scalar transfer function $\dot{x}_1/f_{\rm exc}$, H_{12} represents the scalar transfer function $\dot{x}_1/f_{\rm reb}$, etc. A complete sensor model, however, includes signal conditioning dynamics as shown in Fig. 8.4. For example, the potentials applied to the forcer electrodes are provided by buffer electronics in the blocks labeled H_{d_1} and H_{d_2} in Fig. 8.4. These input dynamics contain smoothing filters and are AC-coupled. The inputs to these buffers are voltages, denoted d_1 and d_2, from the test equipment and the outputs are the electrostatic forces, $f_{\rm exc}$ and $f_{\rm reb}$, applied to the resonator. Similarly, the motion of the resonator is converted into voltages, denoted s_1 and s_2, by the sensor's output buffers. The output buffer dynamics are denoted H_{s_1} and H_{s_2}. Thus, the (measured) transfer function from $[d_1, d_2]^{\rm T}$ to $[s_1, s_2]^{\rm T}$, denoted the *gyro* transfer function $H_{\rm g}$, is given by

$$H_{\rm g} := \begin{bmatrix} H_{s_1} & 0 \\ 0 & H_{s_2} \end{bmatrix} \begin{bmatrix} H_{11} & H_{12} \\ H_{21} & H_{22} \end{bmatrix} \begin{bmatrix} H_{d_1} & 0 \\ 0 & H_{d_2} \end{bmatrix}. \qquad (8.20)$$

In most cases, on the one hand the input signal conditioning dynamics are closely matched to each other so it is reasonable to assume $H_{d_1} = H_{d_2}$. On the other hand, the output signal conditioning dynamics' gains are optimized for the test and control equipment, especially to avoid excessive quantization noise if the controller is implemented with a DSP. Furthermore, in order to maximize the signal-to-noise ratio (SNR) of s_2 , the feedback resistance R_2 associated with the transresistance buffer circuit in Fig. 8.1 should be made as large as possible because the Johnson noise increases in proportion to $\sqrt{R_2}$, whereas the signal gain increases in proportion to R_2 (in other words, every fourfold increase in R_2 yields a doubling of the SNR). On the other hand, the amplitude of the sinusoidal excitation measured by s_1 will always be close to the power supply rails but in this case the gain of the s_1 transresistance op-amp, which is determined by R_1, will be much smaller (an order of magnitude or more) than R_2 because it is desirable to have the physical response of the resonator in the excitation channel be as large a practical since this increases the scale factor of the sensor.

The empirical frequency response in the neighborhood of the coriolis-coupled modes in a DRG with matched output buffer gains is shown in Fig. 8.7. This figure reveals that the two modes have a significant frequency split which, aside from producing a vastly reduced SNR ratio as shown in the Introduction, also have a high degree of cross-channel coupling which would lead to saturation of the higher gain s_2 channel by the d_1 driver. A systematic process for "tuning" these two modes to

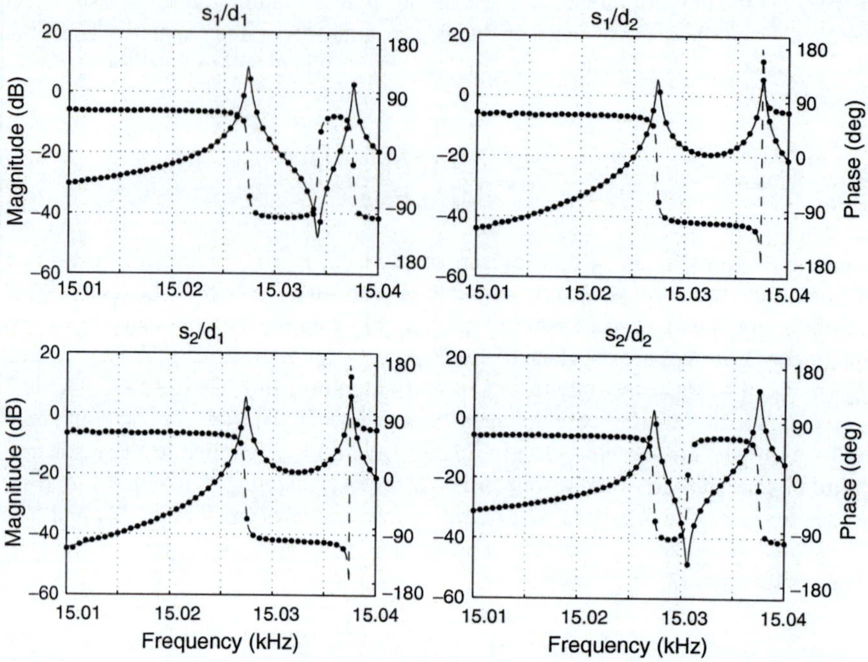

Fig. 8.7 The empirical frequency responses of the four sensor channels $\{s_1/d_1, s_1/d_2, s_2/d_1, s_2/d_2\}$ are shown as the individual points. The magnitude and phase of the identified model (Sect. 8.2.1) are plotted with *solid and dashed lines*, respectively. The input and output units are in volts

near degeneracy is the subject of Sect. 8.3. The frequency response data is obtained by driving a single input channel with a narrow-band periodic chirp, recording the s_1 and s_2 response data and then switching input channels and repeating the data collection. This yields the two-input/two-output empirical frequency responses shown in Fig. 8.7.

8.2.1 Fitting a Model to Frequency Response Data

As a prelude to the frequency tuning method presented in Sect. 8.3 we introduce an analytical model fitting procedure to determine mass, stiffness, and damping matrices from the two-input/two-output frequency response data.

The data shown in Fig. 8.7 represents H_g in (8.20) and provides a complete description of the open-loop dynamics of the sensor, including the electronic buffer dynamics as well as the resonator dynamics. It is desirable to fit a model to this data in order to estimate the mass, stiffness, and damping properties of the resonator.

Unfortunately, (8.20) is overparameterized from the perspective of the frequency response data since the electrical voltage-to-force conversion constants associated with the forcers, and the velocity-to-voltage conversion constants associated with the pickoff electrodes, are not uniquely identifiable from the input–output data. Although it is possible to numerically estimate these constants from the resonator and electrode geometry, it will be evident from the frequency tuning study of Sect. 8.3 that the sensor dynamics can be successfully tuned without knowledge of the conversion constants. One property that can be exploited is the fact that the input dynamics are usually closely matched, that is $H_{d_1} = H_{d_2}$, so that their contribution can be conveniently combined with the output dynamics. Furthermore, given the narrow frequency range of interest, the signal conditioning dynamics can be represented by the first few terms of a power series. Incorporating these ideas, the following model, instead of (8.20), will be fit to the experimental data

$$R(s)Z^{-1}(s), \tag{8.21}$$

where s is the Laplace transform variable and where

$$Z(s) := M_0 s^2 + C_0 s + K_0.$$

The notation Z is used to represent the *impedence* of the sensor. As in (8.1), M_0, C_0, and K_0 represent 2×2, positive definite mass, damping and stiffness matrices, respectively; however, it is to be recognized that these matrices are only proportional to their counterparts in (8.1) and so the subscript is employed to mark this distinction. The signal conditioning dynamics are represented by R. In order to obtain the model parameters, it is assumed that there are a total of m, 2×2, complex-valued frequency response data points denoted $\{\psi_1, \psi_2, \ldots, \psi_m\}$, $\psi_q \in \mathbb{C}^{2\times 2}$, corresponding to the frequencies $\{\omega_1, \omega_2, \ldots, \omega_m\}$. The minimax optimization problem (see [7]) for estimating the parameters of (8.21) is

$$\min_{\substack{M_0 > I, C_0 > 0, K_0 > 0 \\ R_l \in \mathbb{C}^{2\times 2},\, l=0,1,\ldots,n_R}} \; \max_{q=1,\ldots,m} \bar{\sigma}\left(\tilde{R}_q - \psi_q Z(j\omega_q)\right), \tag{8.22}$$

where

$$\tilde{R}_q := \sum_{l=0}^{n_R} R_l \omega_q^l,$$

and where evaluating Z at the qth frequency point yields

$$Z(j\omega_q) := -M_0 \omega_q^2 + K_0 + jC_0 \omega_q.$$

The maximum singular value is denoted $\bar{\sigma}$. The constraint $M_0 > I$ in (8.22) is imposed rather than the typical $M_0 > 0$ because in the latter case all of the free parameters may be scaled by nonzero constant so as to make the cost arbitrarily small without actually changing the model frequency response. Also, note that $\tilde{R}$ is a degree n_R polynomial function of frequency with coefficients in $\mathbb{C}^{2\times 2}$. In fact, $\tilde{R}$ can be viewed as the first few terms of the power series expansion of the frequency response function of the signal conditioning dynamics; however, it also captures sensor-actuator non-colocation effects since $\tilde{R}$ is not constrained to be diagonal.

The model parameters are obtained by recasting (8.22) as the following generalized eigenvalue problem

$$\begin{aligned} \text{min:} \quad & \gamma \\ \text{subject to:} \quad & J_q > 0,\ q = 1, \ldots, m_k \\ & M_0 > I, C_0 > 0, K_0 > 0 \\ & R_l \in \mathbb{C}^{2\times 2}, l = 0, \ldots, n_R. \end{aligned} \tag{8.23}$$

where

$$J_q := \begin{bmatrix} \gamma I & (\tilde{R}_q - \psi_q Z(j\omega_q))^* \\ \tilde{R}_q - \psi_q Z(j\omega_q) & \gamma I \end{bmatrix}.$$

It could be claimed that a more natural formulation of the problem would suggest replacing the objective of (8.22) with

$$\bar{\sigma}\left(\tilde{R}_q Z^{-1}(j\omega_q) - \psi_q\right), \tag{8.24}$$

because this minimizes the largest frequency response error of (8.21). This formulation, however, places too much emphasis on reducing the modeling error at those frequencies and "directions", where $\bar{\sigma}(\psi)$ is relatively large and thereby produces poor fits elsewhere. The frequency response magnitude presented in Fig. 8.7 spans almost three orders of magnitude in a very narrow frequency band – our tests have shown that the objective in (8.22) provides superior matching between the identified model frequency response and empirical data. The frequency response of the model fit to the data in Fig. 8.7 is also shown in this figure.

8.3 Modal Frequency Tuning via Electrostatic Biasing

The model fitting algorithm can be adapted to provide a systematic approach to finding the bias voltages that reduce the modal frequency split to such a degree that the benefits of "tuned" operation are realized. It will be evident that this tuning approach lends itself to automation and requires only a handful of frequency response experiments to determine the final bias voltages. Since the effect of electrostatic biases on the sensor's total stiffness matrix is quadratic, the model can be expanded to include these terms. By perturbing the bias voltages from nominal values and measuring the subsequent frequency response, these "electrostatic stiffness" matrices can be identified along with the original mass, mechanical stiffness, and damping matrices. With these parameters in hand, it is a simple calculation to determine the bias voltages that render equal the generalized eigenvalues of the mass and total stiffness matrix, thus, so as far as the identified model is concerned, the sensor's modal frequencies have been tuned. Additional iterations are possible if the bias voltages in the previous step do not sufficiently reduce the difference between the two modal frequencies.

8.3.1 Modified 2: DOF Model with Electrical "Stiffness" Matrices

For sensors employing electrostatic tuning with dedicated electrodes, (8.1) can be updated to include the dependence of the sensor dynamics on the bias electrodes' voltages to

$$M_0\ddot{\mathbf{x}} + C_0\dot{\mathbf{x}} + \left(K_0 + \sum_{p=1}^{n_e} K_p(v_{res} - v_p)^2\right)\mathbf{x} = \mathbf{f}, \tag{8.25}$$

where the stiffness term is explicitly decomposed into the sum of a positive definite mechanical stiffness matrix, denoted K_0, and n_e negative semidefinite electrostatic stiffness matrices, denoted $K_p, p = 1, \ldots, n_e$, that are associated with the corresponding bias electrodes' potentials, denoted $v_p, p = 1, \ldots, n_e$, where all the bias voltages are defined relative to the sensor's electronic ground. The sensor resonator is held at the constant potential v_{res}. The quadratic appearance of the bias potentials and the fact that $K_p \leq 0$ are due to the fact that the electrostatic force between two plates is an attractive force proportional to the square of the potential difference between the plates. Linearization about nominal plate positions shows that the electrostatics introduces a spring softening effect to the mechanical stiffness. To illustrate the effect of the bias potentials on the sensor dynamics, two – input/two – output empirical frequency response magnitude plots of a DRG at different bias potentials are plotted in Fig. 8.8. The sensor dynamics are quite sensitive to changes in bias potentials by not only shifting the resonant frequencies but also altering the frequency split between the two modes. In general, multiple bias electrodes are necessary to tune the modes to degeneracy. Searching the set of tuning bias potentials, however, is challenging, especially, when there exists a strong coupling between the two modes and the bias potentials (e.g., changing a single bias potential perturbs both modes as clearly shown in Fig. 8.8) and the modal sensitivity to bias potentials indeed varies from sensor to sensor. Thus, a tuning approach must be guided by an accurate estimation of the sensor dynamics from experimental data and a systematic method for tuning the dynamics of electrostatically actuated vibratory gyros is introduced in the following section.

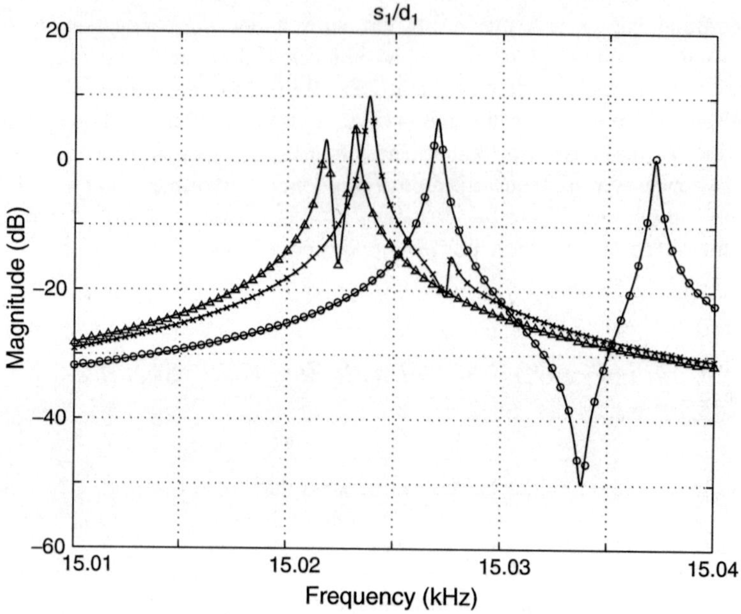

Fig. 8.8 Circles, X's, and triangles are empirical frequency response data generated at $(v_1, v_2) = (0,0)$, (–15,0), and (–15,–15), respectively, when the DRG's resonator is biased at 15 V. The *solid lines* are the frequency response of the single model that is fit to the three two-input/two-output data sets (of which only the s_1/d_1 channel is shown here for clarity)

8.3.2 Electrostatic Tuning Algorithm and Experimental Results

The main idea of the systematic electrostatic tuning algorithm is to directly compute the tuning bias potentials by analyzing the parameters (or their scaled analogs as noted in Sect. 8.2.1) of (8.25) estimated from experimental data. In particular, the individual contributions of the mechanical and electrostatic stiffness matrices to the sensor dynamics can be accurately captured by fitting a single, comprehensive model to multiple empirical frequency response data generated at different bias potentials. From the additive nature of the mechanical and electrostatic stiffness matrices, it is clear that the necessary and sufficient condition for uniquely identifying $K_0, K_1, \ldots, K_{n_e}$ can be stated as

$$\operatorname{rank} \begin{bmatrix} 1 & (v_{res} - v_{1,1})^2 & (v_{res} - v_{2,1})^2 & \ldots & (v_{res} - v_{n_e,1})^2 \\ 1 & (v_{res} - v_{1,2})^2 & (v_{res} - v_{2,2})^2 & \ldots & (v_{res} - v_{n_e,2})^2 \\ \vdots & \vdots & \vdots & \vdots & \vdots \\ 1 & (v_{res} - v_{1,n_{exp}})^2 & (v_{res} - v_{2,n_{exp}})^2 & \ldots & (v_{res} - v_{n_e,n_{exp}})^2 \end{bmatrix} = n_e + 1, \quad (8.26)$$

where the first element of each row is fixed to "1" to represents the fictious potential associate with the mechanical stiffness matrix K_0. A second subscript has been added to the bias potentials to denote the experiment with which they are associated. For example, $v_{2,3}$ represents the the potential applied to the second electrode during the third experiment. In all, we assume there were a total of n_{exp} experiments that were conducted with fixed voltages on the bias electrodes. In order for the rank condition to be satisfied, at least n_e+1 frequency response experiments must be conducted, that is, a necessary condition for the identification of the electrostatic stiffness matrices is $n_{\text{exp}} \geq n_e+1$. Furthermore, we assume that the kth experiment yields m_k frequency response data points $\{\psi_{1,k}, \psi_{2,k}, \ldots, \psi_{m_k,k}\}$, $\psi_{\text{q,k}} \in \mathbb{C}^{2\times 2}$, corresponding to the frequencies $\{\omega_{1,k}, \omega_{2,k}, \ldots, \omega_{m_k,k}\}$.

By incorporating the decomposed stiffness matrices into (8.22), the minimax optimization problem for estimating the model parameters can be stated as

$$\min_{\substack{M_0>I, C_0>0 \\ K_p\leq 0, p=1,\ldots,n_e \\ K_0+\Sigma K_p(v_{\text{res}}-v_{p,k})^2>0 \\ R_l\in\mathbb{C}^{2\times 2}, l=0,1,\ldots,n_R}} \quad \max_{\substack{k=1,\ldots,n_{\text{exp}} \\ q=1,\ldots,m_k}} \overline{\sigma}(\tilde{R}_{\text{q,k}} - \psi_{\text{q,k}} Z(\omega_{\text{q,k}})), \tag{8.27}$$

where evaluating $\tilde{R}$ and Z at qth frequency point associated with the kth experiment produces

$$\tilde{R}_{\text{q,k}} := \sum_{l=0}^{n_R} R_l \omega_{\text{q,k}}^l,$$

and

$$Z(j\omega_{\text{q,k}}) = -M_0\omega_{\text{q,k}}^2 + K_0 + \sum_{p=1}^{n_e} K_p(v_{\text{res}} - v_{p,k})^2 + jC_0\omega_{\text{q,k}}.$$

The generalized eigenvalue problem for estimating parameters is then

$$\begin{aligned} &\text{minimize: } \gamma \\ &\text{subject to: } J_{\text{q,k}} > 0,\ k=1,\ldots,n_{\text{exp}},\ q=1,\ldots,m_k, \\ &\qquad M_0 > I, C_0 > 0, K_0 + \Sigma K_p(v_{\text{res}} - v_{p,k})^2 > 0, \\ &\qquad K_p \leq 0,\ p=1,\ldots,n_e, \\ &\qquad R_l \in \mathbb{C}^{2\times 2},\ l=0,\ldots,n_R \end{aligned} \tag{8.28}$$

where

$$J_{\text{q,k}} := \begin{bmatrix} \gamma I & (\tilde{R}_{\text{q,k}} - \psi_{\text{q,k}} Z(\omega_{\text{q,k}}))^* \\ \tilde{R}_{\text{q,k}} - \psi_{\text{q,k}} Z(\omega_{\text{q,k}}) & \gamma I \end{bmatrix}$$

and $\psi_{\text{q,k}}$ is the qth frequency response at frequency $\omega_{\text{q,k}}$ from the kth experiment.

Once the model parameters are obtained, the tuning bias potentials can be computed in one step by selecting v_p so that the generalized eigenvalues of

$$\omega^2 M_0 - \left(K_0 + \sum_{p=1}^{n_e} K_p (v_{\text{res}} - v_p)^2 \right) \tag{8.29}$$

are equal, that is, the modal frequencies are tuned. Although the identified model is tuned, a frequency response experiment with the predicted tuning bias potentials is prudent in order to verify that the modal frequency split is less than some acceptable criteria. The process can be repeated at the new bias potentials if necessary. Most DRGs can be tuned with $n_e = 2$ (two tuning electrodes are used), however, the tuned frequency cannot be specified. Additional degrees of freedom in which $n_e \geq 3$ makes it possible to satisfy ancillary criteria such as tuning to a specific target frequency or tuning with the smallest maximum bias potentials. This algorithm has been succesfully applied to several MEMS gyro technologies that employ electrostatic tuning including the JPL – Boeing Microgyro[7] and the JPL – Boeing Post-resonator Gyro.

In the remainder of this section, the tuning algorithm is applied to the Boeing DRG. With $n_e = 2$, the DRG requires at least three frequency response experiments that satisfy the rank condition (8.26). The following sets of bias potentials (specified in volts) are used to generate the plots in Fig. 8.8,

$$\begin{aligned} (v_{1,1}, v_{2,1}) &= (0,0), \\ (v_{1,2}, v_{2,2}) &= (-15,0), \\ (v_{1,3}, v_{2,3}) &= (-15,-15). \end{aligned} \tag{8.30}$$

The rank condition (8.26) is satisfied with these bias potentials. Specifying $n_R = 1$, the model parameters $\{M_0, C_0, K_0, K_1, K_2, R_0, R_1\}$ are identified using the three data sets with each data set spanning 15,010 Hz to 15,040 Hz with 0.1 Hz resolution. Figure 8.8 also shows the frequency responses of the identified model evaluated at the bias potentials (8.30), which confirms that the identified model fits the empirical frequency response data extremely well. Solving the following generalized eigenvalue problem yields the bias potentials that tune the modes to degeneracy

$$\begin{aligned} \text{min:} \quad & \gamma \\ \text{subject to:} \quad & \begin{bmatrix} \gamma I & P^* \\ P & \gamma I \end{bmatrix} \geq 0, \\ & \lambda \geq 0, \\ & (v_{\text{res}} - v_p)^2 \geq 0, \; p = 1,2, \end{aligned} \tag{8.31}$$

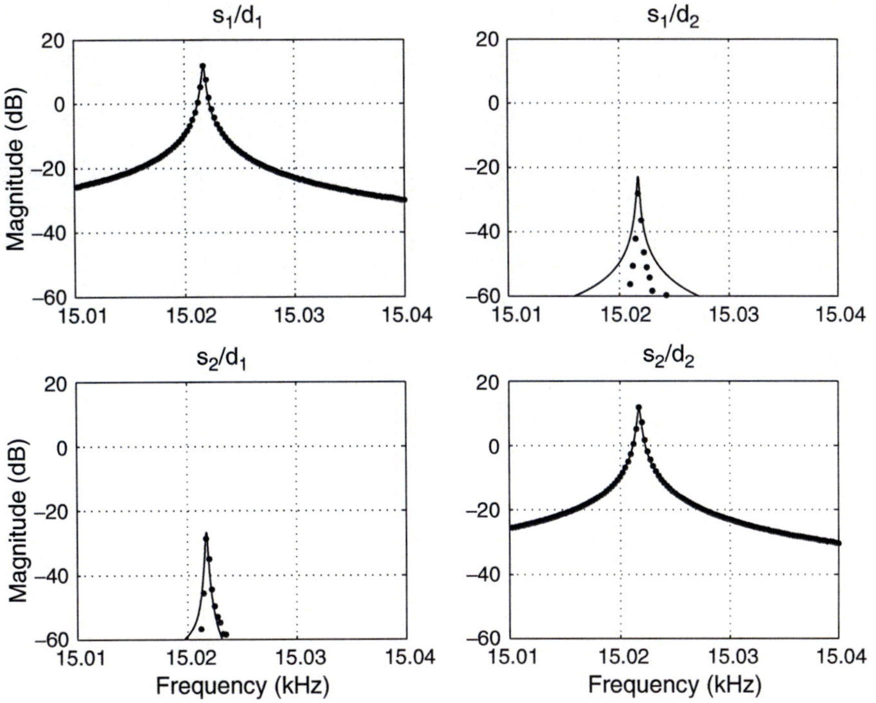

Fig. 8.9 Empirical frequency response (the data points are shown as individual points) of the four sensor channels s_1/d_1, s_2/d_2, etc, with the applied bias voltages that were computed to tune the modes to degeneracy. The frequency response of the identified model (*solid line*) evaluated at the same biases is also shown

where

$$P := \lambda M_0 - \left(K_0 + \sum_{p=1}^{2} K_p (v_{\text{res}} - v_p)^2 \right).$$

This can be efficiently solved by treating $(v_{\text{res}} - v_p)^2$, $p = 1, 2$, as decision variables, that, once computed, yield v_p since v_{res} is known. The predicted tuning bias potentials from (8.31) for the present example yields $(v_1, v_2) = (-16.71, -14.53)$, and 15,021.6 Hz is predicted as a tuned frequency. In order to verify the tuning, the predicted bias potentials are applied to the sensor and its empirical frequency response magnitude is plotted in Fig. 8.9. Only a single peak is evident in each channel and, furthermore, the responses of the off – diagonal channels are significantly reduced which is also expected from the model. A new model fit to the data indicates that the sensor is tuned within 100 mHz. The benefits of the tuned modal frequencies to the sensor performance will be quite evident from the analysis presented in Sect. 8.5.

8.4 Closed-Loop Control Architecture

The Introduction motivated the two control loops that are more or less required components of a vibratory rate sensor: the *excitation loop*, which drives a resonance in the sensor to a stable amplitude, and the *force-to-rebalance* loop, which regulates its input (the s_2 signal Figs. 8.1 and 8.4) to zero. The excitation loop is always present in vibratory rate sensors since it provides the AC waveform onto which Ω, the angular rate, is modulated as the sensor is rotated. The force-to-rebalance loop, however, is only required when the sensor bandwidth exceeds the intrinsic bandwidth of the coriolis-coupled resonances (the inverse of the decay time-constant of the resonances). When the desired sensor bandwidth does exceed the resonance bandwidth, one common approach for achieving the desired bandwidth is to *detune* the modal frequencies by approximately an amount equal to the bandwidth [12]. We show in Sect. 8.5, however, that this approach always leads to an inferior signal-to-noise ratio when compared to increasing the bandwidth of the tuned resonator via feedback.

The starting point for the controller synthesis is the open-loop, tuned, and decoupled DRG dynamics shown in Fig. 8.9. The excitation and force-to-rebalance controller designs can be carried out independently because the off-diagonal frequency response magnitudes are typically at least 30 dB lower than the diagonal channels. The force-to-rebalance loop, or simply "rebalance loop," is considered first. The primary objective of the rebalance loop is to "equalize" the response of the x_2 degree of freedom (refer to Fig. 8.4) to an applied sinusoidal angular rate of constant amplitude but of variable frequency over the desired bandwidth of the sensor. In other words, if the angular rate of the DRG case is given by $\Omega(t) = a_\Omega \sin(\omega_\Omega t)$, where a_Ω and ω_Ω are the amplitude and frequency, respectively, and where ω_Ω is constrained to a value within the desired sensor bandwidth, then it is necessary that the amplitude of x_2 in response to Ω vary in an amount that is consistent with this bandwidth (assuming a_Ω is fixed). The details of the analysis as to why this condition must be satisfied are postponed until Sect. 8.5; however, it is motivated by the desire to have a fixed sensor scale factor that is essentially independent of ω_Ω in the desired bandwidth. At issue is the rapid change in magnitude of H_{22} in a neighborhood of the operating frequency so a controller which "damps" the magnitude is desired and is achieved by implementing a controller that emulates velocity-to-force feedback on the H_{22} channel as shown in Fig. 8.10. This figure shows the loop gain, L, composed of H_{22}, the controller and the analog signal conditioning electronics, that is, $L = H_{s_2} H_{22} H_{d_2} C_{\text{reb}}$. The resonator quality factor is 40 K in this example, which translates to an open-loop bandwidth of approximately 0.19 Hz, which is too small to be useful in most applications. The closed-loop sensor, however, exhibits a much smaller time constant as evident from the complementary sensitivity function in Fig. 8.10. The complementary sensitivity function, $L/(1-L)$, is the transfer function from the rate induced disturbance to the C_{reb} controller output f_{reb} in Fig. 8.4. Note that any sinusoidal input rate Ω with frequency up to 30 Hz falls within the frequency band where the variation of

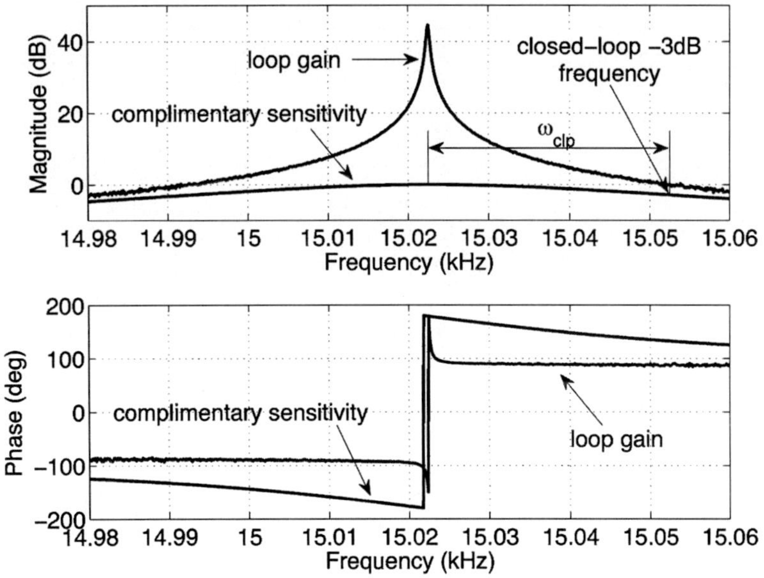

Fig. 8.10 The loop gain $L = H_{s_2} H_{22} H_{d_2} C_{\mathrm{reb}}$ of the force-to-rebalance loop and its complementary sensitivity function $L/(1-L)$. The controller phase is adjusted to dampen the mode. The closed-loop bandwidth, denoted ω_{clp}, is determined by the offset of the -3 dB frequency from the modal frequency and is about 30 Hz in this example

the closed-loop magnitude is less than 3 dB, which is the traditional definition of bandwidth. Thus, the closed-loop bandwidth is 150 times larger than the open-loop case – this opens up a host of vehicle navigation applications that would not be possible without feedback. A discussion of the impact of the pick-off noise n_{s_2} in the closed-loop sensor is postponed until Sect. 8.5.

The rebalance loop controller must not excite other structural modes in the resonator because this can cause saturation of the analog electronics – a condition which would render the DRG useless. If these "spurious" modes are sufficiently far from the modes of interest, however, then a bandpass filter may be adequate to reduce the loop gain magnitude at the spurious modes to a level where their excitation is not possible. An example is shown in Fig. 8.11 which shows the wideband measurement of the loop gain of the rebalance loop. Another benefit of the rebalance loop feedback is to increase the linearity of the sensor's response to angular rate inputs of varying amplitudes. Although the linear model of the sensor mechanics is indispensable for tuning and noise analysis, the DRGs do exhibit amplitude dependent nonlinearities due to electrostatic forces and nonlinear spring rates that give rise to Duffing-type oscillators. The rebalance loop feedback regulates the response of the H_{22} mode to a much smaller range of amplitudes where these nonlinearities are not as significant; however, this puts a greater linearity burden on the forcing electronics which is usually a desirable trade-off.

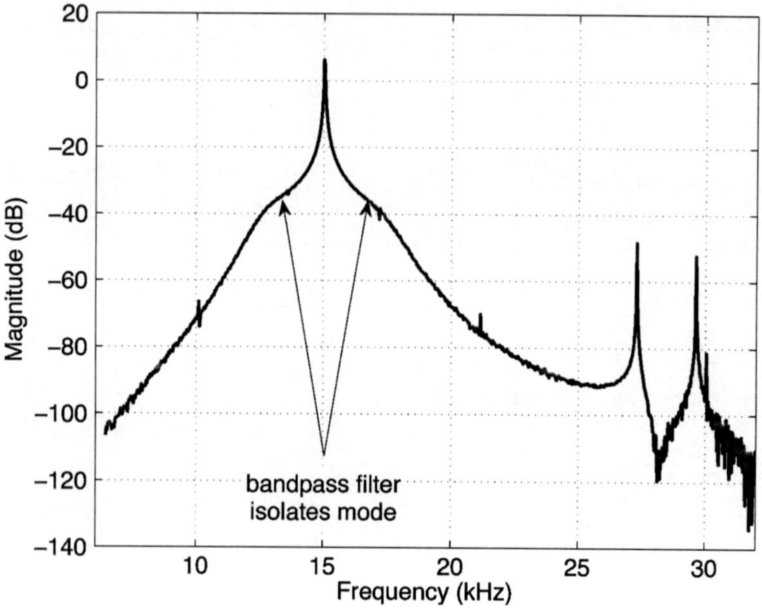

Fig. 8.11 Wideband frequency response of the rebalance loop gain. The mode of interest is near 15 kHz (same mode shown in Fig. 8.10) and is effectively isolated with a bandpass filter such that the higher frequency modes cannot be excited by feedack. Lightly damped modes that are closer to the mode of interest may require phase compensation

The synthesis of the excitation controller is now addressed. The excitation loop is necessary to drive the mode in the H_{11} channel to a constant amplitude. As demonstrated in Sect. 8.1, the $\dot{x}_1$ response provides the carrier signal onto which the angular rate input Ω is modulated, thus, a very stable amplitude is desired since any change in amplitude is reflected as a change in scale factor. Furthermore, the resonator modal frequencies will drift with time due to the fact that the elastic modulus is temperature dependent, so it is necessary for the excitation loop to track the changing modal frequency. One method for achieving these objectives is to use a phase-locked-loop (PLL). The PLL enjoys the frequency tracking property as well as imparting infinite loop gain to both the excitation and rebalance loops at the operating frequency; however, the main disadvantage of the PLL is the requirement of a low phase noise voltage controlled oscillator. Section 8.5 demonstrates that angular rate noise spectra are of interest for frequencies well below 0.1 Hz, which is within the bandwidth of the modal resonance, thus, any noise associated with the PLL oscillator will be passed to the resonator response, that is it will not be filtered. A detailed PLL analysis in the context of vibratory rate sensors is beyond

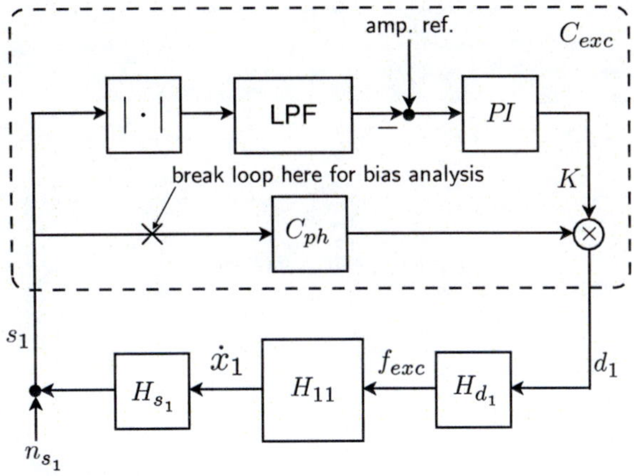

Fig. 8.12 Sensor excitation loop based on *automatic gain control*. The filter C_{ph} adjusts the phase of the feedback signal to achieve the same phase character as the rebalance loop in Fig. 8.10. Although the phase remains fixed, the amplitude of the feedback signal is modulated by the adjustable gain that is the output of the *PI* controller. The input to the *PI* compensation is the amplitude error

the scope of this chapter, especially since we employ an alternative excitation scheme called *automatic gain control* or "AGC." The basic AGC loop is shown in Fig. 8.12 and is essentially damping feedback around an oscillator in which the damping constant (including its polarity) depends on the amplitude error in the oscillator's response. For example, if the oscillator amplitude is smaller than desired, then the AGC sets the damping constant so as to *destabilize* the oscillator, thereby causing its amplitude to increase. Conversely, if the oscillator amplitude is too large, damping is added to the oscillator to decrease its response amplitude. In steady state operation at the desired amplitude, the AGC adds enough energy to the oscillator to counteract its intrinsic dissipation. The disturbance caused by non-zero Ω is not shown in this figure, however, the large rebalance loop loop gain suppresses this disturbance because $\dot{x}_2$ is regulated to be several orders of magnitude smaller then $\dot{x}_1$. The amplitude stability of this scheme as applied to a DRG is shown in Fig. 8.13 and reveals extremely good regulation of the excitation amplitude. The detailed analysis of this loop may be found in several references and so it will not be repeated here [8].

Noise analysis of the angular rate estimate is conducted in Sect. 8.5, where it is assumed that the noise of the excitation loop pick-off can be safely ignored so that only the pick-off noise associated with the rebalance loop need be considered. This assumption can be justified from the analysis of the closed-loop noise spectra but it will not be presented.

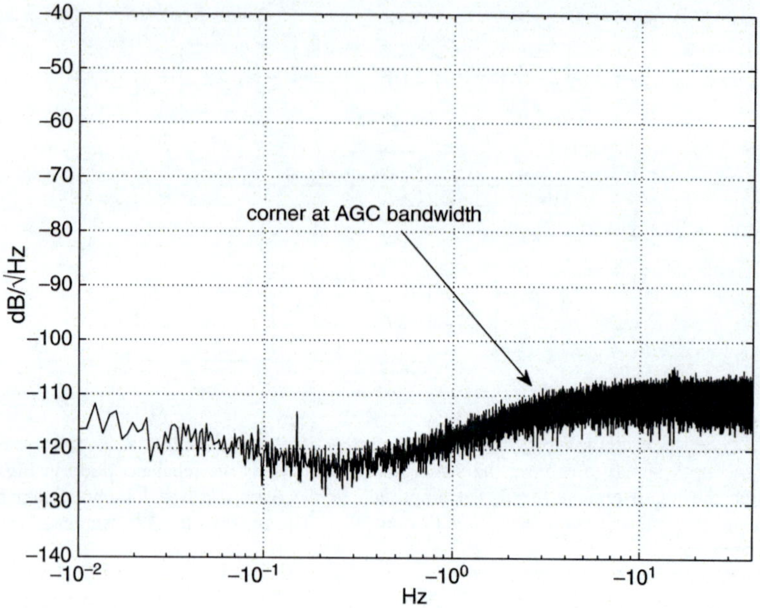

Fig. 8.13 Amplitude stability of s_1 determined by normalizing the spectrum of the amplitude of s_1 (expressed in V/rt-Hz) by the mean value of s_1 (expressed in volts). The main contribution of the amplitude noise is the n_{s_1} pick-off noise. This measurement demonstrates that a properly designed and implemented AGC can hold a very stable amplitude since -110 dB/rt-Hz corresponds to 3 ppm/rt-Hz so the variation is less than 17 ppm over the 30 Hz bandwidth

8.5 Noise Analysis

The angular rate of rotation of the DRG is estimated by demodulating the rebalance loop controller output, d_2, with respect to a phase-shifted copy of s_1 (refer to Fig. 8.4). Due to the stability of s_1 sinusoid, however, the largest noise contribution in the demodulated signal is the noise in the rebalance loop channel which can be produced by a variety of sources including DAC quantization noise if C_{reb} is implemented with a DSP, mechanical–thermal noise of the resonator, and electronic noise associated with the analog buffer electronics. In the DRG, the dominant noise source in the rebalance loop is the electrical noise associated with the rebalance loop pick-off. This noise is produced by the Johnson noise [5] of the feedback resistor in the transresistance buffers in Fig. 8.1. The Johnson noise can be modeled as the additive noise term n_{s_2} in Fig. 8.4 with a flat spectral density given by

$$\sqrt{4k_{\text{B}}TR} \quad [\text{V/rt-Hz}],$$

where k_{B} is Boltzmann's contstant, T is the resistor temperature in Kelvin, and R is the value of the resistor in ohms.

8.5.1 Open-Loop Sense Channel

Before embarking on an analysis of the n_{s_2} pick-off noise on the angular rate estimate, it is useful to have analyzed the case in which $C_{\text{reb}} = 0$, that is, the rebalance loop is left "open" and the s_2 signal (plus noise) is demodulated with respect to s_1. This is referred to as an "open-loop sense channel," and its analysis provides a baseline with which to contrast the case when $C_{\text{reb}} \neq 0$, that is "closed-loop sense channel." Coupling to

the AGC channel is ignored since this typically gives rise to longer-term trends in the rate bias that are not associated with the electrical pick-off noise and that can be treated separately. The open-loop, untuned, sensor dynamics, including signal conditioning, are shown in Fig. 8.7 and denoted H_g (see (8.20)). We will consider the open-loop scenarios in which the two modal frequencies are tuned as well as detuned. In all cases, however, it will be assumed that there is very little cross-coupling between the channels. Cross-coupling is defined by the peak gain in the off-diagonal channels and as can be seen in Fig. 8.7 there is a high degree of cross-channel coupling in the sensors untuned "native" state. The coupling can be vastly reduced, however, by tuning the sensor modes as shown in Fig. 8.9, or, alternatively, by applying a coordinate change such that the new sensor transfer function is given by $O^T H_g O$, where O is an orthonormal matrix. This process is equivalent to creating "virtual" forcers and pick-offs from the electrical input–output signals such that new, virtual electrodes are largely aligned with the anti-nodes of each mode. The decoupling is necessary to reduce bias terms, which can saturate the signal conditioning electronics, especially in the sense channel. Discussion of these bias terms is deferred until Sect. 8.6 and our standing assumption is that the AGC channel and sense channel are decoupled to first order. An example of the transformation-based decoupling is shown in Fig. 8.14, where O is computed from the eigenvectors of the stiffness matrix that is fit to the frequency response data in Fig. 8.7. It is evident that the cross-channel coupling is reduced by two orders of magnitude even though the sensor modes remain detuned.

Noise analysis of the AGC and sense channels can proceed independently because of the decoupled sensor dynamics. When we are forced to confront the fact that there exists finite coupling in Sect. 8.6, a perturbation treatment is quite sufficient to capture its influence on the zero rate bias and quadrature signals. It will also become evident that the key concept is not whether the coriolis-coupled modes are tuned, but whether the excitation frequency, which is denoted ω_0, coincides with the modal frequency of the mode that is used for sensing the angular rate. In MEM gyros, it is often necessary that the excitation frequency be chosen as a modal frequency since it is necessary to achieve a reasonable forced amplitude and, thus, if the excitation frequency coincides with the "sense" mode's modal frequency, this naturally leads to the design of axisymmetric resonators wherein modal degeneracy, if not achieved in practice, is at least the goal. Furthermore, there are additional advantages to operating the excitation loop at a modal frequency: the steep phase curve in a neighborhood of the resonance provides high sensitivity if a

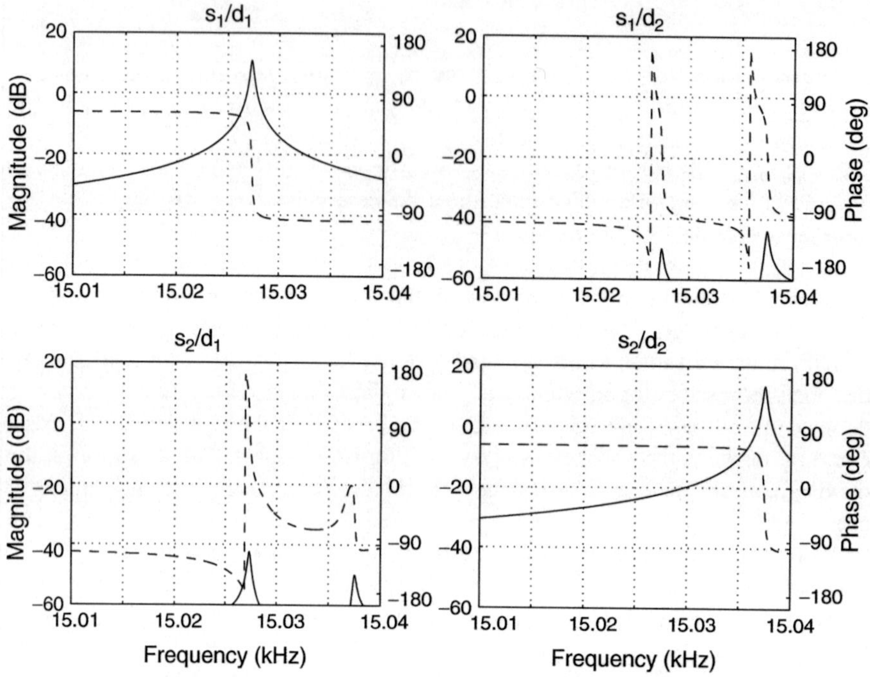

Fig. 8.14 Decoupled sensor dynamics obtained by creating "virtual" pick-offs and forces with an orthonormal transformation at the input and output of the sensor. The modes are decoupled even though they remain detuned. This configuration permits analysis of angular rate noise when the rebalance loop is open

PLL approach is used, or in the case of the AGC, the operating frequency is more robust to phase perturbations introduced by the electronics.

The sense channel dynamics under the assumption of no coupling, that is, Fig. 8.14, may be modeled as

$$m_{22}\ddot{x}_2 + c_{22}\dot{x}_2 - \alpha\Omega\dot{x}_1 + k_{22}x_2 = f_{\text{reb}}, \tag{8.32}$$

where m_{22}, c_{22}, and k_{22} are the effective mass, damping, and stiffness parameters associated with the sense channel mode (the sign change on the α term is for convenience and does not change the conclusions of the analysis). The undamped natural frequency of this resonator is denoted ω_{n},

$$\omega_{\text{n}} := \sqrt{\frac{k_{22}}{m_{22}}}. \tag{8.33}$$

The electrostatic force is produced by H_{d_2} in Fig. 8.4, however, we can assume that the dynamic element H_{d_2} can be replaced by a fixed gain, K_{d_2}, that represents the voltage-to-force conversion that takes place at the sensor input, that is $f_{\rm reb} = K_{d_2} d_2$, where d_2 is the input voltage to H_{d_2}. Similarly, the sense pick-off signal conditioning provides a voltage that is proportional to the resonator velocity at various points in the structure. Thus, the measured voltage is $s_2 := K_{s_2}\dot{x}_2 + n_{s_2}$ where K_{s_2} represents the velocity-to-voltage conversion produced by H_{s_2} and n_{s_2} is the associated pick-off noise. Thus, this model for the measurement does not include output signal conditioning dynamics other than the conversion constant, and the only effect of nonzero phase associated with H_{s_2} is to produce a constant offset from the phase predicted by (8.32). This phase offset does not affect the noise analysis to follow and so the input-output signal conditioning dynamics are not explicitly modeled other than by lumping their effects into the constant gains K_{d_2} and K_{s_2}. Rearranging (8.32) to

$$\ddot{x}_2 + 2\sigma\dot{x}_2 + \underbrace{(\sigma^2 + \omega_d^2)}_{\omega_n^2} x_2 = \frac{\alpha}{m_{22}}\Omega\dot{x}_1 + \frac{K_{d_2}}{m_{22}} d_2, \tag{8.34}$$

where $\sigma = c_{22}/(2m_{22})$ and ω_d are defined such that $\sigma^2 + \omega_d^2 = k_{22}/m_{22}$ shows that the applied input rate may be treated as a disturbance located at the sensor input, which produces a force proportional to the input rate. The coriolis force is summed with the electrostatic force to produce the net force on the x_2 degree of freedom. The open-loop analysis requires $d_2 = 0$, however, in the closed-loop analysis d_2 is specified by a controller that seeks to cancel the coriolis force disturbance.

The transfer function $H_{g,22}$ is defined from the applied forcer voltage d_2 to measurement voltage s_2 (transfer functions from signal u to signal v will often be denoted "v/u"),

$$s_2/d_2 = H_{g,22}(s) := \frac{K_{s_2}K_{d_2}}{m_{22}} \frac{s}{s^2 + 2\sigma s + \sigma^2 + \omega_d^2}. \tag{8.35}$$

The parameters K_{s_2} and K_{d_2} are not measured, although they may be analytically estimated by computing the capacitance of the electrode-resonator gap and its associated signal conditioning circuits. Similarly, m_{22} is not measured directly but may be estimated from finite element analysis of the resonator. Thus, the parameters which may be fit to the frequency response data are ω_d, σ and a gain representing $K_{s_2}K_{d_2}/m_{22}$. Note that the transfer function from the aggregate signal $\Omega\dot{x}_1$ to s_2 is

$$s_2/(\Omega\dot{x}_1) = \frac{\alpha}{K_{d_2}} H_{g,22}. \tag{8.36}$$

The fact that (8.36) is a scaled version of $H_{g,22}$ will be of use in the analysis to follow. It is possible to identify (8.36), though, but it requires subjecting the sensor to an angular rate input. For example, suppose x_1 is driven to a constant amplitude sinusoid at frequency ω_0 (the excitation frequency as defined above) and a constant rate $\Omega(t) = \Omega_0$ is subsequently applied, then the steady-state response of s_2 calibrates (8.36) at $s = j\omega_0$ and so the ratio α/K_{d_2} is determined because $H_{g,22}(j\omega_0)$ is known from the frequency response measurement of s_2/d_2.

The noise analysis in the open-loop case is now derived. It is assumed that the excitation loop establishes a constant amplitude response of the x_1 degree of freedom at frequency ω_0, that is, $x_1(t) = a\cos(\omega_0 t)$. In order to be consistent with the sensing scheme of the DRG, the measurement of s_2 in response to an applied input rate means that we compute the zero state response of $\dot{x}_2$ to an abruptly applied sinusoidal input rate $\Omega(t) = a_\Omega \cos(\omega_\Omega t)$, $t \geq 0$, which is given as follows

$$\dot{x}_2(t) = -\frac{\alpha a a_\Omega \omega_0}{2m_{22}} \Bigg[\lambda D_r(j\lambda)\cos(\lambda t) - \lambda D_i(j\lambda)\sin(\lambda t)$$

$$+ \tilde{\lambda} D_r(j\tilde{\lambda})\cos(\tilde{\lambda} t) - \tilde{\lambda} D_i(j\tilde{\lambda})\sin(\tilde{\lambda} t)$$

$$- \underbrace{\left(\sigma\left(F(\lambda)+F(\tilde{\lambda})\right) + \omega_d\left(G(\lambda)+G(\tilde{\lambda})\right)\right)}_{\mathscr{A}:=} e^{-\sigma t}\sin(\omega_d t)$$

$$+ \underbrace{\left(\omega_d\left(F(\lambda)+F(\tilde{\lambda})\right) - \sigma\left(G(\lambda)+G(\tilde{\lambda})\right)\right)}_{\mathscr{B}:=} e^{-\sigma t}\cos(\omega_d t) \Bigg],$$

where $\lambda = \omega_0 + \omega_\Omega$, $\tilde{\lambda} = \omega_0 - \omega_\Omega$, $D_r(s)$ and $D_i(s)$ are the real and imaginary parts, respectively, of $D(s) := 1/(s^2 + 2\sigma s + \sigma^2 + \omega_d^2)$, $s \in \mathbb{C}$, and $F(\lambda)$ and $G(\lambda)$ are defined as

$$F(\lambda) = \frac{1}{\omega_d}\left(-\sigma D_i(j\lambda) - \lambda D_r(j\lambda)\right),$$

$$G(\lambda) = -D_i(j\lambda),$$

similarly for $F(\tilde{\lambda})$ and $G(\tilde{\lambda})$, and where $\mathscr{A}$ and $\mathscr{B}$ are defined as shown. The measured signal in response to the applied rate is

$$s_2(t) = K_{s_2}\dot{x}_2(t) + n_{s_2}(t),$$

where n_{s_2} is the pick-off noise associated with the signal conditioning electronics. The measured excitation signal is

$$s_1(t) = K_{s_1}\dot{x}_1(t)$$

$$= -K_{s_1} a\omega_0 \sin(\omega_0 t),$$

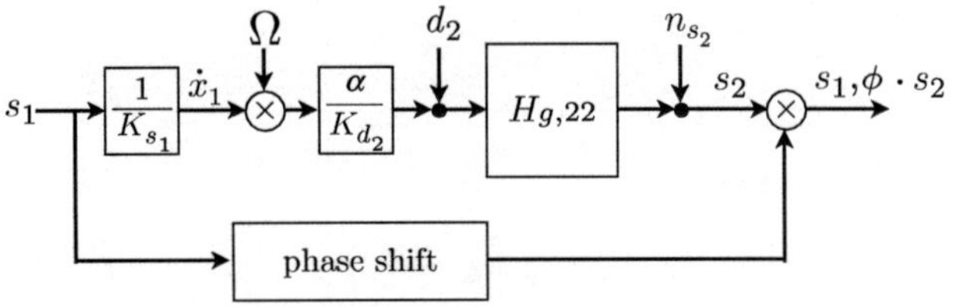

Fig. 8.15 Open-loop operation demodulates the sense pick-off signal s_2 with respect to a phase shifted version of s_1. The phase shift ϕ is chosen to maximize (8.38)

where K_{s_1} is defined in a similar spirit to K_{s_2}. The pick-off noise, n_{s_1}, associated with the measurement electronics can be ignored. The angular rate is estimated by demodulating s_2 with respect to a phase shifted copy of s_1, denoted $s_{1,\phi}$,

$$s_{1,\phi}(t) = -K_{s_1} a\omega_0 \sin(\omega_0 t + \phi),$$

where ϕ represents the phase shift. Figure 8.15 shows a block diagram. The phase shift is chosen to maximize the response of the demodulated signal in response to a constant rate input as will be shown below. The product of $s_{1,\phi}$ and s_2 is

$$\begin{aligned}
s_{1,\phi}(t)s_2(t) = K_{s_1}K_{s_2}&\frac{\alpha a^2 a_\Omega \omega_0^2}{4m_{22}} \\
&\times \Big[\lambda D_r(j\lambda)\big(\sin((2\omega_0+\omega_\Omega)t+\phi) - \sin(\omega_\Omega t - \phi)\big) \\
&- \lambda D_i(j\lambda)\big(\cos(\omega_\Omega t - \phi) - \cos((2\omega_0+\omega_\Omega)t+\phi)\big) \\
&+ \tilde{\lambda} D_r(j\tilde{\lambda})\big(\sin((2\omega_0-\omega_\Omega)t+\phi) + \sin(\omega_\Omega t+\phi)\big) \\
&- \tilde{\lambda} D_i(j\tilde{\lambda})\big(\cos(\omega_\Omega t+\phi) - \cos((2\omega_0-\omega_\Omega)t+\phi)\big) \\
&- \mathscr{A} e^{-\sigma t}\big(\cos((\omega_d-\omega_0)t-\phi) - \cos((\omega_d+\omega_0)t+\phi)\big) \\
&+ \mathscr{B} e^{-\sigma t}\big(\sin((\omega_d+\omega_0)t+\phi) - \sin((\omega_d-\omega_0)t-\phi)\big)\Big] \\
&- K_{s_1} a\omega_0 n_{s_2}(t)\sin(\omega_0 t+\phi).
\end{aligned}$$

This signal is filtered to remove components with frequency near, and above, ω_0,

$$
\begin{aligned}
&\mathrm{LPF}(s_{1,\phi}(t)s_2(t))\\
&\quad = -K_{s_1}K_{s_2}\frac{\alpha a^2 a_\Omega \omega_0^2}{4m_{22}}\Big[\lambda D_r(j\lambda)\sin(\omega_\Omega t-\phi)+\lambda D_i(j\lambda)\cos(\omega_\Omega t-\phi)\\
&\quad -\tilde{\lambda}D_r(j\tilde{\lambda})\sin(\omega_\Omega t+\phi)+\tilde{\lambda}D_i(j\tilde{\lambda})\cos(\omega_\Omega t+\phi)\\
&\quad +\mathscr{A}e^{-\sigma t}\cos((\omega_d-\omega_0)t-\phi)+\mathscr{B}e^{-\sigma t}\sin((\omega_d-\omega_0)t-\phi)\Big]\\
&\quad -\mathrm{LPF}\big((K_{s_1}a\omega_0 n_{s_2}(t)\sin(\omega_0 t+\phi)\big),
\end{aligned}
\tag{8.37}
$$

where the designation "LPF(·)" indicates low, pass filtering of the argument. We assume that the power spectrum of n_{s_2} is independent of frequency within a sufficiently large neighborhood of ω_0. If the intensity of n_{s_2} is μ [V/rt-Hz] in a neighborhood of ω_0, then the intensity of $\mathrm{LPF}\big(n_{s_2}(t)\sin(\omega_0 t+\phi)\big)$ is also μ [V/rt-Hz], independent of frequency, in a low frequency band that includes the origin.

The scale factor is computed by setting $\omega_\Omega = 0$, that is, the angular rate input is constant, and observing the steady-state response which simplifies to (ignoring the noise term for the moment)

$$
\lim_{t\to\infty}\mathrm{LPF}(s_{1,\phi}(t)s_2(t)) = K_{s_1}K_{s_2}\frac{\alpha a^2 a_\Omega \omega_0^3}{2m_{22}}|D(j\omega_0)|\sin(\phi-\angle D(j\omega_0)), \tag{8.38}
$$

where $|D(j\omega_0)|$ and $\angle D(j\omega_0)$ are the magnitude and phase of $D(j\omega_0)$. The demodulation phase, denoted ϕ_d, is chosen to maximize (8.38). In other words, $\phi_\mathrm{d} := \pi/2+\angle D(j\omega_0)$. Note that when $\omega_0 \approx \omega_\mathrm{n}$, the demodulation phase is close to 0°, however, if ω_0 is detuned away from ω_n, the demodulation phase is close to 90°.

The scale factor, denoted γ_sf, is the term multiplying a_Ω in (8.38) when $\phi = \phi_\mathrm{d}$,

$$
\begin{aligned}
\lim_{t\to\infty}\mathrm{LPF}(s_{1,\phi_\mathrm{d}}(t)s_2(t)) &= K_{s_1}K_{s_2}\frac{\alpha a^2\omega_0^3}{2m_{22}}|D(j\omega_0)|a_\Omega\\
&= \frac{K_{s_1}\alpha a^2\omega_0^2}{2K_{d_2}}\left|\frac{K_{s_2}K_{d_2}}{m_{22}}j\omega_0 D(j\omega_0)\right|a_\Omega\\
&= \frac{K_{s_1}\alpha a^2\omega_0^2}{2K_{d_2}}\left|H_{g,22}(j\omega_0)\right|a_\Omega,
\end{aligned}
$$

so

$$
\gamma_\mathrm{sf}(\omega_0) := \frac{K_{s_1}\alpha a^2\omega_0^2}{2K_{d_2}}\left|H_{g,22}(j\omega_0)\right|,
$$

where ω_0 is left as a parameter. Of particular interest is the scale factor when $\omega_0 = \omega_n$. In this case, γ_{sf} reduces to

$$\gamma_{sf}(\omega_n) = \frac{K_{s_1} K_{s_2} \alpha a^2 \omega_n^2}{4\sigma m_{22}}. \tag{8.39}$$

This result will be used when the closed-loop sensor is analyzed.

The scale factor contains constants which are difficult to measure in practice (although they may be estimated from analytical models of the electrode arrangement and finite element analysis of the resonator), so γ_{sf} is typically determined experimentally. Its units are V/deg/s or V/deg/hr. The scale factor is written this way to show that it is proportional to a transfer function which can be measured, that is $H_{g,22}$. Note that if ω_0 is near the modal frequency of $H_{g,22}$, then the scale factor is proportional to the quality factor of the resonator. Thus, increasing the quality factor is a means of increasing the scale factor. The real-time estimate of the rate, denoted Ω_{est}, can be determined by setting $\phi = \phi_d$ in (8.37) and dividing the result by γ_{sf},

$$\begin{aligned}\Omega_{est}(t) &= \frac{1}{\gamma_{sf}(\omega_0)} \mathrm{LPF}(s_{1,\phi}(t) s_2(t)) \\ &= M(\omega_\Omega;\Delta)\cos(\omega_\Omega t - \psi(\omega_\Omega;\Delta)) - \frac{\mathscr{A}}{\gamma_{sf}} \mathrm{e}^{-\sigma t} \cos((\omega_d - \omega_0)t - \phi_d) \\ &\quad - \frac{\mathscr{B}}{\gamma_{sf}} \mathrm{e}^{-\sigma t} \sin((\omega_d - \omega_0)t - \phi_d) - \frac{K_{s_1} a \omega_0}{\gamma_{sf}} \mathrm{LPF}\big(n_{s_2}(t)\sin(\omega_0 t + \phi_d)\big).\end{aligned} \tag{8.40}$$

The magnitude and phase functions associated with harmonic angular rate inputs are denoted M and ψ, respectively, and are functions of the the excitation frequency, ω_0, and the degree of detuning of the excitation frequency from the modal frequency ω_n, which is the parameter $\Delta := \omega_n - \omega_0$.

$$\begin{aligned}M(\omega_\Omega;\Delta) &= \frac{\sqrt{\mathscr{C}^2 + \mathscr{S}^2}}{2\omega_0 |D(j\omega_0)|}, \\ \psi(\omega_\Omega;\Delta) &= \arctan(\mathscr{S}/\mathscr{C}),\end{aligned} \tag{8.41}$$

where

$$\begin{aligned}\mathscr{C} := &(\omega_0 + \omega_\Omega) D_r\big(j(\omega_0 + \omega_\Omega)\big)\cos(\angle D(j\omega_0)) \\ &+ (\omega_0 + \omega_\Omega) D_i\big(j(\omega_0 + \omega_\Omega)\big)\sin(\angle D(j\omega_0)) \\ &+ (\omega_0 - \omega_\Omega) D_r\big(j(\omega_0 - \omega_\Omega)\big)\cos(\angle D(j\omega_0)) \\ &+ (\omega_0 - \omega_\Omega) D_i\big(j(\omega_0 - \omega_\Omega)\big)\sin(\angle D(j\omega_0))\end{aligned} \tag{8.42}$$

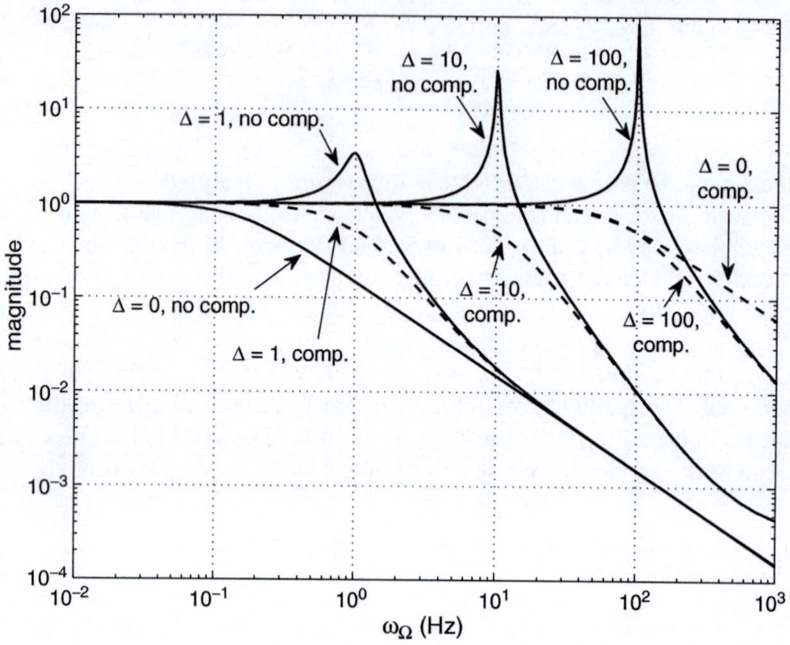

Fig. 8.16 Magnitude function versus ω_Ω for four detuning frequencies $\Delta \in \{0,1,10,100\}$. The uncompensated magnitudes (denoted "no comp.") exhibit resonant peaking, however, mutliplication by a simple notch filter in the cases when $\Delta = \{1,10,100\}$ provides a critically damped 2-pole roll-off (denoted "comp."). A single pole phase lead filter is used to extend the bandwidth of the $\Delta = 0$ case

and

$$\begin{aligned}\mathscr{S} := &-(\omega_0+\omega_\Omega)D_r\big(j(\omega_0+\omega_\Omega)\big)\sin(\angle D(j\omega_0))\\ &+(\omega_0+\omega_\Omega)D_i\big(j(\omega_0+\omega_\Omega)\big)\cos(\angle D(j\omega_0))\\ &+(\omega_0-\omega_\Omega)D_r\big(j(\omega_0-\omega_\Omega)\big)\sin(\angle D(j\omega_0))\\ &-(\omega_0-\omega_\Omega)D_i\big(j(\omega_0-\omega_\Omega)\big)\cos(\angle D(j\omega_0)).\end{aligned} \tag{8.43}$$

The magnitude function is plotted versus ω_Ω for several choices of ω_0 in Fig. 8.16. For consistency with the sensor data presented in this chapter, the following parameters are selected: $\omega_n = 15\,\text{kHz}$, quality factor is $50\,\text{K}$ ($\sigma = 0.94$); ω_Ω is the independent variable. The magnitude response function exhibits a peak at the detuning frequency Δ (for $\Delta \neq 0$) and, furthermore, the zero-state response (8.40) contains transient terms with exponential decay rate equal to σ and frequency equal to Δ. Thus, when $\Delta \neq 0$, processing Ω_{est} with a notch filter of the form

$$\frac{s^2+2\sigma s+\Delta^2}{s^2+2\Delta s+\Delta^2},$$

will equalize the magnitude as shown in Fig. 8.16 and eliminate the slowly decaying transient. In this case, the $-3\,\text{dB}$ bandwidth is approximately 0.6Δ when $\Delta \neq 0$ and follows a critically damped two-pole roll-off. For the case when $\Delta = 0$, in other words, the excitation frequency coincides with the resonant frequency ω_n, no compensation is need to equalize the magnitude, however, the bandwidth is quite small and is equal to σ. As a means of extending the sensor bandwidth when $\Delta = 0$, Ω_{est} can be filtered with a phase lead filter of the form

$$\frac{s/\sigma + 1}{s/\sigma_{\text{clp}} + 1},$$

where σ_{clp} denotes the desired bandwidth. An example is provided in Fig. 8.16 in which $\sigma_{\text{clp}} = 60\,\text{Hz}$ to match the bandwidth of the $\Delta = 100\,\text{Hz}$ case after compensation with the notch filter. The foregoing analysis suggests that detuning the excitation frequency from the modal frequency ω_n is an effective way to change the sensor bandwidth (after suitable filtering with a notch filter). On the other hand, it is also possible to extend the bandwidth by means of a phase lead filter when $\omega_0 = \omega_n$. The question is settled as to which approach is preferred when the rate equivalent noise is considered.

The rate equivalent noise in (8.40) is

$$-\frac{K_{s_1} a \omega_0}{\gamma_{\text{sf}}} \text{LPF}\big(n_{s_2}(t)\sin(\omega_0 t + \phi)\big), \tag{8.44}$$

and has the same units as Ω_{est}, that is deg/hr or deg/sec. Since the noise density of the low-pass filtered term is μ V/$\sqrt{\text{Hz}}$, then the density of (8.44), denoted S_Ω since it is associated with the angular rate estimate, is

$$S_\Omega(\omega_\Omega;\omega_0) = \frac{1}{\gamma_{\text{sf}}(\omega_0)} K_{s_1} a \omega_0 \mu \;\; [\text{deg/hr/rt-Hz}],$$

where it is assumed that γ_{sf} is given in V/deg/hr. Note that S_Ω is independent of frequency. The excitation frequency ω_0 is considered a parameter. If a notch filter or phase lead filter is used to equalize the magnitude of s_2 in response to an angular rate input, however, then the noise density becomes frequency dependent, although in the cases when the notch filter is used the density is still relatively flat up to the bandwidth. In order to compare the rate equivalent noise densities across different cases, we will take as the reference case the constant density of the rate equivalent noise when $\omega_0 = \omega_n$ and no phase lead filter is used thus yielding the following reference value of S_Ω,

$$S_\Omega(\omega_\Omega;\omega_n) = \frac{4\sigma m_{22}}{K_{s_2} \alpha a \omega_n} \mu \;\; [\text{deg/hr/rt-Hz.}] \tag{8.45}$$

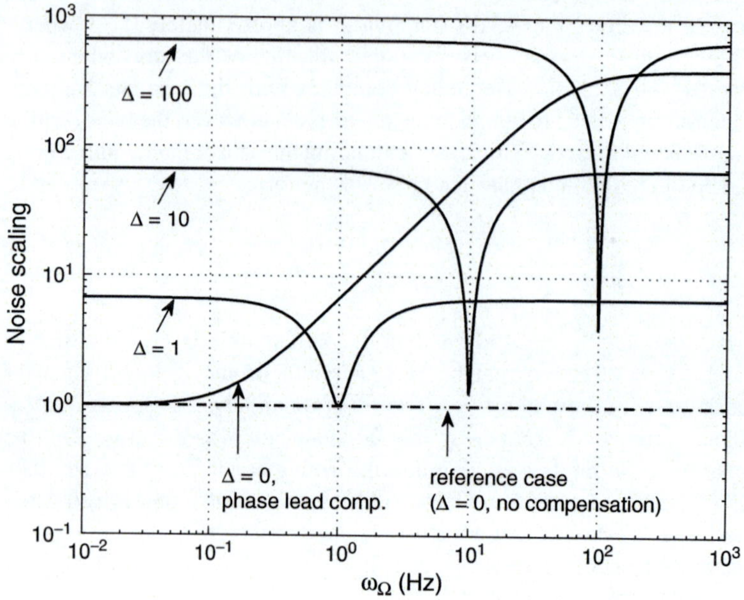

Fig. 8.17 The noise scaling (8.46) for different detuning frequencies. The reference case is one for all frequencies and corresponds to $\Delta = 0$ with no additional filtering (the bandwidth is σ). On the other hand, in order to increase the bandwidth, detuning can be employed at the expense of increasing the noise density. For a given desired bandwidth, the lowest noise is achieved when $\Delta = 0$ and the phase lead filter is used to increase the sensor bandwidth

This reference case is used to normalize the densities for all other cases and yields the following noise scaling,

$$\frac{\text{rate noise with comp.}}{\text{reference noise}} = \frac{|C(j\omega_\Omega)|S_\Omega(\omega_\Omega;\omega_0)}{S_\Omega(\omega_\Omega;\omega_\mathrm{n})} = \left| C(j\omega_\Omega)\frac{D(j\omega_\mathrm{n})}{D(j\omega_0)} \right|, \qquad (8.46)$$

where C denotes the frequency response of the notch or phase lead filter, if used. A plot of the noise scaling for the same cases considered in Fig. 8.16 is shown in Fig. 8.17. The figure reveals that while detuning the excitation frequency from the modal frequency does increase the sensor bandwidth, it is at the expense of a uniform increase in the rate equivalent noise density. In fact, for a given desired bandwidth, the lowest rate equivalent noise is achieved across that bandwidth under the condition when $\Delta = 0$ and a phase lead filter is used in which case the spectrum is

$$|C(j\omega_\Omega)|\,S_\Omega(\omega_\Omega;\omega_\mathrm{n}) = \frac{4\sigma m_{22}}{K_{\mathrm{s}_2}\alpha a \omega_\mathrm{n}}\mu\sqrt{\frac{(\omega_\Omega/\sigma)^2+1}{(\omega_\Omega/\sigma_{\mathrm{clp}})^2+1}}. \qquad (8.47)$$

When comparing the cases in which the sensor bandwidth is about 60 Hz, the low frequency noise density is almost three orders of magnitude smaller when $\Delta = 0$ as compared to $\Delta = 100$. Furthermore, when $\Delta = 0$, the low frequency density matches the noise density of the uncompensated case.

Extending the sensor bandwidth by employing a phase lead filter when $\Delta = 0$ is susceptible to uncertainties in the sensor dynamics. In particular, an accurate model of the damping is required in order to successfully "invert" this aspect of the sensor dynamics. It is possible, however, to achieve the same objectives using feedback as will be demonstrated in the next section.

8.5.2 *Closed-Loop Sense Channel*

Feedback can be used to achieve, in a practical way, the desired sensor bandwidth while minimizing the rate-equivalent noise. The analysis of Sect. 8.5.1 revealed that the optimum open-loop mode of operation is to select the excitation frequency to coincide with the modal frequency of the second degree of freedom, that is, select $\omega_0 = \omega_n$, and then employ a phase lead filter to increase the bandwidth of the demodulated signal. This produced the lowest rate-equivalent noise compared to any detuning scheme and, consequently, it is assumed for the remainder of the chapter that $\omega_0 = \omega_n$, although small deviations from this condition will be analyzed. The filter essentially inverts the dynamics of the second channel s_2/d_2, albeit at "baseband" since the filter is implemented postdemodulation, but this requires precise knowledge of the plant, especially with regard to the damping constant σ of the mode. Feedback, however, can generate an approximate plant inverse that has precisely the same effect as the phase lead filter on the rate-equivalent noise but it can also tolerate deviations of the plant dynamics from a nominal case, deviations that would require a redesign of the phase lead filter. Closing the loop has other advantages as well: running the sensor with s_2 "open" requires that the signal conditioning electronics be designed for larger dynamic range, which may place a limit on the front-end amplifier gain, thereby reducing the SNR ratio, in order to avoid saturation of the electronics. Even more important may be the nonlinearities of the resonant structure that can become significant at larger response amplitudes.

The baseband inversion of the sense channel dynamics corresponds to implementing a classical phase lead filter. The plant inversion can also be achieved by implementing a "wideband" feedback controller in which the lightly damped resonant mode is actively damped. Recall that the transfer function $s_2/\Omega\dot{x}_1$ (see (8.36)) is proportional to $H_{g,22}$ so the objective of the feedback controller is to generate an inverse of $H_{g,22}$ over a limited frequency range since filtering s_2 with this approximate inverse equalizes the response to an applied angular rate. The approximate inverse is generated by the closed-loop transfer function

$$\frac{C_{\text{reb}}}{1 + H_{g,22}C_{\text{reb}}} \approx H_{g,22}^{-1}, \quad \text{when } |H_{g,22}C_{\text{reb}}| \gg 1,$$

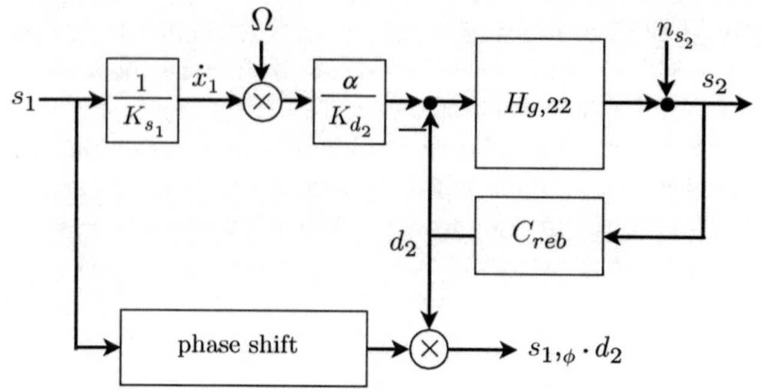

Fig. 8.18 Block diagram corresponding to a closed-loop sense channel. C_{reb} is a high gain controller whose output attempts to cancel the angular rate induced "disturbance" at the input of $H_{g,22}$. This perspective shows that the scale factor is independent of the quality factor of the resonator

where C_{reb} is the so-called *rebalance loop* filter. This transfer function suggests the block diagram in Fig. 8.18 for closed-loop operation. The transfer function from the rate induced signal, injected at the plant input, to the output of C_{reb}, that is, d_2, is approximately "1" under the assumption of large loop gain. Although the applied angular rate "disturbance" is not directly measured, it is inferred from the canceling action taken by the controller. Whether the disturbance at the input of $H_{g,22}$ is due to the device physics, or injected by the engineer, the loop properties may be analyzed by breaking it at d_2. Since $H_{g,22}$ is a sharp resonant peak, C_{reb} is designed so that the loop phase approximates that of velocity feedback – this choice maximizes the damping and minimizes peaking in the sensitivity function. Measurements of the loop gain in a narrow band near the sense channel mode in the DRG are shown in Fig. 8.10. The controller magnitude is selected so that the desired closed-loop bandwidth, denoted ω_{clp}, is achieved but since there may exist other lightly damped out-of-band modes, the controller magnitude is rolled off above and below the band of interest as shown in Fig. 8.11. This figure shows that other lightly damped modes above 20 kHz have been filtered out by the controller. The 3 dB bandwidth is onc-half the interval where the loop gain magnitude is greater than one. Figure 8.10 shows that $\omega_{\text{clp}} = 30$ Hz for this example.

The starting point for rigorous analysis of the closed-loop noise characteristics are the open-loop transfer functions (8.35) and (8.36). As mentioned above, the phase of C_{reb} is chosen so that the loop gain $H_{g,22}C_{\text{reb}}$ emulates velocity-to-force feedback. In practice, this is accomplished by compensating for any phase lag introduced by signal conditioning electronics, and also in the case of digital controller implementation, the delay caused by sampling. The resulting loop gain phase permits the treatment of the dynamic filter C_{reb} as a fixed gain, denoted K_{reb}, for the purposes of noise analysis. The closed-loop transfer function from the signal $\Omega \cdot \dot{x}_1$ to d_2 is

$$d_2/(\Omega \cdot \dot{x}_1) = \frac{\alpha}{K_{d_2}} \frac{C_{\text{reb}} H_{g,22}}{1 + C_{\text{reb}} H_{g,22}} = \alpha \frac{K_{\text{reb}} K_{s_2}}{m_{22}} \frac{s}{s^2 + 2\sigma_{\text{clp}} s + \omega_{\text{n}}^2},$$

where C_{reb} has been replaced by K_{reb} and where closed-loop bandwidth is given by

$$\sigma_{\text{clp}} := \sigma + K_{\text{reb}} \frac{K_{s_2} K_{d_2}}{2m_{22}}.$$

As in the open-loop analysis we assume

$$\dot{x}_1(t) = -a\omega_0 \sin(\omega_0 t)$$

$$\Omega(t) = a_\Omega \cos(\omega_\Omega t).$$

Although in practice $\omega_0 = \omega_{\text{n}}$, ω_0 will be left as a parameter so the effects of detuning can be studied. The *steady-state* response of d_2 to the input $\Omega \cdot \dot{x}_1$ is

$$d_2(t) = -\alpha a a_\Omega \omega_0 \frac{K_{\text{reb}} K_{s_2}}{2m_{22}} \Big[\lambda \hat{D}_{\text{r}}(j\lambda) \cos(\lambda t) - \lambda \hat{D}_{\text{i}}(j\lambda) \sin(\lambda t)$$

$$+ \tilde{\lambda} \hat{D}_{\text{r}}(j\tilde{\lambda}) \cos(\tilde{\lambda} t) - \tilde{\lambda} \hat{D}_{\text{i}}(j\tilde{\lambda}) \sin(\tilde{\lambda} t) \Big] + n_{d_2}(t),$$

where $\lambda := \omega_0 + \omega_\Omega$, $\tilde{\lambda} := \omega_0 - \omega_\Omega$, $\hat{D}_{\text{r}}(s)$ and $\hat{D}_{\text{i}}(s)$ are the real and imaginary parts of $\hat{D}(s) := 1/(s^2 + 2\sigma_{\text{clp}} s + \omega_{\text{n}}^2)$, $s \in \mathbb{C}$, and where the noise term n_{d_2} is due to the pick-off noise n_{s_2}. The transfer function from n_{s_2} to d_2 is

$$d_2/n_{s_2} = \frac{C_{\text{reb}}}{1 + H_{g,22} C_{\text{reb}}}$$

$$= K_{\text{reb}} \frac{s^2 + 2\sigma s + \omega_{\text{n}}^2}{s^2 + 2\sigma_{\text{clp}} s + \omega_{\text{n}}^2},$$

Thus, if the noise density of n_{s_2}, denoted S_{s_2}, is $S_{s_2}(\omega) = \mu$ V/rt-Hz, for all ω in a neighborhood of ω_0, then the noise density of n_{d_2}, denoted S_{d_2}, is

$$S_{d_2}(\omega) = \left| K_{\text{reb}} \frac{\omega_{\text{n}}^2 - \omega^2 + j2\sigma\omega}{\omega_{\text{n}}^2 - \omega^2 + j2\sigma_{\text{clp}}\omega} \right| \mu \;\; [\text{V/rt-Hz}], \tag{8.48}$$

which demonstrates that in the closed-loop case the noise of the signal to be demodulated is frequency dependent and possess a deep notch at ω_{n}.

The rate estimate and scale factor are obtained by demodulating d_2 with respect to a phase shifted copy of s_1, defined as $s_{1,\phi}(t) := -K_{s_1} a\omega_0 \sin(\omega_0 t + \phi)$. Noise associated with s_1 is ignored. The product $s_{1,\phi}(t)\cdot d_2(t)$ is low-pass filtered and yields the following steady-state terms plus noise,

$$\begin{aligned}
&\mathrm{LPF}(s_{1,\phi}(t)\cdot d_2(t))\\
&= \alpha a^2\omega_0^2 K_{\mathrm{reb}}\frac{K_{s_1}K_{s_2}}{4m_{22}}a_\Omega\\
&\quad\Big[-\lambda\hat{D}_{\mathrm{r}}(j\lambda)\sin(\omega_\Omega t-\phi)-\lambda\hat{D}_{\mathrm{i}}(j\lambda)\cos(\omega_\Omega t-\phi)\\
&\quad+\tilde{\lambda}\hat{D}_{\mathrm{r}}(j\tilde{\lambda})\sin(\omega_\Omega t+\phi)-\tilde{\lambda}\hat{D}_{\mathrm{i}}(j\tilde{\lambda})\cos(\omega_\Omega t+\phi)\Big]\\
&\quad-\mathrm{LPF}\left(K_{s_1}a\omega_0 n_{d_2}(t)\sin(\omega_0 t+\phi)\right).
\end{aligned} \tag{8.49}$$

The scale factor and optimum demodulation phase are computed when $\omega_\Omega = 0$,

$$\lim_{t\to\infty}\mathrm{LPF}(s_{1,\phi}(t)\cdot d_2(t)) = \alpha a^2\omega_0^3 K_{\mathrm{reb}}\frac{K_{s_1}K_{s_2}}{2m_{22}}|\hat{D}(j\omega_0)|a_\Omega\sin\left(\phi-\angle\hat{D}(j\omega_0)\right),$$

where the noise term has been momentarily ignored because it has no bearing on the scale factor. The optimal demodulation phase is $\phi_{\mathrm{d}} = \pi/2+\angle\hat{D}(j\omega_0)$ and the scale factor is

$$\gamma_{\mathrm{sf}} = \alpha a^2\omega_0^3 K_{\mathrm{reb}}\frac{K_{s_1}K_{s_2}}{2m_{22}}|\hat{D}(j\omega_0)|.$$

Dividing (8.49) by γ_{sf} yields the steady-state-plus-noise estimate of the applied angular rate,

$$\Omega_{\mathrm{est}}(t) = M(\omega_\Omega;\Delta)\cos\left(\omega_\Omega t-\psi(\omega_\Omega;\Delta)\right)-\frac{K_{s_1}a\omega_0}{\gamma_{\mathrm{sf}}}\mathrm{LPF}\left(n_{d_2}(t)\sin(\omega_0 t+\phi)\right),$$

where the magnitude and phase functions M and ψ are given by the same expressions as those in (8.41), (8.42), and (8.43) except that $\hat{D}$ replaces D in all instances. Ideally, $\Delta = 0$, however, it is left as a parameter to explore the effects of small, unintentional detuning. Figure 8.19 shows four cases of $M(\omega_\Omega;\Delta)$ when $\Delta\in\{0,1,10,100\}$. Note that the bandwidth is quite robust to large detuning. The phase function ψ exhibits similar robustness. Another important fact is that γ_{sf} is essentially constant for these cases since γ_{sf} can be written as

$$\gamma_{\mathrm{sf}} = \alpha a^2\omega_0^2\frac{K_{s_1}}{2K_{d_2}}\underbrace{\left(K_{\mathrm{reb}}\frac{K_{s_2}K_{d_2}}{m_{22}}\omega_0|\hat{D}(j\omega_0)|\right)}_{\approx 1},$$

where the term in parenthesis is approximately one for $\Delta\in[0,10]$. Thus, in contrast to the open-loop case, the closed-loop scale factor is essentially independent of the

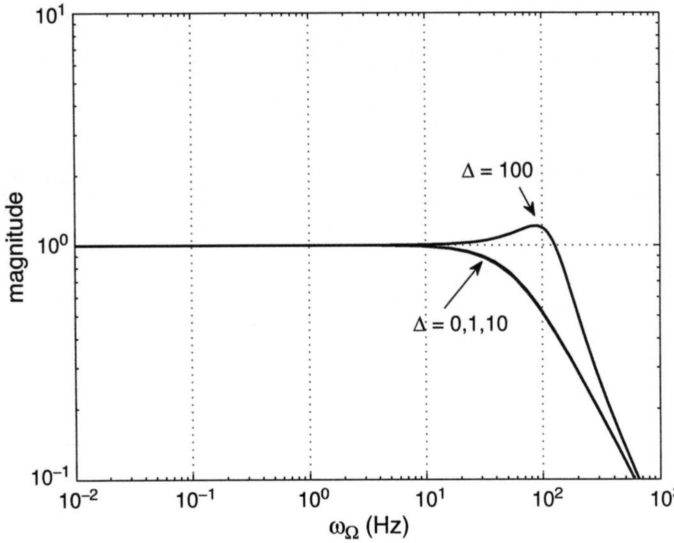

Fig. 8.19 Frequency response associated with closed-loop sensor for detuning frequencies of $\Delta \in \{0, 1, 10, 100\}$ Hz

difference between the excitation frequency and the modal frequency. Furthermore, the scale factor is independent of the modal quality factor as long as the closed-loop bandwidth is larger than the open-loop bandwidth σ.

The trade-off between tuned versus detuned in the closed-loop sensor is revealed by analyzing the rate-equivalent noise that corrupts $\Omega_{\rm est}$, which is given as,

$$-\frac{K_{s_1} a\omega_0}{\gamma_{\rm sf}} \mathrm{LPF}\left(n_{d_2}(t)\sin(\omega_0 t + \phi)\right). \tag{8.50}$$

Before analyzing the characteristics of the noise associated with the demodulated signal, it is useful to consider n_{d_2} prior to demodulation. The density can be given units of deg/hr/rt-Hz by normalizing it with respect to the closed-loop scale factor but since the density is a function of frequency (S_{d_2} has a deep notch at $\omega_{\rm n}$) it is informative to make the comparison with the open-loop case at a particular frequency. The rate-equivalent noise density will be computed at $\omega = \omega_{\rm n}$ and it is also assumed that $\omega_0 = \omega_{\rm n}$, although the preceding analysis has shown that the closed-loop scale factor is essentially independent of ω_0 as long as $\Delta < \sigma_{\rm clp}$, that is, the degree of detuning is less than the closed-loop bandwidth. The normalized density at $\omega = \omega_{\rm n}$ reduces to

$$\frac{1}{\gamma_{\rm sf}} S_{d_2}(\omega_{\rm n}) = \frac{4\sigma m_{22}}{K_{s_1} K_{s_2} \alpha a^2 \omega_{\rm n}^2} \mu \quad \text{(closed-loop).}$$

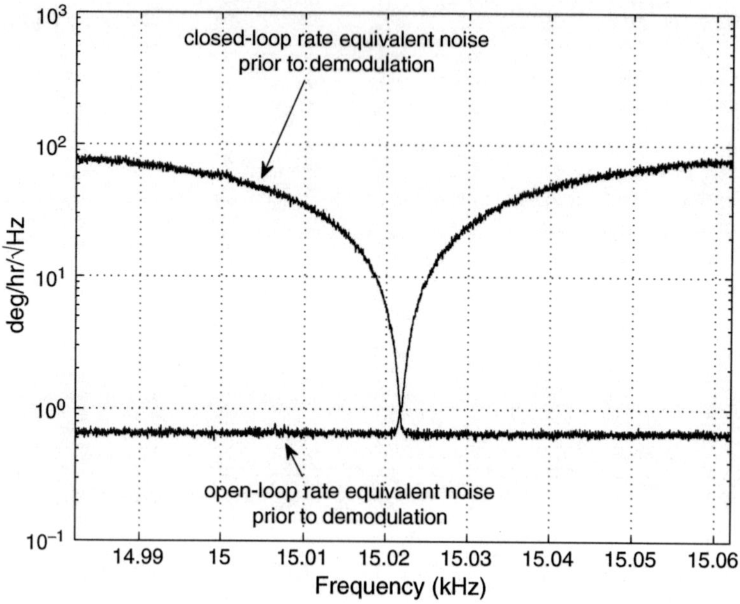

Fig. 8.20 Rate-equivalent noise, prior to demodulation, in the open-loop and closed-loop cases, shown in a neighborhood of $\omega_{\rm n} \approx 15.022\,$kHz. The two noise densities are equal at $\omega_{\rm n}$. The price of increasing the bandwidth is the higher rate-equivalent noise for frequencies not equal to $\omega_{\rm n}$

On the other hand, if the open-loop density S_{s_2} is normalized by the open-loop scale factor when $\omega_0 = \omega_{\rm n}$ (see (8.39)), then the same result is obtained. This produces the spectra shown in Fig. 8.20 where this data was taken from an operational DRG. Note that at $\omega_{\rm n}$ (the bottom of the notch in the closed-loop spectra) the two spectra touch as shown since values of both densities are equal. These spectra are demodulated with respect to a phase shifted copy of s_1 to produce the noise spectra associated with $\Omega_{\rm est}$ in both open- and closed-loop cases.

Returning to the angular rate noise in (8.50), the spectrum of the low-pass filtered term is

$$\frac{1}{\sqrt{2}}\sqrt{S_{d_2}^2\left(\omega_0+\omega_\Omega\right)+S_{d_2}^2\left(\omega_0-\omega_\Omega\right)}\ \ [\mathrm{V/rt\text{-}Hz}],$$

so S_Ω is

$$S_\Omega(\omega_\Omega) = \frac{\sqrt{2}K_{d_2}}{\alpha a \omega_0}\sqrt{S_{d_2}^2\left(\omega_0+\omega_\Omega\right)+S_{d_2}^2\left(\omega_0-\omega_\Omega\right)}\ \ [\mathrm{deg/hr/rt\text{-}Hz}]. \qquad (8.51)$$

In the case where $\omega_0 = \omega_{\rm n}$, S_Ω reduces to

$$
\begin{aligned}
S_\Omega(\omega_\Omega) &= \frac{2K_{d_2}K_{\text{reb}}}{\alpha a \omega_{\text{n}}}\mu \left| \frac{\hat{D}(j(\omega_{\text{n}}+\omega_\Omega))}{D(j(\omega_{\text{n}}+\omega_\Omega))} \right| \\
&= \frac{2K_{d_2}K_{\text{reb}}}{\alpha a \omega_{\text{n}}}\mu \left| \frac{\omega_{\text{n}}^2-(\omega_{\text{n}}+\omega_\Omega)^2+j2\sigma(\omega_{\text{n}}+\omega_\Omega)}{\omega_{\text{n}}^2-(\omega_{\text{n}}+\omega_\Omega)^2+j2\sigma_{\text{clp}}(\omega_{\text{n}}+\omega_\Omega)} \right| \\
&= \frac{2K_{d_2}K_{\text{reb}}}{\alpha a \omega_{\text{n}}}\mu \sqrt{\frac{(2\omega_{\text{n}}\omega_\Omega+\omega_\Omega^2)^2+4\sigma^2(\omega_{\text{n}}+\omega_\Omega)^2}{(2\omega_{\text{n}}\omega_\Omega+\omega_\Omega^2)^2+4\sigma_{\text{clp}}^2(\omega_{\text{n}}+\omega_\Omega)^2}} \quad [\text{deg/hr/rt-Hz}].
\end{aligned}
$$

Although not explicitly stated, the excitation frequency is typically orders of magnitude larger than the input rates we are interested in measuring, that is, $\omega_{\text{n}} \gg \omega_\Omega$. Thus, if we retain only the dominant terms in the previous expression, S_Ω simplifies to

$$
\begin{aligned}
S_\Omega(\omega_\Omega) &\approx \frac{2K_{d_2}K_{\text{reb}}}{\alpha a \omega_{\text{n}}}\mu \sqrt{\frac{(2\omega_{\text{n}}\omega_\Omega)^2+4\sigma^2\omega_{\text{n}}^2}{(2\omega_{\text{n}}\omega_\Omega)^2+4\sigma_{\text{clp}}^2\omega_{\text{n}}^2}} \\
&= \frac{2K_{d_2}K_{\text{reb}}}{\alpha a \omega_{\text{n}}}\frac{\sigma}{\sigma_{\text{clp}}}\mu \sqrt{\frac{(\omega_\Omega/\sigma)^2+1}{(\omega_\Omega/\sigma_{\text{clp}})^2+1}} \\
&= \frac{4\sigma m_{22}}{K_{s_2}\alpha a \omega_{\text{n}}}\mu \underbrace{\sqrt{\frac{(\omega_\Omega/\sigma)^2+1}{(\omega_\Omega/\sigma_{\text{clp}})^2+1}}}_{\text{phase lead magnitude}} \quad [\text{deg/hr/rt-Hz}]. \qquad (8.52)
\end{aligned}
$$

Upon comparing (8.52) with (8.47), we see that the closed-loop rate-equivalent noise under the condition $\omega_0 = \omega_{\text{n}}$ matches the open-loop rate equivalent noise when a phase-lead filter is employed to increase the bandwidth. Thus, the optimum noise properties associated with the "tuned" and compensated open-loop sense channel are in fact achieved with the closed-loop sense channel. The spectral density of the closed-loop DRG rate noise is shown in Fig. 8.21. The parameters associated with this DRG are: $\omega_0 = \omega_{\text{n}} \approx 15\,\text{kHz}$, $\sigma = \omega_{\text{n}}/(2Q) = 0.25\,\text{Hz}$, $\sigma_{\text{clp}} = 30\,\text{Hz}$. Note that the corner frequencies correspond to σ and σ_{clp}.

The effect of detuning ω_0 from ω_{n} can be investigated by returning to (8.51). The detuning is not intentional since the objective of closed-loop operation is to recover the noise characteristics of the tuned open-loop sense channel. In practice, however, some small amount of detuning may be present and we wish to determine its impact on the angular rate estimate with the closed-loop sense channel. Detuning has a detrimental effect on the rate-equivalent noise owing to the deep notch in S_{d_2}. The general effect of detuning is to raise the low frequency noise floor as demonstrated in the Introduction (see Fig. 8.5) by a factor of

$$
\left| \frac{D(j\omega_0)}{D(j\omega_{\text{n}})} \right|. \qquad (8.53)
$$

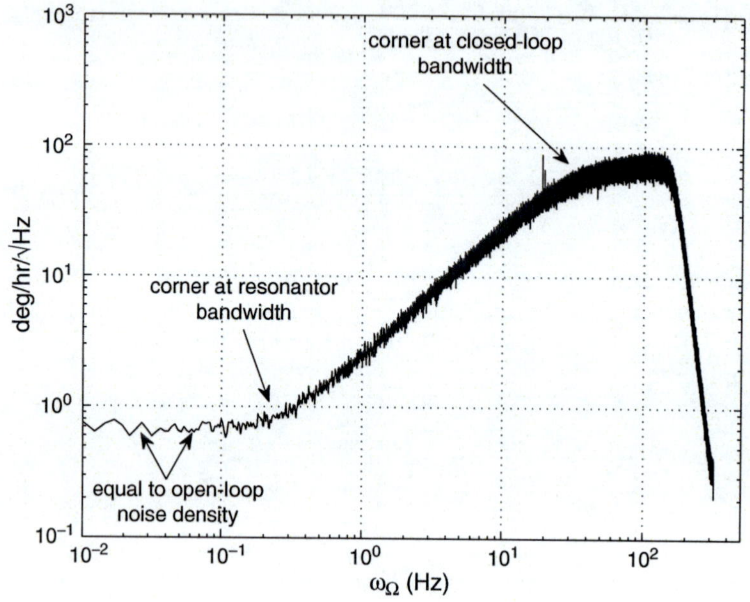

Fig. 8.21 The spectrum of the closed-loop rate-equivalent noise when $\omega_0 = \omega_n$. The lower corner frequency is equal to the open-loop bandwidth of the resonator, the higher corner frequency is equal to the closed-loop bandwidth and the value of the density at low frequencies is equal to open-loop rate-equivalent density. Increasing the resonator Q pushes the lower frequency corner to left and lowers the noise floor. The steep roll-off above 100 Hz is due to anti-alias filtering

Measurements with a DRG (different from the one which produced the data in Fig. 8.21) are compared in Fig. 8.22 when the sensor is tuned and when $\Delta = 1$ Hz.

8.5.3 Angle Uncertainty Due to Angular Rate Noise

The angular rate estimate, Ω_{est}, is often integrated over a fixed interval of time of duration τ seconds to provide an estimate of the *change in angle* experienced by the sensor. Thus, the noise that corrupts the angular rate estimate is integrated and corrupts the estimate of the change in angle. The trends in S_Ω in Fig. 8.21 lead to two distinct trends in the uncertainty attached to the estimate of the change in angle. The low frequency "white" portion of S_Ω is associated with the well-known angle random walk of the angle estimate, while the higher frequency noise in S_Ω where the slope is one produces angle white noise for short integration times.

The uncertainty of the angle estimate can be computed directly using the time series of the angular rate estimate or a frequency domain approach which uses S_Ω. In the time-domain approach, the rate signal is acquired when the sensor is fixed

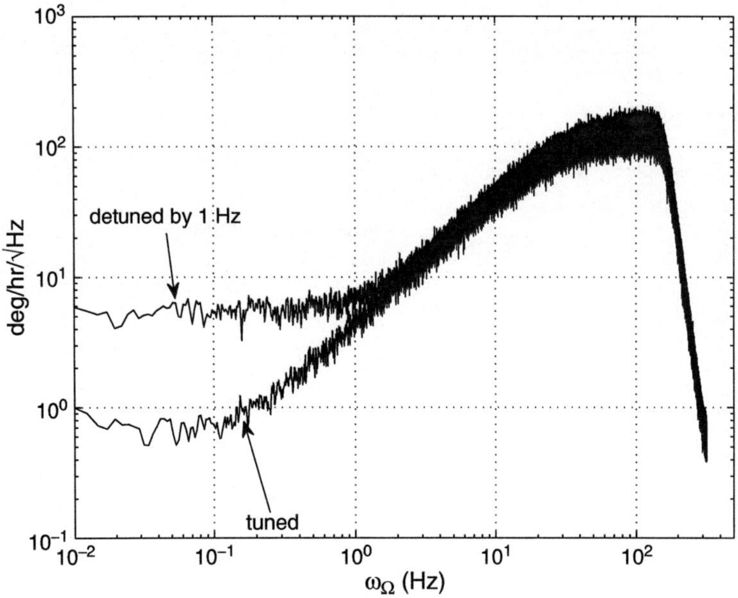

Fig. 8.22 The spectrum of a closed-loop DRG with $\Delta = 1\,\text{Hz}$ compared to the tuned case. Detuning has the effect of raising the low frequency noise floor according to (8.53), which, in the present case with $\omega_n \approx 15\,\text{kHz}$ and $Q = 65\,\text{k}$, computes to 8.7 (the data were taken using a different DRG than the one tested for Fig. 8.21). This value accurately predicts the increase in the measured low-frequency noise floor

$(\Omega(t) = 0)$ and then the bias is removed and the subsequent signal is integrated for τ seconds. The quantity of interest is the value of the integrated signal at the end of the integration interval. This integration process is repeated until a reasonable estimate of the standard deviation of the angle at the end of the integration interval is obtained. The standard deviation associated with each integration interval is then plotted versus τ.

It is also possible to compute the angle standard deviation using S_Ω. The impulse response of the "gated" integrator is given by

$$h_\tau(t) = \begin{cases} 1 & t \in [0, \tau] \\ 0 & t \notin [0, \tau] \end{cases}, \tag{8.54}$$

where τ represents the integration duration in units of seconds. The variance of the angle produced from integrating the rate noise over τ seconds is given by

$$\sigma_\tau^2 := \int_0^\infty S_\Omega^2(\omega)|H_\tau(\omega)|^2 \mathrm{d}\omega, \tag{8.55}$$

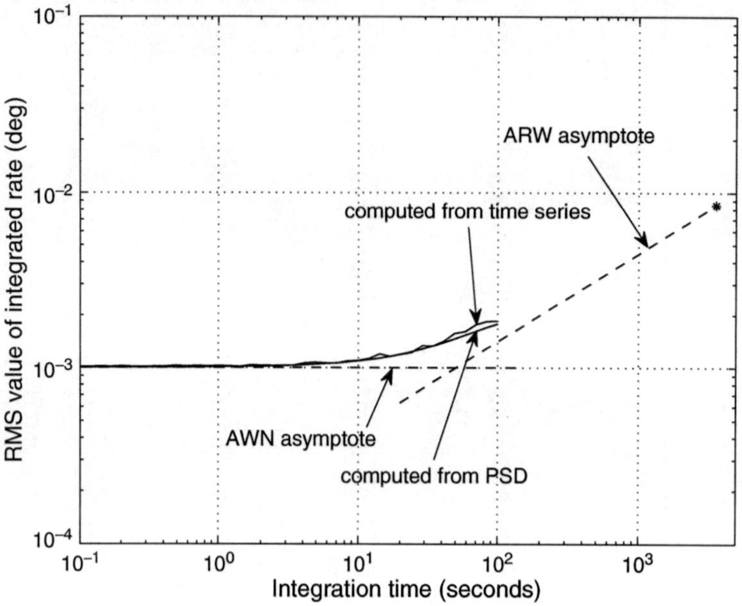

Fig. 8.23 Angle noise computed from the angular rate time series and the angular rate power spectrum in Fig. 8.21. The angle white noise (AWN) and angle random walk (ARW) asymptotes are also shown. The asymptotes are derived from (8.56) and (8.57) and show excellent agreement with the measurement

where S_Ω^2 is the power spectral density of the rate noise expressed in $(\text{deg/s})^2/\text{Hz}$, and H_τ is the Fourier transform of (8.54),

$$H_\tau(\omega) = \frac{1}{\mathrm{j}2\pi\omega}\left(1 - \mathrm{e}^{-\mathrm{j}2\pi\tau\omega}\right), \quad \omega \in (-\infty, \infty),\ \omega \text{ in Hz.}$$

The same data set that was used to produce S_Ω in Fig. 8.21 is analyzed to compute the noise properties of the integrated rate as a function of τ. Figure 8.23 shows the standard deviation of the integrated rate, that is σ_τ, computed by analyzing the time series data as well as the frequency domain data. These two approaches yield nearly identical results. The upper limit of τ is 100 s and is dictated by the length of the data set, that is, enough subrecords are required to obtain a reliable estimate of the standard deviation. The figure shows that for integration times up to about 10 s the standard deviation of the angle remains relatively constant at 0.001 deg. This is *angle white noise*, abbreviated AWN, and is produced from the +1 slope trend in Fig. 8.21. For longer integration times, however, the "flat" trend in the PSD at low frequencies dominates the behavior of the angle uncertainty.

The angle random walk (ARW) asymptote in Fig. 8.23 is an analytical computation of the standard deviation of the angle determined by integrating the rate noise according to (8.55) when S_Ω is set to a constant, denoted c expressed in deg/s/rt-Hz, that matches the value of low frequency density (the low frequency "white" section of S_Ω). Substituting c and the expression for H_τ into (8.55) yields,

$$\sigma_\tau^2 = \int_0^\infty c^2 \left| \frac{1}{\mathrm{j}2\pi\omega} \left(1 - \mathrm{e}^{-\mathrm{j}2\pi\tau\omega}\right) \right|^2 \mathrm{d}\omega = \frac{1}{\pi} c^2 \tau \int_0^\infty \frac{1-\cos\omega}{\omega^2} \mathrm{d}\omega = \frac{1}{2} c^2 \tau.$$

Thus, the expression for the angle random walk is

$$\sigma_\tau = c\sqrt{\frac{\tau}{2}}\ [\mathrm{deg}] \qquad (c \text{ expressed in deg/s/rt-Hz, } \tau \text{ in seconds}). \tag{8.56}$$

At one hour of integration, this expression yields a canonical piece of information associated with gyros used in navigation, and expresses σ_τ like a density,

$$\sigma_{1\mathrm{hr}} = 42.4c\ [\mathrm{deg/rt\text{-}hr}].$$

From the data in Fig. 8.21, the low-frequency noise density is approximately 0.72 deg/hr/rt-Hz so $c = 0.72/3{,}600 = 0.0002$ deg/s/rt-Hz which yields $\sigma_{1\mathrm{hr}} = 0.0085$ deg/rt-hr. This point is shown as the "$\star$" in Fig. 8.23. For comparison, the Honeywell GG1320AN Digital Laser Gyro, the GG1320AN01 Digital Laser Gyro, and the GG5200 MEMS Rate Gyro have a quoted ARWs of 0.0035 deg/rt-hr, 0.01 deg/rt-hr, and 0.2 deg/rt-hr, respectively.

The PSD approach is also useful in explaining the angle white noise trend in the integrated rate data. The +1 trend in rate spectral density can be fit with a simple first-order high-pass filter of the form

$$S_\Omega(\omega) = \beta \left| \frac{\mathrm{j}2\pi\omega}{\mathrm{j}2\pi\omega + 2\pi\sigma_{\mathrm{clp}}} \right|,$$

where σ_{clp} is the closed-loop bandwidth given in Hz, and β is the constant density at frequencies above σ_{clp}, expressed in deg/s/rt-Hz. Thus, (8.55) reduces to

$$\begin{aligned}
\sigma_\tau^2 &= \frac{\beta^2}{4\pi^3\sigma_{\mathrm{clp}}^2} \int_0^\infty (1-\cos(\tau\omega)) \underbrace{\frac{(2\pi\sigma_{\mathrm{clp}})^2}{\omega^2 + (2\pi\sigma_{\mathrm{clp}})^2}}_{NEBW=\pi^2\sigma_{\mathrm{clp}}} d\omega \\
&\approx \frac{\beta^2}{4\pi^3\sigma_{\mathrm{clp}}^2} \int_0^{\pi^2\sigma_{\mathrm{clp}}} (1-\cos(\tau\omega))\, d\omega \\
&= \frac{\beta^2}{4\pi\sigma_{\mathrm{clp}}} \left(1 - \frac{1}{\tau\pi^2\sigma_{\mathrm{clp}}} \sin(\tau\pi^2\sigma_{\mathrm{clp}})\right),
\end{aligned}$$

where the upper integration limit is replaced by the noise-equivalent bandwidth (NEBW) of the low-pass filter. Furthermore, the sinusoidal term has diminishing significance for $\tau \gg 1/(\pi^2 \sigma_{\text{clp}})$, which covers the range of τ of practical interest. Thus, the standard deviation of the angle due to this noise component simplifies to

$$\sigma_\tau = \frac{\beta}{2\sqrt{\pi \sigma_{\text{clp}}}} \text{ [deg]}, \tag{8.57}$$

where σ_{clp} is expressed in Hz. Note that (8.57) is independent of τ: this produces the angle white noise shown in Fig. 8.23. If we carry out the computation with the spectrum in Fig. 8.21 (the same data set that generated Fig. 8.23), we find $\sigma_\tau = 0.001$ degrees because $\beta = 0.02$ deg/s/rt-Hz and $\sigma_{\text{clp}} = 30$ Hz. This prediction of the AWN very closely matches the empirical results. Note although it seems as if σ_τ in (8.57) is inversely proportional to the square root of the sensor bandwidth, when increasing the bandwidth of a given sensor, the noise level β will increase in the same proportion. Thus, we conclude that, for a given sensor, the AWN is proportional to the square root of the sensor bandwidth, that is, a fourfold increase in sensor bandwidth doubles the angle white noise value.

8.6 Analysis of Bias Terms

The preceding section analyzed the features of the estimated rate noise induced by the electronic pick-off noise. In this section, we identify some sources of *zero rate bias* (ZRB) and *quadrature* signals as well as demonstrate that there are two critical dynamic elements whose phases must be accurately identified and rendered stable in order to prevent the "mixing" of the ZRB and quadrature signals. This analysis focuses on elements in Fig. 8.4 other than the sensor dynamics that can produce ZRB and quadrature signals. For vibratory gyros in a tuning fork configuration, the classic references [9, 10] discuss how nonidealities of the vibrating element can produce the these signals. In the generic model (8.1), these nonidealities create the off-diagonal terms in the mass, stiffness, and damping matrices.

Our starting point is to break the loop at the point just upstream of C_{ph} as shown in Fig. 8.12; however, we further imagine that this loop is embedded in the block diagram of Fig. 8.4 since we desire to study the effects of *cross-coupling* in the sensor dynamics – indeed, it is cross-coupling which produces nonzero rate bias and quadrature signals. The input to C_{ph} is denoted u and since we are interested in the bias value of the rate estimate, we can assume $\Omega = 0$. Since the transfer function representation is convenient, we use the ^ notation to denote the Laplace transform of a function. The transfer function from u to $\dot{x}_1$ and $\dot{x}_2$ is computed to be

$$\begin{bmatrix} \hat{\dot{x}}_1 \\ \hat{\dot{x}}_2 \end{bmatrix} = \frac{1}{\Theta} \begin{bmatrix} 1 - H_{s_2} H_{22} H_{d_2} C_{\text{reb}} & H_{s_2} H_{12} H_{d_2} C_{\text{reb}} \\ 0 & 1 \end{bmatrix} \begin{bmatrix} H_{11} H_{d_1} K C_{\text{ph}} \\ H_{21} H_{d_1} K C_{\text{ph}} \end{bmatrix} \hat{u}, \tag{8.58}$$

where

$$\Theta = 1 - H_{s_2} H_{22} H_{d_2} C_{reb},$$

where K represents the AGC "gain" as shown in Fig. 8.12. Under the assumption of large rebalance loop gain, that is,

$$\left| H_{s_2} H_{22} H_{d_2} C_{reb} \right| \gg 1,$$

(8.58) simplifies to

$$\hat{x}_1 = H_{11} H_{d_1} K C_{ph} \left(1 - \frac{H_{12} H_{21}}{H_{11} H_{22}} \right) \hat{u} \tag{8.59}$$

and

$$\hat{x}_2 = -\frac{H_{21} H_{d_1} K C_{ph}}{H_{s_2} H_{22} H_{d_2} C_{reb}} \hat{u}. \tag{8.60}$$

A further simplification is possible in (8.59) because the off-diagonal terms in the sensor transfer function are orders of magnitude smaller than the diagonal components, thus,

$$\left| \frac{H_{12} H_{21}}{H_{11} H_{22}} \right| \ll 1,$$

so we may approximate

$$\hat{x}_1 = H_{11} H_{d_1} K C_{ph} \hat{u}. \tag{8.61}$$

Finally, the return signal at the point where the loop is broken is s_1, thus,

$$\begin{aligned} \hat{s}_1 &= H_{s_1} \hat{x}_1 \\ &= H_{s_1} H_{11} H_{d_1} K C_{ph} \hat{u}. \end{aligned}$$

When the loop is closed, we assume the following steady-state condition is reached

$$s_1(t) = u(t) = A \cos(\omega_0 t),$$

where A is the desired amplitude and ω_0 the operating frequency. At this operating condition, the following expression holds

$$H_{s_1} H_{11} H_{d_1} K C_{ph} \big|_{j\omega_0} = 1, \tag{8.62}$$

where the $|_{j\omega_0}$ notation indicates that the transfer functions are evaluated at $j\omega_0$. Thus, the operating frequency is determined to be the frequency where

$$\mathscr{I} \left(H_{s_1} H_{11} H_{d_1} K C_{ph} \big|_{j\omega_0} \right) = 0,$$

where "$\mathscr{I}$" denotes the *imaginary part* of the expression. In other words, the excitation frequency corresponds to the frequency at which the Nyquist plot of

$H_{s_1}H_{11}H_{d_1}KC_{ph}$ crosses the real axis. The AGC gain magnitude, K, is adjusted such that,

$$K = \frac{1}{H_{s_1}H_{11}H_{d_1}C_{ph}\big|_{j\omega_0}}.$$

The rebalance loop gain in Fig. 8.10 is designed with the same phase target as the excitation loop, namely the peak magnitude occurs at either 0° phase or 180° phase depending on whether the mode is to be stabilized as in the case of the rebalance loop or destabilized as in the case of the excitation loop.

The force-to-rebalance signal, d_2, can be expressed in terms of $\dot{x}_2$, so using (8.60) gives the transfer function from u to d_2,

$$\begin{aligned}\hat{d}_2/\hat{u} &= (\hat{d}_2/\hat{\dot{x}}_2)/(\hat{\dot{x}}_2/\hat{u}) \\ &= (C_{reb}H_{s_2})\left(-\frac{H_{21}H_{d_1}KC_{ph}}{H_{s_2}H_{22}H_{d_2}C_{reb}}\right) \\ &= -\frac{H_{21}H_{d_1}KC_{ph}}{H_{22}H_{d_2}}. \end{aligned} \quad (8.63)$$

The analysis so far has ignored the contribution of the sensor's angular velocity Ω to d_2. This component can be determined by assuming a constant angular rate of rotation $\Omega(t) = \Omega_0$ and determining the $\hat{d}_2/\hat{u}$ transfer function under this condition. Recall that the rate is estimated by demodulating f_{reb} with respect to $\dot{x}_1$, but since we only have access to the signals s_1 and d_2, it will be shown that the dynamics of H_{s_1} and H_{d_2} become important in the angular rate estimate. The angular rate-induced disturbance introduced into the rebalance loop (refer to Fig. 8.4) is approximately cancelled, under the assumption of high gain feedback, by $f_{reb}(t) \approx -\alpha\Omega(t)\dot{x}_1(t)$. Since $\hat{f}_{reb} = H_{d_2}\hat{d}_2$, the component of d_2 in response to an applied constant angular rate of rotation Ω_0 is

$$\begin{aligned}\hat{d}_2 &= \frac{1}{H_{d_2}}\hat{f}_{reb} \\ &= \frac{1}{H_{d_2}}(-\alpha\Omega_0\hat{\dot{x}}_1) \\ &= -\frac{\alpha\Omega_0}{H_{d_2}H_{s_1}}\hat{u}, \end{aligned}$$

where (8.61) was used in the last step. Combining this result with (8.63) gives the expression for d_2 including the disturbance caused by a constant angular rate of rotation

$$\hat{d}_2 = -\frac{H_{21}H_{d_1}KC_{ph}}{H_{22}H_{d_2}}\hat{u} - \frac{\alpha\Omega_0}{H_{d_2}H_{s_1}}\hat{u}.$$

Subsitituting (8.62) into this expression yields

$$\hat{d_2} = -\frac{1}{H_{d_2}H_{s_1}}\left(-\frac{H_{21}}{H_{11}H_{22}} - \alpha\Omega_0\right)\hat{u}.$$

Under the condition that the diagonal terms in the sensor transfer function are dominant, the expression $-H_{21}/(H_{11}H_{22})$ may be approximated as the (2,1) term of H^{-1} which is, given (8.1), the (2,1) term of Z divided by $j\omega_0$. In other words,

$$-\left.\frac{H_{21}}{H_{11}H_{22}}\right|_{j\omega_0} \approx \frac{1}{j\omega_0}\left(-\omega_0^2 m_{21} + j\omega_0 c_{21} + k_{21}\right).$$

Thus, the steady-state response of d_2 to an applied constant rate Ω_0 is

$$d_2(t) = \frac{A}{|H_{d_2}(j\omega_0)H_{s_1}(j\omega_0)|}\Big((c_{21} - \alpha\Omega_0)\cos\left(\omega_0 t - \measuredangle H_{d_2}(j\omega_0) - \measuredangle H_{s_1}(j\omega_0)\right)$$
$$-(m_{21}\omega_0 - k_{21}/\omega_0)\sin\left(\omega_0 t - \measuredangle H_{d_2}(j\omega_0) - \measuredangle H_{s_1}(j\omega_0)\right)\Big).$$

The angular rate is associated with the $\cos\left(\omega_0 t - \measuredangle H_{d_2} - \measuredangle H_{s_1}\right)$ term and the *quadrature* signal is associated with the $\sin\left(\omega_0 t - \measuredangle H_{d_2} - \measuredangle H_{s_1}\right)$ term. The estimated angular rate is obtained by demodulating d_2 with respect to $A\cos(\omega_0 t + \phi)$, which is a copy of s_1 phase shifted by ϕ, and low-pass filtering the result. Thus, the demodulated and filtered signal is

$$\frac{A^2}{2|H_{d_2}(j\omega_0)H_{s_1}(j\omega_0)|}\Big((c_{21} - \alpha\Omega_0)\cos\left(\phi + \measuredangle H_{d_2}(j\omega_0) + \measuredangle H_{s_1}(j\omega_0)\right)$$
$$-(m_{21}\omega_0 - k_{21}/\omega_0)\sin\left(\phi + \measuredangle H_{d_2}(j\omega_0) + \measuredangle H_{s_1}(j\omega_0)\right)\Big).$$

The demodulation phase is ideally chosen as $\phi = -\measuredangle H_{d_2}(j\omega_0) - \measuredangle H_{s_1}(j\omega_0)$ because this eliminates any quadrature signal from the angular rate estimate – note, though, that even if $\Omega_0 = 0$, there is still a spurious rate signal proportional to the cross-axis damping coefficient, c_{21}. Since it is difficult to measure the phases of the H_{d_2} and H_{s_1} elements in practice, a low-frequency angular rate Ω is applied to the gyro and ϕ is adjusted until the quadrature has no correlation with Ω.

Figure 8.24 demonstrates the effect of phase perturbations to H_{s_1}, H_{s_2}, H_{d_1}, and H_{d_2}.

For example, if these elements are fixed at their nominal values with the exception of H_{s_1} whose phase is perturbed by approximately +4.7 degrees, then the zero rate bias is also perturbed along with the operating frequency as shown in Fig. 8.24. In this case, the nonideal demodulating phase incorporates some quadrature into the zero-rate bias. Furthermore, since the phase of the excitation

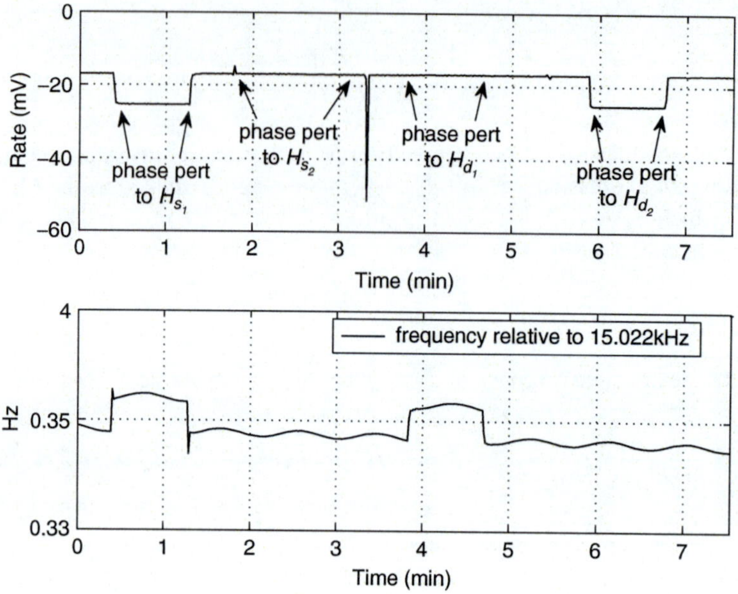

Fig. 8.24 Phase perturbation applied to H_{s_1}, H_{s_2}, H_{d_1}, and H_{d_2}. The top plot shows the effect of these perturbations on a signal proportional to the zero rate bias (the demodulated signal is not multiplied by the scale factor; it is expressed in *mV*). The bottom plot shows the gyro operating frequency. These measurements support the analysis since phase perturbations to H_{s_1} and H_{d_2} are expected to change the zero rate bias, while phase perturbations to H_{s_1} and H_{d_1} are expected to change the operating frequency

loop is perturbed, the operating frequency will undergo a perturbation as well, with the magnitude of the perturbation being determined by the slope of the excitation loop phase curve in a neighborhood of the resonant frequency. This slope is computed to be $2Q/\omega_n$, where Q is the quality factor of the resonator and ω_n is its natural frequency. In the present example $2Q/\omega_n = 2(50{,}000)/(2\pi 15{,}020) \approx 1$ (unit of "seconds"), so the +4.7 degree (+0.082 rad) phase perturbation to H_{s_1} will produce a frequency shift of approximately +0.082 rad/s (+0.013 Hz), which is very close to the measured frequency shift shown in Fig. 8.24. This behavior is contrasted to the same phase perturbation introduced into H_{s_2}: the operating frequency does not shift and the relative phases of s_1 and d_2 are not effected so the zero rate bias is not effected. Introducing the phase perturbation into H_{d_1} will shift the operating frequency as shown in Fig. 8.24 (because the AGC loop phase is perturbed); however, the relative phases between s_1 and d_2 are not effected so the zero rate bias does not change. Finally, if the perturbation is introduced into H_{d_2}, the operating frequency is not effected; however, the relative phase of s_1 and d_2 has changed, thus, a perturbation is introduced into the zero rate bias.

This analysis reveals that H_{d_2} and H_{s_1} are critical dynamic elements whose magnitude and phase must be stabilized over the sensor environment. Any deviation in the magnitude of these elements will cause a change in the sensor scale factor and any deviation in phase will cause the quadrature term to "mix" with the actual angular rate. This mixing is detrimental because even if the gyro dynamics do not change with time, a phase perturbation of the critical elements will create a change in the rate bias, which is indistinguishable from an actual angular rate applied to the sensor. Thus, these phase changes, which typically occur on time scales of minutes, will cause low-frequency noise in the angular rate estimate.

8.7 Conclusion

Vibratory gyros are a technology that could not exist without the use of feedback. The main result of this chapter is a rigorous comparison of open- versus closed-loop operation under the conditions of tuned and detuned coriolis-coupled modal frequencies.

We demonstrated for any desired bandwidth that degenerate modal frequencies yields superior noise properties compared to the detuned case as long as a suitable filter for implementing a "plant" inverse is possible. Indeed, the only practical way to implement the inverse filter is to use feedback which leads us to consider the closed-loop sense channel. It was also shown that the scale factor associated with closed-loop operation is essentially independent of the sense resonance's quality factor (Q) and, furthermore, the scale factor is quite robust to detuning of the excitation frequency from the sense modal frequency. The rate-equivalent noise, however, is very sensitive to detuning in the closed-loop sensor and a challenge of resonator design and sensor packaging is to minimize any inadvertent detuning. It should be noted that the benefits of "high" Q are only realized when the sharp resonance occurs within the desired bandwidth. In other words, a more accurate statement that captures the essence of what is important is that the open-loop resonator *time constant* be much larger than the time constant associated with the closed-loop bandwidth. The quality factor is a less-useful metric because of its dependence on the modal frequency. For example, the ratio of time constants exceeds 100 for the DRG prototypes discussed in this chapter.

The challenge of tuning high Q resonators was also addressed. The objective was to develop an approach that lends itself to automation. Although the practiced engineer can often tune the modes in an ad hoc manner, this cannot be used in a production environment. Our proposed approach, which has been validated on numerous vibratory gyro designs, uses multi-input/multi-output frequency response data to fit a sensor model that includes the sensitivities of the sensor dynamics to changes in the bias electrode voltages. These models are subsequently used for determining the bias voltages that render the modes degenerate. There have been exciting recent developments in resonator tuning that rely on perturbing the resonator mass matrix instead of electrostatically changing the stiffness matrix. This may ultimately be the preferred approach since the complicated electronics that are associated with the electrostatic tuning method are not required. The mass matrix perturbation results were not reported in this chapter due to length limitations;

however, they use the same fundamental sensor model based on frequency response data to guide the tuning process [11].

It was also shown that the price of increasing the sensor bandwidth relative to the intrinsic resonator bandwidth is the appearance of angle white noise in the angle measurement produced by integrating the rate signal. This trend seems to be unfamiliar to most engineers who are users of rate gyros for navigation because the current technologies, such as untuned, open-loop vibratory gyros or ring laser gyros, are dominated by angle random walk. Nevertheless, we demonstrate that the angle random walk associated with tuned, open-loop vibratory sensor is asymptotically recovered by the closed-loop sensor as the integration time increases. In other words, the uncertainty associated with the angle measurement is not impacted by the fact that rebalance feedback adds more noise to the rate estimate because the added noise is at frequencies above the resonator bandwidth. It is the authors' hope that the analysis in this chapter dispels some misconceptions associated with the use of feedback in vibratory gyros.

Acknowledgement The authors thank Dorian Challoner at the Boeing Satellite Development Center for his enthusiastic support. This work was supported in part by the National Science Foundation (ECS 0601622) and the UC MICRO program in conjunction with The Boeing Company.

Disclosure The authors provide consulting services to Boeing on the development of MEMs gyro technology.

References

1. Y.-C. Chen, R.T. M'Closkey, T.A. Tran, and B. Blaes. A control and signal processing integrated circuit for the JPL-Boeing micromachined gyroscopes. *IEEE Trans Cntrl Sys Tech*, vol. 13, no. 2:286–300 (2005) .
2. K. Fearnside and P.A.N. Briggs. The mathematical theory of vibratory angular tachometers. *IEE Monograph*, no. 264:155–166 (1957).
3. T.B. Gabrielson. Mechanical-thermal noise in micromachined acoustic and vibration sensors. *IEEE Trans Elec Dev*, vol. 40, no. 5:903–909 (1993).
4. J. Hale. *Ordinary differential equations*. Wiley, New York (1969).
5. J.B. Johnson. Thermal agitation of electricity in conductors. *Phys Rev*, vol 32:97–109 (1928).
6. D.W. Jordan and P. Smith. *Nonlinear ordinary differential equations*, 2nd ed. Oxford University Press, Oxford (1987).
7. D.-J. Kim and R.T. M'Closkey. A systematic method for tuning the dynamics of electrostatically actuated vibratory gyros. *IEEE Trans Cntrl Sys Tech.*, vol. 14, no. 1:69–81 (2006).
8. R.T. M'Closkey, A. Vakakis, and R. Gutierrez. Mode localization induced by a nonlinear control loop. *Nonlinear Dynamics*, vol. 25, no. 1:221–236 (2001).
9. C.T. Morrow. Zero signals in Sperry tuning fork gyrotron. *J. Acoust Soc Am*, vol 27:581–585 (1955).
10. G.C. Newton Jr. Theory and practice in vibratory rate gyros. *Control Eng*, vol 10:95–99 (1963).
11. D. Scwhartz, D.-J. Kim, and R.T. M'Closkey. Frequency tuning of a disk resonator gyroscope via mass matrix perturbation. *ASME J. Dyn Sys Meas Cntrl*, vol. 131, no. 6 (2009).

12. A. Sharma, M.F. Zaman, and F. Ayazi. A sub-0.2°/hr bias drift micromechanical silicon gyroscope with automatic CMOS mode-matching. *IEEE Trans Sld St Crct*, vol. 44, no. 5:1593–1608 (2009).
13. A.C. To, W.K. Liu, G.B. Olson, et al. Materials integrity in microsystems: a framework for a petascale predictive-science-based multiscale modeling and simulation system. *Comput Mech* (2007). doi: 10.1007/s00466-008-0267-1

Chapter 9
Feedback Control of Microflows

Mike Armani, Zach Cummins, Jian Gong, Pramod Mathai, Roland Probst, Chad Ropp, Edo Waks, Shawn Walker, and Benjamin Shapiro

9.1 Introduction

Microfluidics refers to fluid flow inside systems whose features range in size from millimeters to micrometers. This length scale matches the size of biological entities. Consequently, many microfluidic systems are aimed at biochemical applications and some of these have now progressed to medical and clinical use. Under development

M. Armani
Fischell Department of Bioengineering, University of Maryland, and Pathogenetics Unit, Laboratory of Pathology, National Cancer Institute, Bethesda, MD.
e-mail: armanimd@mail.nih.gov

Z. Cummins • R. Probst
Fischell Department of Bioengineering, University of Maryland, College Park MD, USA

J. Gong
Micromanufacturing Laboratory, University of California, Los Angeles, CA, USA

P. Mathai
Aerospace Engineering, University of Maryland, College Park MD, USA

C. Ropp
Electrical Engineering, University of Maryland, College Park MD, USA

E. Waks
Electrical Engineering & Institute for Research in Electronics and Applied Physics (IREAP), University of Maryland, College Park MD, USA

S. Walker
Department of Mathematics & Center for Computation and Technology (CCT), Louisiana State University, Baton Rouge, LA, USA

B. Shapiro (✉)
Fischell Department of Bioengineering & Institute for Systems Research,
University of Maryland, College Park MD, USA
e-mail: benshap@umd.edu

J.J. Gorman and B. Shapiro (eds.), *Feedback Control of MEMS to Atoms*,
DOI 10.1007/978-1-4419-5832-7_9, © Springer Science+Business Media, LLC 2012

and demonstrated biomedical applications include microarrays for rapid analysis of DNA [1, 2], analysis and detection of proteins [3], monitoring and analysis of cells [4], and implantable drug injection systems [5].

Creating, or fabricating, microscale and microfluidic systems is a large and active research area with significant portions of journals (e.g., the *Journal of Micro-Electro-Mechanical-Systems* and *Lab-on-a-Chip*), conferences (Micro-Total Analysis Systems [μTAS]; Hilton Head Sensors, Actuators and Micro-Systems; and the MEMS conference), and books [6–8] devoted to it. MEMS fabrication methods are usually based on lithography – the process of shining radiation (most easily light) through masks onto photosensitive materials to then etch out or build up material layers of the system [9]. The wavelength of light ($\lambda \approx 0.5\,\mu m$) limits the minimum size of the features that can be produced in this way, thus the term *micro* in micro-electro-mechanical-systems. Shorter wavelength radiation (such as electron beams) or other fabrication methods (such as controlled atomistic growth for carbon nanotubes [10] or self-assembly [11, 12]) can enable fabrication of nanoscale features. Generally, lithography fabrication techniques are grouped into methods for rigid substrates, e.g., for silicon and glass (e.g., see [9]), versus methods for polymer materials (soft lithography) [8, 13, 14]. Fabrication often requires deep expertise and dedicated clean-room facilities with expensive machines for each aspect of a fabrication process yet, in some cases, it can be achieved by nonexperts working on a bench top. In fact, there is a spectrum of fabrication capabilities from one to the other, with the former usually needed for smallest feature sizes, hard materials, mirror smooth finishes, and high aspect ratio (thin and deep) features; the latter applying more to soft materials, larger features, and inexpensive (e.g., disposable) systems.

The physics inside microfluidic devices is diverse. Even though momentum effects are usually negligible in microfluidic systems, which means the computationally complex Navier–Stokes equations [15] reduce to the far simpler (linear) Stokes equations (see [16–20]), and even though noncontinuum effects are not yet evident in bulk microscale flows (for example, the mean free path of water is <1 nm [21] which is still negligible compared to $\geq 1\,\mu m$ device length scales), this does not mean microscale fluid dynamics is easier to understand, model, or quantify than macroscale fluid flow. The complexity is just in different areas: it is in the boundary conditions (the actuation of a fluid by electrically modifying surface tension), in the mix between continuum and discrete elements (cells or DNA chains undergoing Brownian motion in a moving continuous bulk fluid), in the complexity of the bulk fluid itself (in the non-Newtonian behavior of blood), and in the interaction of hundreds or thousands of different fluid samples on a single chip which microfluidic systems allow. Even behavior that is simpler on the microscale takes some time getting used to. Experiments that show that there is no convective (turbulent) mixing on the microscale, the classic T-junction experiments where a green and blue fluid enter each of the T inputs and exit the T output as unmixed half green/half blue [22], are still counter-intuitive even to microfluidic experts. Mixing, which is achieved naturally on the macroscale, becomes an issue that must be solved artificially on the microscale [23–25].

Feedback control is needed in microfluidic applications for the same reasons that it is required on the macroscale [26–28]: to create new capabilities and to enable high performance in the face of uncertainty. Microsystems often operate in largely unknown environments and can have significant geometric, parametric, and dynamic uncertainty. Outside environments may contain unknown biochemical species (as in sensing applications where the presence of rare species must be reliably detected against a background of other common, diverse, and widely varying species); biological fluid samples have a large degree of variability (urine samples vary with disease, with hydration, and from patient to patient), and the characteristics of specific entities inside the samples can vary (cells of the same type will have different shapes and properties). Device geometric uncertainty is created by fabrication limits: the wavelength of light limits lithography resolution to $\sim$0.5 µm, hence devices with 5 µm sized features will have a >10% variability in geometry. Finally, mathematical models that characterize biochemical behavior (such as models of surface tension boundary conditions, reaction rates, diffusion, migration, species adsorption, and desorption) contain uncertain parameters and unmodeled effects. The system design that necessarily relies on these models must be made insensitive to the errors that they contain. Feedback is required to address all of these uncertainties and to create robust behavior, enable new tasks, and improve system performance.

To illustrate results and challenges this chapter includes two broad examples. The first deals with control of fluid packets on chip using electrically modulated surface tension forces (in collaboration with UCLA). The second provides results on steering of individual particles (cells and quantum dots) by microflow control. In both cases, we show how feedback control can improve performance and enable new capabilities.

These two examples illustrate common challenges encountered by us, in these and other projects, and they closely match the challenges encountered by others (as evidenced by the recurrence of the same issues from chapter to chapter in this book). Broadly, these challenges are: choice of problem (which need should the microscale system address?), fabrication (build it), physics (which phenomena occur?), modeling (describing the physics by equations and then quantifying their solution by numerical methods), control (problem definition, the more critical aspect, and subsequently problem solution), experimental verification, and validation.

A thread that runs through all of the preceding aspects is multidisciplinary communication, or, more accurately, multidisciplinary training and collaboration. We have found time and again that we can only create working systems once we have deeply understood the needs, physics, numerics, and experiments for that application, or, alternately, once each of those areas of expertise is represented in the research team. Being a controls expert just *talking* to a MEMS expert or a clinician is almost always not good enough. The clinician must be willing to become a part of the team. The reverse is likely also true: being a MEMS expert and talking to a controls expert is insufficient, rather, the microsystems expert or clinician should find the right controls person and make him or her a part of the research program.

9.2 Control for Electrowetting Actuation

Electrowetting refers to the local modification of surface tension by electric actuation to manipulate liquid droplets on the microscale [29–41] (Fig. 9.1). By applying electric fields via actuating electrodes, surface tension and electrical effects compete [29, 42, 43], and this competition can create forces that vary in both space and time and that can be used to move, split, merge, shape, and mix fluids in microscale devices. The size of the manipulated droplets usually roughly matches the size of the electrodes (as shown below), or is larger, to permit each droplet to contact multiple electrodes and thus be actuated in different directions.

Applications of electrowetting include re-programmable lab-on-a-chip systems [30, 31], auto-focus cell phone lenses [37] (commercially marketed by Varioptic), and the development of colored oil pixels for laptop screens and flexible video-speed "smart paper" [34–36] (under development by Liquavista).

9.2.1 Modeling

Our control design for electrowetting has benefited dramatically from, and in fact has been permitted by, the system modeling that we have carried out. The modeling that is needed must be sufficiently rich to capture the basic physics but sufficiently compact (or, more subtly, in a form that is suitable) for control design – the models must be *useable* in conjunction with available control analysis and synthesis techniques [27, 28, 46, 47]. Modeling is essential and still useful for our control design even though we work with complex and messy systems that we know we cannot fully capture mathematically – both because we do not know all the physics and because each physical phenomena will have aspects (such as the detailed chemistry of surface tension) which are outside our reach. Feedback allows us to manage the uncertainty in our models. Essentially, our models must be good enough to tell us how to actuate to make things better – to push the fluid from where it is to closer to where it should be. In both the electrowetting example and in the electro-osmotic flow control that follows, this is sufficient to control the fluid in dramatic ways and the objects within it to startling accuracy.

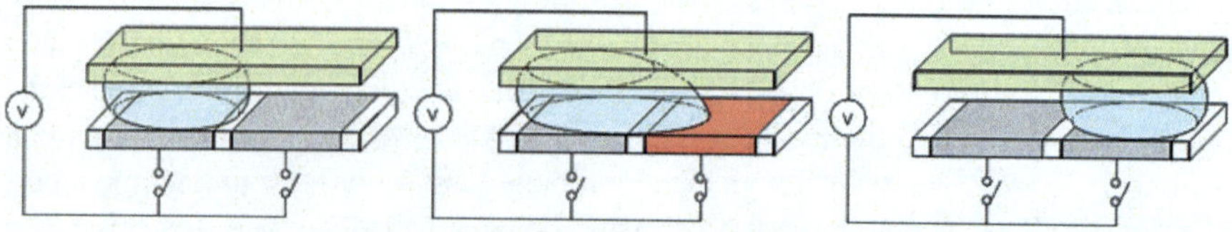

Fig. 9.1 An example of electrowetting actuation (schematic). The activated electrode (*red pad*) effectively and locally decreases the surface tension of the liquid above it, causing it to move to the right [44, 45]. (Used with permission. Copyright John Wiley & Sons, Ltd; American Institute of Physics)

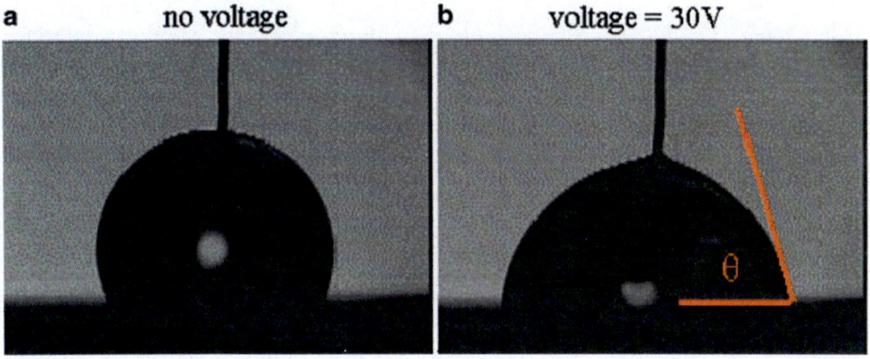

Fig. 9.2 In the simplest example of electrowetting, an applied voltage changes the shape of a liquid droplet. The figure shows the induced change in the contact angle of a droplet of deionized, distilled water (pH 6.5) on 50 nm thick Teflon AF coating a 120 nm thick silicon dioxide dielectric layer. An applied voltage of 30 V between the inserted platinum wire and underlying gold electrode causes the droplet to flatten reversibly [44]. (Used with permission. Copyright John Wiley & Sons, Ltd)

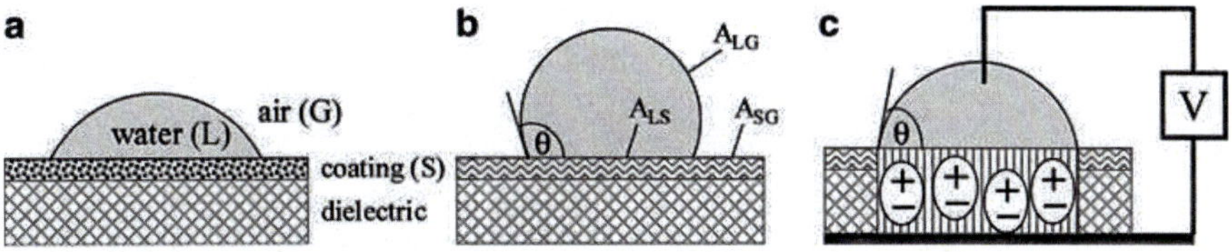

Fig. 9.3 Electrowetting shape change actuation: a competition between surface tension and dielectric energy. (**a**) A drop of water on a hydrophilic surface. The liquid/solid energy per unit area is low, and so the drop prefers a shape with a high (low penalty) liquid/solid area and a lower (higher penalty) liquid/gas area at its energy minimum (equilibrium). (**b**) The same drop on a hydrophobic surface. The contact angle θ and the liquid/solid, liquid/gas, and solid/gas interface areas are marked. (**c**) The drop of (b) with voltage actuation. The resulting charged solid volume is marked by the $\pm$ dipoles. Surface tension penalizes the liquid/solid area, but the dielectric energy in the charged volume (and the voltage source) rewards an increase in this area. The $\pm$ charges in the dielectric lead to a surface charge (here $+$) at the solid/liquid surface. (A schematic two-dimensional slice is shown. The figure is not to scale, the solid is very thin)

At the core of electrowetting is a competition between surface tension and dielectric energy. As noted in the review by Mugele and Baret [29], there is now a basic agreement that this is the dominant physical cause, and, at equilibrium, this competition can be quantified either by classical thermodynamics, an energy minimization, or a force balance electromechanical approach. All of these yield the same classical Young–Lippmann equation which predicts, to first order, the shape (contact angle θ, Fig. 9.2) of a single droplet as a function of the applied voltage [48]. We briefly summarize the energy minimization argument.

Figure 9.3 shows a schematic of the setup and physics for the experiment of Fig. 9.2. At equilibrium, an un-actuated droplet will minimize its total surface tension potential energy (the energy due to gravity is negligible on the microscale).

This energy is composed of the liquid/gas, liquid/solid, and solid/gas interface energies, each of which is given by a surface tension coefficient times that interfacial area [49, 50]. In this view, a water droplet on a hydrophobic surface beads up because its liquid/solid surface tension coefficient is high compared to its liquid/gas coefficient, and the droplet shrinks its L/S area at the expense of a larger L/G area.

When a voltage V is applied, then in addition to surface tension energy there is also an electrical energy. If the liquid is conducting, it stores no electric energy, the applied voltage is transferred to the liquid/solid contact area, and the solid material underneath the liquid is dielectrically charged (Fig. 9.3c). The solid dielectric energy scales as the charged volume, which is equal to the L/S area times the (constant) material thickness. Along with this energy stored in the material, there is also an energy stored in the voltage source. The total energy now becomes [29, 43, 51]:

$$U = \left(\sigma_{\mathrm{LS}} - \sigma_{\mathrm{SG}} - \varepsilon_{\mathrm{S}} V^2/2h\right) A_{\mathrm{LS}} + \sigma_{\mathrm{LG}} A_{\mathrm{LG}}, \tag{9.1}$$

where σ_{XY} are the surface tension coefficients, A_{XY} are the interface areas, L, S, and G denote liquid, solid, and gas, ε_{S} is the dielectric constant of the solid and h is its thickness. This is the new energy that must be minimized by the liquid shape and it alters the classical Young energy minimum [43, 49, 50] to the Young–Lippmann minimum [48]:

$$\cos\theta = -\left(\sigma_{\mathrm{LS}} - \sigma_{\mathrm{SG}} - \varepsilon_{\mathrm{S}} V^2/2h\right)/\sigma_{\mathrm{LG}}. \tag{9.2}$$

In this way the applied voltage V changes the contact angle θ, i.e., the shape, of the liquid droplet.

Our interest is in controlling the *dynamics* of electrowetting, so beyond equilibrium, we require models for the fluid dynamics. As stated earlier, our modeling must capture the essential physics but still be tractable (computationally cheap, of a suitable form) for control design. Our prior dynamic modeling efforts [19, 45, 52] have focused on the UCLA electrowetting system, a planar liquid-in-air electrowetting-on-dielectric (EWOD) system [30, 31, 39] (Fig. 9.4). (Also see Lu et al. [53].) For dynamic modeling of EWOD, the critical aspects are the low Reynolds number fluid dynamics of the bulk liquid (water, glycerine, etc.), the liquid/air boundary conditions, numerical methods for tracking the moving liquid/air interfaces and enforcing appropriate boundary conditions, as well as incorporating loss mechanisms such as contact angle saturation, contact line hysteresis and pinning/de-pinning which limit system performance [29].

We model the bulk fluid flow by simplifying the Navier–Stokes equations. The continuum assumptions behind the NS equations remain valid because the micrometer device length scales are still far greater than the mean free path of both air and water molecules [16]. Because we are modeling the flow of water or glycerine, incompressibility and Newtonian fluid assumptions also hold [15]. At low Reynolds numbers, for flow of liquid surrounded by air between two narrowly spaced plates (Fig. 9.5), the Navier–Stokes equations reduce to the Hele-Shaw equations [54, 55], with a pressure boundary condition given by the Young–Laplace

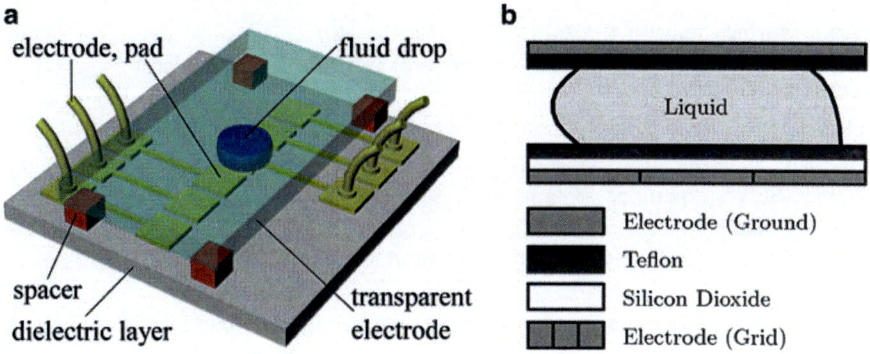

Fig. 9.4 The UCLA EWOD system. (**a**) Schematic. (**b**) Cross-section view [19,45]. (Used with permission. Copyright IEEE, American Institute of Physics)

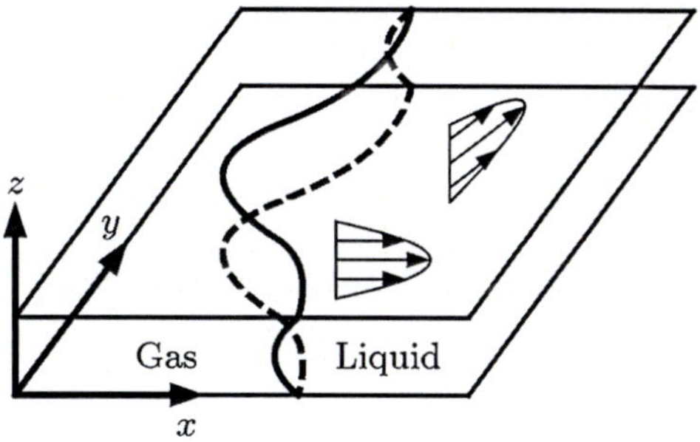

Fig. 9.5 Liquid flow in a Hele-Shaw cell: the fluid velocity field is assumed to have a quadratic profile in the z-direction [19]. (Used with permission. Copyright IEEE)

relation at the liquid/gas interface. Thus the two-dimensional fluid equations inside (the possibly many) droplets actuated by EWOD are given in nondimensional form as (see [19,45,52] for details):

$$\alpha \frac{\partial \vec{u}}{\partial t} + \beta \vec{u} + \nabla p = 0, \quad \text{in } \Omega$$
$$\nabla^2 p = 0, \quad \text{in } \Omega, \tag{9.3}$$

where Ω is the domain of the liquid, $\vec{u}$ is the two-dimensional vector velocity field (in the plane of the device), and p is the pressure. The nondimensional constants α and β depend on the fluid parameters and the geometry of the device:

$$\alpha = \left(\frac{\rho L U_0}{\mu}\right) \mathrm{Ca}, \qquad \beta = 12 \left(\frac{L}{H}\right)^2 \mathrm{Ca}, \qquad \mathrm{Ca} = \frac{\mu U_0}{\sigma_{\mathrm{LG}}},$$

where ρ is the fluid density, H is the height between the parallel plates, L is the planar liquid length scale (e.g., the pitch of the EWOD electrodes), U_0 is the velocity scale, μ is the dynamic viscosity, Ca is the capillary number (the ratio of viscosity versus surface tension forces), and σ_{LG} is the surface tension coefficient of the liquid/air interface. The time derivative term in (9.3) is nonstandard, usually it is negligible in microflows, but it is included here because fast actuation of the EWOD electrodes can lead to an imposed fast time-scale making local momentum effects (the $\partial\vec{u}/\partial t$ term) appreciable. The convective momentum term $(\vec{u}\cdot\nabla)\,\vec{u}$ in the Navier–Stokes equations remains negligible even with fast EWOD actuation (see [19] for details) and so does not appear in our model. Crucially for control design (see the next section) this means the bulk fluid equations are linear from pressure to velocity.

Boundary conditions at the liquid/air interface drive the bulk flow. These conditions include surface tension, electrowetting actuation (which is really a competition between dielectric and surface tension energy as described above but effectively acts at the liquid/gas interface), electrowetting loss-phenomena (such as contact angle saturation), and pinning and de-pinning of the triple line (the moving liquid/gas/solid interface). A portion of these effects can be written in a straightforward way, in particular surface tension [49, 50] and ideal electrowetting actuation [29, 48] have standard descriptions, but interface loss-phenomena, such as contact angle saturation and triple line pinning, are subtle, depend on fine scale physics and chemistry, vary from system to system, and remain the subjects of fierce debate.

Surface tension creates a pressure jump across curved interfaces [49, 50], this pressure jump is quantified by the classic Young–Laplace relation [55]:

$$\Delta\tilde{p} = \sigma_{LG}(\kappa_1 + \kappa_2), \tag{9.4}$$

where κ_1 and κ_2 denote the principal curvatures [56] (we have written this equation in dimensional form). In the planar EWOD devices, we can take the principal curvatures to have one direction in the plane of the device and the other along the channel height. After nondimensionalizing and setting the arbitrary outside reference pressure to zero, the nondimensional pressure just inside the liquid/air interface is given by [19]:

$$p = \kappa + \frac{L}{H}\kappa_z, \quad \text{on } \Gamma \tag{9.5}$$

where $p = \Delta\tilde{p}/\Delta P_0$, κ is the curvature of the liquid/air interface in the plane (the xy curvature), κ_z is the curvature of the interface in the vertical z-direction, and $\Delta P_0 = \sigma_{LG}/L$ is the dimensional pressure scale.

Electrowetting actuation modifies the pressure of (9.5) by bending the interface – it changes κ_z. For a small vertical gap H, which allows us to assume that the liquid/air interface is circular in the vertical direction, there is a direct geometric relation between κ_z and the local top and bottom contact angles θ_t and θ_b. In the UCLA devices, electrodes are placed on the bottom and so it is θ_b that is actuated – whenever an electrode is turned on the local contact angle above it decreases (the liquid spreads out). For applied voltages that cause small or medium changes in contact angle, the relation between the cosine of the angle and the applied voltage is quadratic ((9.2), the Young–Lippmann relation). But relation (9.2) does not hold indefinitely, at some point further increasing the actuation voltage brings into play non-ideal phenomena that prevent a further decrease in contact angle – this is known as contact angle saturation and is one of the fundamental limits of electrowetting performance. Causes of contact angle saturation are under debate, they vary from system to system [29], and they are difficult to predict from first principles. Thus, in our modeling, we have quantified the relation between the applied voltage V and the resulting contact angle $\theta(V)$ by fitting to experimental data for the UCLA system (see [19,45]).

In addition to contact angle saturation, there is also contact line pinning. Line pinning is the phenomenon that prevents droplets from sliding down a vertical window pane, it can oppose gravity at zero velocity, its force is not proportional to a fluid shear, and it is, therefore, not a viscous effect per se. Rather, it is a kind of molecular adhesion that occurs at the triple contact line of the droplet [57]. In electrowetting devices it can prevent droplets from returning to a perfectly circular shape when electrowetting actuation is turned off, which is what they would do under surface tension in the absence of pinning. Like contact angle saturation, the physics of contact line pinning is complex and under debate. It is often modeled by fine scale atomistic simulations which are too computationally expensive to be included in models for control. Recently, we proposed a modeling paradigm that includes a simple description and can numerically track pinning [45] but we have not yet carried out control studies for this new modeling capability.

Solving the bulk flow equations with the boundary conditions above (equation (9.5)) yields a flow velocity at each time, and this flow velocity convects the fluid/gas interface Γ as follows. For each point $\vec{x}$ on the interface:

$$\frac{\partial \vec{x}}{\partial t} = \vec{u}(\vec{x}) = \left[\vec{u}(\vec{x}) \cdot \hat{n}(\vec{x})\right] \hat{n}(\vec{x}), \tag{9.6}$$

the velocity of that point is given by the flow velocity at its location. However, since the change in droplet shape is due only to the normal component of the velocity, the second form in (9.6) is also correct, where $\hat{n}$ is the unit outward normal.

To summarize the model, the planar bulk fluid dynamics of the liquid is described by (9.3), the air is ignored, and the liquid/air boundary conditions that drive the bulk flow are given by (9.5). Here κ_z depends on the voltage applied at the electrode underneath that portion of the interface and its dependence on that voltage is identified from experiments. This gives a complete set of equations for the pressure and fluid velocity at each time. The fluid velocity then updates the shape of the liquid by (9.6).

A sound numerical implementation of this model is difficult. The hard part is good numerical tracking of the moving interfaces, which move quickly and undergo topological changes (split and merge events), and accurately computing and applying interface boundary conditions. In particular, surface tension boundary conditions require a clean and robust computation of interface curvature: this

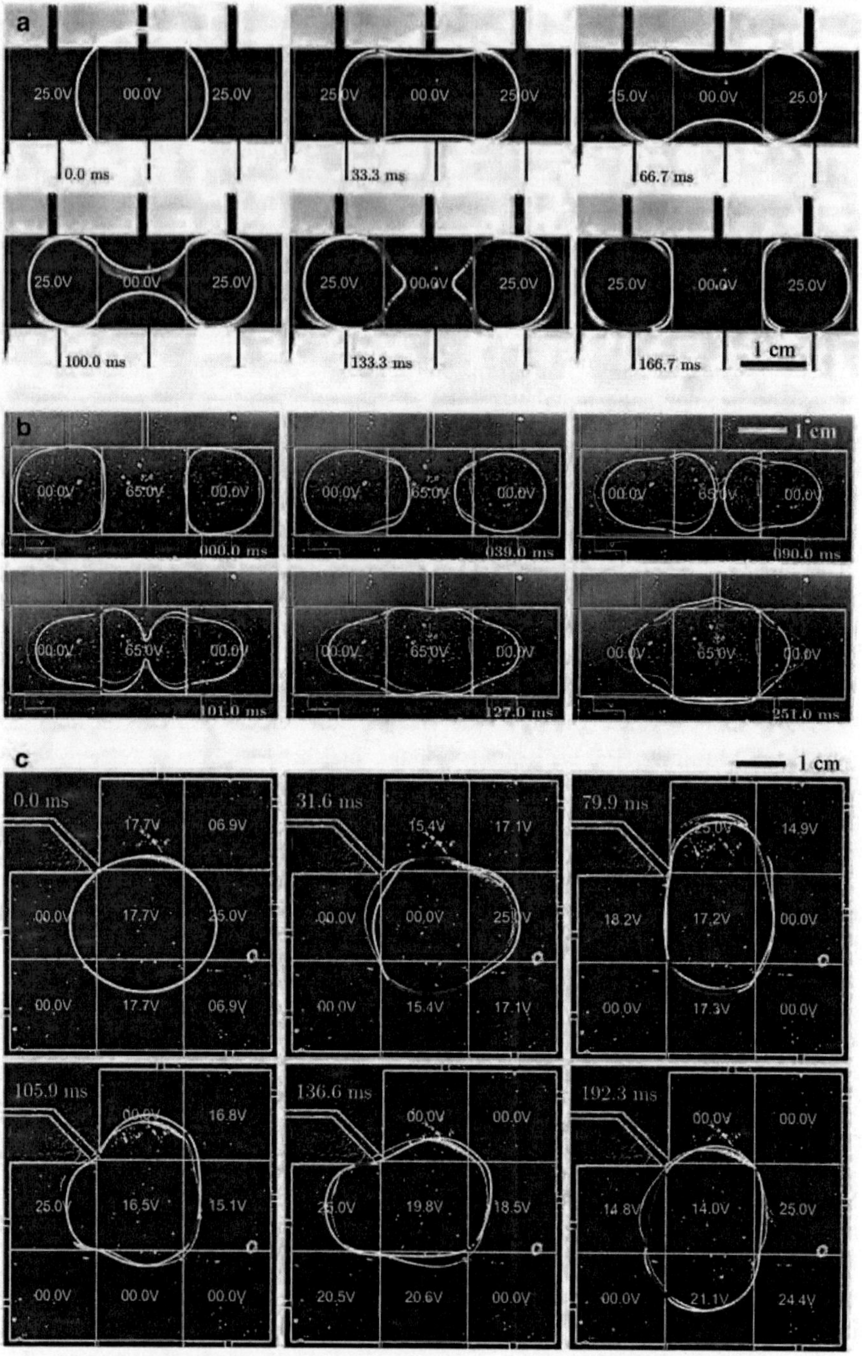
a
25.0V
00.0V
25.0V
0.0 ms
33.3 ms
66.7 ms
100.0 ms
133.3 ms
166.7 ms
1 cm
b
1 cm
00.0V
65.0V
00.0V
000.0 ms
039.0 ms
090.0 ms
101.0 ms
127.0 ms
251.0 ms
c
1 cm
0.0 ms
17.7V
06.9V
00.0V
17.7V
00.0V
17.7V
06.9V
31.6 ms
15.4V
17.1V
00.0V
00.0V
00.0V
15.4V
17.1V
79.9 ms
14.9V
18.2V
17.2V
00.0V
00.0V
17.3V
00.0V
105.9 ms
16.8V
16.5V
15.1V
00.0V
00.0V
00.0V
136.6 ms
00.0V
00.0V
19.8V
18.5V
20.5V
20.6V
00.0V
192.3 ms
00.0V
00.0V
14.8V
14.0V
25.0V
00.0V
21.1V
24.4V

involves a second derivative in space and can lead to large numerical noise if not handled correctly. In our first numerical implementation, we tracked interfaces by a level-set approach [19]. The level-set method implicitly tracks the liquid boundaries as the zero level-set of a scalar function defined on the plane [58–60]. This function is convected by the fluid velocity and so deforms and changes shape – its zero level corresponds to the liquid/gas boundary. In level-set approaches there are issues with computing curvature and enforcing mass conservation. Using an explicit calculation of curvature requires restricting the time-step to be less than the square of the mesh spacing, which can be expensive [58,59]. Computing the curvature implicitly can be done in the level-set framework [61] but is more expensive. Mass conservation is an issue because the standard level-set schemes are not globally conservative, thus they tend to lose mass over the course of a simulation [58,62,63]. This can be alleviated by other techniques [64–66] but they further complicate the method.

In our more recent work [45,52,67], we use a variational front-tracking approach that represents the interface explicitly and solves the underlying PDEs using a variational formulation which is discretized by a finite element method. Our new method has the advantage that it is globally conservative, so mass conservation is not an issue. Here we discretize curvature using a semi-implicit method that is straightforward to implement in our variational approach. In this new method we have to deal with distortion of the underlying finite element mesh and this is especially important in the case of topological changes. But we handle this by a hybrid variational front-tracking level-set approach that is able to take explicit meshes through a pinching/merging event without too much computational overhead [67]. The primary advantage of the level-set method is enabling topological changes and we use it in this way only when we need it (i.e., only where and when a pinch or merge event occurs). Another advantage that our variational approach has is in modeling and computing the effects of contact line pinning which we describe in detail in [45]. Our contact line pinning hybrid method uses a variational inequality to capture the pinning effect – it is not at all obvious how this could be implemented in a pure level-set framework.

Our numerical models evaluate in minutes on a laptop computer and accurately capture behavior observed experimentally in the UCLA EWOD systems. Figure 9.6 shows a representative sample of results (taken from [45]). Our models are fast yet accurate; this makes them amenable to the control design discussed next.

Fig. 9.6 A sample of comparisons between our EWOD model and experimental data from UCLA (from [45]). In all panels, the simulated interface is the *solid curve* (*white* for free, *gray* for locally and transiently pinned), and the experimental interface is visible as a *thin line* that is sometimes motion blurred. The numbers show the voltage applied at that electrode pad. The view is from the top through the top transparent electrode of Fig. 9.4. From top to bottom: (**a**) drop being split into two, (**b**) two drops joining into one, and (**c**) a drop being moved along a complex path. The model includes a simple force–threshold contact line pinning description that enables us to capture, to a degree, the final noncircular pinned shape of the droplet [45]. (Used with permission. Copyright American Institute of Physics)

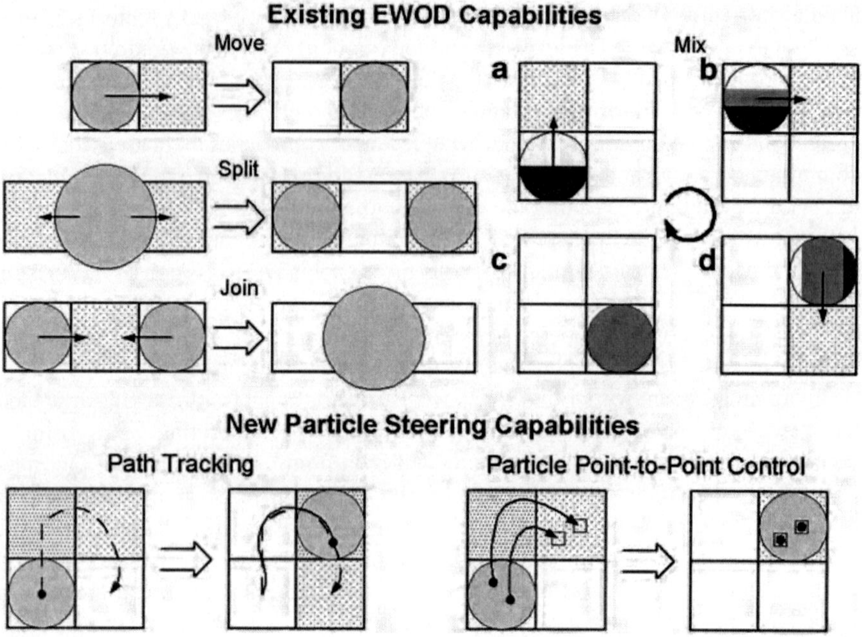

Fig. 9.7 Existing (move, split, join, and mix) capabilities of electrowetting devices are shown schematically above (see [24,30,40,68–71]) alongside the new particle steering capability enabled by the control methods described next. The view is from the top. *Shaded circles* represent droplets of liquid. *Squares* are electrodes where the *dotted hatching* indicates the electrode is on. *Directed lines* specify the direction of motion. The *multishaded* droplet shows the diffusion and mixing of two chemicals, here mixing is enhanced by the fluid dynamics created inside the droplet due to its imposed motion [72]. (Used with permission. Copyright Royal Society of Chemistry)

9.2.2 *Control*

Current electrowetting systems use simple control scripts but can already perform the key operations outlined in Fig. 9.7 – they can move, split, and join fluid droplets and effectively mix chemicals inside them. Feedback control can improve precision and robustness and our specific results below could enable manipulation of individual particles within single EWOD droplets – a new capability for electrowetting. The control algorithms presented next are based on the EWOD model developed above ((9.3) through (9.6)) but without contact line pinning.

9.2.2.1 Control for Particle Steering

Steering of multiple particles inside EWOD driven droplets, using actuators already available in standard EWOD devices, requires more sophisticated control of the electrode voltages. The voltages directly influence, through the boundary conditions, the pressure gradient field inside the droplet (see (9.5)). Hence, by manipulating the voltages, we can control the fluid flow fields (9.3), and thereby control the velocities and positions of particles inside the liquid droplets. We consider neutral (uncharged) particles that are simply carried along by the (vertically averaged) planar fluid flow. Thus a particle at the location (x, y) will simply follow the velocity of the fluid at its location:

$$\dot{x} = u(x,y), \qquad \dot{y} = v(x,y), \tag{9.7}$$

where (u, v) is the flow field from (9.3) and the dots denote derivatives with respect to time. Therefore, our control problem is to find electrode voltage sequences that create temporally and spatially varying flow fields that will carry the target particles along their desired trajectories.

The control problem described above is a trajectory-tracking problem: we seek to find the control inputs that will cause the system (in this case the particle positions) to follow a desired trajectory. A naïve inspection of the equations of motion, especially (9.3) for the particle dynamics, would suggest that the control problem is standard in linear control theory and that a linear quadratic regulator (LQR) tracking controller [73] could be used. However, the particle motion depends on the droplet shape and on the number of electrodes that the droplet overlays at any given moment. This information is not known a priori, which means that an LQR approach cannot be used. For this reason, we do local estimation and control at each time-step of our simulation using a least-squares framework to compute the necessary pressure boundary conditions, and then find the electrode voltages that will achieve these boundary pressures. Any particle deviation from the desired trajectory that may arise from thermal fluctuations, external disturbances, and actuation errors is corrected using feedback of the particle's position. Figure 9.8 gives a diagram of the needed closed-loop feedback architecture.

For a single particle, the control algorithm would proceed as follows:

1. Initialization: represent trajectory as a set of points connected by straight lines.
2. Find the particle position and the location of the droplet boundary.
3. Find the closest trajectory point to the particle.
4. Set the particle's desired direction of motion to be toward a nearby next trajectory point.
5. Solve a least-squares problem for the necessary voltage actuation to induce a pressure gradient field that will move the particle along the desired direction of step 4.
6. Apply control voltages, solve for the resulting pressure and fluid velocity, and update the position of all the particles. Advance to the next time-step of the simulation. Go back to step 2 and iterate.

The control algorithm details are described next.

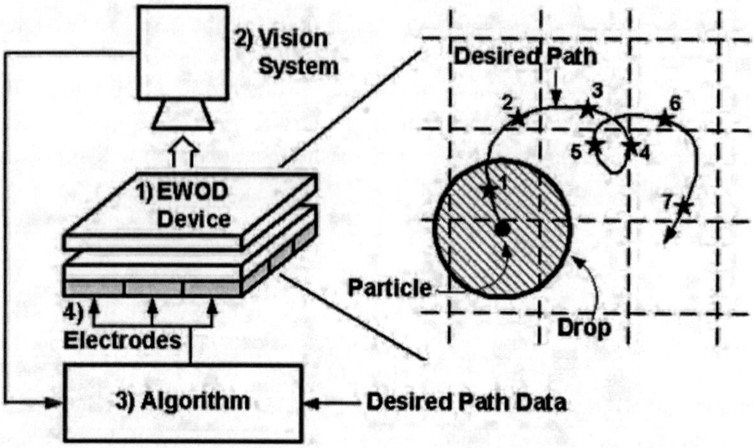

Fig. 9.8 Particle steering closed-loop feedback control architecture. (*1*) The EWOD device will be observed by (*2*) an image system (a microscope/camera or an on-chip contact imager) which transmits information to (*3*) a computer or chip that contains (*3*) an image processing algorithm to identify droplet shapes and the location of the particles and a control algorithm that computes the actuator voltages that will move the particles from where they are to where they should be, and (*4*) these actuation voltages are then applied on the EWOD device. The loop would repeat at each time step to steer the particles along their desired trajectories. The zoomed top view of the EWOD device shows a single droplet with one particle floating inside. The *curvy line* indicates the desired path of the particle. In our simulated control algorithm, we sample the trajectory by many points (only seven points are shown here; see the numbered *stars 1–7*)

Algorithm Initialization

We represent the desired trajectory curve for each particle as a fine sampling of points connected by straight lines. The points are indexed in the order in which the particles should follow them (i.e., the trajectory is parameterized; see Fig. 9.8). Complicated trajectories are broken up into separate segments for ease of particle tracking. For simplicity, only one particle and trajectory is considered in the following sections. Simultaneous multiple particle steering is discussed in the least-squares step.

Particle Position and Droplet Boundary Sensing

We need to know the shape and position of the droplet as well as the position of each particle in order to apply our control algorithm. At the beginning of each time step, we obtain the position of the particle and the location of the droplet boundary using feedback through a vision system (see Fig. 9.8). The issues of integrating a vision system with an EWOD device are not considered here. For the purposes of this chapter, the particle positions and droplet shape information are taken directly from the simulation.

Compute the Desired Direction of Particle Motion

Next, the desired direction of motion for the particle is chosen to be a unit vector that points from the particle's current coordinates toward one of the trajectory points. Since maximum forcing of the pressure gradient is used to drive the particle in the desired direction (see Fig. 9.10), it is necessary to choose a trajectory point that is just out of reach of the particle for the current time step. Otherwise, it is possible that the particle could overshoot trajectory points and trace out an unwanted zigzag path around the trajectory.

Hence, we find the target trajectory point by first finding the closest trajectory point to the particle. Then, using the trajectory parameterization (i.e., the index list; see Fig. 9.8), we look ahead after the closest point and choose the target to be the first trajectory point that is at least one grid spacing away. This ensures the particle will move forward along the trajectory. It also guarantees that the target point is out of reach because the time steps of our simulation are chosen by the CFL criterion [74], which says that no particle can move more than a grid step at each time-step. If the closest trajectory point is the last point of the trajectory, then the particle aims for the last point.

For a self-intersecting or extremely curvy trajectory, it is possible that the particle could become stuck in a loop and not travel the entire trajectory. We resolve this issue by breaking the trajectory into smooth segments that do not intersect and only allow the particle to see one segment at a time. As a result, the particle follows one piece of the trajectory until it reaches the end, where our algorithm switches to the next segment. Therefore, without loss of generality, we assume in the following subsections that the trajectory consists of just one segment.

The forcing of the particle is created by the pressure gradient. And the desired unit vector discussed above determines the direction of forcing. This unit vector is used in the next section to calculate the pressure boundary conditions needed to realize the pressure gradient that will move the particle in the desired direction.

Least-Squares Solution of the Required Pressure Boundary Conditions

Figure 9.9 shows a top view of a sample droplet in the EWOD device containing a single particle. The current drop shape overlaps four electrodes; hence four actuators are available to move the single particle. In each of the four cases, only one electrode is on; the rest are off. The arrows inside the droplet show the fluid flow for each of the four voltage actuations. The black dot represents the particle with a thick arrow indicating the negative direction of the pressure gradient at the particle location (note that the fluid flows opposite to the pressure gradient).

Our algorithm centers on the idea of taking the right linear combination of pressure gradients in Fig. 9.9 to make the particle (or particles) move in the direction(s) we want at a particular time step. This will directly correspond to finding the right combination of electrode voltages at every time step to realize the desired particle motion (or motions).

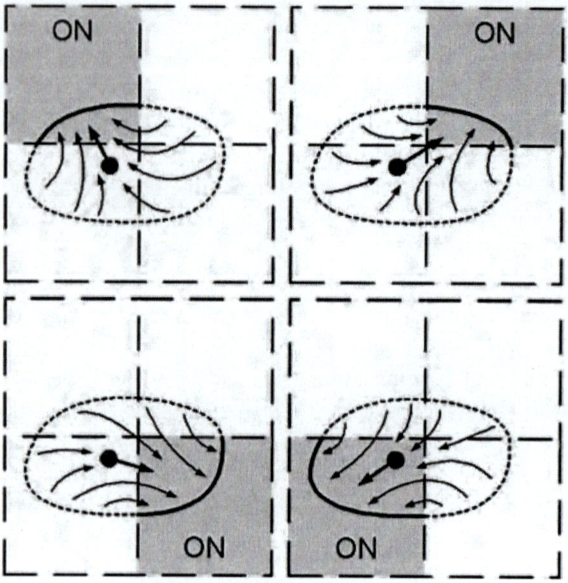

Fig. 9.9 Linear combination of pressure gradients for a single droplet overlaying four electrodes (*small dashed squares*). The diagram above shows a droplet in an EWOD system with four different instances of voltage actuation. In each instance, only one of the four electrodes is on. The particle floating inside the droplet (*black dot*) has a *thick arrow* indicating its direction of motion for each single electrode actuation. These *arrows* actually represent the opposite direction of the pressure gradient when a unit pressure boundary condition is set on the *thick curve* that overlays the *shaded* electrode, with zero pressure boundary conditions everywhere else. The *thin curvy arrows* show the fluid flow inside the droplet. Since the pressure equation (second equation in (9.3)) is linear, we can make the particle move in any desired direction by taking an appropriate linear combination of the four possible boundary conditions given above [72]. (Used with permission. Copyright Royal Society of Chemistry)

First, given the current droplet configuration, we solve the pressure equation in (9.3) for the pressure field inside the droplet for a single active electrode. The pressure boundary conditions are defined to be one on the droplet boundary that lies over the active electrode and zero everywhere else (see Fig. 9.9). From the pressure solution, the pressure gradient at each particle's position is computed. After repeating this for each electrode, we obtain a matrix of pressure gradients:

$$G = -\begin{bmatrix} \nabla P_1(x_1,y_1) & \cdots & \nabla P_N(x_1,y_1) \\ \vdots & \ddots & \vdots \\ \nabla P_1(x_m,y_m) & \cdots & \nabla P_N(x_m,y_m) \end{bmatrix}, \tag{9.8}$$

where $(x_j,\, y_j)$ are the coordinates for the jth particle. Each column of pressure gradients $\nabla P_k(x_j, y_j)$ in the matrix corresponds to a single active electrode; each row to a single particle. The total number of particles is m and the number of available electrodes is N. The minus sign accounts for the direction of particle motion.

Next, given the desired pressure gradient at each particle's location in the droplet, we wish to find the appropriate boundary conditions to realize it. Since Laplace's equation for the pressure in (9.3) is linear regardless of the droplet shape, solutions for single active electrodes can be combined linearly to obtain the pressure gradient field due to many active electrodes. This reduces our problem to solving a linear system:

$$G\alpha = b,\ \alpha = \begin{bmatrix} \alpha_1 \\ \vdots \\ \alpha_N \end{bmatrix}, \quad b = \begin{bmatrix} \nabla P_{\mathrm{D}}(x_1, y_1) \\ \vdots \\ \nabla P_{\mathrm{D}}(x_m, y_m) \end{bmatrix}, \tag{9.9}$$

where $\nabla P_{\mathrm{D}}(x_j,\ y_j)$ is a 2×1 vector representing the desired pressure gradient at the jth particle and α is the vector of boundary values that will achieve b. We set $\nabla P_{\mathrm{D}}(x_j,\ y_j)$ equal to the unit vector that represents the desired direction of motion for the jth particle. If $2m \geq N$, the number of particle degrees of freedom is greater than the available actuators and (in general) (9.9) cannot be solved exactly. Then, a least-squares solution is needed to obtain the best fit of actuations α. Otherwise, it is a pseudo-inverse problem, which has a solution as long as the matrix G has full row rank [75].

We solve (9.9) for α using singular value decomposition (SVD) [75]. In addition, each component of the solution vector must be made to satisfy an inequality constraint:

$$\alpha_{\min} \leq \alpha_j \leq \alpha_{\max}, \qquad 1 \leq j \leq N, \tag{9.10}$$

where $\alpha_{\min}$ and $\alpha_{\max}$ are the minimum and maximum values that the pressure boundary condition can be for any electrode. These constraints come from the limitations of varying the contact angle (i.e., contact angle saturation). Hence, $\alpha_{\min}$ and $\alpha_{\max}$ are related to the maximum and minimum contact angles achievable in the EWOD device. In order to satisfy (9.10), we take the solution α to (9.9) and transform each of its components so that the full dynamic range of boundary forcing is utilized (see Fig. 9.10).

With this new transformed α, we know what the pressure boundary values should be to realize the desired pressure gradient field. But it is not possible to exactly enforce α because we cannot directly control the planar curvature term κ in (9.5). For a circular droplet, the planar curvature term is constant and has no effect on the pressure gradient field [76]; hence, it can be ignored. Using (9.5), it is straightforward to compute the contact angles needed to implement α. For noncircular droplets, we still use the same procedure. It is not reasonable to use the planar curvature term in our control algorithm because it involves 2nd derivatives of data that cannot be accurately measured in experiments [77]. Instead, we view it as a small error to the desired directional forcing of the particles. This error grows as the droplet deviates from being a circle. This is not a problem for particle steering for two reasons. First, the linear transformation of the boundary conditions in Fig. 9.10 ensures maximum forcing of the particles. Thus, the relative magnitude of the error

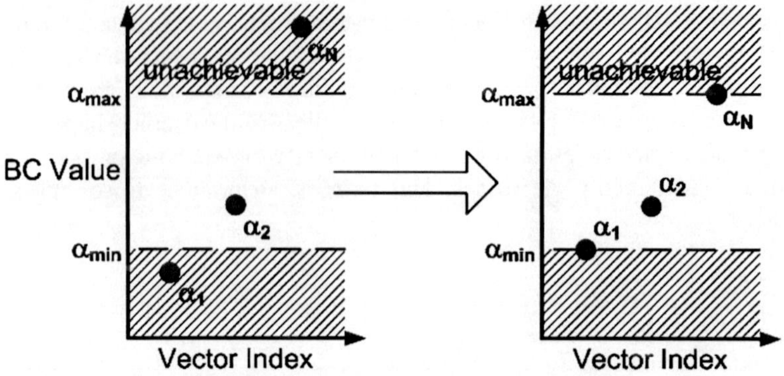

Fig. 9.10 Linear transformation of boundary conditions. An example of satisfying the boundary condition constraints is shown above. On the *left*, the components of the solution to (9.9) are plotted with the maximum and minimum constraint bounds denoted by *dashed lines* (see (9.10)). On the *right*, the components have been linearly mapped to enforce the constraints. This introduces a scaling factor into (9.9), which affects the magnitude of the pressure gradient b vector (i.e., the magnitude of the force acting on the particles). In effect, this causes the particle to be forced as much as possible in the desired direction – it imposes a limit on the maximum velocities that can be applied

due to the xy planar curvature is minimized. Second, any particle trajectory tracking errors that may occur are corrected through our feedback system (see the numerical simulations in the next section). However, the planar curvature does limit the type of trajectories that the particles can follow and this is also discussed in the next section.

Finally, the electrode voltages needed to actuate the contact angles corresponding to the pressure boundary vector α are computed by inverting the curve-fitted data of the contact angle versus voltage function $\theta(V)$.

Apply Voltages, Update Particle Position, Advance to the Next Time-Step

Our simulation advances to the next time step after using the voltages computed above to solve for the induced pressure and velocity fields. The velocity field is then used to update the position of the particle (see Fig. 9.11). The scaling described in Fig. 9.10 ensures the particle will be forced as fast as possible along the desired direction. Our algorithm runs by repeating this process for each time step.

Multiple particle steering is easily handled by applying the above discussion to each particle and its respective trajectory. The only change is that the linear system above has more rows to accommodate the extra particles. If the number of electrodes is limited, then this can adversely affect the controllability we have. In a single

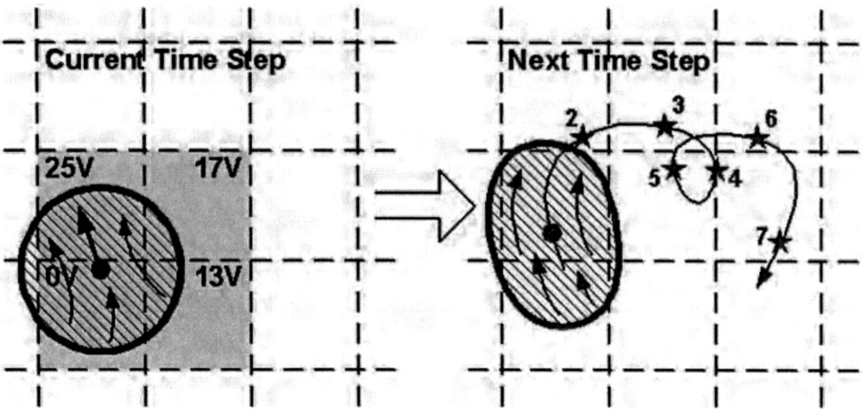

Fig. 9.11 EWOD particle steering control algorithm update. The droplet configuration from Fig. 9.8 is shown in the diagram above. The direction of motion for the particle is toward the trajectory point that is just out of reach for the current time-step. This control strategy ensures the particle will move as fast as possible and stay close to its desired trajectory. On the *left*, the *shaded* electrodes contain the voltages needed to move the particle in the desired direction. These are computed by the least-squares solution discussed above and by the inversion of the contact angle versus voltage curve-fit θ(V). The varying voltage grid induces a pressure gradient field inside the droplet such that the pressure gradient at the particle is pointing along the desired direction of motion. This moves the droplet and particle along the trajectory to the next time-step

small droplet, a single particle can be made to track interesting trajectories as long as the droplet overlaps enough electrodes (see Figs. 9.12 and 9.13). Also in a small droplet, two particles can be controlled for simple trajectories as shown in Fig. 9.14. For more than two particles in small droplets, all but the simplest trajectories (i.e., straight lines) cannot be tracked. This is a consequence of the number of actuators (N, which is typically around four for small droplets that only touch neighboring electrode pads) needing to exceed the number of particle degrees of freedom ($2m$) for the inverse problem to have an exact solution. For larger droplets that overlay more electrodes, control of more particles should be feasible.

9.2.2.2 Simulation Results and Discussion

In this section, we present some results that demonstrate basic electrowetting particle steering control using our experimentally validated simulations. A 3×3 electrode grid is used to actuate and control the droplet and each square electrode is 1.4 mm on a side. We present four cases that are controllable and three cases that are not and then discuss the possibilities and limits of our method. The voltages generated by our algorithm are reasonable and are within the limits of the UCLA device discussed in [39].

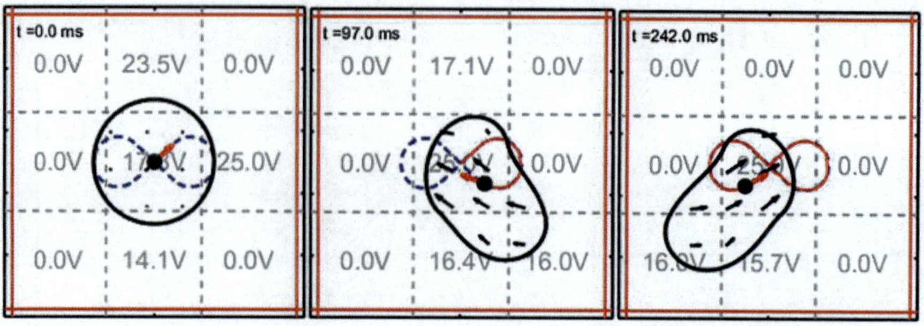

Fig. 9.12 Particle following a figure "8" path. In the simulation results above, we have a droplet (denoted by the *thick black curve*) lying on a 3 × 3 grid of electrodes (denoted by the *dashed lines*). The *blue dashed curve* is the desired figure "8" path and a *black dot* represents the particle with a *thick red arrow* pointing in the desired direction of travel. The *red curve* is the actual path of the particle. The *black arrows* inside the droplet denote the fluid velocity field inside the droplet. The voltages on the grid are time varying in such a way as to keep the particle moving along the path and are computed using the control method above, (9.8)–(9.10) [72]. (Used with permission. Copyright Royal Society of Chemistry)

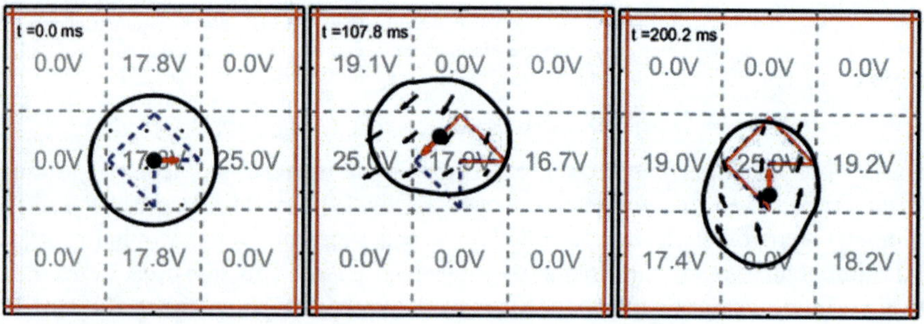

Fig. 9.13 Particle following an angular path (same format as in Fig. 9.12). The particle is able to track the trajectory very well, including at the corners

Controllable Cases

Figure 9.12 shows a droplet moving in a way that makes a particle floating inside follow a figure "8" path. A circular droplet starts on the center electrode with a particle resting in the center of the droplet. The blue dashed curve represents the desired trajectory, which is made up of a fine sampling of points. Two segments are used to represent the trajectory because of the self-intersection. The voltages on the electrode grid are actuated using the control algorithm above, which causes the particle to move forward along the trajectory. For this case, the droplet always overlaps enough electrodes to allow it to be controlled in a way that keeps the particle moving on the figure "8" path. The particle never deviates more than 20 micrometers from its desired trajectory.

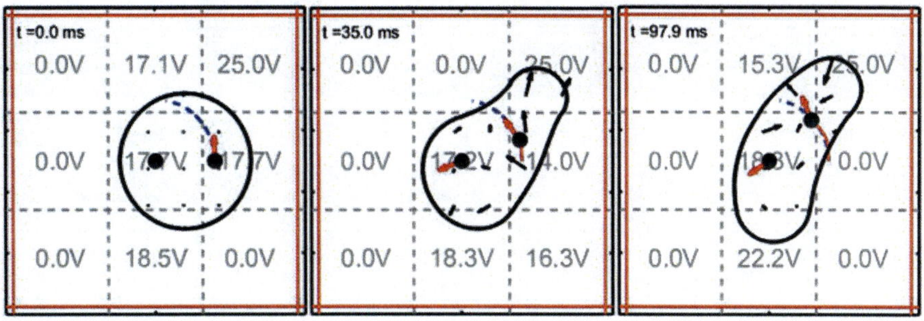

Fig. 9.14 Two-particle control: one particle moves on a quarter circle, the other is stationary (same format as in Fig. 9.12). The stationary particle's trajectory is a single point. As the particle on the right follows the circular arc, the droplet distorts to accommodate both particle motions

In Fig. 9.13, a particle is shown following an angular path that is represented by five separate straight line segments. This is to prevent the particle from rounding off the corners as it travels along the trajectory. Just as in Fig. 9.12, the droplet always overlaps enough electrodes to keep the particle on the path, with a maximum deviation error of 25 μm.

An example of two-particle control is shown in Fig. 9.14. One particle is held stationary while the other moves along a circular arc. The trajectory for the stationary particle consists of a single point, which ensures that it stays close to that point. As the particle on the right follows the circular arc trajectory, the stationary particle oscillates around its desired position to within 10 μm. The droplet itself becomes deformed because of the limited actuators and the restrictive task of moving one particle and holding another still. This also prevents the particle on the circular arc from moving past the point shown in the last frame of Fig. 9.14 and completing the circle.

In Fig. 9.15, we demonstrate particle separation. A droplet starts in the first panel with two particles spaced 0.31 mm apart. Both particles follow separate diverging trajectories designed to stretch the droplet and separate the particles. Once the particles are near the ends of their trajectories (see the third frame), our control algorithm turns off and we command an open-loop voltage of 25 V on the middle left and right electrodes and zero volts everywhere else. This causes the droplet to split into two smaller drops, each of which contains a single particle. The reason for not using our control algorithm to complete the split is because of numerical instability. When both particles are in the lobes of the dumbbell shape of the pinching droplet, the available forcing at the particles' positions is fairly weak. This would cause the condition number of the G matrix in (9.8) to degenerate and produce errors in the least-squares solution. Therefore, we avoid this by commanding open-loop voltages that we know will split the droplet (see Fig. 9.6a). Also, see Fig. 9.17 for an example of how this numerical instability can affect particle control.

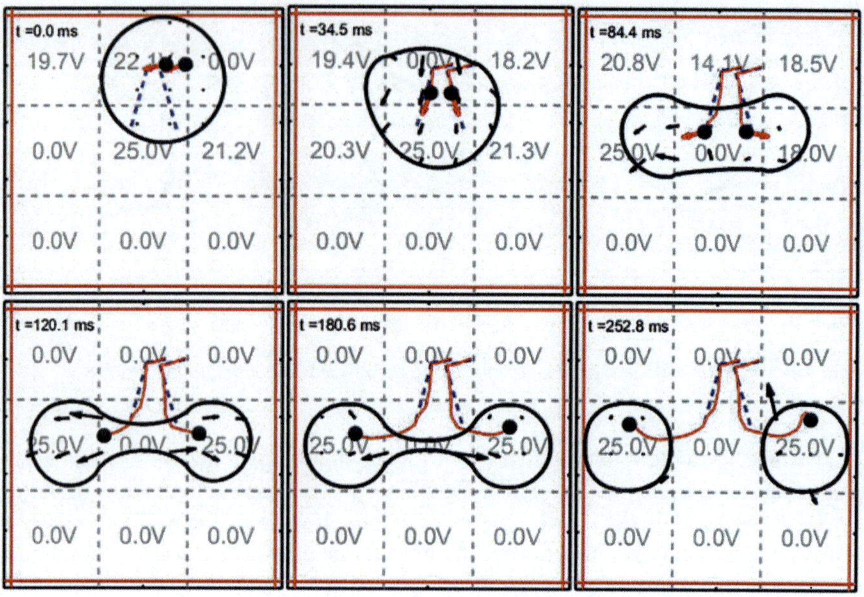

Fig. 9.15 Two-particle separation into two satellite drops (same format as in Fig. 9.12). Each particle first follows a trajectory that takes them away from each other. When there is sufficient distance between the two particles, our control algorithm turns off and the separation is completed in the usual way by applying the open-loop voltages used in the experimental splitting example (Fig. 9.6a) [72]. (Used with permission. Copyright Royal Society of Chemistry)

Uncontrollable Cases

We now show some cases that cannot be effectively controlled. In Fig. 9.16, a particle is shown trying to track a sine wave path. The particle is able to track the trajectory very well until near the end where there is a kink in the particle's path. The loss of tracking is because the droplet's shape and position at that moment are such that the number of available electrodes is very limited. It becomes impossible to create a pressure gradient field that will continue moving the particle in the tangential direction of the desired trajectory. Hence, the particle drifts away from the trajectory by more than 100 μm. This situation corresponds to (9.9) having no exact solution, which means only a least-squares best fit of the desired pressure gradient can be computed. Eventually, however, the particle is able to reacquire the trajectory.

Figure 9.17 shows two, initially separate, particles trying to come together and touch. The desired motion of the particles induces the droplet to try and pinch together in an effort to have the particles touch. However, when the particles begin to near each other, the droplet ceases its splitting action. Instead, the droplet holds the necking region and begins to oscillate up and down. This is because we are

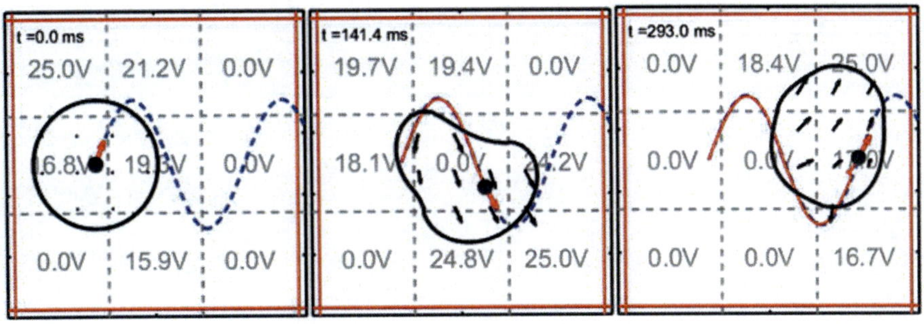

Fig. 9.16 Particle traveling on a sine wave (same format as in Fig. 9.12). The particle is able to track the sine wave path until the last time frame where the particle drifts away from the desired trajectory (see the kink in the particle's path)

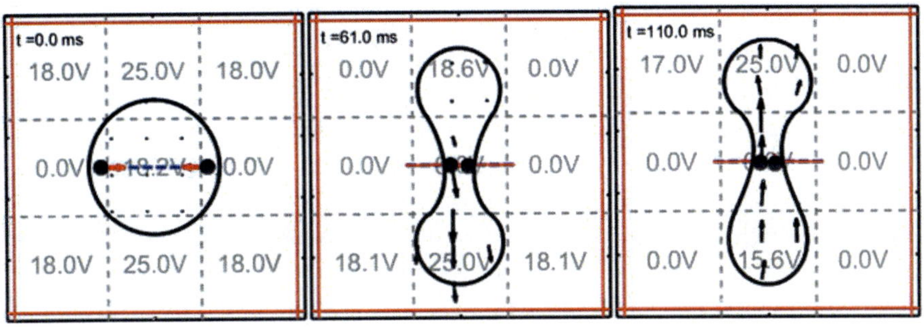

Fig. 9.17 Two particles trying to come together and pinch a droplet (same format as in Fig. 9.12). The particles travel on two separate trajectories that would, ideally, bring them together. However, as they come together, numerical instabilities in (9.9) cause random variations in the control voltages. This causes the droplet to hold its shape and move up and down in an undesirable way

trying to specify two opposite directions of motion at points that are very close together, which leads to a numerical instability in solving (9.9). As the particle positions get closer together, the condition number of the matrix G degenerates causing spurious oscillations in the control voltages. The droplet is unable to bring the particles together, much less pinch, because of the randomly varying electrode voltages.

Figure 9.18 shows two particles trying to follow diverging paths. At first the droplet is able to deform enough to keep the two particles on their respective trajectories but this quickly fails. The droplet is unable to continue deforming in a way that keeps both particles on track and moving forward. Since the trajectories are just straight lines represented by two points each, the control algorithm keeps the particles moving forward while trying to force them toward the endpoints of the trajectories. The end result is both particles stay roughly parallel with each other and are unable to recover their trajectories. This stems from a lack of available electrodes and the limitations imposed by contact angle saturation.

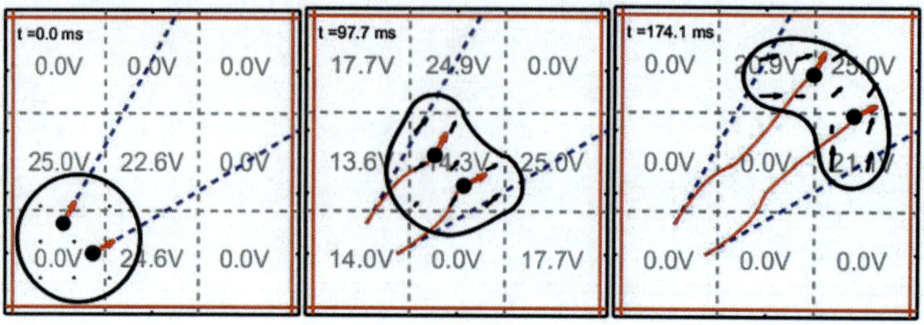

Fig. 9.18 Two particles on diverging paths (same format as in Fig. 9.12). Each particle is attempting to follow separate trajectories, both of which lead away from each other. Due to limitations of the pressure boundary actuation, and a lack of electrodes, the control algorithm is unable to keep both particles moving on their respective paths

The limitations of achievable electrowetting particle control arise from having a small number of electrodes available for actuation and from contact angle saturation. Moving several particles in different directions requires many degrees of freedom in adjusting the pressure boundary conditions. As the droplet moves, it must overlap enough electrodes to allow the realization of the pressure gradient field needed to push the individual particles along their trajectories. Hence, a finer electrode grid would allow more precise control of more particles simultaneously (not surprisingly, it is more challenging to fabricate electrowetting systems with a finer grid of electrodes). Also, some trajectories will require the droplet to become extremely distorted and may require it to split into several pieces. To do this, one needs enough dynamic range in the boundary forcing to overcome the droplets natural tendency to remain in a circular shape (see the *xy* planar curvature term in (9.5)). Contact angle saturation limits the boundary forcing and the degree of droplet deformation, which can cause controllability to be lost and particles to drift off their desired trajectories (see Figs. 9.16 and 9.18). In addition, if two particles are very close together, it is not possible to force them in arbitrary directions. The limits of boundary forcing and the numerical instability that enters into solving (9.9) inhibit close-particle control (as in Fig. 9.17) no matter how many actuators are present.

As of today, EWOD devices employ an electrode pitch and are then used to manipulate droplet of about that size (if the electrodes are made smaller, then smaller droplets can be used). This means that there are only a few actuators per droplet and this allows control of only one or two particles per droplet. Nevertheless, it is both interesting and surprising that existing electrowetting systems already have enough control authority to steer single particles along complex trajectories and to steer two particles along simple paths – usually it is assumed that additional types of control (e.g., laser tweezers, magnetic forces, etc.) are required to control single particles inside EWOD systems. In our next example, which uses electro-osmotic or electrophoretic control, it is possible to control particles with more freedom, to do so to nanometer precision, and to control particles that try to swim away (we control swimming bacteria). We also have initial results on three-dimensional control and controlling the orientation of objects by creating flows with the right amount of shear at the objects location.

9.3 Manipulating Objects by Flow Control and/or Electrophoresis

As in macroscale technologies and applications, there is a need to put things where they need to go (cells into testing chambers or to sensor locations, quantum dots into photonic cavities), and this is difficult to do on the microscale. We have developed, and experimentally demonstrated, a suite of techniques based on feedback control of the surrounding flow and/or electric fields to steer, place, and hold objects in microfluidic systems. Flow control methods to individually position *and* orient micro- and nanoscale objects, such as nanorods, are being demonstrated next.

Our approach has advantages over laser tweezers and optoelectronic methods [78–81], which are the current state-of-the-art approaches for manipulating micro- and nanoscale objects. Our method is simpler and cheaper. We can control any kind of visible objects in liquid solutions, not only objects with the right dielectric properties to permit force trapping by optical or optoelectric means [82]. Our method can be integrated into a hand-held system, and position error correction is implemented over a large working area instead of relying on particle capture into a small optical trap [83] thus allowing robust manipulation over a large region. And our method has a more favorable scaling with object size [84] – optical forces scale with the volume of the object making it difficult to control very small objects [85], fluid control forces scale with the object diameter [15] so we get bigger forces more easily at the nanoscale. Our large control working region has allowed us to steer and hold swimming bacteria (we continuously bring them back as they try to swim away) and the more favorable force scaling has allowed us to manipulate single quantum dots to nanometer precision for as long as they remain visible [84] without using high power lasers that can damage the particles they are meant to control. Our method also has limitations compared to optical methods: laser tweezers can control more particles at once [82] and they can more readily be used to quantitatively measure particle-to-particle interactions [86, 87] (for us to measure such forces would require precision inversion of a fluid dynamic model that has uncertainties in it that will degrade the inversion). Laser tweezers are also routinely used for three-dimensional manipulation whereas we have only recently demonstrated three-dimensional control in simulations [88].

Current applications for our method include manipulation of cells on chip for basis science biology studies and for lab-on-a-chip applications such as sample preparation (e.g., sorting out cells of interest, such as bacteria, stem cells, or circulating tumor cells, from human samples), and positioning quantum dots on photonic crystals for creating multidot quantum information systems [89]. We envision being involved in many additional applications now that the method is mature and has been experimentally demonstrated to be flexible, robust, and nanoprecise.

9.3.1 System Setup and Device Fabrication

Our basic system to manipulate micro- and nanoscale objects by flow or electrical control consists of a microfluidic device, a microscope and a camera to observe the location of objects inside the device in real-time, actuating electrodes powered by a digital to analog converter, and a control algorithm on a standard personal computer (Fig. 9.19). The microfluidic device is made out of a soft polymer (polydimethylsiloxane (PDMS)) and is fast and easy to fabricate. It can be laid on top of other devices, e.g., on top of a glass device with patterned chemical features, on top of a silicon device with other MEMS capabilities, or on top of a photonic crystal for our quantum dot placement project. Details on system setup are given in [83]. More advanced capabilities, to manipulate swimming cells, to steer and trap multiple particles at once, and to place single quantum dots to nanometer precision on chip are described in [83, 84, 89].

9.3.2 Physics and Modeling

Our system can actuate micro- and nanoscale objects in one of two ways. It can either move the fluid in the device by electro-osmotic actuation (described next) to carry particles along, this works for both neutral and charged particles; or, if a particle is charged, then it can be actuated by an electric field which applies an electrostatic (Coulomb) force and moves the particle relative to the surrounding fluid (electrophoretic actuation) [49, 50]. Particles often acquire a surface charge through weak chemical interactions with the surrounding fluid, for example the polystyrene beads we used in [83] have a surface charge in water as do the yeast cells we also controlled. Thus charged particles are the norm rather than the exception but the amount of charge can vary depending on the chemistry of the object and the surrounding medium.

Electro-osmotic actuation of flow is routine in microfluidic devices, e.g. [90–92]. Here an applied electric field electrophoretically moves a thin layer of charges that form naturally at the fluid/device interface. Typically, these charges are ions present in the liquid that migrate to the solid/liquid boundary to shield stationary charges formed there, for example, by weak acid/base chemistry occurring at the interface (the same type of chemical mechanisms also lead to charge formation on the surfaces of particles). Which charges (positive or negative) and how much they accumulate inside the liquid immediately adjacent to the device surfaces depends on the chemistry of the liquid and solid materials, on the pH, the amount and type of dissolved ions, surface treatments, and many other factors [93–95]. The electric

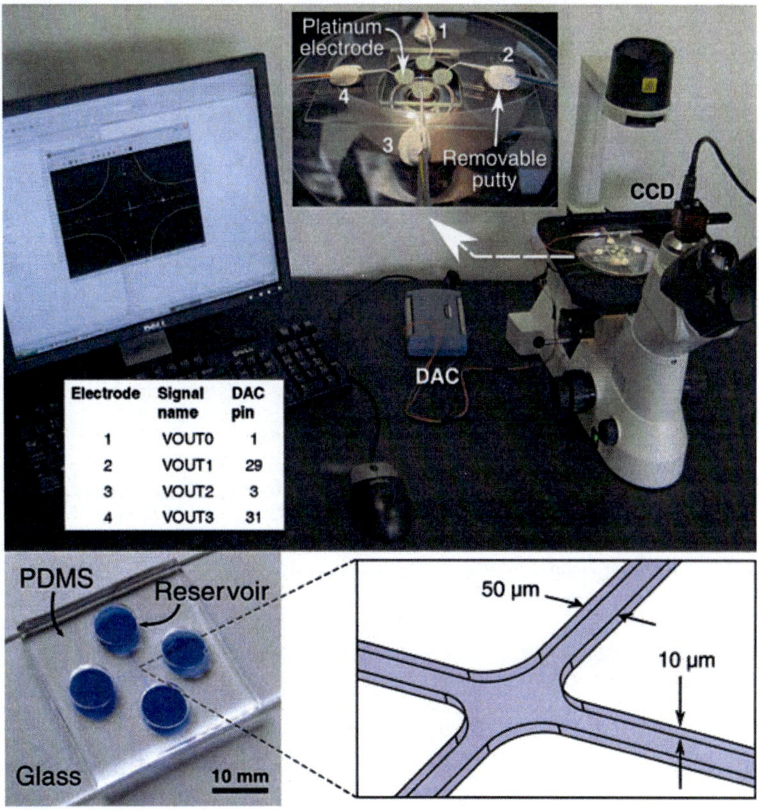

Fig. 9.19 Our flow control system for a single particle. *Top*: Photograph of the experimental setup, the flow control device is on the right on top of the inverted microscope which is connected to a CCD camera. *Bottom left*: Photograph of a four-channel PDMS on a glass device filled with blue food coloring to clearly show the microfluidic channels and reservoirs. Each microchannel is 10 mm long, 50 μm wide close to the particle steering intersection region and 300 μm wide otherwise, and 10 μ m deep. *Bottom right*: Schematic of the channel intersection and the 100 μm × 100 μm cell steering control area. The corresponding system closed-loop block diagram is shown in Fig. 9.21

field applied by the electrodes moves these free charges (the Debye layer) in one predominant direction. This thin moving layer of charges then drags the rest of the fluid along by viscous forces, the electro-osmotic actuation (Fig. 9.20). (Charges in the interior of the fluid do not cause a net fluid motion, since there is essentially an equal number of positive and negative ions. Only a small fraction of ions of one type are taken away into the Debye layer. The remaining interior charges create equal and opposite electrical forces on the fluid in the channel center, their only net effect is to move through the fluid and heat it.) A more detailed description and analysis of the physics of electro-osmotic actuation can be found in [49, 50, 96].

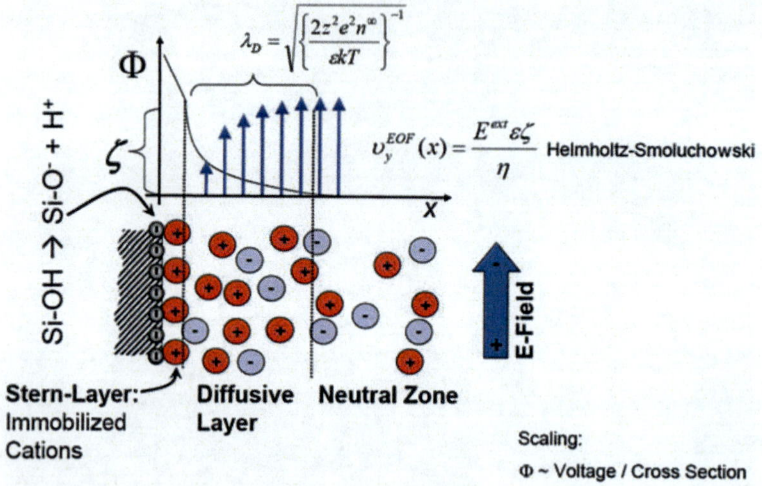

Fig. 9.20 The physics of electro-osmotic actuation. A schematic side view through a microfluidic channel is shown (the channel wall is on the left side, the flow is being electro-osmotically actuated up the channel by the applied electric field). The *minus* signs represent the fixed charges at the solid/liquid interface, *large circles* (+ or −) show ions naturally found in the liquid (e.g., in water). These ions accumulate to shield the surface charges forming a thin Debye layer that has a predominant charge (here mostly positive, on the left). The electric field moves this layer and it drags the fluid in the channel by viscous forces. Charges in the interior of the channel (the "neutral zone") remain essentially balanced (only a small fraction of the charge goes to the surfaces) and so they create no net fluid motion effect [96]. (Used with permission. Copyright COMSOL)

In electro-osmotic flow the fluid is dragged by moving charges that are actuated by the applied electric field. In our planar devices this means that the flow will follow the electric field that is present at the floor and ceiling of the device. The electric field we apply is uniform in the vertical direction but it can have complex patterns in the horizontal *xy* plane. The resulting microflow will exhibit these same complex horizontal patterns. It is possible to show this rigorously starting from the Navier–Stokes equations, as we do in [97], the end result is that the fluid velocity follows the applied electric field essentially instantaneously (with a microsecond time constant) [98, 99]. Thus, see also [49],

$$\vec{V}(x,y,z,t) = (\varepsilon\xi/\eta)\,\vec{E}(x,y,t) = -(\varepsilon\xi/\eta)\nabla\phi(x,y,t), \tag{9.11}$$

where $\vec{V}$ is the electro-osmotic fluid velocity, $\vec{E}$ is the applied electric field which is uniform in the vertical direction, ϕ is the electric potential as created by the actuators of Fig. 9.19, ε is the permittivity of the liquid, η is its dynamic viscosity, and ξ is the zeta potential (essentially the voltage) at the liquid/solid interface [49, 50]. Electric fields are governed by Laplace's equation, the electrostatic limit of Maxwell's equations [51], with boundary conditions at the electrodes set by the voltages that we apply there.

In the above it is ξ which quantifies the amount of charge that is contained in the Debye layer. Since this value depends on the details of the surface chemistry and cannot be predicted a priori, it is usually inferred from experiments by applying a known electric field and measuring the resulting flow velocity. The chemistry that happens at the solid/liquid interface is complicated and so the above discussion of electro-osmotic actuation should be understood as a first order simplified explanation (further explanations can be found in [100, 101]). Although the underlying chemical principles of electro-osmosis are still not well understood, that does not prevent us from using it to precisely control microscopic and nanoscopic particles as we show in the remainder of this chapter.

Neutral particles are carried along by the created electro-osmotic flow. In addition, these particles experience Brownian motion. When the particles are comparable in size to the channel height, as for example the yeast cells that are $\sim 5\,\mu\text{m}$ in diameter compared to the $11\,\mu\text{m}$ high channels we used in [83], then the channel floor and ceiling constrain vertical diffusion. When the particles are small, e.g., the nanoscopic quantum dots, then they diffuse in all three directions. In either case, we only control their motion in the xy plane leaving their motion to be free in the z-direction.

Thus, in the plane, the particle positions are governed by $\dot{\vec{P}}_j = \vec{V}(\vec{P}_j) + \vec{w}$, where $\vec{w}$ is Brownian noise and $\vec{P}$ is the vector of particle x and y positions. The electric potential is described by Laplace's equation $\nabla^2\phi = 0$ with Dirichlet boundary conditions at the electrode boundaries $\phi(\partial D_j) = u_j$, where ∂D_j denotes the liquid/electrode interface location and u_j is the jth applied voltage. Insulating Neumann conditions hold at other surfaces. The solution of Laplace's equation is linear in the applied voltages so:

$$\dot{\vec{P}} = \vec{V}\left(\vec{P}\right) + \vec{w} = c\vec{E}\left(\vec{P}\right) + \vec{w} = -c\nabla\phi\left(\vec{P}\right) + \vec{w} = -c\sum_{j=1}^{n}\nabla\phi_j\left(\vec{P}\right)u_j + \vec{w}, \tag{9.12}$$

where $c = \varepsilon\xi/\eta$ is the electro-osmotic mobility, ϕ_j is the solution to Laplace's equation when electrode j has a unit applied voltage and all other electrodes are at zero voltage, and $\vec{u}$ is the time-varying vector of applied voltages. Note that the velocities of the particles are in the direction of the locally applied electric field and so depend on where they are with respect to the electric potential $\phi(x, y)$. For the same set of voltages, two different particles in two different locations can be actuated in different directions. In summary, the equations to be controlled for m neutral particles are linear in the control and nonlinear in the particle positions, they are:

$$\dot{\vec{P}} = A\left(\vec{P}\right)\vec{u} + \vec{w}, \tag{9.13}$$

where $\vec{P} = (x_1, y_1, x_2, y_2, \ldots, x_m, y_m)$ is the position vector for the planar location of the m particles of interest and the A matrix contains spatial information about the electric fields originating from each electrode.

If the particles are charged then there is an added electrostatic force that also points with the electric field – either along it for a positively charged particle or directly opposite it for a negatively charged particle. This can be incorporated into the A matrix by modifying the mobility coefficient for each particle. Variations in the electro-osmotic zeta potential and the amount of charge on the particles can change these mobility coefficients, but the control algorithm is robust to these variations – the control basically sets the direction of particle motion at the location of each particle. So long as the sign of the mobility coefficient for that particle does not flip (a rare occurrence) the control works. To further improve performance, we usually identify the mobilities of the particles of interest before starting an experiment by applying a known electric field and observing their resulting velocity through our vision system. Our particle steering experiments in [83] function to 1 μm precision even though the polystyrene particle and cell mobilities in that case are only known to within ±50%. Our quantum dot experiments show 45 nm accuracy even though the charge on the QD also varies.

9.3.3 Feedback Control

Figure 9.21 shows the basic control idea for a single particle: a four channel microfluidic device, an optical observation system, and a computer with a control algorithm are connected in a feedback loop. The vision system locates the position of the particle in real-time, the computer then compares the current position of the particle with the desired (preprogrammed or user input) particle position, the control algorithm computes the necessary actuator voltages that will create the electric field, or the fluid flow, that will carry the particle from where it is to where it should be, and these voltages are applied at electrodes in the microfluidic device. For example, if the particle is currently South/East of its desired location, then a North/West flow is created. The process repeats at each time instant and forces the particle to follow the desired path (see [83, 102] for details).

Surprisingly, it is also possible to steer multiple particles independently using microflow control [20]. A multielectrode device is able to actuate multiple fluid flow or electric field modes. Different modes cause particles in different locations to move in different directions. By judiciously combining these modes, it is possible to move all the particles in the desired directions. We note here that this kind of flow control, where we control the fluid so precisely that we can hold or steer multiple objects at once in different locations, is not possible in macroscale fluid dynamics. Here we are exploiting the linear nature of the electrostatic equations and Stokes flow (the nonlinear fluid momentum terms, the "Navier" part, are negligible on the microscale) to be able to invert the problem to achieve control. We certainly would not be able to invert a high Reynolds number or turbulent flow in the same fashion since it would amplify small changes in actuation to large errors in particle motion.

The multiparticle steering control algorithm is more sophisticated than the single-particle algorithm: its operation relies on inversion of the flow and electric fields

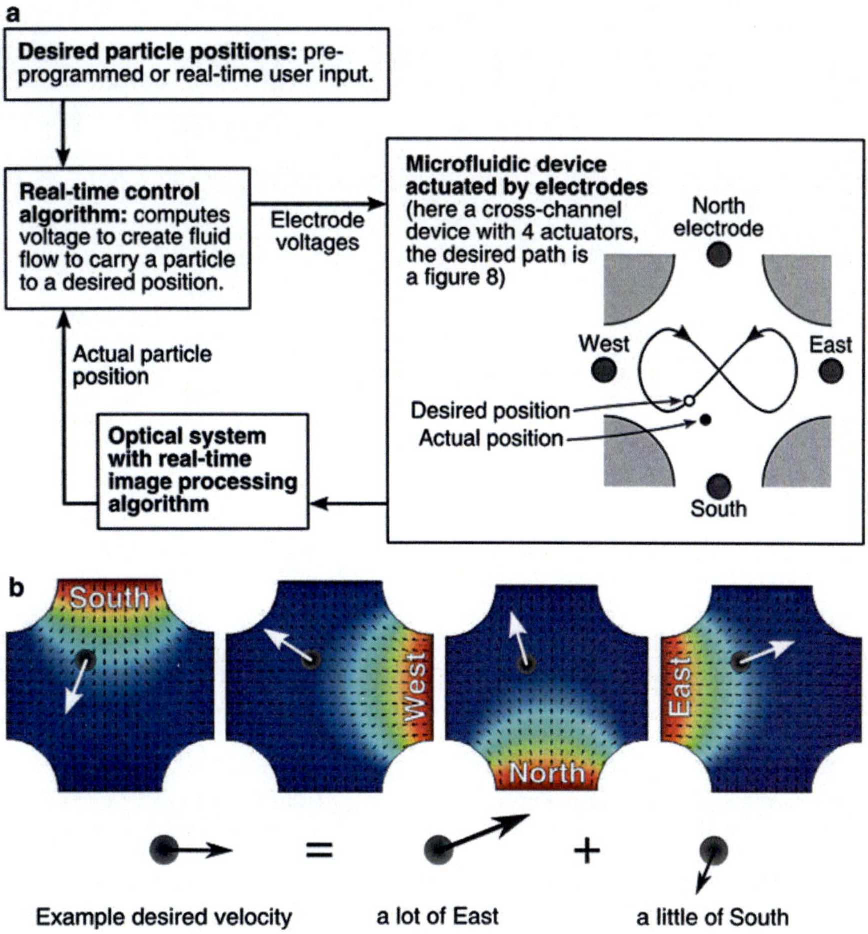

Fig. 9.21 (**a**) Feedback control steering approach for a single particle. A microfluidic device with electro-osmotic actuation is observed by a vision system that informs the control algorithm of the current particle position. The control algorithm compares the actual position against the desired position and finds the actuator voltages that will create a fluid flow, at the particle location, to steer that particle from where it is to where it should be. The process repeats continuously to steer the particle along its desired path. (**b**) Four basic flows that can be generated by applying a voltage to each electrode individually (from simulations). By actuating these four flows together correctly, it is possible to generate an electrokinetic (electro-osmotic + electrophoretic) velocity at the chosen particles location in any desired direction to always carry that particle from where it is to closer to where it should be [83]. (Used with permission. Copyright IEEE)

predicted by the model. An eight-electrode device, as in Fig. 9.22, can create seven independent electric/fluid modes (one of the eight electrodes acts as ground, or, equivalently, if the electrodes float, raising or lowering all of them by a constant voltage does not impact the electric field, so only seven degrees of freedom remain).

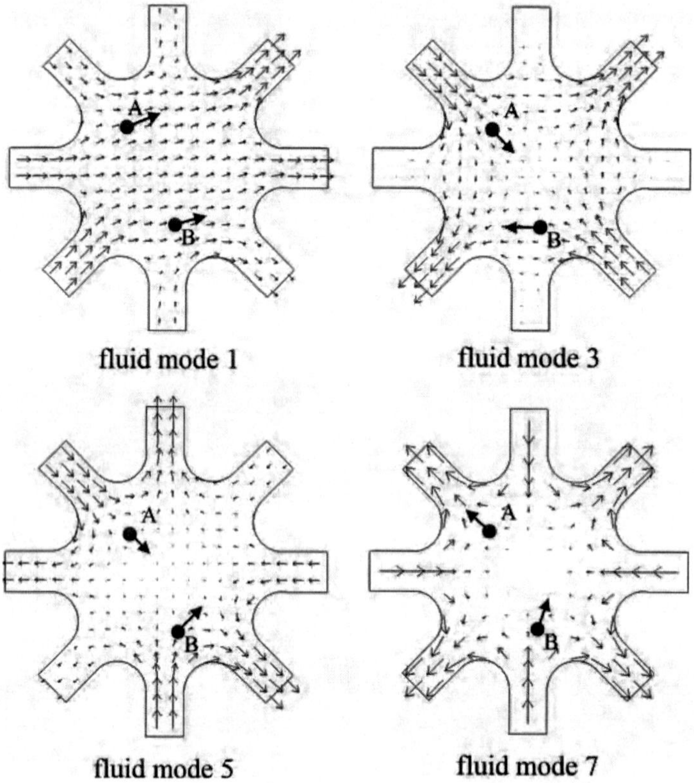

Fig. 9.22 Electro-osmotic microflow modes for an eight-electrode device. The above figure shows the first, third, fifth, and seventh modes computed from the model stated above (also see [20, 83]). The two example neutral particles A and B (shown as *black dots* above) will then experience the velocities shown by the *arrows* [83]. (Used with permission. Copyright IEEE)

Four of these seven modes are shown above. The key point is that the different modes force particles at different locations in different directions (see particles A and B in Fig. 9.22): by intelligently actuating a combination of modes, we can force all the particles toward the right locations at each instant in time. Since each particle has two degrees of freedom (an x and a y position), an eight-electrode device can precisely control up to three particles (particle degrees of freedom $3 \times 2 = 6 \leq 7$ actuation degrees of freedom).

In its simplest incarnation, the multiparticle control algorithm works as follows (details in [20]). We define a desired correction velocity vector between where all the particles of interest are observed to be versus where we would like them to be at the current time:

$$\vec{v}_{\text{correction}} = k\left(\vec{P}_{\text{desired}} - \vec{P}_{\text{observed}}\right), \tag{9.14}$$

here k is the control gain. Our task is now to choose the voltages at the electrodes to create a velocity as close to this desired correction velocity as possible. Since, by (9.13), there is a linear relation between the control and the velocity (we know the particle positions since the camera can see them), and since this velocity is achieved essentially instantaneously as soon as we apply the voltages, we can solve a static linear problem to determine the needed set of electrode voltages. Specifically, as in the EWOD problem, we solve a least-squares problem to find the set of actuator voltages that will create velocities at all the particles of interest as close as possible to the desired correction velocities. The other particles (the particles not of interest) are actuated in some random way that depends on the electric fields they will see at their locations. This gives the feedback control:

$$\begin{aligned}\vec{u}^* &= \left[A^T\left(\vec{P}\right)A\left(\vec{P}\right)\right]^{-1}A^T\left(\vec{P}\right)\vec{v}_{\text{correction}} \\ &= k\left[A^T\left(\vec{P}\right)A\left(\vec{P}\right)\right]^{-1}A^T\left(\vec{P}\right)\left(\vec{P}_{\text{desired}}-\vec{P}_{\text{observed}}\right). \end{aligned} \tag{9.15}$$

For the case where there are more actuation than particle degrees of freedom ($n-1 \geq 2m$), the A matrix typically has full row rank (unless two particles are at the same location) and the above least-squares answer achieves the desired velocity with minimum control effort (with minimum $\|\vec{u}\|_2$) [75]. For cases where we try to control more particle degrees of freedom than we have actuators, the experimental performance rapidly degrades to unusable. For example, four particles (eight degrees of freedom) can be controlled badly by eight electrodes (seven degrees of freedom, one electrode is ground), but five particles cannot. Since it is possible to fabricate devices with many electrodes, the real limit to the number of particles that can be controlled is the condition number of the matrix A as discussed below.

We pre-compute the electric fields that make up the matrix A ahead of time, this means we can use a lookup table to determine A for any particle positions $\vec{P}$ seen by the camera. We then compute the pseudo-inverse $(A^TA)^{-1}A^T$ in real-time, in milliseconds, as the control proceeds. It is convenient to carry out this calculation in the coordinate system of the fluid modes of Fig. 9.22 (the singular values modes of the matrix A evaluated on a fine grid of points). The dominant (lower spatial frequency) modes are the ones that are better conditioned: at the higher spatial modes very high voltages are required to create even small fluid velocities. Thus we truncate our matrix A onto these first modes and compute the pseudo-inverse above for that well conditioned matrix. It is in fact this conditioning that sets how many particles we can control at once. For our experimental image sensing and actuation errors, we can robustly access just over the first ten or so modes which means we have been able to control up to five particles simultaneously in experiments. There are also other issues, such as a limit to the voltage that can be applied at the electrodes. Too high a voltage causes electrolysis [103], a chemical reaction that creates bubbles, and must be avoided – this voltage limit depends on buffer and electrode chemistry, for us it is around 10 V. We have treated this actuation limit in two ways: either by turning down the control gain per particle as we approach this limit or, more rigorously, by phrasing a linear programming constrained optimization to choose the gain per particle to maximize performance but not exceed actuator limits. These two approaches both work equally well in experiments.

Our control works robustly across the entire control region – so long as we have done the singular value mode conditioning above there are no regions or combinations of particle locations where we cannot reliably pseudo-invert A. The only time the inversion fails is if two particles are right on top of each other but we are trying to move them in different directions (this is physically impossible since we have to create two different fluid flow directions at the same location). Indeed, our particles can be controlled very close together – in experiments we have shown an ability to steer particles to within 8 μm of each other.

9.3.4 Experimental Results

9.3.4.1 Control of Single Particles to Micrometer Precision

Experimental results for manipulation of one particle, first reported in [102], needed only a simple control algorithms (if the particle was North/West of its target, we created a South/East flow in the entire device) but required solution of practical issues such as device fabrication, fast and reliable vision sensing, operating in a regime of reliable electro-osmotic actuation but with no unwanted chemical reactions (no electrolysis), and prevention of device fouling and particle sticking. Smoothing out of the control algorithms and optimization of the vision system enabled us to control single particles, e.g., polystyrene beads and yeast cells, to single micrometer accuracy (Fig. 9.23). The achieved single micrometer resolution was set by the 1 μm field of view that corresponded to each camera pixel – so we controlled as well as we could see, to single pixel accuracy. One micrometer also roughly corresponds to a more fundamental vision sensing limit, the wavelength of visible light, which sets the absolute minimum on how close two features can be before they can no longer be distinguished one from the other. We discuss how it is possible to bypass this sensing limit for particle control in the section on controlling single quantum dots to nanometer precision.

9.3.4.2 Control of Multiple Particles to Micrometer Precision

Control of more than one particle at the same time requires the more sophisticated pseudo-inverse control algorithm described previously. Below we show results for steering three particles at once using eight electrodes, all to 1 μm accuracy (again

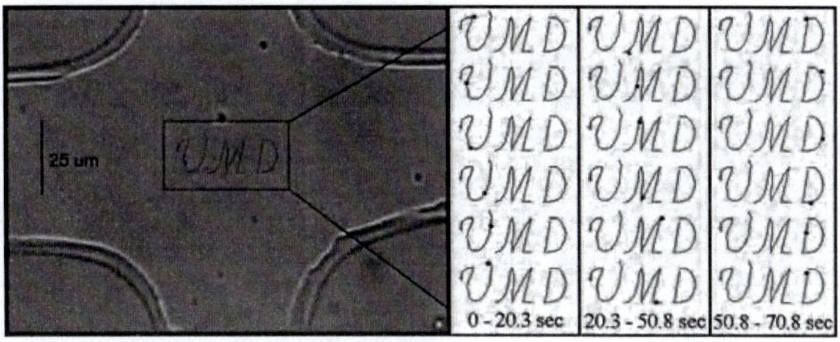

Fig. 9.23 Steering of a slightly charged yeast cell along a UMD path. The cell had an approximate electrophoretic mobility of $c_{\text{ep}} = (-23.3 \pm 6.9) \times 10^{-9}\text{m}^2\text{V}^{-1}\text{s}^{-1}$. By comparison, the electro-osmotic mobility of our PDMS devices was $c_{\text{eo}} = (36.5 \pm 3.6) \times 10^{-9}\text{m}^2V^{-1}\text{s}^{-1}$. *Left*: Close-up photograph of the microfluidic devices with the desired cursive "UMD" path overlaid on the image. *Right*: The actual path of the chosen 5 μm yeast cell (Red Star® Yeast) (*black dot*) in the feedback control experiment. Snapshots are shown at six equally spaced times for each letter. The yeast cell follows the required trajectory to within 1 μm [83]. (Used with permission. Copyright IEEE)

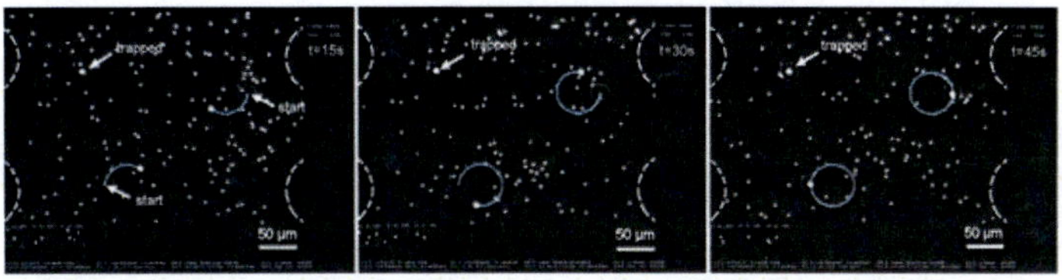

Fig. 9.24 Steering of two fluorescent beads (2.2 μm diameter, Duke Scientific) around two circles while a third bead is held stationary. In the experiment, the fluorescent beads appear as *small green dots* on a black background and the device geometry, which does not fluoresce, is not visible. Here, the *white dots* are the beads (enlarged), the *solid curves* are the actual trajectories that the target beads have traced out (overlaid), and the *dashed white curves* (also overlaid) show the geometry of the channels and the particle control chamber. Snapshots are shown at three time-steps. The two beads are being steered to within an accuracy of one pixel (corresponding to less than 1 μm). The desired paths are not shown because, at this image resolution, they would perfectly underlay the actual paths. The trapped bead is marked by an *arrow*, and is trapped by the control algorithm to an accuracy of better than 1 μm. Every time the bead deviates from its desired position, a flow is created that pushes the bead back toward its desired location [83]. (Used with permission. Copyright IEEE)

from [83]). We have also demonstrated control of five particles at once but the accuracy is degraded away from 1 μm. This level of control, that it is possible to actuate multiple objects at once in the interior by actuating a fluid by electrodes on its boundary, was and is surprising to the microfluidics community. It is a concrete example, experimentally demonstrated, that shows control theory can enable simple microfluidic systems to perform complex and precise tasks (Figs. 9.24 and 9.25).

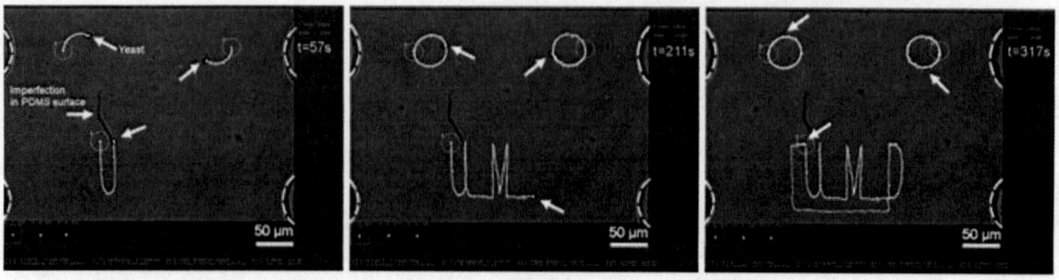

Fig. 9.25 Steering of three yeast cells (5 µm diameter) with small surface charge (electrophoretic mobility $c_{ep} = (-23.3 \pm 6.9) \times 10^{-9} m^2 V^{-1} s^{-1}$) around two circles and a "UMD" path. The cells do not fluoresce. In these images there is no high-pass filter before the camera and the raw images are shown. The yeast cells are visible as small *black dots with a white center* (the three target cells are marked with a *white arrow* in each image), and the *white curves* are the trajectories that the target cells have traced out. The three cells are being steered to within an accuracy of one pixel (corresponding to less than 1 µm) [83]. (Used with permission. Copyright IEEE)

9.3.4.3 Control of Live Swimming Cells

Compared to laser tweezers and optoelectronic techniques [78–81], our technique has the big advantage that it works over a large control area (as shown in Fig. 9.26). This means it is easier for us to manipulate swimming cells: every time they swim away we bring them back (as opposed to moveable trap methods where the bacteria may exceed the optical forces, swim out of the optical trap, and thus escape its intended manipulation). So long as our control can correct the location of the microbe faster than that microbe can swim away it will be effective in trapping and steering it. Whether this can be done or not depends on both the swim speed of the microbe and its preferred swim patterns – fast swimming microbes that like to swim in small circles can be controlled because, even though they swim fast, they do not swim far away; in contrast, medium speed microbes that swim out in straight lines in random directions get further away and are harder to bring back. Below we show initial results for manipulation of medium speed (<10 µm/s) swimmers (Fig. 9.27). We plan to improve our slow control update (every 1/30th of a second) to 300 Hz, this should allow us to control even fast swimmers.

The target applications here involve preparation of biological samples that contain moving organisms, e.g., precisely removing motile bacteria from human samples, steering them to chambers for sensing and subsequent analysis, and (when we achieve control of multiple swimming organisms at once) testing the reaction of one swimming organism against another. Faster hardware (currently we operate at a slow 30 Hz) will allow us to control more often per second and will thus give the microbe less time to escape between control corrections. We also plan to develop smarter control algorithms that will detect and exploit the properties of the specific microbe we are trying to control.

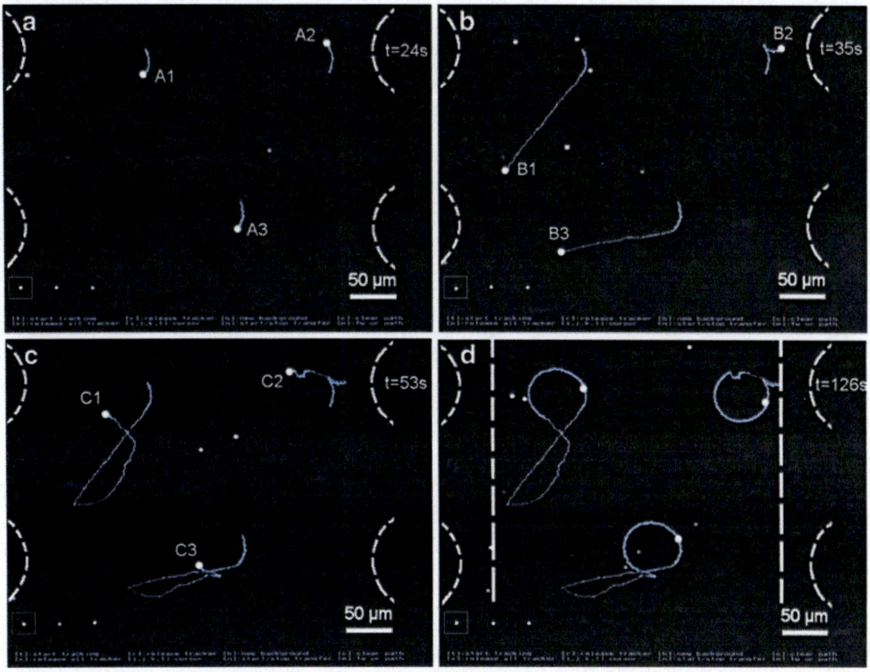

Fig. 9.26 The control algorithm is globally stable and can correct for large errors in particle positions. This figure shows steering of three fluorescent beads (2.2 μm diameter, Duke Scientific) around three circles. At time $t = 24$ s, corresponding to bead positions marked A_1, A_2, and A_3, the control was turned off for 11 s, allowing the particles to drift away (primarily due to the slow parasitic flow inside the device caused by surface tension forces at the reservoirs) by up to 150 μm. The control was then turned back on at $t = 35$ s (B_1, B_2, and B_3), and the control algorithm steered the three original beads back to their desired positions (C_1, C_2, and C_3). Four time instants are shown: (**a**) right before control is turned off, (**b**) right before control is turned back on (the three beads have drifted away a large distance), then (**c**) at a time when the beads are back on track, and (**d**) the final time when the beads have completed the remainder of their three circular paths (again to an accuracy of better than 1 μm). The two straight lines in the last image illustrate the left and right boundaries of the control region [83]. (Used with permission. Copyright IEEE)

9.3.4.4 Control of Single Quantum Dots to Nanometer Precision

We end this experimental section by showing manipulation of a single nanoscopic particle, a quantum dot, to nanometer precision by flow control. This is needed for creating nanophotonic and nanoelectronic devices, in that situation there is a need to place multiple quantum dots in the high electric field regions of nanophotonic [104–106] and plasmonic [107, 108] structures and this has not been done in any other way. The high field regions of photonic cavities are small, approximately 250 nm in size, so nanometer placement is necessary [109]. (Once one dot has been placed, it is possible to fix it in place by a chemical binding reaction to the surface [110] or

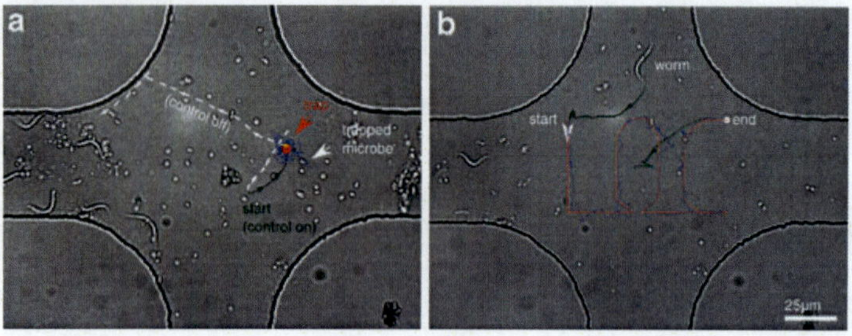

Fig. 9.27 (**a**) A swimming microbe found in river water was moved to an arbitrary trapping location and trapped for 30 s until being released from control. Uncontrolled swimming is shown by a *dashed line*, initial control to the trap or path is shown by a *thin line with arrow heads*, and the controlled motion is shown by a thin line without arrow heads – as is evident, the microbe swims away after the control is turned off which means it was not harmed by the control. (**b**) A worm was steered around a trajectory spelling "LOC" (for lab-on-a-chip)

by solidifying the surrounding fluid [111, 112], thus allowing placement and fixing of one QD after another to make multidot devices.)

In the experimental results reported earlier, our vision sensing accuracy had limited the control precision. That limit was both experimental setup specific (camera pixel size) and more fundamental (the wavelength of light limitation). It is possible to improve the sensing to determine the location of a particle to well below the wavelength of light, in real-time. The key is to realize that a nanoscopic particle, such as a quantum dot, appears as a diffraction limited spot under a microscope, and this spot spans many pixels (see the inset in Fig. 9.29a). By averaging correctly over the many pixels it is possible to infer the center of the diffraction pattern to better than single pixel resolution, a technique known as subpixel averaging [113], which we perform in real-time.

The errors in quantum dot placement are now a combination of vision sensing errors (which can be driven down to tens of nanometers) and diffusion between control updates (which can be reduced by doing control updates more often and by using a higher viscosity fluid – we added a polymer to water that increases its viscosity). Quantum dots also presented other problems that had to be solved to achieve nanometer precision. QDs blink on and off: they blink in and out of view. We pause our control actuation when a QD blinks off and continue actuating when it blinks back on. The QD position is controlled in the horizontal plane but the QD still diffuses in the vertical direction. This diffusion makes the QD leave the focal plane of the microscope and causes a defocusing which hurts our sensing accuracy. Thus we wrapped in a second control loop that uses the variance of the QD image as its metric and drives this metric to a minimum by moving the microfluidic device up or down using a piezo stage. The problem is that going up out of the focal plane and going down below it both look the same, so for this second control loop we introduce a small jitter and then check if the dot looks more focused when going

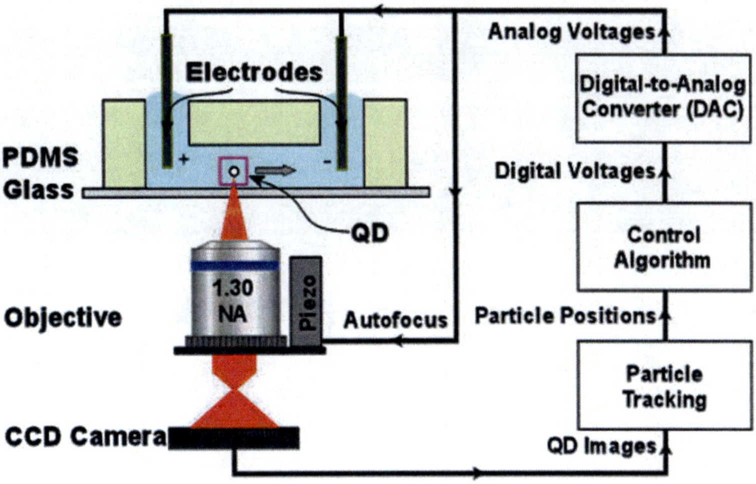

Fig. 9.28 Illustration of the optical and electronic setup for tracking and feedback control of QDs. A CCD camera images the QD and sends the information to a tracking algorithm that uses subpixel averaging to accurately determine the current position of the QD. The control algorithm uses this information to determine the proper voltages to apply to the electrodes in order to move the QD to its desired position. A second feedback loop moves the imaging objective in the z-direction using a piezo stage to keep the QD in focus [84,89]. (Used with permission. Copyright American Chemical Society.)

up or down. This tells us if we are above or below the focal plane and we then use a Newton-bracketing algorithm to steer to the minimum image variance. This inner loop runs slowly compared to the main *xy* control loop. The end result is higher accuracy control in the *xy* plane. The vertical loop also tells us where the QD is with respect to the bottom photonic crystal so we can wait until the QD diffuses to the bottom to freeze it in place. Chemistry is also an issue. We had to create a fluid that could be actuated by electro-osmosis, that had a high viscosity, and that was compatible with our device (with PDMS) and with the QDs (would not cause them to fall out of solution). With our colleagues, we are now further creating fluids that satisfy all of the above criteria and, in addition, can be solidified to nanometer precision (by two-photon absorption) [114] to allow us to fix a QD at a photonic crystal cavity by solidifying just a small amount of fluid around it.

Our current single QD positioning results are reported in [84,89] which includes all details on the experimental setup (Fig. 9.28), the error analysis, and an optical autocorrelation measurement that proves we are indeed controlling just one single quantum dot. We were able to hold a QD at a single location to 45 nm accuracy and steer it along a path with an average deviation of 120 nm (Fig. 9.29). The dot was controlled for 1 hour, its useable (i.e., visible) lifetime.

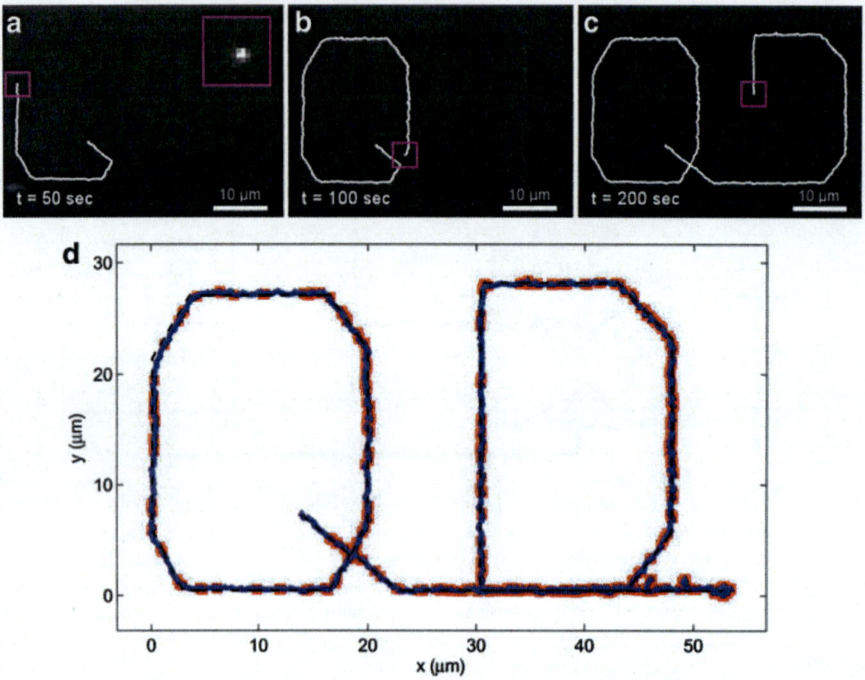

Fig. 9.29 Single quantum dot trajectory. (**a–c**) Time stamped CCD camera images of a single quantum dot being steered along the desired trajectory. The *white trace* shows the measured path of the quantum dot up until its current location. The *square magenta box* shows the subpixel averaging window used to determine the current position of the QD. The *inset* in panel (**a**) shows a closeup of the subpixel averaging window which contains the QD near its center. (**d**) Plot of quantum dot position along its trajectory. The *dotted black line* shows the desired trajectory programmed into the controller. The actual measured QD trajectory is shown in blue. The *solid red squares* depict when the quantum dot blinks off. At the end of the trajectory the QD is held in place for 2 min. The deviation of the QD from the desired trajectory was measured to be 104 nm [84]. (Used with permission. Copyright American Chemical Society.)

9.3.5 Ongoing Research: Toward Three-Dimensional Control and Control of Object Orientation

Control in the third dimension is also possible [88]. A microfluidic device with multiple levels (as shown in Fig. 9.30) can create fluid flows or electric fields with up and down components, in addition to the prior horizontal actuation directions. For example, an actuation from the top North electrode to the bottom South-West electrode will create both a Southwards flow as well as a downward component. As before, different actuation modes move different particles in different directions, and using the same least-squares control algorithm as before these modes can be judiciously combined to create particles velocities as close as possible to desired three-dimensional velocity vectors.

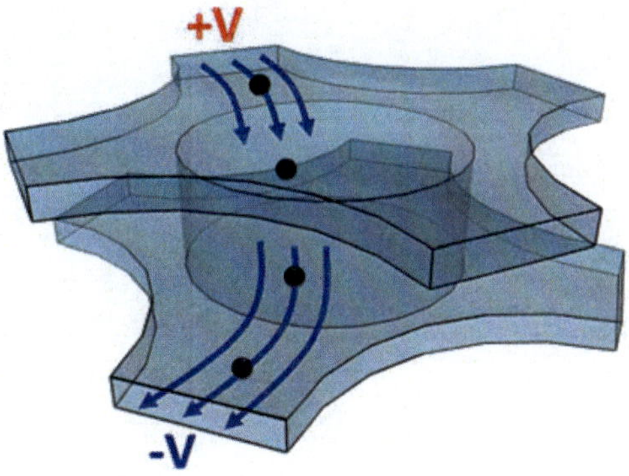

Fig. 9.30 Sample device design for three-dimensional particle control. By placing electrodes in a top and bottom layer, a flow or electric field actuation component can be created from top to bottom or vice versa in the central control region [88]. (Used with permission. Copyright Institute of Physics (IOP) Publishing.)

The device above with eight electrodes can readily control one and two particles in all three dimensions (Fig. 9.31). As previously [83], effective control remains possible in the presence of noise and is still accurate even if the properties of the particles (and the device) are not known perfectly. Control of two 10 nm diameter particles (whose Brownian motion is significant in water) is shown in Fig. 9.32 along two orthogonal and self-intersecting circles. In this case we assumed that the control algorithm does not accurately know the charge on these particles – it believes their charge is $\pm 50\%$ of the true value. In this uncertain case, the simulation shows that manipulation can be achieved with a precision of $2\,\mu$m.

In addition to controlling the position of objects, it is also possible to control their orientation. The discussion below is stated back in two spatial dimensions, but the same method can be used in three dimensions as well. The idea is that now, in addition to creating a translating flow, a flow shear is also created to turn the object. Understanding how to create the right flow is subtle. It is not possible to create a flow rotation: the flow follows the applied electric field and the electric field is irrotational ($\nabla\times \vec{E} = \nabla \times (-\nabla\phi) = 0$).

It is, however, possible to create irrotational flows with shear. Only some types of shear flows can be made. It is not possible to only create the shear flow in one direction as shown in the first panel of Fig. 9.33. The illustrated $\partial u/\partial y$ horizontal flow shear (clockwise rotation) must be exactly cancelled by an equal and opposite vertical flow shear $\partial v/\partial x = \partial u/\partial y$ (counterclockwise rotation) as follows immediately from the zero curl equation for the electric field, or equivalently fluid velocity vector field, $\nabla\times \vec{E} = \nabla\times \vec{V} = \partial v/\partial x - \partial u/\partial y = 0$. But it is possible to create saddle flows, with two balancing shears in opposite directions, as shown in the second and third panels.

If this saddle flow is chosen correctly with respect to the object – here if the shear that will rotate the ellipsoid clockwise is oriented to work on its long axis

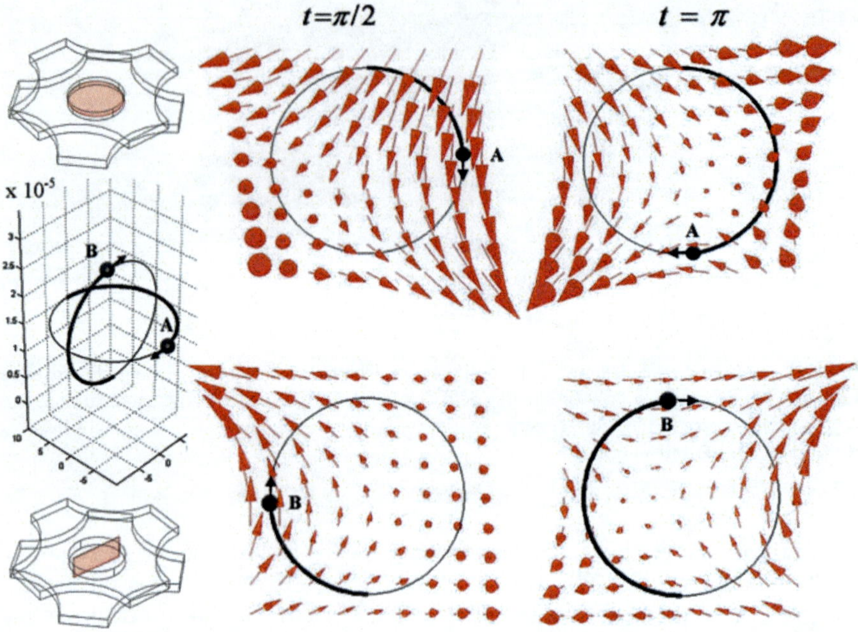

Fig. 9.31 Two charged particles controlled simultaneously on two orthogonal circular paths. The horizontal and vertical paths are shown at the *top* and the *bottom* of the figure, respectively. The desired path of the two particles (*A* and *B*) is in *thin black* and the achieved path is in *thick black*. The (*red*) *arrows* show the created electric field at the two time instants (arrows that appear as *round dots* show flow coming out of that plane) [88]. (Used with permission. Copyright Institute of Physics (IOP) Publishing.)

while the opposing counterclockwise shear only has the short axis to work with – then one rotation will win over the other and the object can be turned clockwise in a controlled fashion. This works for any object that is not fully symmetric. For example, a sphere, which is fully rotationally symmetric, will not be turned as depicted in the second panel of the figure. However, the ellipsoid, shown in the third panel, can be turned by an irrotational saddle flow.

In Fig. 9.34 we show initial results for position and orientation control of an ellipsoidal object in the plane in simulations. The fluid dynamics in the device is the same as before. Also, as before, there is a linear mapping from electrode actuations to object configuration velocity, here to its translation and rotation velocities. For any location and orientation of the ellipsoid, this mapping can be inverted by least-squares to find the electrode actuations that will move and rotate the object from where it is to where it should be. For even less symmetric objects than ellipsoids, like helixes, there will be coupling between translational and rotational motion. In that case a linear mapping between the applied voltage and particle velocity still holds in principle, and consequently control should still be possible in a

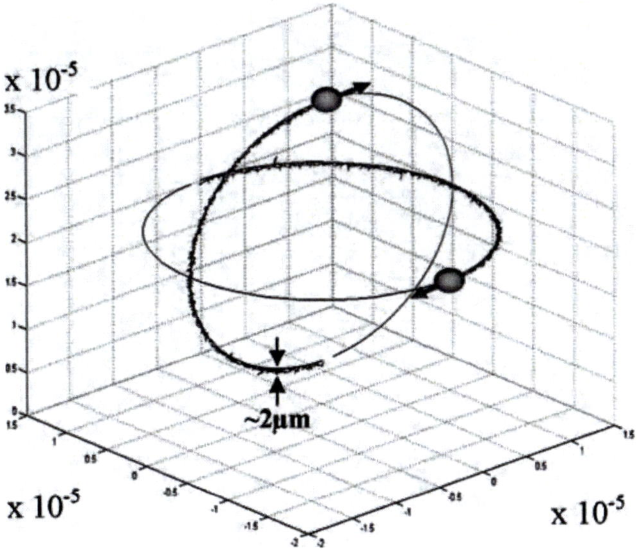

Fig. 9.32 Two nanoparticles (diameter 10 nm) controlled simultaneously in the presence of Brownian motion and a 50% charge uncertainty [88]. (Used with permission. Copyright Institute of Physics (IOP) Publishing.)

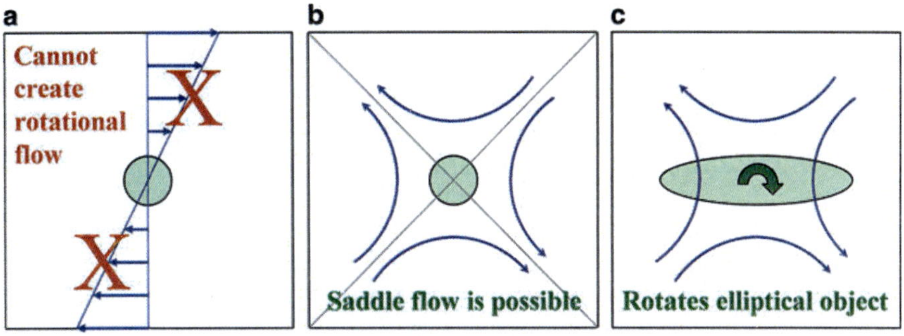

Fig. 9.33 Flow actuation to turn a nonspherical object (the shown flow would be in addition to flow being used to translate the object). (**a**) It is not possible to create the illustrated unidirectional shear flow since that flow is rotational. In the devices the flow follows the electric field which is always irrotational. However, a saddle flow can be created in the device. A saddle flow will not turn a fully rotationally symmetric object like a sphere (**b**) but it will turn an object with less rotational symmetry like the ellipsoid (**c**)

similar least-squares inversion fashion. A further description of our rotation control simulation efforts can be found in [115]. Experiments to test flow control of object translation and orientation are currently underway and will be reported in future publications.

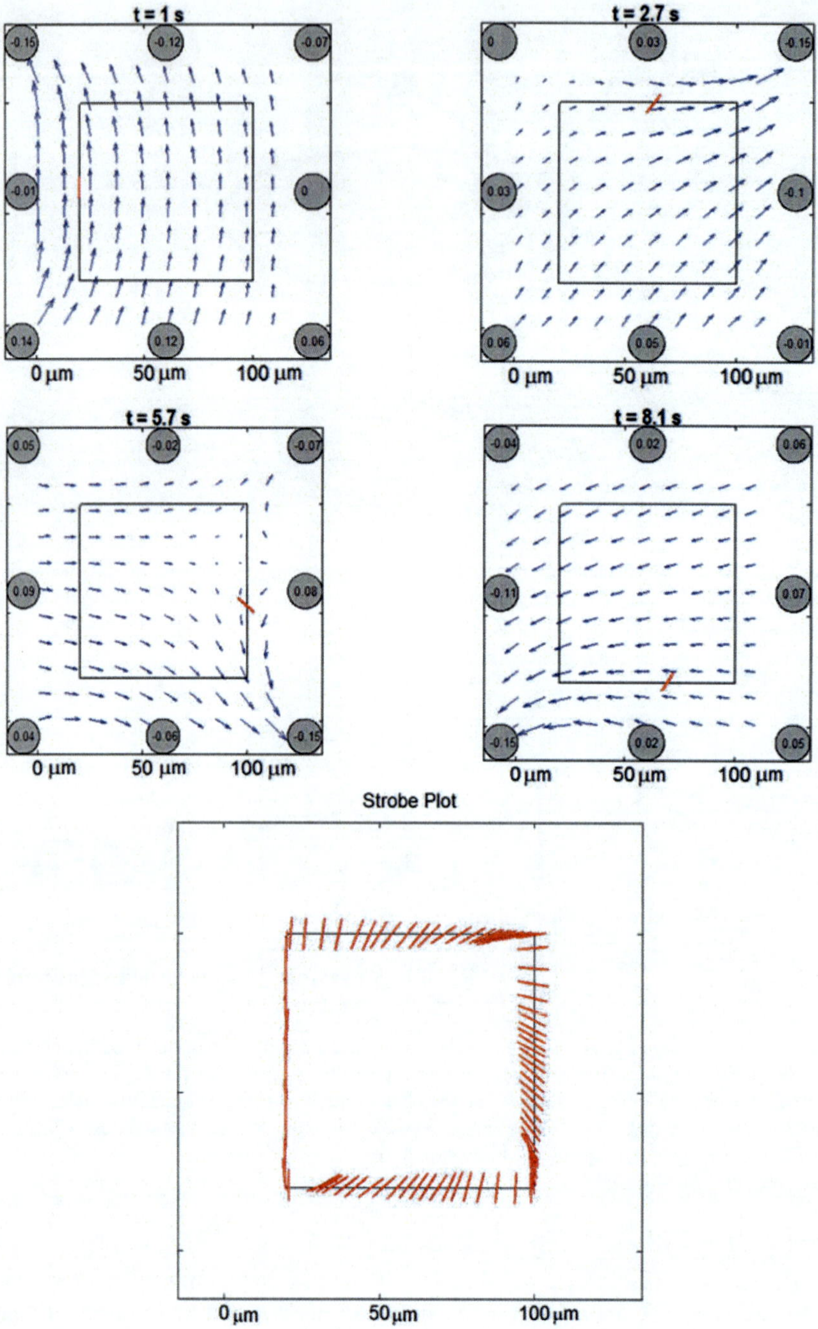
t = 1 s
-0.15
-0.12
-0.07
-0.01
0
0.14
0.12
0.06
0 µm
50 µm
100 µm
t = 2.7 s
0
0.03
-0.15
0.03
-0.1
0.06
0.05
-0.01
0 µm
50 µm
100 µm
t = 5.7 s
0.05
-0.02
-0.07
0.09
0.08
0.04
-0.06
-0.15
0 µm
50 µm
100 µm
t = 8.1 s
-0.04
0.02
0.06
-0.11
0.07
-0.15
0.02
0.05
0 µm
50 µm
100 µm
Strobe Plot
0 µm
50µm
100µm

9.4 Conclusion

Feedback control has enabled microfluidic devices, here electrowetting-on-dielectric as well as simple PDMS devices, to carry out new tasks robustly and with unexpected precision. We show in simulations that smart control can enable EWOD systems to manipulate single particles, and we show that cheap and easy to fabricate PDMS devices with standard electro-osmotic actuation can steer and trap one and multiple particles experimentally. Our control results have enabled nanometer precision placement of quantum dots on photonic crystals for creating multidot quantum information devices, something that has not been achieved using any other particle manipulation technique.

All our control results, for the two examples in this chapter and for other examples in our research (e.g., magnetic control for directing drugs to tumors [116–118]), have been and are being enabled by detailed physical modeling. Especially for new physical situations, this modeling is difficult and time consuming (our modeling effort for electrowetting has continued over many years), but in every case it has enabled us to create controllers that far exceed the performance that would have been possible without modeling. In situations where we deal with chemistry, new physics, and complex samples (fluids that can be solidified with light, living cells, and human samples), we have to choose carefully what to model. Often we are not able to list, let alone mathematically describe in detail, all the relevant physical phenomena; yet we must model enough key physics so that the control algorithm can know how to make things better at each time. To identify the key physics and find the right modeling balance is one of our major challenges.

In terms of control algorithm design, defining a tractable mathematical control problem is the most critical step. It is easy to state a control problem that is clearly useful and we would like to solve, e.g., control of nonlinear partial differential equations through their moving boundary conditions, but that will not admit a useable solution in the foreseeable future. Instead, we try to define more specific problems that are still relevant but that can be solved, and then to build up our expertise to more general domains. For example, for control of electrowetting, the critical insight was that there is a linear mapping from the pressures created by the electrode pads to the particle velocities. This linear mapping reduced the control problem to a least-squares inversion of the small linear matrix map

Fig. 9.34 Position and orientation control of an ellipsoid in the plane by electro-osmotic flow control using eight electrodes. The ellipsoid is controlled to start at the bottom left corner of the desired trajectory, trace the square path, and then return to the bottom left corner. Along each of the four segments of the desired trajectory, the orientation task is to align the major axis of the ellipsoid along that segment by the time it reaches the end of that segment. The ellipsoid is perturbed by translational and rotational thermal (Brownian) motion. *Top*: Four time snapshots are shown (electrode actuation voltages are shown by the values in the *gray circles*, the resulting EO flow field is shown by the *arrows*). *Bottom*: The resulting sequence of ellipsoid positions and orientations is shown for 95 times. (Used with permission. Copyright Institute of Physics (IOP) Publishing.)

from the pressures to the velocities (9.9), after which we corrected for the static nonlinear relationship between the applied voltages and the pressures they create. The least-squares matrix problem, unlike a more general nonlinear PDE control with moving boundary conditions problem, is tractable and can be solved in real-time with minimal computational power. It made control implementation practical for electrowetting.

As in the electrowetting example, we are always trying to map from application needs through our modeling to available or possible control design schemes. For example, an application task (such as putting these living cells here) must be translated through the language of modeling into tractable control schemes (e.g., least-squares, feedback linearization). In cases where existing control schemes remain insufficient for all reasonable formulations of the problem, as has turned out to be the case for focusing of magnetic drugs to deep tumors, we have to invent new control methods. In this case we must define a new control question that we believe has a hope of being answered tractably. (For magnetic drug targeting we have whittled the drug focusing goal down to a sequence of quadratic maps from magnet control inputs to desired drug distributions: now semidefinite programming tools can be used to find an optimal control at each time [119].) Achieving the right balance between the needs of the application and tractable control approaches is our second great challenge.

Finally, our third and most important challenge has been learning to communicate and effectively interact with microfabricators, chemists, physicists, biologists, clinicians, and doctors. Without them the results above would not have been possible and, more importantly, would have been without purpose.

Acknowledgments This research was supported by funding from the National Science Foundation (CAREER award), DARPA, and AFOSR. There is also current funding support from the National Institutes of Health to begin to transition these results from laboratory to medical research and clinical use. Success in this area relies on many collaborators that we would like to acknowledge and thank: Pamela Abshire, Andrew Berglund, Michael Emmert-Buck, Robin Garrell, "CJ" Kim, Alex Liddle, Andreas Lubbe, Ricardo Nochetto, Juan Santiago, Elisabeth Smela, and Ian White.

References

1. M.A. Northrup, et al. *A Miniature analytical instrument for nucleic acids based on micromachined silicon reaction chambers. Analytical Chemistry*, 1998. **70**(5): p. 918–922.
2. R. Jabeen, et al. *Capillary electrophoresis and the clinical laboratory. Electrophoresis*. **27**(12): p. 2413–2438.
3. D. Figeys. *Adapting arrays and Lab-on-a-chip technology for proteomics. Proteomics*, 2002. **2**(4): p. 373–382.
4. J. Seo, et al. *Integrated multiple patch-clamp array chip via lateral cell trapping junctions. Applied Physics Letters*, 2004. **84**(11): p. 1973–1975.
5. G. Spera. *Implantable pumps improve drug deliver, strengthen weak hearts*. Medical Devices & Diagnostic Industry Magazine, 1997.

6. M. Gad-el-Hak, ed. *MEMS: Design and fabrication* 2ed. The MEMS handbook. Vol. 2. 2006, CRC Press, Taylor and Francis Group. 664.
7. C.T. Leondes, ed. *Fabrication Techniques for MEMS/NEMS*. MEMS/NEMS handbook. 2006, Springer: New York, NY.
8. K.E. Herold and A. Rasooly, eds. *Fabrication and microfluidics*. Lab on a chip technology. Vol. 1. 2009, Caister Academic Press Norfolk, UK. 409.
9. C.A. Mack. *Fundamental principles of optical lithography: The science of microfabrication*. 2008, West Sussex, England: John Wiley and Sons Ltd. 534.
10. M. Meyyappan. *A review of plasma enhanced chemical vapour deposition of carbon nanotubes. Journal of Physics D-Applied Physics*, 2009. **42**(21).
11. R. Bogwe. *Self-assembly: a review of recent developments. Assembly Automation*, 2008. **28**(3): p. 211–215.
12. J.C. Huie. *Guided molecular self-assembly: a review of recent efforts. Smart Materials & Structures*, 2003. 12(2): p. 264–271.
13. S.D. Minteer, ed. *Microfluidic Techniques: Reviews and protocols (Methods in molecular biology)*. Methods in molecular biology. Vol. 321. 2005, Humana Press: Clifton, New Jersey. 247.
14. P. Kim, et al. *Soft lithography for microfluidics: a review. Biochip Journal*, 2008. **2**(1): p. 1–11.
15. R.L. Panton. *Incompressible flow*. 2 ed. 1996, New York, NY: John Wiley & Sons, Inc.
16. G.E. Karniadakis and A. Beskok. *Micro flows: Fundamentals and simulation*. 2001, New York, NY: Springer Verlag.
17. A. Beskok. *Physical challenges and simulation of micro fluidic transport*. in *39th Aerospace Sciences Meeting & Exhibit*. 2001. Reno, Nevada: AIAA.
18. M. Gad-el-Hak. *The fluid mechanics of microdevices – The freeman scholar lecture. Journal of Fluids Engineering*, 1999. **121**: p. 5.
19. S. Walker and B. Shapiro. *Modeling the fluid dynamics of electro-wetting on dielectric (EWOD). Journal of Micro-Electro-Mechanical Systems*, 2006. **15**(4): p. 986–1000.
20. S. Chaudhary and B. Shapiro. *Arbitrary steering of multiple particles at once in an electroosmotically driven microfluidic system. IEEE Transactions on Control Systems Technology*, 2006. **14**(4): p. 669–680.
21. L. Pauling. *General chemistry*. 1970, New York: Dover Publications, Inc.
22. S.R. Quake and T.M. Squires. *Microfluidics: Fluid physics at the nanoliter scale. Reviews of Modern Physics*, 2005. bf 77(3): p. 977–1026.
23. M.H. Oddy, J.G. Santiago, and J.C. Mikkelson. *Electrokinetic instability micromixing. Analytical Chemistry*, 2001. **73**(24): p. 5822–32.
24. J. Fowler, H. Moon, and C.J. Kim. *Enhancement of mixing by droplet-based microfluidics*. IEEE Conf. MEMS, Las Vegas, NV, 2002: p. 97–100.
25. I. Glasgow and N. Aubry. *Enhancement of microfluidic mixing using time pulsing. Lab on a Chip*, 2003. **3**(2): p. 114–120.
26. K.J. Astrom and R.M. Murray. *Feedback systems: An introduction for scientists and engineers*, 2006. Princeton University Press. 2008.
27. J.C. Doyle, B.A. Francis, and A.R. Tannenbaum. *Feedback control theory*. 1992, New York, NY: Macmillan Publishing Company.
28. R.M. Murray, et al. *Control in an information rich world*. 2002, Air Force Office of Scientific Research (AFOSR).
29. F. Mugele and J. Baret. *Electrowetting: from basics to applications. journal of physics: condensed matter*, 2005. **17**: p. R705–R774.
30. J. Lee, et al. *Electrowetting and electrowetting-on-dielectric for microscale liquid handling. Sensor Actuat. A-Phys*, 2002. **95**: p. 269.
31. S. Fusayo, et al. *Electrowetting on dielectrics (EWOD): Reducing voltage requirements for microfluidics. Polymeric Materials: Science & Engineering*, 2001. **85**: p. 12–13.
32. H. Ren, et al. *Dynamics of electro-wetting droplet transport. Sensors and Actuators B-Chemical*, 2002. **87**(1): p. 201–206.

33. E. Seyrat and R.A. Hayes. *Amorphous fluoropolymers as insulators for reversible low-voltage electrowetting. Journal of Applied Physics*, 2001. **90**(3).
34. R.A. Hayes and B.J. Feenstra. *Video-speed electronic paper based on electrowetting. Nature*, 2003. **425**(6956): p. 383–385.
35. T. Roques-Carmes, R.A. Hayes, and L.J.M. Schlangen. *A physical model describing the electro-optic behavior of switchable optical elements based on electrowetting. Journal of Applied Physics*, 2004. **96**(11): p. 6267–6271.
36. T. Roques-Carmes, et al. *Liquid behavior inside a reflective display pixel based on electrowetting. Journal of Applied Physics*, 2004. **95**(8): p. 4389–4396.
37. B. Berge and J. Peseux. *Variable focal lens controlled by an external voltage: An application of electrowetting. European Physical Journal E*, 2000. **3**(2): p. 159–163.
38. C. Quilliet and B. Berge. *Electrowetting: a recent outbreak. Current Opinion in Colloid & Interface Science*, 2001. **6**(1): p. 34–39.
39. S.K. Cho, H. Moon, and C.J. Kim. *Creating, transporting, cutting, and merging liquid droplets by electrowetting-based actuation for digital microfluidic circuits. Journal of Microelectromechanical Systems*, 2003. VOL. **12**(NO. 1): p. 70–80.
40. M.G. Pollack, R.B. Fair, and A.D. Shenderov. *Electrowetting-based actuation of liquid droplets for microfluidic applications. Applied Physics Letters*, 2000. **77**(11).
41. H.J.J. Verheijen and W.J. Prins. *Reversible electrowetting and trapping of charge: Model and experiments. Langmuir*, 1999. **15**: p. 6616–6620.
42. B. Berge. *Electrocapillarity and wetting of insulator films by water. Comptes Rendus de l Academie des Sciences Series II*, 1993. **317**(2): p. 157–163.
43. B. Shapiro, et al. *Equilibrium behavior of sessile drops under surface tension, applied external fields, and material variations. Journal of Applied Physics*, 2003. **93**(9).
44. M. Armani, et al. *Control of micro-fluidic systems: Two examples, results, and challenges. International Journal of Robust and Nonlinear Control.*, 2005. **15**(16): p. 785–803.
45. S.W. Walker, B. Shapiro, and R.H. Nochetto. *Electrowetting with contact line pinning: Computational modeling and comparisions with experiments. Physics of Fluids*, 2009. **23**(10): p. 102103.
46. K. Zhou, J.C. Doyle, and K. Glover. *Robust and optimal control*. 1996: Prentice Hall, New Jersey.
47. A. Isidori. *Nonlinear control systems*. 3 ed. Communications and control engineering. 1995: Springer.
48. G. Lippmann. *Relation entre les phenomenes electriques et capillaires. Ann. Chim. Phys.*, 1875. **5**: p. 494–549.
49. R.F. Probstein. *Physicochemical Hydrodynamics: An introduction*. 2 ed. 1994, New York: John Wiley and Sons, Inc.
50. P.C. Hiemenz and R. Rajagopalan. *Principles of colloid and surface chemistry*. 3 ed. 1997, New York, Basel, Hong Kong: Marcel Dekker, Inc.
51. R.P. Feynman, R.B. Leighton, and M. Sands. *The feynman lectures on physics*. 1964: Addison-Wesley Publishing Company.
52. S.W. Walker. *Modeling, simulating, and controlling the fluid dynamics of electro-wetting on dielectric, in aerospace engineering*. 2007, University of Maryland: College Park.
53. H.W. Lu, et al. *A diffuse interface model for electrowetting droplets in a hele-shaw cell. J. Fluid Mech*, 2007. **590**: pp. 411–435.
54. H.S. Hele-Shaw. *The flow of water. Nature*, 1898. **58**: p. 34–35.
55. G.K. Batchelor. *An Introduction to fluid dynamics*. 1967: Cambridge University Press.
56. M.P. do Carmo. *Differential geometry of curves and surfaces*. 1976, Upper Saddle River, New Jersey: Prentice Hall.
57. T.D. Blake. *The physics of moving wetting lines. Journal of Colloid and Interface Science*, 2006. **299**: p. 1–13.
58. S. Osher and R. Fedkiw. *Level set methods and dynamic implicit surfaces*. 2003: Springer-Verlag New York.

59. S.A. Sethian. *Level set methods & fast marching methods*. 2 ed. 1999: Cambridge University Press.
60. R. Caiden, R. Fedkiw, and C. Anderson. *A numerical method for two phase flow consisting of separate compressible and incompressible regions. J. Comput. Phys.*, 2001. **166**: p. 1–27.
61. P. Smereka. *Semi-implicit level set methods for curvature and surface diffusion motion. Journal of Scientific Computing*, 2003. **19**(1–3): p. 439–456.
62. W.J. Rider and D.B. Kothe. *Stretching and tearing interface tracking methods, in 12th AIAA CFD Conference*. 1995.
63. H.C. Kuhlmann and H.J. Rath. *Free surface flows 1ed*. CISM courses and lectures, international centre for mechanical sciences. Vol. **39**, New York, NY: Springer. 300.
64. D. Enright, R. Fedkiw, J. Ferziger, I. Mitchell. *A hybrid particle level set method for improved interface capturing. Journal of Computational Physics*, 2002. **183**: p. 83.
65. F. Losasso, F. Gibou, R. Fedkiw. *Simulating water and smoke with an octree data structure. in ACM Trans. Graph. (SIGGRAPH Proc.)*. 2004.
66. X. Yang, et al. *An adaptive coupled level-set/volume-of-fluid interface capturing method for unstructured triangular grids. Journal of Computational Physics* 2006. **217**(2): p. 364–394
67. S.W. Walker. *A hybrid variational-level set approach to handle topological changes, in mathematics*. 2007, University of Maryland: College Park.
68. C.J. Kim. *Micropumping by electrowetting. in Int. mechanical engineering congress and exposition, New York, NY, IMECE2001/HTD-24200*. 2001.
69. V. Srinivasan, V. Pamula, M. Pollack, and R. Fair. *A digital microfluidic biosensor for multianalyte detection*. in *Proceedings of the IEEE 16th Annual International Conference on Micro Electro Mechanical Systems*. 2003.
70. P. Paik, V.K. Pamula, and R.B. Fair. *Rapid droplet mixers for digital microfluidic systems. Lab on a Chip*, 2003. **3**: p. 253–259.
71. S.K. Cho, H. Moon, J. Fowler, S.K. Fan, and C.J. Kim. *Splitting a liquid droplet for electrowetting-based microfluidics. in Int. mechanical engineering congress and exposition, New York, NY, IMECE2001/MEMS-23831*. 2001.
72. S. Walker and B. Shapiro. *A control method for steering individual particles inside liquid droplets actuated by electrowetting. Lab on a Chip*, 2005. **12**(1): p. 1404–1407.
73. F.L. Lewis, and V.L. Syrmos. *Optimal control*. 2nd ed. 1995, New York, NY: Wiley-Interscience. 560.
74. K.W. Morton, and D.F. Mayers. *Numerical solution of partial differential equations*. 1994: Cambridge University Press. 239.
75. G. Strang. *Linear algebra and its applications*. 3 ed. 1988, New York, NY: Brooks Cole. 520.
76. E.C. Zachmanoglou and D.W. Thoe. *Introduction to partial differential equations with applications*. 1986, New York, NY: Dover Publications, Inc.
77. A. Ralston, and P. Rabinowitz. *A first course in numerical analysis*. 2nd ed. 2001, Mineola, NY: Dover.
78. L. Jauffred, A.C. Richardson, and L.B. Oddershede. *Three-dimensional optical control of individual quantum dots. Nano Letters*. **8**(10): p. 3376–3380.
79. A. Ashkin, et al. *Observation of a single-beam gradient force optical trap for dielectric particles. Optics Letters*. **11**(5): p. 288–290.
80. C. Pei-Yu, et al. *Light-actuated AC electroosmosis for nanoparticle manipulation. Microelectromechanical Systems, Journal of*. **17**(3): p. 525–531.
81. A.H.J. Yang, et al. *Optical manipulation of nanoparticles and biomolecules in sub-wavelength slot waveguides. Nature*. *457*(7225): p. 71–75.
82. A. Ashkin. *History of optical trapping and manipulation of small-neutral particles, atoms, and molecules. IEEE Journal on Selected Topics in Quantum Electronics*, 2000. **6**(6): p. 841–856.
83. M. Armani, et al. *Using feedback control and micro-fluidics to independently steer multiple particles. Journal of Micro-Electro-Mechanical Systems*, 2006. **15**(4): p. 945–956.
84. C. Ropp, et al. *Manipulating quantum dots to nanometer precision by control of flow. Nano Letters*, 2010. **10**(7): p. 2525–2530.

85. K.C. Neuman and A. Nagy. *Single-molecule force spectroscopy: optical tweezers, magnetic tweezers and atomic force microscopy. Nat Meth.* **5**(6): p. 491–505.
86. H. Zhang and K.K. Liu. *Optical tweezers for single cells. Journal of the Royal Society, Interface/the Royal Society.* **5**(24): p. 671–690.
87. F. Qian, et al. *Combining optical tweezers and patch clamp for studies of cell membrane electromechanics. Review of Scientific Instruments.* **75** (9): p. 2937–2942.
88. R.Probst, B.Shapiro, *3-dimensional electrokinetic tweezing: Device design, modeling, and control algorithms*, Journal of Micromechanics and Microengineering, 2011 **21**(2)
89. C. Ropp, et al. *Positioning and immobilization of individual quantum dots with nanoscale precision. Nano Letters*, 2010. **10** p. 4673–4679.
90. L.E. Locascio, C.E. Perso, and C.S. Lee. *Measurement of electroosmotic flow in plastic imprinted microfluid devices and the effect of protein adsorption on flow rate. Journal of Chromotography A*, 1999. **857**: p. 275–284.
91. D.J. Harrison, et al. *Micromachining a miniaturized capillary electrophoresis-based chemical-analysis system on a chip. Science*, 1993. **261**(5123): p. 895–897.
92. A. Manz, et al. *Electroosmotic pumping and electrophoretic separations for miniaturized chemical-analysis systems. Journal of Micromechanics and Microengineering*, 1994. **4**(4): p. 257–265.
93. W. Korohoda and A. Wilk. *Cell electrophoresis — a method for cell separation and research into cell surface properties. Cellular & Molecular Biology Letters.* **13**(2): p. 312–326.
94. J.N. Mehrishi and J. Bauer. *Electrophoresis of cells and the biological relevance of surface charge. Electrophoresis.* **23**(13): p. 1984–1994.
95. G.G. Slivinsky, et al. *Cellular electrophoretic mobility data: A first approach to a database. Electrophoresis.* **18**(7): p. 1109–1119.
96. F. Schonfeld. *CμFD – Case study: Flow patterning by phase-shifted electroosmotic flows, in comsol user's conference* 2006: Frankfurt, Germany.
97. S. Chaudhary and B. Shapiro. *Arbitrary steering of multiple particles at once in an electroosmotically driven micro fluidic system. IEEE Trans. on Control Systems Technologies*, 2005: 14669–680.
98. T. Zhou, et al. *Time-dependent starting profile of velocity upon application of external electrical potential in electroosmotic driven microchannels. Colloids and Surfaces A: Physicochemical and Engineering Aspects*, 2006. **277** (1–3): p. 136–144.
99. D. Yan, et al. *Visualizing the transient electroosmotic flow and measuring the zeta potential of microchannels with a micro-PIV technique. The Journal of Chemical Physics.* **124**(2): p. 021103–4.
100. M.A. Henderson. *The interaction of water with solid surfaces: fundamental aspects revisited. Surface Science Reports*, 2002. **46**(1–8): p. 1–308.
101. J. Lyklema, et al. *Fundamentals of interface and colloid science.*
102. M. Armani, et al. *Using feedback control and micro-fluidics to steer individual particles. in 18th IEEE International Conference on Micro Electro Mechanical Systems.* 2005. Miami, Florida.
103. I. Rodriguez and N. Chandrasekhar. *Experimental study and numerical estimation of current changes in electroosmotically pumped microfluidic devices. Electrophoresis*, 2005. **26**(6): p. 1114–1121.
104. A. Badolato, et al. *Deterministic coupling of single quantum dots to single nanocavity modes. Science.* **308**(5725): p. 1158–1161.
105. K. Hennessy, et al. *Quantum nature of a strongly coupled single quantum dot-cavity system. Nature. 445*(7130): p. 896–899.
106. D. Englund, et al. *Controlling cavity reflectivity with a single quantum dot. Nature.* **450**(7171): p. 857–861.
107. A.V. Akimov, et al. *Generation of single optical plasmons in metallic nanowires coupled to quantum dots. Nature.* **450**(7168): p. 402–406.
108. D.E. Chang, et al. *A single-photon transistor using nanoscale surface plasmons. Nat Phys.* **3**(11): p. 807–812.

109. Y. Akhane, et al. *High-Q photonic nanocavity in a two-dimensional photonic crystal. Nature*, 2003. **425**: p. 944–947.
110. Q. Zhang, et al. *Large ordered arrays of single photon sources based on II-VI semiconductor colloidal quantum dot. Opt. Express*. **16**(24): p. 19599, 19592–19599, 19592.
111. J. Liu, et al. *Controlled photopolymerization of hydrogel microstructures inside microchannels for bioassays. Lab on a Chip*. **9**(9): p. 1301–1305.
112. J.T. Fourkas and L. Li. *Multiphoton polymerization. Mater. Today*, 2007. **10**(6): p. 30–37.
113. R.E. Thompson, D.R. Larson, and W.W. Webb. *Precise nanometer localization analysis for individual fluorescent probes. Biophysical Journal*, 2002. **82**(5): p. 2775–2783.
114. L.J. Li and J.T. Fourkas. *Multiphoton polymerization. Materials Today*, 2007. **10**(6): p. 30–37.
115. P. Mathai, A. Berglund, A. Liddle, B. Shapiro, *Simultaneous positioning and orientation of a single nano-object by flow control*, New Journal of Physics, **13**: p. 013027, 19 January 2011.
116. B. Shapiro, et al. *Control to concentrate drug-coated magnetic particles to deep-tissue tumors for targeted cancer chemotherapy. in 46th IEEE Conference on Decision and Control*. 2007. New Orleans, LA.
117. B. Shapiro, et al. *Dynamic control of magnetic fields to focus drug-coated nano-particles to deep tissue tumors, in 7th International Conference on the Scientific and Clinical Applications of Magnetic Carriers*. 2008: Vancouver, British Columbia.
118. B. Shapiro. *Towards dynamic control of magnetic fields to focus magnetic carriers to targets deep inside the body. Journal of Magnetism and Magnetic Materials*, 2009. **321**(10): p. 1594–1599.
119. A.Komaee and B.Shapiro, *Magnetic steering of a distributed ferrofluid towards a deep target with minimal spreading, In 50th IEEE Conference on Decision and Control and European Control Conference (CDC-ECC)*. Dec 2011. Orlando, FL.

Chapter 10
Problems in Control of Quantum Systems

Navin Khaneja

10.1 Introduction

The chapter describes some differential equation models that arise in the control and manipulation of quantum mechanical phenomena. Control of spin dynamics in nuclear magnetic resonance (NMR) spectroscopy [2–4] is used as a paradigm to outline general principles in the control of quantum systems and describe some common characteristic phenomenon encountered in control of these physical systems.

The defining equation for the state of a quantum mechanical system is the Schröedinger equation

$$|\dot{\psi}\rangle = -i\left[H_0 + \sum_{j=1}^{n} u_j H_j\right]|\psi\rangle, \tag{10.1}$$

where the state of the quantum system is represented by a vector $|\psi\rangle \in \mathscr{H}$, a suitable Hilbert space. H_0 and H_j are Hermitian operators, representing Hamiltonians of the system and $u_j(t)$ are time-varying functions that represent controls in the system dynamics. For models discussed in this chapter, $\mathscr{H}$ is finite dimensional and in a chosen basis, H_0 and H_j are Hermitian matrices. We assume that $\mathscr{H}$ is finite, unless stated explicitly. Modulating u_j affects the Hamiltonian of the system and therefore affects the evolution of the state of the system. Equation (10.1) has the familiar form of a bilinear control system

$$\dot{x} = \left(A + \sum_{j=1}^{n} u_j B_j\right)x, \tag{10.2}$$

N. Khaneja (✉)
School of Engineering and Applied Sciences, Harvard University, Cambridge, MA 02138, USA
e-mail: navin@hrl.harvard.edu

J.J. Gorman and B. Shapiro (eds.), *Feedback Control of MEMS to Atoms*,
DOI 10.1007/978-1-4419-5832-7_10,

where in the present context, A, B_j are skew Hermitian matrices and x is $|\psi\rangle$. The evolution preserves the norm of the state $|\psi\rangle$. The evolution is unitary and the state vector at time t is related to the initial state vector by a unitary transformation $U(t)$, such that

$$|\psi(t)\rangle = U(t)|\psi(0)\rangle, \tag{10.3}$$

where $U(t)$ satisfies the differential equation

$$\dot{U} = -i\left[H_0 + \sum_{j=1}^{n} u_j H_j\right]U, \;\; U(0) = \mathbf{1}, \tag{10.4}$$

where $\mathbf{1}$ is the identity matrix. A textbook example of a system where such a bilinear control model arises is the evolution of the magnetic moment of a spin $\frac{1}{2}$ in a magnetic field B. Spin, like charge, is a physical property of elementary particles. It is a measure of their intrinsic angular momentum, and the state of a spin $\frac{1}{2}$ is represented by a complex vector of dimension 2. The Hamiltonian generates rotations on the state space of a quantum system. The Hamiltonian of a spin $\frac{1}{2}$ can be written in terms of the generators of rotations on a two-dimensional space and these are the Pauli matrices $-i\sigma_x, -i\sigma_y, -i\sigma_z$, where,

$$\sigma_z = \frac{1}{2}\begin{bmatrix} 1 & 0 \\ 0 & -1 \end{bmatrix}; \;\; \sigma_y = \frac{1}{2}\begin{bmatrix} 0 & -i \\ i & 0 \end{bmatrix}; \;\; \sigma_x = \frac{1}{2}\begin{bmatrix} 0 & 1 \\ 1 & 0 \end{bmatrix}. \tag{10.5}$$

Note

$$[\sigma_x, \sigma_y] = i\sigma_z, \;\; [\sigma_y, \sigma_z] = i\sigma_x, \;\; [\sigma_z, \sigma_x] = i\sigma_y, \tag{10.6}$$

where $[A, B] = AB - BA$ is the matrix commutator and

$$\sigma_x^2 = \sigma_y^2 = \sigma_z^2 = \frac{\mathbf{1}}{4}. \tag{10.7}$$

The classical energy E of the magnetic moment μ of a spin in a magnetic field is $E = -\mu \cdot B$. The magnetic moment of a spin is proportional to its angular momentum, given by $\mu = \gamma L$, where γ is the gyromagnetic ratio (a characteristic property of the nucleus) and L is the angular momentum operator. Therefore, the Hamiltonian of spin $\frac{1}{2}$ is

$$H = -\gamma[B_x L_x + B_y L_y + B_z L_z],$$

where L_x, L_y, L_z are now operators, representing angular momentum in the x, y, z direction, respectively.

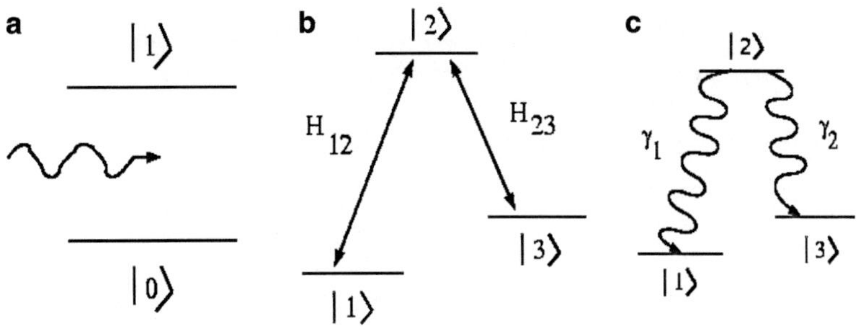

Fig. 10.1 (**a**) Shows the simplest of the quantum objects, a two level system being probed with an electromagnetic field. (**b**) Shows the energy level diagram of a three level Lambda system often studied in the area of laser spectroscopy. (**c**) Shows spontaneous decay of the population in state $|2\rangle$ to energy levels $|1\rangle$ and $|3\rangle$

Since angular momentum is a generator of rotation, the angular momentum operators L_x, L_y, L_z are identified with the Pauli matrices $\sigma_x, \sigma_y, \sigma_z$, the generators of rotation in a two-dimensional Hilbert space. The Schröedinger equation then takes the form

$$|\dot{\psi}\rangle = i\gamma[\sigma_z B_0 + \sigma_y B_y(t) + \sigma_x B_x(t)]|\psi\rangle, \tag{10.8}$$

where we use $B_0 = B_z$, $\omega_0 = -\gamma B_0$, and $(u(t), v(t)) = -\gamma(B_x(t), B_y(t))$. The above equation is then rewritten as

$$|\dot{\psi}\rangle = \frac{-i}{2}\begin{bmatrix} \omega_0 & u - iv \\ u + iv & -\omega_0 \end{bmatrix}|\psi\rangle. \tag{10.9}$$

The eigenstates of σ_z, labeled, $|0\rangle = \begin{bmatrix} 1 \\ 0 \end{bmatrix}$ and $|1\rangle = \begin{bmatrix} 0 \\ 1 \end{bmatrix}$, with eigenvalues $\frac{1}{2}$ and $-\frac{1}{2}$, correspond to the state of the spins oriented along or opposite to the magnetic field B_0. Equation (10.9) represents the most basic of all quantum mechanical objects, a two level system being manipulated by an external field. A schematic of such a system is shown in Fig. 10.1a.

The differential equation model (10.9) describes the dynamics of spin $\frac{1}{2}$ in the NMR experiments when manipulated by transverse magnetic fields $(B_x(t), B_y(t))$ which manifest themselves as control inputs $(u(t), v(t))$. The primary goal of these experiments is to accurately measure ω_0 by manipulating or probing the system with control inputs $(u(t), v(t))$, which provides a wealth of information on chemistry and structure of molecules carrying spins, as detailed subsequently. NMR experiments are performed on an ensemble of spins. All the members of the ensemble may not have identical state vectors. In which case, the description of a quantum system is a density matrix as described by

$$\rho = \sum_j p_j |\psi_j\rangle\langle\psi_j|, \tag{10.10}$$

where p_j is the proportion of ensemble elements ($\sum_j p_j = 1$) in the state $|\psi_j\rangle$ (The notation $\langle\psi_j|\psi_j\rangle$ and $|\psi_j\rangle\langle\psi_j|$, denote inner and outer product of vector $|\psi_j\rangle$ with itself, respectively). In an ensemble of spin $\frac{1}{2}$, with $\frac{1}{3}$ of spins in the state $|0\rangle$ and $\frac{2}{3}$ of spins in the state $|1\rangle$, the density matrix is

$$\rho = \begin{bmatrix} \frac{1}{3} & 0 \\ 0 & \frac{2}{3} \end{bmatrix}.$$

By the postulates of quantum mechanics, the observable quantities are represented by self-adjoint operators. The expected value of an observable $\mathbf{O}$, when the quantum system is in the state $|\psi\rangle$ is simply $\langle\psi|O|\psi\rangle$, where O is the operator that represents the observable $\mathbf{O}$. Therefore, for an ensemble of quantum systems, the expected value of an observable $\mathbf{O}$ is

$$\langle O\rangle = \sum_j p_j\langle\psi_j|O|\psi_j\rangle = tr(\rho O).$$

The density matrix ρ of the quantum system evolves as

$$\dot{\rho} = -i\left[H_0 + \sum_{j=1}^{n} u_j H_j, \rho\right], \tag{10.11}$$

where $[A,B]$, as before, is the matrix commutator. This follows from simply differentiating (10.10), where each $|\psi_j\rangle$ satisfies the same (10.1). Some properties of the density matrix ρ are evident from its construction. It is a Hermitian, positive semidefinite and satisfies $tr(\rho) = 1$. The evolution of the density operator is

$$\rho(t) = U(t)\rho(0)U^\dagger(t), \tag{10.12}$$

where $U(t)$ is the unitary transformation in (10.4). Also by construction, $tr(\rho^2) \le 1$, with equality holding only if only one of the p_j in (10.10) is nonzero and equal to 1. For an ensemble of spin $\frac{1}{2}$, the density matrix ρ is a 2×2, Hermitian matrix, which can be decomposed as

$$\rho = \frac{1}{2}\mathbf{1} + m_x\sigma_x + m_y\sigma_y + m_z\sigma_z. \tag{10.13}$$

Therefore, for the density matrix in (10.13), we obtain that the expected value of the angular momentum along x, y, z, represented by operators σ_x, σ_y, and σ_z, is then simply proportional to m_x, m_y, m_z. Then, (10.9) implies

$$\frac{\mathrm{d}}{\mathrm{d}t}\underbrace{\begin{bmatrix} m_x \\ m_y \\ m_z \end{bmatrix}}_{M} = \begin{bmatrix} 0 & -\omega_0 & v(t) \\ \omega_0 & 0 & -u(t) \\ -v(t) & u(t) & 0 \end{bmatrix}\begin{bmatrix} m_x \\ m_y \\ m_z \end{bmatrix}, \tag{10.14}$$

where the vector $M = (m_x, m_y, m_z)'$ is a measure of the net magnetic moment or magnetization of the spin ensemble and the above equation is the well-studied Bloch equation, which describes the precession of the magnetic moment in a magnetic field and can be concisely written as $\dot{M} = \gamma M \times B$, where $B = (B_x(t), B_y(t), B_0)'$ is the magnetic field vector as defined before. Observe, (10.14) evolves on a sphere, and we normalize the norm of M to 1. Equation (10.14) is at the heart of the subject of NMR spectroscopy, where a typical task is to engineer $(u(t), v(t))$ to manipulate or steer the vector M to estimate the parameter ω_0. In the following subsection, we describe some characteristic features of the control inputs for the manipulation of (10.14).

10.1.1 Control of Bloch Equations

Note, (10.14) can be written as

$$\dot{M} = (\omega_0 \Omega_z + u(t)\Omega_x + v(t)\Omega_y)M, \tag{10.15}$$

where

$$\Omega_x = \begin{bmatrix} 0 & 0 & 0 \\ 0 & 0 & 1 \\ 0 & -1 & 0 \end{bmatrix}, \ \Omega_y = \begin{bmatrix} 0 & 0 & 1 \\ 0 & 0 & 0 \\ -1 & 0 & 0 \end{bmatrix}, \ \Omega_z = \begin{bmatrix} 0 & -1 & 0 \\ 1 & 0 & 0 \\ 0 & 0 & 0 \end{bmatrix}. \tag{10.16}$$

A typical control problem is to steer the system from its equilibrium state $M(0) = (0,0,1)$ to a terminal state on the equator. A salient feature of such problems is that the external excitations $(u(t), v(t))$ are typically significantly smaller in strength as compared to the natural dynamics represented by ω_0, which is four to five orders of magnitude larger in the NMR experiments. Therefore, for the external control to be most effective in manipulating the system, it is essential that the control be oscillatory (see subsequent remarks). To fix ideas, let

$$(u(t), v(t)) = A(t)(\cos(\omega_1 t + \theta(t)),\ \sin(\omega_1 t + \theta(t))) \tag{10.17}$$

and consider the control problem of steering M in (10.15) from an initial point $(0,0,1)'$ to the final target state $(1,0,0)'$. Observe, by transforming to a coordinate system such that $X = \exp(-\omega_1 \Omega_z t)M$, we obtain that

$$\dot{X} = (\omega \Omega_z + u_1(t)\Omega_x + v_1(t)\Omega_y)X, \tag{10.18}$$

where $\omega = \omega_0 - \omega$, and $(u_1(t), v_1(t)) = (A(t)\cos\theta(t), A(t)\sin\theta(t))$. Now, since $X(0) = M(0)$, by simply choosing $\omega = \omega_0$, $\theta(t) = \theta$, and $A(t) = A$, as constants, we drive $X(0)$ to $(\sin\theta, -\cos\theta, 0)$ in $T = \frac{\pi}{2A}$ units of time. When $\omega = \omega_0$, (10.18) has no natural dynamics. Therefore, it does not matter how weak $A(t)$ is, given sufficient

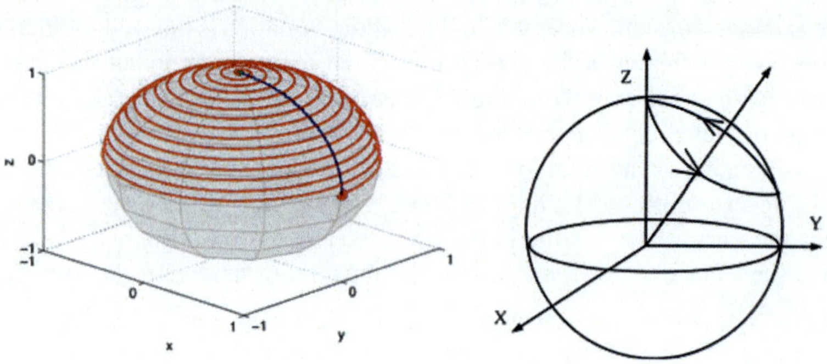

Fig. 10.2 The figure [1] shows the rotation of the vector X in (10.18), around a tilted axis, when $\Delta\omega$ is comparable to the strength of control A

time, $X(0)$ can be steered to the transverse plane. $X(T)$ can be put anywhere on the transverse plane by appropriate choice of θ. The left panel in Fig. 10.2 shows this transfer, both in the rotating frame as in (10.18) (the winding curve) and the laboratory frame (10.15) (geodesic curve). However, a choice of constant control $(u(t), v(t)) = (A\cos\theta, A\sin\theta)$, would be completely ineffective, if applied to the original system in (10.15), as the net motion is then a precession around the axis $(A\cos\theta, A\sin\theta, \omega_0)$, as shown in Fig. 10.2 and therefore significantly falls short of the desired transfer when $A \ll \omega_0$. Therefore, it is desirable to use a control input as in (10.17), which is oscillatory at the frequency of natural oscillation of the system. It is not difficult to show that the choice of such a control input is the minimum energy control for driving the system in (10.15) to the transverse plane. Applying such a control input corresponds to exciting the system at its resonance, with a very weak field, the system can be driven far from equilibrium. $A(t)$, $\theta(t)$ and ω_0 are naturally termed amplitude, phase and carrier frequency of the applied radio-frequency field as in radio communication. Design of appropriate control inputs $(u(t), v(t))$ is in fact the design of appropriate amplitude and phase modulations.

After the magnetic moment is driven to the transverse plane by choice of an appropriate control input, the oscillating control is switched off and the magnetic moment M precesses around the static magnetic field B_0 with a frequency ω_0. This is just the evolution in (10.17), after the controls are switched off. This precessing magnetic moment by Faraday's law induces an oscillating current in the nearby placed receiver coil and is termed as free induction decay (FID) (Top right of Fig. 10.3 shows the FID). This FID, when Fourier transformed, shows a peak at ω_0. At the magnetic field strength of $B_0 = 14$ Tesla, ω_0 for hydrogen nuclei is 600 MHz, for carbon is 150 MHz, and for nitrogen is 60 MHz (negative sign in definition of ω_0 is absorbed in clockwise rotation). The frequency ω_0 of an atomic nuclei is also dependent on its chemical/electronic environment in a molecule. The secondary magnetic fields produced on an atomic nuclei by its electronic environment results in a shift of the frequency ω_0 to $\omega_0(1-\sigma)$, where σ is specific to the chemical environment of the nuclear spins and is usually of the order 10^{-6},

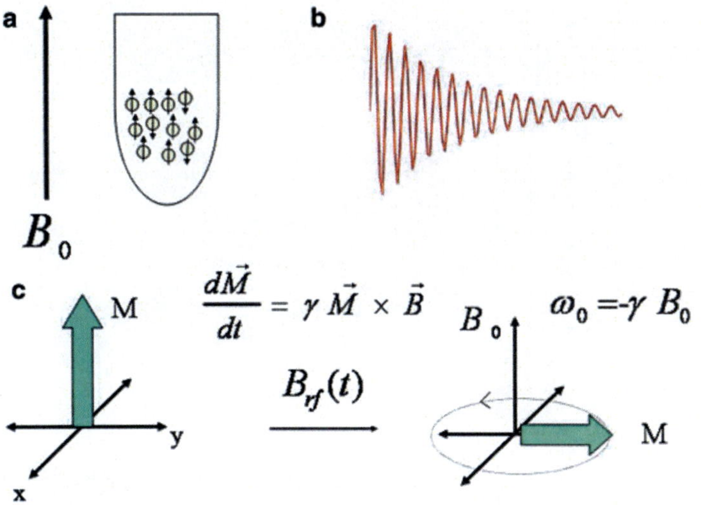

Fig. 10.3 The top figure [1] shows the basic features of an NMR experiment. Panel (**a**) depicts use of a field B_0 to polarize the sample. Panel (**c**) shows the use of pulsed magnetic fields to steer the net magnetization and generate FID. Panel (**b**) shows the profile of a free induction decay

so when ω_0 is around 500 MHz, the shift $\omega_0\sigma$ is in KHz. The Fourier transform of the FID signal then shows many peaks, corresponding to different nuclei with their chemical environment-specific characteristic shifts. Figure 10.5 shows a typical proton NMR spectra from two different size molecules. NMR (Fig. 10.4) is therefore an important analytical tool in chemistry as the peaks in the NMR spectrum serve as a characteristic finger print of a molecule. Starting as a tool for characterization of organic molecules, the use of NMR has spread to areas as diverse as pharmaceutics, medical diagnostics (medical resonance imaging) and structural biology [5, 6]. The principles of NMR have served as a paradigm for other physical methods that rely on interaction between radiation and matter. It is therefore not surprising that experiments in NMR also serve as good model problems in control of quantum systems.

Equation (10.17) gives the wrong impression that the magnetic moment on the transverse plane will continue to precess for ever. Overtime, the magnetic moments of spins making the magnetization vector M experience local fluctuations in the ambient field B_0, causing them to precess differently and hence lose coherence (decoherence). This gives the FID, a decaying envelop (see Fig. 10.5). This phenomenon, termed decoherence, is described in detail subsequently and explicit models are derived to analyze the effect. This leads to the study of open quantum systems where a quantum system interacts with the external environment, but one is only interested in the dynamics of the quantum system of interest. Additional terms need to be incorporated in (10.11) to account for this effect. Before, we describe the dynamics of open quantum systems, a few general comments about the oscillatory control described before are in order.

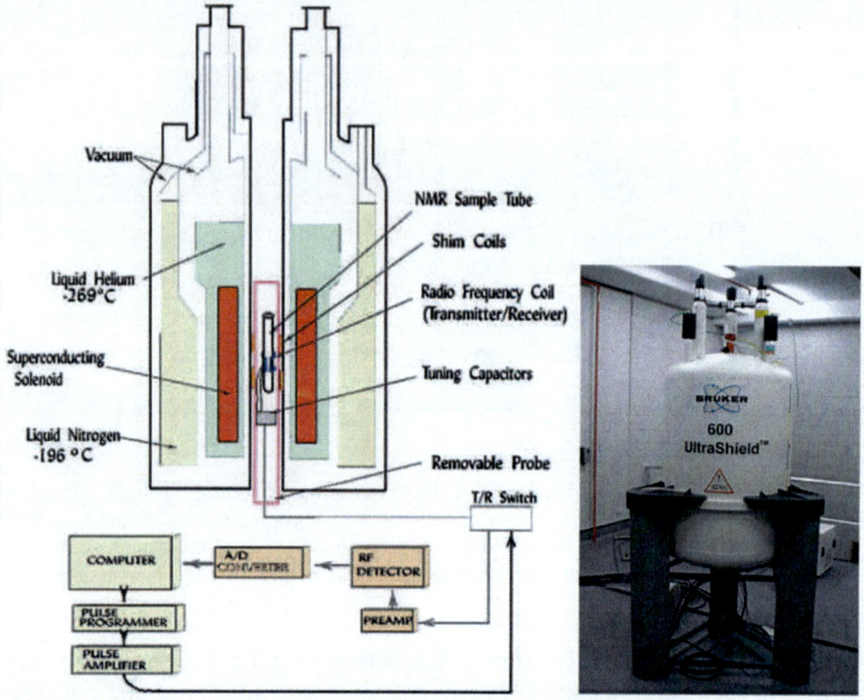

Fig. 10.4 The above figure shows the schematic of a high field NMR instrument

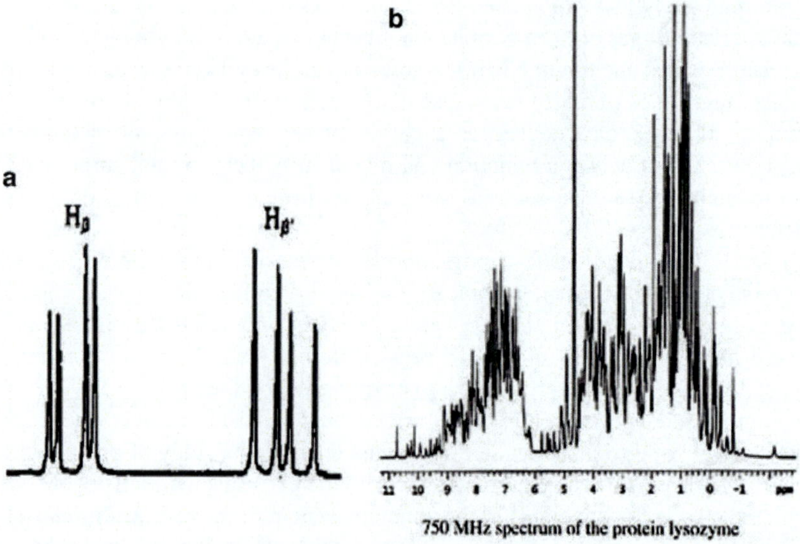

Fig. 10.5 The figure [1] shows a typical proton NMR spectra of a small- and medium-sized molecule, shown in the *left* and *right* panel, respectively

The oscillatory control described in (10.17) consists of irradiating the spin ensemble with an oscillating field along, say, the x and y direction. In practice, the same effect can be obtained by simply having a single oscillating field along, for example, the x direction. This corresponds to

$$(u(t),v(t)) = (2A(t)\cos(\omega t+\theta),0).$$

Such a control can be written as the superposition of two control inputs $(u_1,v_1) = A(t)(\cos(\omega_0 t+\theta), \sin(\omega_0 t+\theta))$ and $(u_2,v_2) = A(t)(\cos(\omega_0 t+\theta), -\sin(\omega_0 t+\theta))$. Transforming again into a rotating frame results in the equation

$$\dot{X} = (A(t)(\cos\theta+\cos(2\omega_0 t+\theta))\Omega_x + A(t)(\sin\theta-\sin(2\omega_0 t+\theta))\Omega_y)X.$$

Since $\omega_0 \gg A(t)$, the oscillating terms average out, giving identical equation as in (10.18). This averaging of the fast oscillating terms is often termed the rotating wave approximation.

10.1.2 Oscillatory Control

The control inputs described in the previous section have an oscillatory character. In general, consider an n dimensional quantum system where the control u is used to modulate the Hamiltonian H_1 in

$$\dot{\rho} = -i[H_0 + uH_1, \rho]. \tag{10.19}$$

Now, since H_0 is a Hermitian operator, let $|\phi_j\rangle$ denote the orthogonal eigenvectors, with eigenvalues ω_j. Then $|\phi_j\rangle$ diagonalize H_0, i.e. we rewrite $H_0 = \sum_j \omega_j |\phi_j\rangle\langle\phi_j|$, and then note $|\phi_j\rangle\langle\phi_j|$ all commute and $\exp(-i\omega_j|\phi_j\rangle\langle\phi_j|t) = \exp(-i\omega_j t)|\phi_j\rangle\langle\phi_j|$, implying that

$$\exp(-iH_0 t) = \sum_j \exp(-i\omega_j t)|\phi_j\rangle\langle\phi_j|.$$

Now, transforming (10.19) into a rotating frame

$$\rho_r = \exp(iH_0 t)\rho\exp(-iH_0 t)$$

gives

$$\dot{\rho}_r = -i[u(t)\exp(iH_0 t)H_1\exp(-iH_0 t), \rho_r]. \tag{10.20}$$

Now, let $h_{jk} = \langle\phi_j|H_1|\phi_k\rangle$. This, then gives that

$$u(t)\exp(iH_0 t)H_1\exp(-iH_0 t) = u(t)\sum_{jk} h_{jk}\exp(-i\omega_{jk} t)|j\rangle\langle k|,$$

where $\omega_{jk} = \omega_k - \omega_j$ and we assume ω_{jk} are all distinct, such that $|u| \leq |\omega_{jk} - \omega_{lm}|$. If u is modulated at one of the ω_{jk}, i.e., $u(t) = A(t)\cos(\omega_{jk}t + \theta(t))$, where the variation in $A(t)$ and $\theta(t)$ is assumed to be much slower than ω_{jk}, then the resulting Hamiltonian in (10.19) averages to

$$\dot{\rho}_{\mathrm{r}} = -i\frac{A}{2}[h_{jk}\exp(i\theta(t))|\phi_j\rangle\langle\phi_k| + \exp(-i\theta(t))h_{kj}|\phi_k\rangle\langle\phi_j|, \rho_{\mathrm{r}}], \tag{10.21}$$

where $h_{jk} = h^*_{kj}$. By modulating the Hamiltonian at the frequency of the difference of the energies of the eigenstates $|\phi_k\rangle$ and $|\phi_j\rangle$, we obtain effective Hamiltonians

$$H_{jk} = h_{jk}|\phi_j\rangle\langle\phi_k| + h_{kj}|\phi_k\rangle\langle\phi_j|, \tag{10.22}$$

$$G_{jk} = -ih_{jk}|\phi_j\rangle\langle\phi_k| + ih_{kj}|\phi_k\rangle\langle\phi_j|, \tag{10.23}$$

which induces a transition from state j to state k and vice versa. If all ω_{jk} are distinct, one can synthesize Hamiltonians H_{jk} independently by simply choosing the frequency of the control $u(t)$. Therefore, one can write an effective control system for (10.19), which takes the form

$$\dot{\rho}_{\mathrm{r}} = -i\left[\sum_{jk} u_{jk}H_{jk} + v_{jk}G_{jk}, \rho_{\mathrm{r}}\right], \tag{10.24}$$

where u_{jk}, v_{jk} are controls that can be turned on and off. Some of the h_{jk}, and therefore H_{jk}, might be zero and hence H_1 cannot induce a transition between the eigenstates $|\phi_j\rangle$ and $|\phi_k\rangle$ directly. These constraints are often termed as the selection rules in physics. Figure 10.1 shows the energy level diagram of the so-called Lambda system studied in laser spectroscopy. There is no direct transition between states $|1\rangle$ and $|3\rangle$, but there is an indirect transition through the state $|2\rangle$.

Of fundamental interest is to know whether the system in (10.4) can be driven between states of interest. This is the standard problem of controllability of bilinear systems. Therefore, the standard techniques [7–9] for studying controllability of systems evolving on compact Lie groups can be directly applied [13]. The main result being that if the Lie algebra $\{-iH_0, -iH_j\}_{LA}$, spanned by $\{-iH_0, -iH_j\}$, is the Lie algebra $su(n)$ of the state space of the system in (10.4), then the system is controllable. Then, checking for controllability reduces to checking the Lie algebraic rank condition. For example, although there is no direct transition between states $|1\rangle$ and $|3\rangle$ in Fig. 10.1b, the system is controllable. The unitary propagator for the effective control system in (10.24), evolves as

$$\dot{U} = -i\begin{bmatrix} 0 & \Omega_c(t) & 0 \\ \Omega_c^*(t) & 0 & \Omega_p(t) \\ 0 & \Omega_p^*(t) & 0 \end{bmatrix} U, \tag{10.25}$$

where $\Omega_c(t)$ and $\Omega_p(t)$ are complex valued controls that induce transitions between $|1\rangle$ and $|2\rangle$ and $|3\rangle$ and $|2\rangle$, respectively. The subject of explicit synthesis of the control laws for control of (10.4) has received significant attention recently in the context of control of spin systems [10–12]. We will discuss some of these results subsequently.

10.2 Open Quantum Systems

10.2.1 Master Equations

Equation (10.11) describes the evolution of a closed quantum system. We now derive an equation for the dynamics of open quantum systems. The derivation is not the most general, but captures the essence of how such a model is usually arrived at [4, 14, 15]. The effect of the environment on the system is modeled by an Hamiltonian H_1, which randomly fluctuates with time.

$$\dot{\rho} = -i[H_0 + f(t)H_1, \rho]. \tag{10.26}$$

The state at time dt is related to time 0 by

$$|\psi(t)\rangle = \exp(-i(H_0 dt + H_1 dW))|\psi(0)\rangle, \tag{10.27}$$

where dW is a Brownian increment distributed $dW \sim N(0, dt)$. Then we have, $\rho(dt) = E(|\psi(dt)\rangle\langle\psi(dt)|)$

$$\rho(dt) = E(\exp(-i(H_0 dt + H_1 dW))\underbrace{|\psi(0)\rangle\langle\psi(0)|}_{\rho(0)}\exp(i(H_0 dt + H_1 dW)). \tag{10.28}$$

Using the Baker-Campbell Hausdorff formula and above definitions of expectations,

$$\rho(dt) = \rho(0) - i[H_0 dt + H_1 dW, \rho(0)] - \frac{1}{2}[H_0 dt + H_1 dW[H_0 dt + H_1 dW, \rho(0)]] + \ldots, \tag{10.29}$$

Using $E(dW) = 0$ and $E(dW^2) = dt$

$$\rho(dt) = \rho(0) - i[H_0, \rho(0)]dt + \frac{1}{2}[-iH_1[-iH_1, \rho(0)]dt + o(dt^2), \tag{10.30}$$

we obtain,

$$\frac{d\rho}{dt} = -i[H_0, \rho] + \frac{1}{2}[iH_1[iH_1, \rho]. \tag{10.31}$$

If we let $\tilde{\rho(t)} = E(\rho(t))$, where expectation is over various initial states of the ensemble, this then gives us that the evolution of the density operator for the open quantum system is no longer isospectral. The effect of the term $L(\rho)$ is then to reduce the value of $tr(\rho^2)$. For instance,

$$\frac{\mathrm{d}\, tr(\rho^2(t))}{\mathrm{d}t} = tr([H_1,\rho]^2), \tag{10.32}$$

where $tr([H_1,\rho]^2) \leq 0$, as $[H_1,\rho]$ is skew Hermitian (implying that $tr(\rho^2)$ decreases with time). The effect of the coupling to an external heat bath is to transform a pure state into a mixed state.

A more general form of $L(\rho)$ is

$$L(\rho) = \sum_j k_j[H_j,[H_j,\rho]] \tag{10.33}$$

arising because of random modulations of Hamiltonians H_j with fluctuations that are uncorrelated. There are many interesting problems involving control of open quantum systems in the presence of dissipation. The operator $L(\rho)$ is a negative definite operator, such that $\frac{dTr(\rho^2)}{dt} = Tr(\rho L(\rho)) \leq 0$. If we measure the entropy of the ensemble of quantum systems by

$$S = 1 - tr(\rho^2),$$

also termed Renyi entropy, then observe that pure states have entropy 0 and the effect of the decoherence is to increase the entropy of the system.

In NMR experiments, fluctuations $f(t)$ in (10.26) arise because the magnetic field B seen by the spins fluctuates with time due to coupling of the spin ensemble with an external bath. We will study the source of these fluctuations subsequently. The equation for the density matrix of the 2×2 spin system then takes the form

$$\frac{\mathrm{d}\rho}{\mathrm{d}t} = -i[\omega_0\sigma_z + f(t)\sigma_z + u\sigma_x + v\sigma_y, \rho]. \tag{10.34}$$

The resulting master equation is then

$$\frac{\mathrm{d}\rho}{\mathrm{d}t} = -i[\omega_0\sigma_z + u\sigma_x + v\sigma_y, \rho] - k[\sigma_z,[\sigma_z,\rho]]. \tag{10.35}$$

When the above equation is written as a Bloch equation, the evolution of the Bloch equation takes the form

$$\frac{\mathrm{d}}{\mathrm{d}t}\begin{bmatrix} m_x \\ m_y \\ m_z \end{bmatrix} = \begin{bmatrix} -k & -\omega_0 & v(t) \\ \omega_0 & -k & -u(t) \\ -v(t) & u(t) & 0 \end{bmatrix}\begin{bmatrix} m_x \\ m_y \\ m_z \end{bmatrix}. \tag{10.36}$$

The constant k is called the transverse relaxation rate and is responsible for the decay of the FID signal with time. Equation (10.35) is, however, not a complete description, because eventually M returns back to the original state $(0,0,1)'$. A more general model for the Lindblad equations is [15]

$$L(\rho) = \sum_k \left[A_k \rho A_k^\dagger - \frac{1}{2}\left\{A_k^\dagger A_k, \rho\right\}\right], \tag{10.37}$$

where $\{A,B\} = AB + BA$ is the anticommutator. If A_k are Hermitian operators, then $L(\rho)$ reduces to the familiar form $\sum_k [A_k[A_k,\rho]]$. However, A_k in general can have both Hermitian and non-Hermitian parts. If we take

$$A_1 = \kappa_1 \begin{bmatrix} 0 & 0 \\ 1 & 0 \end{bmatrix}; \; A_2 = \kappa_2 \begin{bmatrix} 0 & 1 \\ 0 & 0 \end{bmatrix}; \; A_3 = \kappa_3 \begin{bmatrix} 1 & 0 \\ 0 & -1 \end{bmatrix},$$

such that $\kappa_1 > \kappa_2$, then the system in (10.38) will follow the equation

$$\frac{\mathrm{d}}{\mathrm{d}t}\begin{bmatrix} m_x \\ m_y \\ m_z \end{bmatrix} = \begin{bmatrix} -k & -\omega_0 & v(t) \\ \omega_0 & -k & -u(t) \\ -v(t) & u(t) & -(T_1)^{-1} \end{bmatrix}\begin{bmatrix} m_x \\ m_y \\ m_z \end{bmatrix} + \begin{bmatrix} 0 \\ 0 \\ T_1^{-1} m_0 \end{bmatrix}. \tag{10.38}$$

Here k and T_1 in the above equation depend on κ_1, κ_2 and κ_3 as in (10.37). Then after u, v is switched off, m_z eventually returns to $0 \leq m_0 \leq 1$, at a characteristic time T_1, also called the longitudinal relaxation rate.

Although, the effect of the Lindblad operator $L(\rho)$ in (10.33) is to increase the entropy of the system, the most general form of the Lindblad equations as in (10.37) can lead to a decrease in the entropy of the quantum system when mixed suitably with external control. An important application of this feature is in the field of laser cooling [16], where an interplay between unitary control and Lindblad terms, as in (10.37), is used to decrease the entropy of the quantum system. We study a concrete example to understand the basic ideas in this subject in [17].

10.2.2 Laser Cooling

Consider again, the three-level Λ system as depicted in Fig. 10.1b. The evolution of the density matrix of the three-level Λ system is given by

$$\dot{\rho} = -i[H(t),\rho] + \gamma_1\left(E_1\rho E_1^\dagger - \frac{1}{2}\left\{E_1^\dagger E_1, \rho\right\}\right) + \gamma_2\left(E_2\rho E_2^\dagger - \frac{1}{2}\left\{E_2^\dagger E_2, \rho\right\}\right), \tag{10.39}$$

where $E_1 = |1\rangle\langle 2|$ and $E_2 = |3\rangle\langle 2|$.

If the initial state of the system represented by its density matrix is diagonal, say

$$\rho = \begin{bmatrix} \lambda_1 & 0 & 0 \\ 0 & \lambda_2 & 0 \\ 0 & 0 & \lambda_3 \end{bmatrix},$$

where λ_1, λ_2 and λ_3 describe the population distribution in three energy states. Then, in the absence of any external controls, the density matrix stays diagonal and the diagonal entries evolve as

$$\frac{\mathrm{d}}{\mathrm{d}t}\begin{bmatrix} \lambda_1 \\ \lambda_2 \\ \lambda_3 \end{bmatrix} = \underbrace{\begin{bmatrix} 0 & \gamma_1 & 0 \\ 0 & -(\gamma_1+\gamma_2) & 0 \\ 0 & \gamma_2 & 0 \end{bmatrix}}_{A} \begin{bmatrix} \lambda_1 \\ \lambda_2 \\ \lambda_3 \end{bmatrix}. \tag{10.40}$$

We assume that the system is completely controllable and any unitary rotation $U(t)$ on the three level system in (10.25) can be synthesized in arbitrary small time. In particular, consider the unitary transformation

$$P = \begin{bmatrix} 0 & 0 & 0 \\ 0 & 0 & 1 \\ 0 & 1 & 0 \end{bmatrix},$$

which swaps the population states 2 and 3. The effect of this unitary transformation on the diagonal of the density matrix is

$$P\begin{bmatrix} \lambda_1 \\ \lambda_2 \\ \lambda_3 \end{bmatrix} = \begin{bmatrix} \lambda_1 \\ \lambda_3 \\ \lambda_2 \end{bmatrix}.$$

Then, consider the following sequence of operations,

$$\exp(At_n)P\exp(At_{n-1})\dots P\exp(At_1),$$

where t_i are chosen long enough so that as λ_2 in (10.40) decays below the value λ_3, for example $\lambda_2 = \frac{\lambda_3}{2}$, these operations keep the density matrix diagonal and

$$\exp(At_n)P\exp(At_{n-1})\dots P\exp(At_1)\begin{bmatrix} \lambda_1 \\ \lambda_2 \\ \lambda_3 \end{bmatrix} \sim \begin{bmatrix} 1 \\ 0 \\ 0 \end{bmatrix}. \tag{10.41}$$

It is clear that the population in state 1 keeps building up, while the population from states 2 and 3 is eventually drained off. The cooling strategy consists of a sequence of dissipative evolutions and unitary control (with electromagnetic fields) to synthesize the Hamiltonians H_{23} in Fig. 10.1b. Therefore, by an interplay of external control and evolution of the natural dynamics, all the population is eventually driven to state 1, although one starts with a state where the population is distributed across all the states [16].

10.3 Control of Ensembles with Parametric Inhomogeneities and Dispersions

We now return to the Bloch equations in (10.18). As discussed earlier, in NMR experiments, there is dispersion in the parameter ω, as the chemical shifts σ of the nuclear spins are dispersed over a certain range. In practice, there is another source of the dispersion. The applied radio-frequency field is not uniform on the whole sample but is dispersed over a range, captured by the parameter $\varepsilon \in [1-\delta, 1+\delta]$. In this case, (10.18) is then modified to

$$\dot{X} = (\omega\Omega_z + \varepsilon u_1(t)\Omega_x + \varepsilon v_1(t)\Omega_y)X. \tag{10.42}$$

Dispersion in the parameters in the system dynamics poses some interesting questions on controllability and control design. Equation (10.42) represents a continuum of systems parametrized by ω and ε. Figure 10.2 shows that application of the control input $(u_1(t), v_1(t))$ as in (10.18), results in poor transfer to the transverse plane for spins with ω comparable or greater than $A_{\max}$, where $\sqrt{u_1^2(t)+v_1^2(t)} \leq A_{\max}$ (the net rotation is around a tilted axis B_{r} as shown in the picture). The control challenge is to steer the ensemble of inhomogeneous systems, to a desired target state, inspite of variation in their internal dynamics, by application of the same control law $(u_1(t), v_1(t))$. We say that the system in (10.42) is ensemble controllable if the system can be steered from an initial state of the ensemble described by vector valued function $X_0(\omega, \varepsilon)$ arbitrarily close to the target state $X_F(\omega, \varepsilon)$ (where the distance $||X_F(\omega, \varepsilon) - X_0(\omega, \varepsilon)||$ is measured by, say, an L_2 distance $\int\int |X_F(\omega, \varepsilon) - X_0(\omega, \varepsilon)|^2 \mathrm{d}\omega \mathrm{d}\varepsilon$ between functions).

This problem represents a typical problem in the control of quantum systems, with dispersion or uncertainty in the parameters governing the dynamics, using the same control field. The problem of designing excitations that can steer an ensemble and be robust and immune to dispersion in the dynamics of the spin system is a well-studied subject in NMR spectroscopy, and extensive literature exists on the subject of so-called composite pulses that can correct for dispersion in system dynamics [18]. In many cases of practical interest, one wants to find control inputs that prepare the final state as some desired function of the parameter. For example, slice selective excitation and inversion pulses in magnetic resonance imaging (MRI) [19, 21, 22]. Only recently, these problems have been understood and posed as questions in controllability of infinite dimensional systems [23–25]. A principled study of the controllability of these systems reveals aspects of system dynamics, which makes it possible to engineer excitations that can steer a quantum ensemble and be robust to the dispersion in the system dynamics. These problems therefore motivate development of new methods and techniques for studying controllability and constructive controllability of a class of infinite dimensional nonlinear systems.

To fix ideas, we first set the dispersion ω in (10.42) to zero, and only consider dispersion arising due to an inhomogeneous RF field on the sample, measured by the parameter ε. Rewriting (10.42), we obtain

$$\dot{X} = (\varepsilon u_1(t)\Omega_x + \varepsilon v_1(t)\Omega_y)X. \tag{10.43}$$

We now summarize the basic ideas [23] that make it possible to engineer input excitations that can steer the whole ensemble uniformly and be immune to the dispersion in the value of ε. Observe for small dt, the evolution

$$U_1^x(dt) = \exp\left(\varepsilon\Omega_y\sqrt{dt}\right)\exp\left(\varepsilon\Omega_x\sqrt{dt}\right)\exp\left(-\varepsilon\Omega_y\sqrt{dt}\right)\exp\left(-\varepsilon\Omega_x\sqrt{dt}\right) \tag{10.44}$$

to leading order in ε is $I+\varepsilon^2[\Omega_y,\Omega_x]dt+o(dt^{\frac{3}{2}})$, where $o(dt^{\frac{3}{2}})$ represents a term of order $dt^{\frac{3}{2}}$, i.e., we can synthesize the generator $[\varepsilon\Omega_x,\varepsilon\Omega_y]=\varepsilon^2\Omega_z$, by back and forth maneuver in the directly accessible directions Ω_x and Ω_y. Similarly, we can synthesize higher order Lie brackets like $[\varepsilon\Omega_y,[\varepsilon\Omega_x,\varepsilon\Omega y]]=\varepsilon^3\Omega_x$. By successive Lie brackets, terms of the type $\varepsilon^{2k+1}\Omega_x$ can be synthesized to leading order.

One such construction is for $k>0$ given by

$$U_k^x(dt) = I+\varepsilon^{k+1}ad_{\Omega_y}^k(\Omega_x)\,dt^{\alpha_k}+o\left(dt^{\frac{3}{2}}\right),$$

we have, for $\gamma_{k+1}=\frac{3}{4}-\frac{\alpha_k}{2}$, gives

$$U_{k+1}^x(dt) = \exp(\varepsilon\Omega_y dt^{\gamma_{k+1}})\;U_k^x(dt)\exp(-\varepsilon\Omega_y dt^{\gamma_{k+1}})\,U_k^{-x}(dt),$$
$$U_{k+1}^x(dt) = I+\varepsilon^{k+2}ad_{\Omega_y}^{k+1}(\Omega_x)\,dt^{\alpha_{k+1}}+o\left(dt^{\frac{3}{2}}\right)$$

with $\alpha_{k+1}=\frac{3}{4}+\frac{\alpha_k}{2}$. Note $ad_{\Omega_y}^{k+1}(\Omega_x)$, is the leading order term $(ad_{\Omega_y}^{k+1}(\Omega_x)=[\Omega_y,ad_{\Omega_y}^k(\Omega_x)])$.

Now using $\{\varepsilon\Omega_x,\varepsilon^3\Omega_x,\ldots,\varepsilon^{2n+1}\Omega_x\}$ as generators, we can produce an evolution

$$\exp\left\{\sum_{k=0}^{n} c_k\varepsilon^{2k+1}\Omega_x\right\},$$

where n, and the coefficients c_k can be so chosen so that

$$\sum_{k=0}^{n} c_k\varepsilon^{2k+1}\approx\theta$$

for all $\varepsilon\in[1-\delta,1+\delta]$. Therefore, an evolution $\exp(\theta\Omega_x)$ can be synthesized for all values of ε to any desired accuracy. Therefore, one achieves robustness with dispersion to ε by generating effective generators with arbitrary high powers of the dispersion parameter ε.

Contrast to the situation in (10.43), with the following control system, the well-studied non-holonomic integrator,

$$\frac{\mathrm{d}}{\mathrm{d}t}\begin{bmatrix}x\\y\\z\end{bmatrix}=\varepsilon u\begin{bmatrix}1\\0\\-y\end{bmatrix}+\varepsilon v\begin{bmatrix}0\\1\\x\end{bmatrix}. \tag{10.45}$$

If ε is fixed, then the system in (10.45) is controllable as the vector fields

$$f = \begin{bmatrix} 1 \\ 0 \\ -y \end{bmatrix}; \; g = \begin{bmatrix} 0 \\ 1 \\ x \end{bmatrix}$$

generate the vector field $[f,g] = [0,0,1]'$. The three vector fields $f, g, [f,g]$, then span the three-dimensional space. However, the Lie algebra generated by f, g is nilpotent and therefore $[\cdot,[f,g]] = 0$. The dispersion parameter ε cannot be raised to higher powers by iterated brackets and therefore such an ensemble of inhomogeneous nilpotent systems is not ensemble controllable. On the contrary, the Lie algebra $\mathfrak{g} = so(3)$ generated by Ω_x and Ω_y in (10.43) is semi-simple (implying $[\mathfrak{g},\mathfrak{g}] = \mathfrak{g}$) and the iterated Lie brackets of Ω_x and Ω_y never terminate allowing for design of robust input excitations. Similarly, linear systems

$$\frac{\mathrm{d}X}{\mathrm{d}t} = AX + \varepsilon Bu \tag{10.46}$$

cannot be steered uniformly by application of the same control input $u(t)$, as the output, for $X(0) = 0$ is

$$X(t) = \varepsilon \int_0^t \exp(A(t-\tau))B(\tau)u(\tau)\mathrm{d}\tau,$$

which is just a linear function of the input. No matter how $u(\tau)$ is modulated, the output depends linearly on the input.

Interesting control design problems arise in the manipulation of inhomogeneous quantum ensembles. To provide a flavor for such problems, we describe one synthesis method [26] for designing input excitations that robustly steer the inhomogeneous ensemble in (10.43). This synthesis method has recently been used in the design of RF pulse sequences in NMR spectroscopy that are robust to RF inhomogeneity. Consider the following rotations obtained by alternate rotations around x- and y-axis for appropriate durations. Let

$$U_1 = \exp(k\pi\Omega_x\varepsilon)\exp\left(\frac{\beta_k}{2}\Omega_y\varepsilon\right)\exp(-k\pi\Omega_x\varepsilon), \tag{10.47}$$

$$U_2 = \exp(-k\pi\Omega_x\varepsilon)\exp\left(\frac{\beta_k}{2}\Omega_y\varepsilon\right)\exp(k\pi\Omega_x\varepsilon). \tag{10.48}$$

Now by choosing β_k small enough, we have

$$V_k = U_2U_1 \sim \exp(\varepsilon\beta_k\Omega_y\cos(k\pi\varepsilon)).$$

Consider, a sequence of transformations

$$\Pi_k(V_k)^{n_k} \sim \exp\left(\varepsilon\sum_k \alpha_k\cos(k\pi\varepsilon)\,\Omega_y\right), \tag{10.49}$$

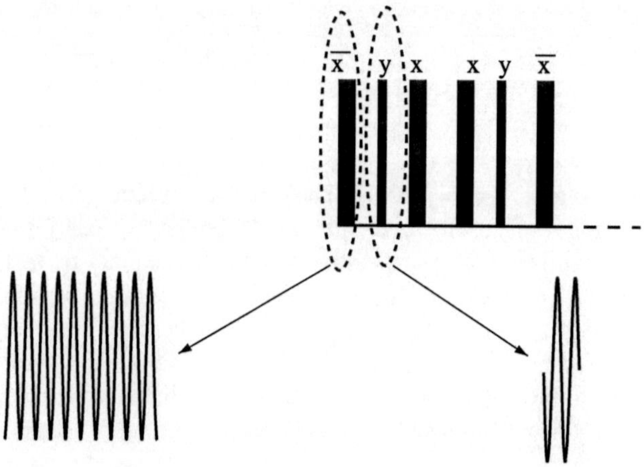

Fig. 10.6 The figure [1] shows the sequence of pulses with alternate phases as described in (10.47), which forms the building block of a composite pulse train as in (10.49). Each pulse is an oscillatory control input the phase of which is changed from pulse to pulse

where $\alpha_k = n\beta_k$. Now, the coefficients α_k can be so chosen that

$$\sum_k \alpha_k \cos(k\pi\varepsilon) = \frac{\theta}{\varepsilon}$$

for $1-\delta \leq \varepsilon \leq 1$, with $0 < \delta < 1$.

Therefore,

$$\Pi_k (V_k)^{n_k} \sim \exp(\theta). \tag{10.50}$$

The actual control input to (10.15) consists of an oscillatory input $(u,v) = (A\cos(\omega_0 t + \phi), 0)$, where the phase ϕ is switched between 0, $\frac{\pi}{2}$ and π to achieve rotations around x, y, and $-x$ axes, respectively. Figure 10.6 shows this control input and its pictorial depiction as a pulse sequence.

10.3.1 Controllability of Bloch Equations in the Presence of Frequency Dispersion

Now, consider the Bloch equations as in (10.18), with dispersion in the Larmor frequencies. We now like to show that this system is ensemble controllable with respect to the dispersion in the parameter ω. This is an ubiquitous problem in NMR spectroscopy, where one wants to excite a broad range of frequencies with limits on RF-power/amplitude, which translates to limits on $A(t)$ in (10.18).

There is an important conceptual issue that emerges in the controllability analysis of such problems. In studying controllability of control systems of the kind

$$\dot{x} = (A + uB)x$$

evolving on compact Lie groups, it is possible to synthesize commutators of the kind $[A,B]$ to leading order by an evolution

$$\exp(-A\mathrm{d}t)\exp(-B\mathrm{d}t)\exp(A\mathrm{d}t)\exp(B\mathrm{d}t), \qquad (10.51)$$

where the backward evolution $\exp(-A\mathrm{d}t)$ is generated by letting the forward map $\exp(At)$ evolve for a sufficient period of time. The free evolution on a compact group almost returns back after sufficient time. However, the situation is different for a continuum of such systems as in (10.18). In the presence of a continuum of frequencies $\omega \in [-B,B]$, given small time $\mathrm{d}t$, there is no forward evolution time T, such that $\exp(\omega T) = \exp(-\omega\ \mathrm{d}t)$, for all $\omega \in [-B,B]$. However, using control, we can synthesize an effective backward evolution. Two limits are of particular interest here.

We first assume that our control inputs in (10.18) are unbounded a priori ($A_{\max} \gg \omega$). Note, because of the assumption of strong fields, we can reverse the evolution of the drift term in (10.18),

$$\exp(\pi\Omega_x)\exp(\omega\Omega_z dt)\exp(-\pi\Omega_x) = \exp(-\omega\Omega_z dt), \qquad (10.52)$$

where $\omega \in [-B,B]$, and the π rotations like $\exp(\pi\Omega_x)$ can be produced in negligible time. Now, we consider the case when the controls u and v are bounded, i.e., $\sqrt{u^2(t)+v^2(t)} \leq A_{\max}$ for all t, so that we cannot produce rotations of the type $\exp(-\Omega_x\pi)$ in an arbitrarily small time as in (10.52).

Nonetheless, the system is ensemble controllable as will be shown below. The key to showing this is to produce the backward evolution of the drift term, $\exp(-\omega\Omega_z dt)$. This helps us to generate higher-order Lie brackets with the drift term containing higher powers of the dispersion parameters ω, which can be combined to produce an evolution that is robust to ω. Our construction initially uses the well-known construction in the physics literature called adiabatic following, which helps to synthesize an evolution $\exp(-\omega\Omega_z dt)$. This construction can be used to show ensemble controllability with respect to both Larmor dispersion and RF inhomogeneity in the Bloch equations (10.43). Adiabatic following is a technique widely used in a variety of experiments involving control of quantum systems as it is robust to inhomogeneity in the system dynamics. It is of independent interest from the perspective of nonlinear control.

10.3.1.1 Adiabatic Following

Consider the Bloch equations with only Larmor dispersion as in (10.42), which we rewrite to reflect ensemble of systems with ω dependence.

$$\dot{X}(t,\omega) = \Big[\omega\Omega_z + u_1\Omega_y + v_1\Omega_x\Big]X(t,\omega), \qquad (10.53)$$

where $\omega \in [-B,B]$. Let $(u_1(t), v_1(t)) = A(t)(\cos\phi(t),\ -\sin\phi(t))$, where $A = \sqrt{u_1^2(t)+v_1^2(t)}$. We then slowly vary $\dot{\phi}(t)$ from an initial value $\dot{\phi}(0) \ll -B$ to

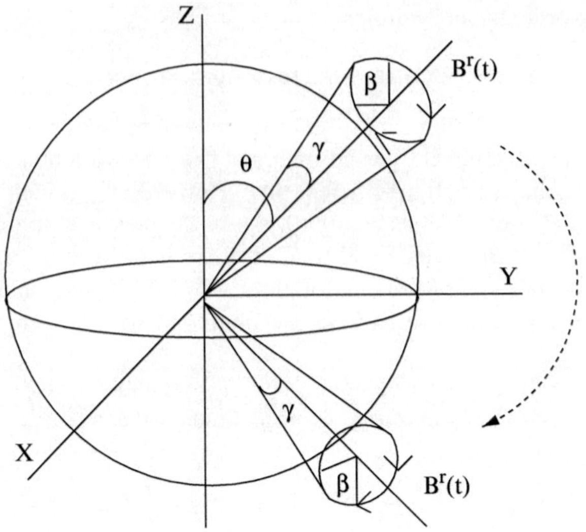

Fig. 10.7 The figure [1] shows how the vector X in (10.15) can be dragged from $(x,y,z)=(0,0,1)$ to $(0,0,-1)$, independent of the value of ω, by slowly varying $\tilde{\omega}(t)$ in (10.54)

$\dot{\phi} \gg B$. We show that if the change in $\dot{\phi}(t)$ is slow enough, all systems as in (10.53) can be steered from $(0,0,1)^T$ to $(0,0,-1)^T$. We first make a change of coordinates

$$Y(t,\omega) = \exp[-\phi(t)\Omega_z]X(t,\omega).$$

The resulting system then takes the form

$$\dot{Y}(t,\omega) = ([\omega - \tilde{\omega}(t)]\,\Omega_z + A\Omega_y)Y(t,\omega),$$

where $\tilde{\omega}(t) = \dot{\phi}(t)$. Thus, the effective generator of motion is

$$(\omega - \tilde{\omega}(t))\Omega_z + A\Omega_y.$$

In standard physics terminology, the Bloch vector $Y(t,\omega)$ rotates around the effective field $B^{\mathrm{r}} = A\mathbf{j} + [\omega - \tilde{\omega}(t)]\,\mathbf{k}$ (see Fig. 10.7) and has the net magnitude of rotation

$$|\, B^{\mathrm{r}} \,| = \sqrt{(A)^2 + [\omega - \tilde{\omega}(t)]^2} = A\sqrt{1+\cot^2\theta}.$$

The angle θ through which B^{r} is tilted with respect to A is defined by

$$\cot\theta = \frac{\omega - \tilde{\omega}(t)}{A}. \tag{10.54}$$

By differentiating (10.54), we get the rate of change for the angle $\theta(t)$

$$\dot{\theta} = \frac{\dot{\tilde{\omega}}(t)}{A}\sin^2\theta.$$

The maximum value of the right hand side (RHS) of the above expression is obtained when $\theta = \frac{\pi}{2}$ and we have

$$| \dot{\theta} |_{\max} = \frac{| \dot{\tilde{\omega}}(t) |}{A}.$$

In addition, the smallest rate of rotation of X around B^{r} is A. This happens when $\tilde{\omega}(t) = \omega$, i.e., θ in (10.54) is 0. If we vary $\tilde{\omega}(t)$ slowly enough so that $| \dot{\theta}_{\max} | \ll A$, i.e.,

$$| \dot{\tilde{\omega}}(t) | \ll A^2$$

from $\theta(0) = 0$, to the final state π such that the variation is slow, then $X(t,\omega)$ for all ω follows the effective field (remains locked around B^{r}) from $(0,0,1)^T$ to $(0,0,-1)^T$ simultaneously. This can be seen by the following averaging argument. Observe that in Fig. 10.7, the rate of change of the angle γ at time t is a function of $\dot{\theta}$ and β, i.e.,

$$\frac{\mathrm{d}\cos\gamma}{\mathrm{d}t} = h(\dot{\theta}, \theta, \beta),$$

where the angles γ and β are defined in Fig. 10.7. Because β changes at a much faster rate compared to θ, i.e., $\dot{\beta} \gg \dot{\theta}$, the time scale separation gives

$$\cos\gamma(t+\tau) - \cos\gamma(t) = \int_0^{\tau} h(\dot{\theta}(t), \beta(t+\sigma))\mathrm{d}\sigma \approx 0,$$

with error propotional to $\frac{\dot{\tilde{\omega}}}{A^2}(\theta(t+\tau) - \theta(t))$ (with $\dot{\tilde{\omega}}$ constant), and τ is the period for β to rotate by 2π over which $\dot{\theta}(t)$ is supposed to be constant. Therefore, we can maintain $\gamma(t)$ very small throughout, i.e., $0 \leq \gamma(t) \leq \varepsilon$ for all t, and ε can be controlled by the rate $\dot{\theta}(t)$. Now note that $\theta(T) \approx \pi$ and hence $X(T,\omega) \approx -X_0$ for $\omega \in [-B,B]$, where $X_0 = (0,0,1)^T$. As a result, there exists a net evolution $U(\omega)$ for all $\omega \in [-B,B]$ such that

$$U(\omega)X_0 \approx -X_0. \tag{10.55}$$

Therefore, by doing an Euler angle decomposition, we can decompose

$$U(\omega) = \exp(f(\omega)\Omega_z)\exp(\pi\Omega_x)\exp(g(\omega)\Omega_z),$$

where $f(\omega)$ and $g(\omega)$ are some functions of ω. Then, observe,

$$U^2(\omega) \sim \exp(f(\omega)\Omega_z)\exp(-g(\omega)\Omega_z)\exp(-f(\omega)\Omega_z)\exp(g(\omega)\Omega_z) = \mathbf{1}\,,$$

$$U(\omega)\exp(\omega\Omega_z t)U(\omega) = \exp(-\omega\Omega_z t). \tag{10.56}$$

This propagator $U(\omega)$ can be used to reverse the direction of drift in (10.56). Now, constructions as described before can be used to produce any rotation in (10.42) as a function of ω. This approximation in (10.55) is in L_2 sense as described earlier and can be made arbitrarily good by regulating how slowly $\dot{\theta}$ is changed. In fact, it is possible to write down the explicit time-dependent control law that will transfer X_0 to $-X_0$. This is the well-studied complex hyperbolic secant pulse [19] and is very interesting from the perspective of nonlinear control.

We have sketched the basic ideas required to show that the Bloch equation (10.18) can be steered to a target state that has the desired dependence of the drift parameter ω. In many applications in NMR and MRI, one requires input control design that only excites spins with specific value of ω to the equator, with a final state that depends in a specified way on ω and leaves other spins invariant.

10.3.2 Ensemble Control by Method of Multiply Rotating Frames

We again consider the problem of broadband excitation. Consider the unitary transformation (we use I_α to denote the Pauli matrix such that the frobenius norm $|I_\alpha| = \frac{1}{\sqrt{2}}$)

$$\dot{U}(\omega) = -i\{\omega I_z + u(t)I_x + v(t)I_y\}U(\omega), \tag{10.57}$$

where $\omega \in [-c_0, c_0]$. The goal is to design $(u(t), v(t)) = A(t)(\cos\phi(t), \sin\phi(t))$, which will synthesize the propagator $\exp(-i\frac{\pi}{2}I_y)$, uniform for all $\omega \in [-c_0, c_0]$ for a given value of $\frac{c_0}{A} = \alpha_0$, where $max\{A(t)\} \leq A$. We show that it is possible to perform a sequence of coordinate transformations

$$\Theta_n(\omega,t) = \exp\left(i\upsilon_n I_{z_n(\omega)}t\right)\ldots\exp\left(i\upsilon_k I_{z_k(\omega)}t\right)\ldots\exp\left(i\upsilon_1 I_{z_1(\omega)}t\right)U(\omega,t), \tag{10.58}$$

such that (for notation simplicity, we suppress the time index t, where it is obvious)

$$\dot{\Theta}_n(\omega) = -i\left\{f_n(\omega)I_{z(\omega)} + \frac{w_n}{2^n}I_y + a(t)\right\}\Theta_n(\omega), \tag{10.59}$$

such that $|f_n(\omega)| \ll c_0$, $z(\omega) \in x-z$ plane and $a(t)$ captures oscillating components, with zero time average, such that their assumed effect is small, and the system can be approximated by

$$\dot{\Theta}_n(\omega) = -i\left\{f_n(\omega)I_{z(\omega)} + \frac{w_n}{2^n}I_y\right\}\Theta_n(\omega), \tag{10.60}$$

$\frac{|2^n f_n(\omega)|}{w_n} \ll 1$, for all $\omega \in [-c_0, c_0]$. Then, we can perform a $\exp(-i\frac{\pi}{2}I_y)$ rotation, with high fidelity in time $T = \frac{2^{n-1}\pi}{w_n}$. Furthermore, things can be arranged such that $\upsilon_k T = 2n_k\pi$ for integer n_k. This ensures that $U(\omega,T) \sim \Theta_n(\omega,T) \sim \exp(-i\frac{\pi}{2}I_y)$. To achieve this, we choose the following control and resulting Hamiltonian

$$H_0(t) = \omega I_z + w_0 I_x + \underbrace{(w_1 \sin\upsilon_1 t + w_2\cos\upsilon_1 t\sin\upsilon_2 t + w_3\cos\upsilon_1 t\cos\upsilon_2 t\sin\upsilon_3 t + \ldots)}_{A_0(t)}I_y, \tag{10.61}$$

which we rewrite as

$$H_0(t) = \tilde{\omega}I_{z_1(\omega)} + A_0(t)I_y, \tag{10.62}$$

where spread of frequencies is $\tilde{\omega} \in [w_0, \sqrt{w_0^2 + c_0^2}]$. Now by choosing υ_1 in the first transformation as exactly the center of this spread, we get

$$H_1(t) = \underbrace{\underbrace{(\tilde{\omega} - \upsilon_1)}_{f_1(\omega)} I_{z1(\omega)} + \frac{w_1}{2} I_{x_1} + A_1(t) I_y}_{H_1'(t)} + H_1''(t), \tag{10.63}$$

where $H_1''(t)$ is a fast oscillating part that we neglect for now. The new frequency $f_1(\omega) \in [-c_1, c_1]$, where $c_1 < c_0$, A_1, and H_1'' are written in their general form,

$$A_k(t) = \frac{1}{2^k} \left\{ \sum_{m=k+1}^{n-1} w_m \prod_{i=k+1}^{m-1} \cos(\upsilon_i t) \sin(\upsilon_m t) + w_n \prod_{i=k+1}^{n} \cos(\upsilon_i t) \right\} I_y, \tag{10.64}$$

$$\begin{aligned} H_k' &= f_k(\omega) I_{zk(\omega)} + \bar{\omega}_k I_{xk} + A_k(t) I_y. \\ H_k''(t) &= -\frac{w_k}{2^k} \exp\left(i 2\upsilon_k I_{z_k} t\right) I_{x_k} \exp\left(-i 2\upsilon_k I_{z_k} t\right) \\ &\quad + A_k(t) \exp\left(i 2\upsilon_k I_{z_k} t\right) I_y \exp\left(-i 2\upsilon_k I_{z_k} t\right), \end{aligned} \tag{10.65}$$

$$c_{k+1} = \frac{\sqrt{c_k^2 + \bar{w}_k^2} - \bar{w}_k}{2}; \quad \upsilon_{k+1} = \frac{\sqrt{c_k^2 + \bar{w}_k^2} + \bar{w}_k}{2}, \tag{10.66}$$

where $\bar{w}_k = 2^{-k} w_k$. The system obtained after first coordinate transformation has the desired feature that the ratio of chemical shift spread to control strength $\frac{c_1}{w_1/2}$ is reduced over $\frac{c_0}{w_0}$ for the original system. We can now iterate the above construction.

At the kth stage, ($k > 0$), the system takes the form

$$H_k(t) = H_k'(t) + H_k''(t). \tag{10.67}$$

Combining (10.58) and (10.59), we write the total evolution as

$$\begin{aligned} U(\omega, t) &= \exp\left(-i\upsilon_1 I_{z_1(\omega)} t\right) \dots \exp\left(-i\upsilon_k I_{z_k(\omega)} t\right) \\ &\quad \dots \exp\left(-i\upsilon_n I_{z_n(\omega)} t\right) \exp\left(-i\bar{w}_n I_y t\right) \Theta_n'(\omega, t), \end{aligned} \tag{10.68}$$

where

$$\dot{\Theta}_n'(\omega) = -i\left\{a'(t,\omega)\right\} \Theta_n'(\omega), \tag{10.69}$$

and Θ_n' is the evolution of Θ_n in the frame of $\exp(-i\bar{w}_n I_y t)$. The oscillating terms in the above equation can be written as

$$a'(t,\omega) = \sum_k \underbrace{c_k \cos \nu_k t \; I_{k\alpha} + s_k \sin \nu_k t \; I_{k\beta}}_{g_k(t)}, \tag{10.70}$$

where the frobenius norm of the Pauli-matrices $|I_{k\alpha}| = |I_{k\beta}| = \frac{1}{\sqrt{2}}$. Define $G_k = \sqrt{c_k^2 + s_k^2 + 2|c_k s_k \cos\theta_k|}$, where θ_k is the angle between Pauli matrices $I_{k\alpha}$ and $I_{k\beta}$. We now calculate, how well the oscillating terms $a'(t,\omega)$ will average out by bounding the deviation of $\Theta_n'(\omega)$ from the identity.

Let U_k denote the evolution,

$$\dot{U}_k = -i \sum_{j=k}^{n} g_k(t) U_k. \tag{10.71}$$

To evaluate U_k, we transform into the frame of U_{k+1}, i.e., $V_k = U_{k+1}' U_k$, which gives

$$\dot{V}_k = -i\, U_{k+1}' g_k U_{k+1}\, V_k. \tag{10.72}$$

We now evaluate the Peano Baker series for V_k over a period $\tau_k = \frac{2\pi}{v_k}$. Let $\tilde{H}_k = U_{k+1}' g_k(t) U_{k+1}$,

$$V_k(\tau_k) = I + \int_0^{\tau_k} \tilde{H}_k(\tau)\mathrm{d}t + \underbrace{\int_0^{\tau_k}\int_0^{\tau} \tilde{H}_k(d\tau)\tilde{H}_k(d\sigma)\mathrm{d}\tau\mathrm{d}\sigma + \dots}_{\Delta}. \tag{10.73}$$

For $G_k\tau_k < 1$, we have $|\Delta| \leq \sqrt{2}\frac{G_k^2\tau_k}{v_k}$.
Let

$$D_k(t) = \int_0^t g_k(\sigma)\mathrm{d}\sigma. \tag{10.74}$$

Then note $D_k(n\tau_k) = 0$. Then, using integration by parts, we write

$$\int_0^{\tau_k} U_{k+1}^\dagger g_k(t) U_{k+1}(t)\mathrm{d}t = U_{k+1}^\dagger D_k(t) U_{k+1}(t)|_0^{\tau_k}$$
$$-i\int_0^{\tau_k} U_{k+1}^\dagger \left[\sum_{j=k+1}^{n} g_j, D_k(t)\right] U_{k+1}\mathrm{d}\tau. \tag{10.75}$$

We can bound the second integral on the RHS by

$$\int_0^{\tau_k} |\, U_{k+1}^\dagger \left[\sum_{j=k+1}^{n} g_j,\, D_k(t)\right] U_{k+1}\, |\mathrm{d}\tau \leq \sqrt{2}\frac{G_k \sum_{j=k+1}^{n} G_j}{v_k}\tau_k, \tag{10.76}$$

where

$$D_k(t) = \frac{1}{v_k}\{c_k I_{k\alpha}\sin v_k\tau - s_k I_{k\beta}\cos v_k\tau\}|_0^t. \tag{10.77}$$

This gives that

$$|V_k(t) - I| \leq \sqrt{2}\left\{\frac{G_k \sum_{j=k+1}^{n} G_j}{\nu_k} + \frac{G_k^2}{\nu_k}\right\} t. \tag{10.78}$$

We evolve U_1 in the frame of U_2, followed by U_2 in the frame of U_3 and so on. Then we write the total evolution

$$\Theta_n'(\omega,t) = V_n \dots V_1 = (I + E_n \Delta_n) \dots (I + E_1 \Delta_1), \tag{10.79}$$

where $|\Delta_k| = 1$ and

$$E_k \leq \sqrt{2}\left\{\frac{\sum_{j=k+1}^{n} G_j}{\nu_k} G_k + \frac{G_k^2}{\nu_k}\right\} t. \tag{10.80}$$

The total error $E = |\Theta_n'(\omega,t) - I|$ is written as

$$E = \sum_k E_k. \tag{10.81}$$

Two limits are of interest. When all $\nu_k = \nu$, then

$$\frac{E}{\sqrt{2}} \leq \frac{1}{2\nu}\left\{\left(\sum_{k=1}^{n} G_k\right)^2 + \sum_{k=1}^{n} G_k^2\right\} t, \tag{10.82}$$

when $G_k \gg \sum_{j=k+1}^{n} G_j$ then we approximate,

$$\frac{E}{\sqrt{2}} \leq \sum_k \frac{G_k^2}{\nu_k} t. \tag{10.83}$$

Remark 10.1. We rewrite (10.66) as

$$c_{n-(k+1)} = 2\sqrt{c_{n-k} \upsilon_{n-k}}; \quad \bar{w}_{n-(k+1)} = \upsilon_{n-k} - c_{n-k}. \tag{10.84}$$

Define $\frac{c_k}{\upsilon_k} = \alpha_k$. Then, choosing $\upsilon_{n-(k+1)} = 2\upsilon_{n-k}$, we write

$$\alpha_{n-(k+1)} = \sqrt{\alpha_{n-k}}. \tag{10.85}$$

Let $q_k = \log_2 \alpha_k$. Then, taking log of the above equation, we get

$$q_{n-(k+1)} = \frac{1}{2} q_{n-k}. \tag{10.86}$$

The above equation gives

$$\log \frac{c_{n-(k+1)}}{c_{n-k}} - 1 = \left(\frac{1}{2^k}\right)\left(\log \frac{c_{n-1}}{c_n} - 1\right). \tag{10.87}$$

Let $a = (\log \frac{c_{n-1}}{c_n} - 1)$. Adding $k+1$ such equations, and exponentiating both sides, we get

$$\frac{c_{n-(k+1)}}{c_n} = 2^{k+1} 2^{a \sum_{j=0}^{k} \frac{1}{2^j}}. \tag{10.88}$$

Then we have,

$$\frac{c_{n-(k+1)}}{c_n} = 2^{k+1} 2^{a\left(2-\frac{1}{2^k}\right)}. \tag{10.89}$$

By choosing,

$$\bar{w}_{n-1} = 1 - \frac{c_{n-1}^2}{4}. \tag{10.90}$$

We obtain $c_n = \frac{c_{n-1}^2}{4}$, which gives, $c_n 4^a = 1$, with $\upsilon_n = 1$. Then, we have

$$c_{n-(k+1)} = 2^{k+1} 2^{-\frac{a}{2^k}} \tag{10.91}$$

and

$$\bar{w}_{n-(k+2)} = 2^{k+1}\left(1 - 2^{-\frac{a}{2^k}}\right) = 2^{k+1}\left(1 - \mathrm{e}^{-\ln 2 \frac{a}{2^k}}\right) < \underbrace{2a\ln 2}_{\Delta}, \tag{10.92}$$

where, for $x \geq 0$, we have $\mathrm{e}^{-x} \geq 1 - x$ and $\mathrm{e}^{-x} \leq 1 - x + \frac{x^2}{2}$. Note $\Delta = \ln c_n^{-1}$. Then $w_{n-k} \leq 2^{n-k}\Delta$, for $k \geq 2$. Note $w_{n-1} = 2^{n-1}(1 - c_n)$. We choose $w_n = \frac{2^n}{4}$ and a typical value of $c_n^{-1} = 100$, so that $\alpha_n = .04$.

The term

$$w_n \prod_{k=1}^{n} \cos(\upsilon_k t) = \frac{w_n}{2^n} \sum_{i=1}^{2^n} \cos(f_i t), \tag{10.93}$$

where f_i are the sum and differences of the frequencies υ_k. This choice of the RF-field has rms power that scales like $2^{n+1}\Delta^2$ and an rms amplitude A_e that scales like $2^{\frac{n+1}{2}}\Delta$. Let $A_e = 2^{\frac{n=1}{2}}\Delta$, we define,

$$(\tilde{u}(\tau), \tilde{v}(\tau)) = A_\mathrm{e}^{-1}\left(u\left(A_\mathrm{e}^{-1}\tau\right), v\left(A_\mathrm{e}^{-1}\tau\right)\right),$$

then this control input has rms amplitude 1 and will perform a uniform excitation for $c_0 = \Delta^{-1} 2^{\frac{n-1}{2}}$ in time $\tilde{t}_f \leq \frac{2^{\frac{n-1}{2}} \pi \Delta}{\bar{w}_n}$. The ratio $\frac{c_0}{t_f} \sim \kappa$, a constant, which suggests a linear scaling of time-bandwidth for a constant rms amplitude (Fig. 10.8).

In summary, in this section, we presented a new method for performing broadband rotations on an inhomogeneous spin ensemble with dispersion in their natural

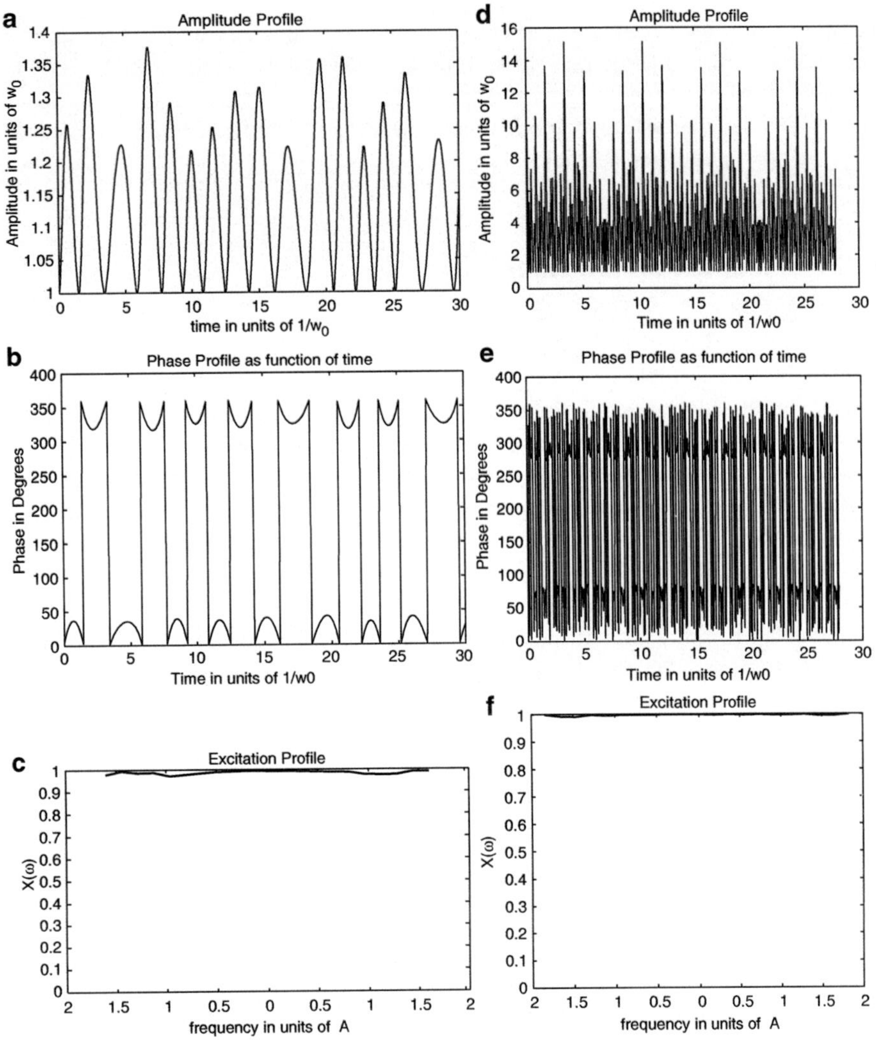

Fig. 10.8 Panel (**a**) shows a segment of the amplitude profile of the RF-field in units of $\frac{1}{w_0}$. Panel (**b**) shows a segment of the phase profile of the RF-field in units of $\frac{1}{w_0}$. Panel (**c**) shows the excitation profile, the x coordinate of the Bloch equation at time $t_f = 2,662/w_0$, starting from $z = 1$ for a range of frequencies expressed in the units of the maximum amplitude of the available RF-field. The control input takes the form expressed in (10.61). We start with an initial $\frac{c_0}{w_0} \sim 2$ and in the final frame $\frac{c_7}{w_7} = 0.04$. We choose $\upsilon_k = 3\upsilon_{k+1}$. The maximum amplitude of the RF-field, $A = 1.38w_0$, while root mean square amplitude $A_{\mathrm{m}} = 1.15w_0$. Panels (**d–f**) depict the same for $\upsilon_k = 2\upsilon_{k+1}$. In this case, $t_f = 27.9/w0$. The maximum amplitude of the RF-field, $A = 15w_0$, while root mean square amplitude $A_{\mathrm{m}} = 5.1w_0$. The excitation profile in units of A_{m} is much broader in this case

frequencies. We derived an upper bound for the error in performing a $\frac{\pi}{2}$ pulse. The simulations show that the performance of the method is far superior than what the bound was when applied to the parameters of the problem would suggest. This opens many new and interesting methodological challenges in understanding the effect of the multiply modulated RF-field.

10.4 Coupled Spin Dynamics

Until now, we have described bilinear control systems that arise in the control of spin $\frac{1}{2}$ or an ensemble of spin $\frac{1}{2}$. A rich class of model control problems arise, when one considers dynamics of two coupled spin $\frac{1}{2}$. The dynamics of two coupled spins forms the basis for the field of quantum information processing and computing and is fundamental in multidimensional NMR spectroscopy experiments as detailed subsequently. Let $|0\rangle$ and $|1\rangle$ represent a choice of the orthogonal basis for the Hilbert space of state of the spin $\frac{1}{2}$, for example, the eigenstates of the operator σ_z, with eigenvalues $\frac{1}{2}$ and $-\frac{1}{2}$, respectively. The joint Hilbert space of the coupled spin system is the tensor product of the individual one of these. A possible choice of the basis for the joint Hilbert space is the tensor product of basis for each individual space (also termed the product operator basis), and we represent these basis as $|00\rangle$, $|01\rangle$, $|10\rangle$, and $|11\rangle$, where

$$|00\rangle = \begin{bmatrix} 1 \\ 0 \end{bmatrix} \otimes \begin{bmatrix} 1 \\ 0 \end{bmatrix} = \begin{bmatrix} 1 \\ 0 \\ 0 \\ 0 \end{bmatrix}.$$

An arbitrary vector in this space takes the form

$$a|00\rangle + b|01\rangle + c|10\rangle + d|11\rangle. \tag{10.94}$$

Not all the vectors in the joint Hilbert space can be written as the tensor product $|\phi_1\rangle \otimes |\phi_2\rangle$. Vectors that can be decomposed in this way are called separable states and those that cannot are termed entangled states. For example, the states

$$|\psi_A \pm\rangle = \frac{|00\rangle \pm |11\rangle}{\sqrt{2}}, \tag{10.95}$$

$$|\psi_B \pm\rangle = \frac{|01\rangle \pm |10\rangle}{\sqrt{2}} \tag{10.96}$$

are examples of entangled states and are given the special name of the Bell states.

The Hamiltonian for a system of two coupled spins then takes the general form

$$H_0 = \sum a_\mu \sigma_\mu \otimes \mathbf{1} + \sum b_\nu \mathbf{1} \otimes \sigma_\nu + \sum J_{\mu\nu} \sigma_\mu \otimes \sigma_\nu, \tag{10.97}$$

where $\mu, \nu \in \{x, y, z\}$. The Hamiltonians $\sigma_\mu \otimes \mathbf{1}$ and $\mathbf{1} \otimes \sigma_\nu$ are termed local Hamiltonians and the Hamiltonian

$$H_c = \sum c_{\mu\nu} \sigma_\mu \otimes \sigma_\nu, \tag{10.98}$$

the coupling or interaction Hamiltonian. The local Hamiltonians only operate on one of the spins. For example, $\sigma_\mu \otimes \mathbf{1}$ only transforms the first spin (labeled as I)

$$\sigma_\mu \otimes \mathbf{1} \, |\phi_1\rangle \otimes |\phi_2\rangle = \left(\sigma_\mu |\phi_1\rangle\right) \otimes |\phi_2\rangle. \tag{10.99}$$

Similarly $\mathbf{1} \otimes \sigma_\mu$ only transforms the second spin (labeled as S).

The following notation is therefore common place in the NMR literature.

$$I_\mu = \sigma_\mu \otimes \mathbf{1} \;\; ; \;\; S_\nu = \mathbf{1} \otimes \sigma_\nu. \tag{10.100}$$

The operators I_μ and S_ν commute and therefore

$$\begin{aligned} \exp(-i\sum_\mu a_\mu I_\mu + \sum_\nu b_\nu S_\nu) &= \exp\left(-i\sum_\mu a_\mu I_\mu\right) \exp\left(-i\sum_\nu b_\nu S_\nu\right) \\ &= \left(\exp\left(-i\sum_\mu a_\mu \sigma_\mu\right) \otimes \mathbf{1}\right) \otimes \left(\mathbf{1} \otimes \exp\left(-i\sum_\nu b_\nu \sigma_\nu\right)\right) \end{aligned} \tag{10.101}$$

and therefore

$$\begin{aligned} \exp\left(-i\sum_\mu a_\mu I_\mu + \sum_\nu b_\nu S_\nu\right) |\phi_1\rangle \otimes |\phi_2\rangle &= \left(\exp(-i\sum_\mu a_\mu \sigma_\mu)|\phi_1\rangle\right) \\ &\quad \otimes \left(\exp\left(-i\sum_\nu b_\nu \sigma_\nu\right)|\phi_2\rangle\right) \end{aligned}$$

implying that the evolution of local Hamiltonians preserves separable states. The unitary transformations of the kind

$$\exp\left(-i\sum_\mu a_\mu \sigma_\mu\right) \otimes \exp\left(-i\sum_\nu b_\nu \sigma_\nu\right)$$

obtained by evolution of the local Hamiltonians are called local unitary transformations.

Entangled states can be generated starting from separable states by letting the coupling Hamiltonian evolve. The coupling Hamiltonian can be written as

$$H_c = \sum J_{\mu\nu} I_\mu S_\nu. \tag{10.102}$$

Written explicitly, some of these matrices take the form

$$I_z = \sigma_z \otimes \mathbf{1} = \frac{1}{2}\begin{bmatrix} 1 & 0 & 0 & 0 \\ 0 & 1 & 0 & 0 \\ 0 & 0 & -1 & 0 \\ 0 & 0 & 0 & -1 \end{bmatrix} \tag{10.103}$$

and

$$I_z S_z = \sigma_z \otimes \sigma_z = \frac{1}{2}\begin{bmatrix} 1 & 0 & 0 & 0 \\ 0 & -1 & 0 & 0 \\ 0 & 0 & -1 & 0 \\ 0 & 0 & 0 & 1 \end{bmatrix}. \tag{10.104}$$

The 15 operators,

$$-i\{I_\alpha, S_\beta, I_\alpha S_\beta\}$$

for $\alpha, \beta \in \{x,y,z\}$, form the basis for the Lie algebra $\mathfrak{g} = su(4)$, the 4×4, skew Hermitian matrices. For the coupled two spins, the generators $-iH_0, -iH_j \in su(4)$ and the evolution operator $U(t)$ in (10.4) is an element of $SU(4)$, the 4×4, unitary matrices of determinant 1. The density matrix for a two spin system is then a 4×4 Hermitian matrix with trace 1 and can be written as

$$\rho = \frac{\mathbf{1}}{4} + \sum_\mu a_\mu I_\mu + \sum_\nu b_\nu S_\nu + \sum_{\mu\nu} J_{\mu\nu} I_\mu S_\nu. \tag{10.105}$$

It is customary to omit $\mathbf{1}$ in the formula for (10.105), as it does not transform under a unitary transformation. The various terms in the decomposition of the density matrix have a special meaning. A density matrix $\rho_I = \frac{\mathbf{1}}{4} + \alpha_I I_z$ corresponds to the state of the spin ensemble, where there are excess of spins I oriented along the z axis, the B_0 field direction, while there is no preferred orientation for spin S. Similarly, a density matrix $\rho_S = \frac{\mathbf{1}}{4} + \alpha_S S_z$ corresponds to the state of the spin ensemble, when there are excess of spins S oriented along the B_0 field direction, while there is no preferred orientation for spins I.

Numerous experiments in NMR spectroscopy involve synthesizing unitary transformations that require interaction between the spins (evolution of the coupling Hamiltonian). These experiments involve transferring, for example, the initial state of the spin ensemble represented by a density operator of the kind ρ_I to a final density operator of the kind ρ_S and involves evolution of interaction Hamiltonians. Such transfer experiments are used to improve the sensitivity of the measurement and will be discussed subsequently. Similarly, many protocols in quantum communication and information processing [36] involve synthesizing entangled states as in (10.95) starting from the separable states. This again requires evolution of interaction Hamiltonians between the spins.

A typical feature of many of these problems is that evolution of interaction Hamiltonians takes significantly longer than the time required to generate local unitary transformations. Local unitary transformations on spins are obtained by application of RF-pulses, whose strength may be orders of magnitude larger than the couplings between the spins. Given the unitary evolution

$$\dot{U} = -i\left[H_c + \sum_{j=1}^{n} u_j H_j\right] U,\ \ U(0) = I, \tag{10.106}$$

where H_c represents a coupling Hamiltonian as in (10.98), we ask what is the minimum time required to synthesize any unitary transformation in the coupled spin system, when the control generators H_j are local Hamiltonians and are much stronger than the coupling between the spins. Design of time optimal RF-pulse sequences is an important research subject in NMR spectroscopy and quantum information processing as minimizing the time to execute quantum operations can reduce relaxation losses which are always present in an open quantum system as described in Sect. 10.2.1. This is the problem of time optimal control of bilinear control systems, as in (10.2), evolving on compact Lie groups. The present problem has a special mathematical structure that helps to characterize all the time optimal trajectories [10, 34]. The special mathematical structure manifested in the coupled two spin system motivates a broader study of control systems with the same properties.

The Lie algebra $\mathfrak{g} = su(4)$ has a decomposition $\mathfrak{g} = \mathfrak{p} \oplus \mathfrak{k}$, where

$$\mathfrak{k} = -i\{I_\mu, S_\nu\},\ \ \mathfrak{p} = -i\{I_\mu S_\nu\}. \tag{10.107}$$

Here $\mathfrak{k}$ is a subalgebra of $\mathfrak{g}$ made from local Hamiltonians. It is easy to verify that

$$[\mathfrak{k},\mathfrak{k}] \subset \mathfrak{k},\quad [\mathfrak{k},\mathfrak{p}] \subset \mathfrak{p},\quad [\mathfrak{p},\mathfrak{p}] \subset \mathfrak{p}. \tag{10.108}$$

This decomposition of a real semi-simple Lie algebra $\mathfrak{g} = \mathfrak{p} \oplus \mathfrak{k}$ satisfying (10.108) is called the Cartan decomposition of the Lie algebra $\mathfrak{g}$ [35].

The coupling Hamiltonian $-iH_c \in \mathfrak{p}$ in (10.106), while the control Hamiltonians $-iH_j \in \mathfrak{k}$. We will assume that the Lie algebra generated by the control terms $-iH_j$ span the whole $\mathfrak{k}$, i.e., $\{-iH_j\}_{LA} = \mathfrak{k}$. Under this assumption, a computation shows that the system in (10.106) is controllable for any $-iH_c \in \mathfrak{p}$. Let $K = \exp(\mathfrak{k})$. We assume that control amplitudes are unbounded a priori, and any element of the subgroup K of transformations can be synthesized in arbitrarily small time. This is typical and will be elaborated in the context of NMR applications and quantum information processing, where any local unitary transformation can be produced in negligible time compared to the evolution of the couplings.

The Cartan decomposition of the Lie algebra $\mathfrak{g}$ in (10.108) leads to the decomposition of the Lie group G [35]. Let $\mathfrak{a}$ denote the largest abelian subalgebra contained inside $\mathfrak{p}$. Then, any arbitrary element of the group $G = SU(4)$ can be written as

$$G = K_1 \exp(a_1) K_2, \tag{10.109}$$

where $K_1, K_2 \in K$, and $a_1 \in \mathfrak{a}$. Furthermore, the Cartan decomposition entails that for $-iH_c \in \mathfrak{p}$, and $K_1 \in K$, we have $\underbrace{K_1(-iH_c)K_1^\dagger}_{Ad_{K_1}(-iH_c)} \in \mathfrak{p}$.

Example 10.1. For $\mathfrak{g} = su(4)$, as in (10.107), one choice of $\mathfrak{a}$ is

$$\mathfrak{a} = -i\{I_\alpha S_\alpha\}; \ \alpha \in \{x,y,z\}. \tag{10.110}$$

Note $\mathfrak{a}$ is three dimensional. Then, any arbitrary element of any element $U \in SU(4)$ can then be written explicitly as

$$G = \underbrace{\exp\left(-i\sum_\mu c_\mu I_\mu + \sum_\nu d_\nu S_\nu\right)}_{K_1} \exp\left(-i\sum_\alpha J_\alpha I_\alpha S_\alpha\right) \underbrace{\exp\left(-i\sum_\mu a_\mu I_\mu + \sum_\nu b_\nu S_\nu\right)}_{K_2} \tag{10.111}$$

for appropriate choice of coefficients a_μ, b_ν, c_μ, d_ν, J_α etc.

Example 10.2. For $\mathfrak{g} = su(n)$ and $\mathfrak{k} = so(n)$, and $\mathfrak{p} = -iA$, where A is a traceless symmetric matrices, the decomposition $\mathfrak{g} = \mathfrak{p} \oplus k$ is a Cartan decomposition. Let $\mathfrak{a}$ be the space of all traceless diagonal matrices, where

$$\mathfrak{a} = \left\{ -i \begin{bmatrix} \lambda_1 & 0 & 0 & 0 & 0 \\ 0 & \lambda_2 & 0 & 0 & 0 \\ 0 & 0 & \ddots & 0 & 0 \\ 0 & 0 & 0 & \lambda_{n-1} & 0 \\ 0 & 0 & 0 & 0 & \lambda_n \end{bmatrix} \right\}.$$

Then any element of $U \in SU(n)$ can be written as $U = K_1 \exp(D) K_2$, where $K_1, K_2 \in SO(n)$ and $D \in \mathfrak{a}$ is a diagonal matrix as above.

Theorem 10.1. [10] For the control system in (10.106), all the elements of G that can be reached starting from $U(0) = I$ in time $T > 0$, denoted as $R(I,T)$, are characterized by its closure as

$$\bar{R}(I,T) = K_1 \exp\left(T \sum_j \alpha_j Z_j\right) K_2,$$

where $K_1, K_2 \in K$, $\alpha_j \geq 0$ and $\sum_j \alpha_j = 1$ and $Z_j \in Ad_K(-iH_c) \cap \mathfrak{a}$. The points Z_j are called the Weyl points, the set of points where the orbit $Ad_K(-iH_c)$ intersects the Cartan subalgebra $\mathfrak{a}$. Let $\mathfrak{c}(-iH_c)$ denote the convex hull of the Weyl points $Ad_K(-iH_c) \cap \mathfrak{a}$, then the reachable set can also be written as

$$\bar{R}(I,T) = K_1 \exp(T\mathfrak{c}(-iH_c)) K_2. \tag{10.112}$$

Remark 10.2. In essence, the KAK decomposition of the group G allows us to write any $U \in G$ as $U = K_1 \exp(Y) K_2$ with $Y \in \mathfrak{a}$ and the minimum time T to synthesize U is to find the smallest time T such that Y/T lies in the convex hull of the Weyl points Z_j. Given that T is the minimum time such that $Y/T = \sum_j \alpha_j \underbrace{Ad_{K_j}(-iH_c)}_{Z_j}$, with $\alpha_j > 0$, $\sum_j \alpha_j = 1$ and $\underbrace{Ad_{K_j}(-iH_c)}_{Z_j} \in \mathfrak{a}$, we synthesize $\exp(Y)$ as

$$\exp(Y) = \prod_{j=1}^{n+1} K_j \exp(-iH_c t_j) K_j^{\dagger}, \tag{10.113}$$

where $K_j \in K$, and therefore take negligible time to synthesize. The optimal trajectory consists of a sequence of fast control rotations, interspersed with the periods of free evolution.

Example 10.3. In example 2, the Weyl points Z_j are

$$-i \begin{bmatrix} \lambda_{\sigma(1)} & 0 & 0 & 0 & 0 \\ 0 & \lambda_{\sigma(2)} & 0 & 0 & 0 \\ 0 & 0 & \ddots & 0 & 0 \\ 0 & 0 & 0 & \lambda_{\sigma(n-1)} & 0 \\ 0 & 0 & 0 & 0 & \lambda_{\sigma(n)} \end{bmatrix},$$

the various permutations of the eigenvalues of H_c. The closure of the reachable set in time T is all matrices of the form

$$K_1 \exp\left(-i \begin{bmatrix} \mu_1 & 0 & 0 & 0 & 0 \\ 0 & \mu_2 & 0 & 0 & 0 \\ 0 & 0 & \ddots & 0 & 0 \\ 0 & 0 & 0 & \mu_{n-1} & 0 \\ 0 & 0 & 0 & 0 & \mu_n \end{bmatrix} \right) K_2,$$

where $\mu = (\mu_1, \mu_2, \ldots, \mu_n)'$ satisfies $\mu \prec T\lambda$, where $\lambda = (\lambda_1, \lambda_2, \ldots, \lambda_n)'$, and the symbol $\prec$ stands for majorization, i.e., μ lies in convex hull of the vector $T\lambda$ and its various permutations.

Example 10.4. For the coupled spins, as in example 1, if we choose $\mathfrak{a} = -i\{I_\alpha S_\alpha\}$, then the Weyl points Z_j have the form $c_x I_x S_x + c_y I_y S_y + c_z I_z S_z$, where

$$(c_x, c_y, c_z) \in \left\{ \varepsilon_1 c_{\sigma(1)}, \varepsilon_2 c_{\sigma(2)}, \varepsilon_3 c_{\sigma(3)} | \varepsilon_i = \pm 1, \prod_i \varepsilon_i = 1 \right\},$$

where $(c_{\sigma(1)}, c_{\sigma(2)}, c_{\sigma(3)})$ are various permutations of (c_1, c_2, c_3), and $c_1 \geq c_2 \geq |c_3|$. Then the reachable set for the system in time T in example 1 is now described by Theorem 10.1 and can be explicitly written as

$$\bar{R}(I,T) = K_1 \exp(b_1 I_x S_x + b_2 I_y S_y + b_3 I_z S_z) K_2, \tag{10.114}$$

such that $b_1 \geq b_2 \geq |b_3|$, and $b_1 \leq Tc_1$ and $b_1 + b_2 \pm |b_3| \leq T(c_1 + c_2 \pm |c_3|)$. As before $K_1, K_2 \in K = SU(2) \otimes SU(2)$.

Theorem 1 gives a complete characterization of the reachable set for a coupled qubit system. The results derive from the Cartan decomposition of the group $G = SU(4)$ in terms of the associated subgroup $K = SU(2) \otimes SU(2)$, where elements of K can be synthesized in arbitrarily small time. Until now we have only talked about two coupled qubits. Experiments in quantum information processing and NMR spectroscopy involve control of dynamics of multiple coupled spins. For a system of n spin $\frac{1}{2}$, the Hilbert space is 2^n dimensional. Unitary transformation on such a space belong to the group $G = SU(2^n)$. The control Hamiltonians for such a system generate the subgroup $K = SU(2) \otimes SU(2) \ldots SU(2)$, the group of local unitary transformations that affects individual spins. Lie group decompositions such as the KAK decomposition can be used to decompose any unitary transformation $U \in G$ as

$$U = K_{n+1} \exp(-iH_c t_n) \ldots \exp(-iH_c t_1) K_1,$$

where $K_i \in K$ are local rotations and are interspersed with the evolution of the coupling Hamiltonian $-iH_c$ for appropriate time. These decompositions then provide explicit synthesis methods for generating unitary transformations in G.

There are numerous beautiful control problems of efficient synthesis of unitary transformations belonging to $SU(2^n)$, using the coupling Hamiltonian between the spins and the control subgroup K [11]. Finding time optimal control for synthesizing unitary transformations in the big space G can be reduced to problems in Sub-riemannian geometry and have been recently studied [30]. Characterization of time optimal trajectories for multiple spin systems, however, remains largely open.

10.5 Control of Coupled Spin Dynamics in the Presence of Relaxation

Many experiments in coherent spectroscopy and quantum information processing require transfer between different states of coupled spin system. The presence of decoherence arising due to coupling to the environment limits how close the state of a spin system can be driven to a target state. In the previous section, we described problems in time optimal control of coupled spin dynamics with the goal of minimizing decoherence effects by reducing the time to perform

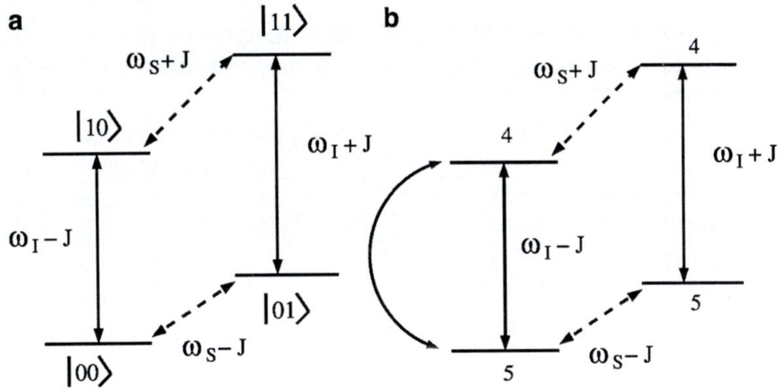

Fig. 10.9 (**a**) Shows the eigenstates of the Hamiltonian $H_0 + H_c$ for the two spin system as in (10.115). The energies are in frequency units. (**b**) Corresponds to an ensemble where there are excess of spin I oriented along the B_0 field direction. The populations in various states are shown below the energy bars [1]

quantum operations. In this section, we describe some problems of optimal design of trajectories of coupled spin evolution so that they suffer minimum decoherence loss. We show that by exploiting explicit models for decoherence, represented by Lindblad operators as described in (10.33), it is possible to design trajectories of the coupled spin system so that they suffer minimum decoherence loss [31–33].

We consider a coupled spin system consisting of spin I and S. The Hamiltonian for the spin system takes the form

$$H(t) = \underbrace{2\pi\nu_I I_z + 2\pi\nu_S S_z}_{H_0} + \underbrace{\pi J 2 I_z S_z}_{H_c} + \underbrace{2\pi A\cos(\omega t + \theta(t))}_{u_1(t)}\,\underbrace{(I_x + S_x)}_{H_1}\,. \quad (10.115)$$

The first two terms of H_0 represent energy of the spins I and S in a static magnetic field along the z direction. The term $2I_zS_z$ corresponds to the interaction Hamiltonian, which gives a positive contribution when spins are oriented alike and negative contribution when the spins are oriented opposite to each other. The control consists of an oscillating magnetic field along the x direction, whose amplitude, frequency, and phase, given by $A(t)$, ω, and ϕ, can be varied. In these experiments, $J \ll A \ll \nu_I, \nu_S$. Typical values of J and A are in Hz and kHz, respectively, while ν_I and ν_S at B_0 field strength of Tesla are hundreds of megahertz.

The eigenstate of the Hamiltonian are the product operator basis $|00\rangle$, $|01\rangle$, $|10\rangle$, and $|11\rangle$, where $|0\rangle$ and $|1\rangle$ are eigenstates of σ_z, with eigenvalues $\frac{1}{2}$ and $-\frac{1}{2}$. Therefore, the energies of these eigenstates are $-\frac{(\nu_I+\nu_S-J/2)}{2}$, $\frac{(\nu_I-\nu_S-J/2)}{2}$, $\frac{(-\nu_I+\nu_S-J/2)}{2}$, and $\frac{(\nu_I+\nu_S+J/2)}{2}$. These energies are depicted in the energy level diagram in Fig. 10.9a. Observe that the difference in the energies of the $|1\rangle$ and $0\rangle$ states of spin I depend on whether the S spin is in $|0\rangle$ or $|1\rangle$ state and corresponds to transitions I and II in Fig. 10.9a. The corresponding energies are $\nu_I - \frac{J}{2}$ and $\nu_I + \frac{J}{2}$.

Therefore, if one performs an NMR experiment as described earlier in Fig. 10.3 on a spin ensemble of coupled spin I and S where I spins have the Larmor frequency ω_{I}, then one observes two resonances, one at $\nu_I - \frac{J}{2}$ and one at $\nu_I + \frac{J}{2}$.

Figure 10.9a shows the state of a spin ensemble with the population in different states written below the energy bar. The proportion of the ensemble when spin I is in the state $|0\rangle$, vs when the spin I is in the state $|1\rangle$, is $5/4$, while the ensemble has equal number of spin S in $|0\rangle$ and $|1\rangle$ states. Writing down a density matrix for this system then gives,

$$\rho = \frac{5}{18}|00\rangle\langle 00| + \frac{5}{18}|01\rangle\langle 01| + \frac{4}{18}|10\rangle\langle 10| + \frac{4}{18}|11\rangle\langle 11|. \tag{10.116}$$

We obtain

$$\rho = \frac{1}{4}\mathbf{1} + \frac{1}{18}I_z,$$

which signifies that we have an ensemble of spins with an excess of spins I oriented along the z direction. An important experiment in NMR spectroscopy is to synthesize unitary transformations that will transform an ensemble of the kind

$$\frac{1}{4}\mathbf{1} + \alpha_{\mathrm{I}} I_z + \alpha_{\mathrm{S}} S_z,$$

where $\alpha_{\mathrm{I}} > \alpha_{\mathrm{S}}$ into an ensemble that looks like $\frac{1}{4}\mathbf{1} + \alpha_{\mathrm{S}} I_z + \alpha_{\mathrm{I}} S_z$. If the gyromagnetic ratio $\gamma_{\mathrm{I}} > \gamma_{\mathrm{S}}$, then in thermal equilibrium spins I are more polarized than spin S and therefore $\alpha_{\mathrm{I}} > \alpha_{\mathrm{S}}$. By transforming the ensemble so that more of the spins S get more polarized compared to their equilibrium state, it is possible to improve the sensitivity of NMR experiments that determine the Larmor frequency of spins S. This experiment is called the transfer of polarization experiment. To make matters more transparent, we assume $\alpha_{\mathrm{S}} = 0$ and drop the factor $\frac{1}{4}\mathbf{1}$ as this part of the density matrix does not transform under rotations. We consider operations that will transform the spin ensemble from the initial state

$$I_z \to S_z. \tag{10.117}$$

One method for performing this manipulation is to first perform a rotation on spin I conditioned on the state of spin S, so that $|10\rangle \leftrightarrow |00\rangle$, while $|01\rangle$ and $|11\rangle$ is unperturbed. In the language of quantum information processing, this is so-called a controlled not (CNOT) operation, and the corresponding unitary transformation denoted U_{cnot} inverts the state of spin I, conditioned on state of spin S being $|0\rangle$. This is depicted by arc I in Fig. 10.9b. Now, we can perform a CNOT operation on spin S, such that the S spin is inverted if spin I is 0. As a result of the first CNOT operation, the ensemble in (10.116) transforms to

$$\rho = \frac{5}{18}|10\rangle\langle 10| + \frac{5}{18}|01\rangle\langle 01| + \frac{4}{18}|00\rangle\langle 00| + \frac{4}{18}|11\rangle\langle 11| = \frac{1}{4}\mathbf{1} + \frac{1}{18}2I_z S_z. \tag{10.118}$$

As a result of the second CNOT operation, the ensemble in (10.118) transforms to

$$\rho = \frac{5}{18}|10\rangle\langle 10| + \frac{5}{18}|00\rangle\langle 00| + \frac{4}{18}|01\rangle\langle 01| + \frac{4}{18}|11\rangle\langle 11| = \frac{1}{4}\mathbf{1} + \frac{1}{18}S_z. \quad (10.119)$$

Another operation of fundamental importance in quantum information processing is to transform a separable state $|00\rangle$ to an entangled state of the form $\frac{|00\rangle+|11\rangle}{2}$. Entangled states are useful resources in many quantum information processing protocols. One mechanism of performing such an operation is to start with an initial state $|00\rangle$ and transform it as

$$|00\rangle \rightarrow |0\rangle \frac{|0\rangle + |1\rangle}{\sqrt{2}}.$$

Such a transformation simply involves doing a local unitary transformation of the type $\exp(-i\frac{\pi}{4}\mathbf{1} \otimes \sigma_x)$ and can be obtained by evolution of local Hamiltonians as described in the previous section. Such operations are significantly faster than the evolution of the coupling Hamiltonians. Now, by performing the unitary transformation, U_{cnot}, on this state, where the state of the first spin I is inverted conditioned on the state of the spin S, such that $|00\rangle \rightarrow |00\rangle$ and $|01\rangle \rightarrow |11\rangle$, we obtain

$$|0\rangle \frac{|0\rangle + |1\rangle}{\sqrt{2}} \rightarrow \frac{|00\rangle + |11\rangle}{2}.$$

In the presence of decoherence or dissipation in the system, the desired transfer cannot be performed with complete fidelity. Interesting optimal control problems arise with the goal of maximizing the fidelity of the desired transformations in the presence of decoherence as described subsequently. The control system describing the transfer is obtained by first writing the Schröedinger equation of the coupled spin system in terms of the product basis $|00\rangle$, $|01\rangle$, $|10\rangle$, and $|11\rangle$. This gives us that

$$\frac{\mathrm{d}}{\mathrm{d}t}\begin{bmatrix}\psi_1\\ \psi_2\\ \psi_3\\ \psi_4\end{bmatrix} = \frac{-i}{2}\begin{bmatrix} -\omega_I - \omega_S + J & u & u & 0 \\ u & -J + \omega_S - \omega_I & 0 & u \\ u & 0 & \omega_I - \omega_S + J & u \\ 0 & u & u & \omega_I + \omega_S + J \end{bmatrix}\begin{bmatrix}\psi_1\\ \psi_2\\ \psi_3\\ \psi_4\end{bmatrix}. \quad (10.120)$$

We add decoherence into our system model by introducing fluctuations into the system Hamiltonian $H(t)$ in (10.120). The resulting density matrix equation then takes the form

$$\dot{\rho} = -i[H(t) + f_1(t)I_z + f_2(t)S_z + f_3(t)I_zS_z, \rho], \quad (10.121)$$

where $f_1(t)$, $f_2(t)$, and $f_3(t)$ are assumed to be uncorrelated fluctuations such that $E[f_i(t+\tau)f_j(t)] = \delta_{ij}\delta(\tau)k_i$. This captures the fact that various terms contribute to the Hamiltonian fluctuating. This leads to the master equation

$$\dot{\rho} = -i[H(t),\rho] + \underbrace{k_1[iI_z[iI_z,\rho] + k_2[iS_z[iS_z,\rho] + k_3[iI_zS_z[iI_zS_z,\rho]}_{L(\rho)}. \tag{10.122}$$

Now, by choosing $u(t) = 2\cos\omega_I t$, where ω_I is the resonance frequency of spin I, and transforming into a rotating frame described by taking the density matrix

$$\rho(t) \to \exp(iH_0t)\rho(t)\exp(-iH_0t),$$

we obtain that

$$\dot{\rho} = -i[2JI_zS_z + \underbrace{A\cos\phi(t)}_{u(t)}I_x + \underbrace{A\sin\phi(t)}_{v(t)}I_y,\ \rho] + L(\rho). \tag{10.123}$$

We can rewrite the corresponding density equation as

$$\frac{\mathrm{d}}{\mathrm{d}t}\begin{bmatrix} z_1 \\ y_1 \\ x_1 \\ x_2 \\ y_2 \\ z_2 \end{bmatrix} = \begin{bmatrix} 0 & u & -v & 0 & 0 & 0 \\ -u & -k & -J & 0 & 0 & 0 \\ v & J & -k & 0 & 0 & 0 \\ 0 & 0 & 0 & -k & -J & v \\ 0 & 0 & 0 & J & -k & -u \\ 0 & 0 & 0 & -v & u & 0 \end{bmatrix}\begin{bmatrix} z_1 \\ y_1 \\ x_1 \\ x_2 \\ y_2 \\ z_2 \end{bmatrix}, \tag{10.124}$$

where (x_1,y_1,z_1) is a Bloch vector associated with the two level system $|00\rangle$ and $|01\rangle$. Similarly, (x_2,y_2,z_2) is the Bloch vector associated with the two level system $|10\rangle$ and $|11\rangle$.

The goal is then to synthesize $u(t)$ and $v(t)$ that transfer

$$\begin{bmatrix} 1 \\ 0 \\ 0 \\ 0 \\ 0 \\ 1 \end{bmatrix} \to \begin{bmatrix} -1 \\ 0 \\ 0 \\ 0 \\ 0 \\ 1 \end{bmatrix}.$$

This would correspond to the selective inversion of the transition I in Fig. 10.9a.

We reexpress the above equations with coordinates $Z_1 = \frac{z_1+z_2}{2}$ and $Z_2 = \frac{z_1-z_2}{2}$. Similarly, we define X_1, X_2, Y_1, Y_2. Then, we obtain the following control system:

$$\frac{d}{dt}\begin{bmatrix} Z_1 \\ X_1 \\ X_2 \\ Z_2 \end{bmatrix} = \begin{bmatrix} 0 & -u(t) & 0 & 0 \\ u(t) & -k & -J & 0 \\ 0 & J & -k & v(t) \\ 0 & 0 & -v(t) & 0 \end{bmatrix} \begin{bmatrix} Z_1 \\ X_1 \\ X_2 \\ Z_2 \end{bmatrix}. \tag{10.125}$$

Now, the goal is to steer the above system from $(1,0,0,0)'$ to $(0,0,0,\eta)'$, maximizing the value of η. Here the parameters J and k represents the coupling and relaxation in the system. Equation (10.125) represents a typical problem in the control of quantum systems in the presence of decoherence where one requires natural dynamics, represented by the parameter J, to steer the system between points of interest and the natural dynamics is dissipative, as represented by the parameter k. When the strength J is comparable to the parameter k, one necessarily dissipates, resulting in $\eta < 1$. Since the controls can be made much larger than the natural parameters in the system, we define $r_1 = \sqrt{Z_1^2 + X_1^2}$, $r_2 = \sqrt{Z_1^2 + X_1^2}$, $\tan\theta_1 = \frac{Z_1}{X_1}$, and $\tan\theta_2 = \frac{Z_2}{X_2}$. Writing an equation for r_1 and r_2 gives us

$$\frac{d}{dt}\begin{bmatrix} r_1 \\ r_2 \end{bmatrix} = \begin{bmatrix} -ku_1^2 & -Ju_1u_2 \\ Ju_1u_2 & -ku_2^2 \end{bmatrix}, \tag{10.126}$$

where $u_1(t) = \cos\theta_1(t)$ and $u_2(t) = \cos\theta_2(t)$. The goal is that for $0 \le u_1(t), u_2(t) \le 1$, find the maximum possible transfer to the final state r_2, starting from the initial state $(r_1, r_2) = (1,0)$. Now, this problem can be solved by direct application of the maximum principle.

Let (λ_1, λ_2) represent the costate variable for the system in (10.126). Along the optimal trajectory, the Hamiltonian

$$H(u_1,u_2) = \begin{bmatrix} \lambda_1 & \lambda_2 \end{bmatrix} \begin{bmatrix} -ku_1^2 & -Ju_1u_2 \\ Ju_1u_2 & -ku_2^2 \end{bmatrix} \begin{bmatrix} r_1 \\ r_2 \end{bmatrix}$$

should be maximized. The Hamiltonian can then be written as

$$H(u_1,u_2) = \begin{bmatrix} u_1 & u_2 \end{bmatrix} \underbrace{\begin{bmatrix} -k\lambda_1 r_1 & J\frac{\lambda_2 r_1 - \lambda_1 r_2}{2} \\ J\frac{\lambda_2 r_1 - \lambda_1 r_2}{2} & -k\lambda_2 r_2 \end{bmatrix}}_{B} \begin{bmatrix} u_1 \\ u_2 \end{bmatrix}.$$

Then the optimal (u_1^*, u_2^*) should satisfy that $H(u_1^*, u_2^*) = 0$. This then implies that $\det B = 0$ and $B\begin{bmatrix} u_1^* \\ u_2^* \end{bmatrix} = 0$. Then substituting $\det B = 0$ and letting $a = \frac{\lambda_2}{\lambda_1}$, $b = \frac{r_2}{r_1}$ and $\xi = \frac{k}{J}$, we obtain that

$$\sqrt{\frac{b}{a}} = \sqrt{1+\xi^2} - \xi.$$

Now using the condition that $B\begin{bmatrix} u_1^* \\ u_2^* \end{bmatrix} = 0$, implies that $\frac{u_1^*}{u_2^*} = \frac{a-b}{2\xi}$, resulting in

$$\frac{u_2^* r_2}{u_1^* r_1} = \sqrt{1+\xi^2} - \xi. \tag{10.127}$$

The optimal feedback control law for (10.125) entails that $\frac{X_2}{X_1} = \sqrt{1+\xi^2} - \xi$. This policy leads to an optimal value of $\eta = \sqrt{1+\xi^2} - \xi$ in (10.125) and this is the largest possible value of r_2 in (10.127). Infact, it is now straightforward to write down an optimal return function $V(r_1, r_2)$, representing the maximum possible achievable value of r_2 starting from arbitrary value of r_1 and r_2 and it turns out to be

$$V(r_1, r_2) = \sqrt{\eta^2 r_1^2 + r_2^2}.$$

In (10.121), we assumed that fluctuations $f_i(t)$ are uncorrelated. Interesting model systems arise when we consider correlations between various noise mechanisms [32]. Suppose, we assume that $E(f_1(t) f_3(t+\tau)) = k_c \delta(\tau)$, which represents interference effect between noise mechanisms in NMR spectroscopy [32]. The following transfer problem arises in this case, which is of fundamental and practical interest. Given the control system

$$\frac{\mathrm{d}}{\mathrm{d}t}\begin{bmatrix} Z_1 \\ Y_1 \\ X_1 \\ X_2 \\ Y_2 \\ Z_2 \end{bmatrix} = \begin{bmatrix} 0 & u(t) & -v(t) & 0 & 0 & 0 \\ -u(t) & -k_a & -J & 0 & k_c & 0 \\ v(t) & J & -k_a & k_c & 0 & 0 \\ 0 & 0 & k_c & -k_a & -J & v(t) \\ 0 & k_c & 0 & J & -k_a & -u(t) \\ 0 & 0 & 0 & -v(t) & u(t) & 0 \end{bmatrix}\begin{bmatrix} Z_1 \\ Y_1 \\ X_1 \\ X_2 \\ Y_2 \\ Z_2 \end{bmatrix}, \tag{10.128}$$

find optimal $(u(t), v(t))$ such that starting from $(Z_1, Y_1, X_1, X_2, Y_2, Z_2) = (1,0,0,0,0,0)$, what is the largest value of $(0,0,0,0,0,\eta)$?

The above optimal control problem can be solved in closed form. Consider the vectors $l_2 = (X_2, Y_2)$ and $l_1 = (X_1, Y_1)$. The optimal solution is then given by the following two invariants of motion. The ratio

$$\frac{l_2}{l_1} = \sqrt{1+\xi^2} - \xi = \eta; \;\; \xi = \sqrt{\frac{k_a^2 - k_c^2}{k_a^2 + J^2}} \tag{10.129}$$

is maintained constant and the angle between vectors l_2 and l_1 is maintained constant. The maximum transfer of efficiency is then η.

It is worthwhile to point out that researchers in magnetic resonance have developed novel pulse sequences that have improved the transfer described in (10.128); however, the fundamental limits of the transfer described here was not

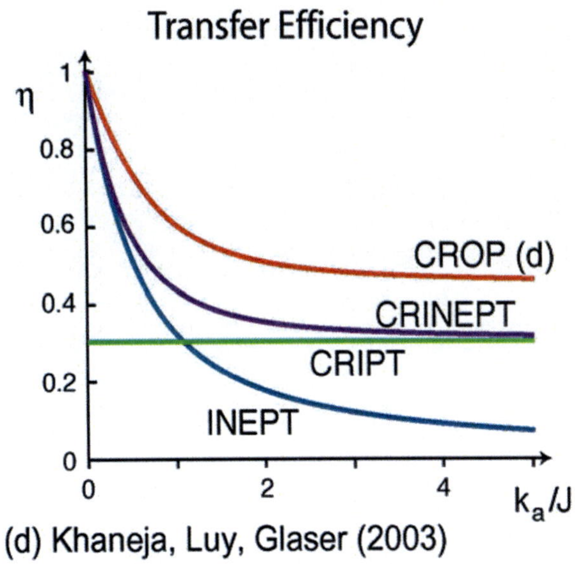

Fig. 10.10 The figure shows efficiency of various state of the art pulse sequences as a function of $\frac{k_a}{J}$ for the transfer in (10.128) for $k_c = 0.75$. The CROP pulse sequences developed using optimal control of system in (10.128) provide the optimal transfer [32]

known. Figure 10.10 shows a plot of transfer efficiency of various state-of-the-art pulse sequences as a function of the ratio $\frac{k_a}{J}$ for $k_c = 0.75$. The CROP pulse sequence obtained by solving the above transfer problem using methods of optimal control performs better than all state-of-the-art methods and provide significant improvement in sensitivity. Furthermore, methods of optimal control help to state limits on how close a quantum dynamical system can be driven to a target state.

In summary, in this section, we provide some concrete examples of state transfer problems involving control of coupled spin dynamics in the presence of decoherence or relaxation. We showed how optimal control of these dissipative bilinear systems can lead to design of better experiments. A systematic study of the controllability and optimal control of problems related to Lindblad equations of open quantum systems is expected to have immediate impact in areas of coherent spectroscopy and quantum information processing.

Acknowledgments N. Khaneja would like to thank Sinbae Kim and Paul Coote for discussions, and acknowledge NSF-0724057, ONR 38A-1077404, and AFOSR FA9550-05-1-0443 for supporting this work. Figure 10.4 is taken from the website "http://lucas.lakeheadu.ca" and "http://chem.mq.edu.au."

References

1. N. Khaneja "On some model problems in quantum control," Commun. Inf. Syst. Volume 9, Number 1, 1-40 (2009).
2. R. R. Ernst, G. Bodenhausen, A. Wokaun, *Principles of nuclear magnetic resonance in one and two dimensions* (Clarendon Press, Oxford, 1987).
3. C. P. Slichter, *Principles of magnetic resonance* (Springer Verlag 1978).
4. A. Abragam, *The principles of nuclear magnetism* (Oxford University Press 1978).
5. K. Wüthrich, *NMR of proteins and nucleic acids* (Wiley & Sons 1986).
6. J. Cavanagh and W. J. Fairbrother and A. G. Palmer and N. J. Skelton, *Protein NMR spectroscopy, principles and practice* (Academic Press, 1996).
7. R. W. Brockett, "System theory on group manifolds and coset spaces," SIAM Journal of Control, **10**: 265-284 (1972).
8. V. Jurdjevic and H. Sussmann, "Control systems on lie groups," Journal of Differential Equations, **12**: 313-329 (1972).
9. R. W. Brockett, "Nonlinear control theory and differential geometry," Proceedings of the International Congress of Mathematicians, 1357-1367, (1984).
10. N. Khaneja, R.W. Brockett and S.J. Glaser, "Time optimal control of spin systems", Phys. Rev. A **63**, 032308 (2001).
11. N. Khaneja and S.J. Glaser, "Cartan decomposition of $SU(2^n)$ and control of spin systems", Chemical Physics **267**, 11-23, (2001).
12. D. D'Alessandro, "Constructive controllability of one and two spin 1/2 particles," *Proceedings 2001 American Control Conference*, Arlington, Virginia, June 2001.
13. V. Ramakrishna, H. Rabitz, M. V. Salapaka, M. Dahleh and A. Peirce, "Controllability of the molecular systems," Phys. Rev. A **51**:960-66 (1995).
14. A. G. Redfield, "The theory of relaxation processes," Adv. Magn. Reson. **1**, 1-32 (1965).
15. G. Lindblad, "On the generators of quantum dynamical semigroups," Commun. Math. Phys. **48**, 199 (1976) .
16. H. J. Metcalf and P. Straten, *Laser cooling and trapping* (Springer 1999).
17. S.E. Sklarz, D.J. Tannor and N. Khaneja, "Optimal control of quantum dissipative dynamics: analytic solution for cooling a three level Lambda system", Phys. Rev. A **69**, 053408 (2004).
18. M. H. Levitt, Prog. NMR Spectrosc. **18**, 61 (1986).
19. M. S. Silver, R. I. Joseph, C.-N. Chen, V. J. Sank, D. I. Hoult, Nature. **310**, 681 (1984).
20. David E Rourke, Ph.D. Thesis (1992).
21. M. Shinnar, S. Eleff, H. Subramanian, J. S. Leigh, Resonance Med. **12**, 74-80 (1989).
22. P. Le Roux, Proc. 7th SMRM 1049 (1988).
23. J.S. Li and N. Khaneja, "Control of inhomogeneous quantum ensembles", Phys. Rev. A, **73**, 030302 (2006).
24. J.S. Li and N. Khaneja, "Ensemble controllability of the Bloch equations" IEEE Conference on Decision and Control, San-Diego (2006).
25. J. S. Li, "Control of inhomogeneous ensembles" Ph.d. Thesis, Harvard University (2006).
26. B. Pryor and N. Khaneja "Fourier synthesis techniques for control of inhomogeneous quantum systems" IEEE Conference on Decision and Control, San Diego, December (2007).
27. T.E. Skinner, T. Reiss, B. Luy, N. Khaneja and S.J. Glaser, J. Magn. Reson., **163**, 8, (2003).
28. K. Kobzar, B. Luy, N. Khaneja, S. J. Glaser, "Pattern pulses: design of arbitrary excitation profiles as a function of pulse amplitude and offset", J. Magn. Reson. **173**, 229-235 (2005).
29. K. Kobzar, T. E. Skinner, N. Khaneja, S. J. Glaser, B. Luy, "Exploring the limits of broadband excitation and inversion pulses", J. Magn. Reson. **170**, 236-243 (2004).
30. N. Khaneja, S.J. Glaser and R.W. Brockett, "Sub-Riemannian geometry and optimal control of three spin systems ", Phys. Rev. A **65**, 032301 (2002).
31. N. Khaneja, T. Reiss, B. Luy, S. J. Glaser, "Optimal control of spin dynamics in the presence of relaxation", J. Magn. Reson. **162**, 311-319 (2003).

32. N. Khaneja, B. Luy, and S.J. Glaser, "Boundary of quantum evolution under decoherence", Proceedings of National Academy of Sciences **100**, no. 23, 13162-66 (2003).
33. N. Khaneja, Jr. Shin Li, C. Kehlet, B. Luy, S.J. Glaser, "Broadband relaxation optimized polarization transfer in magnetic resonance", Proceedings of National Academy of Sciences, USA. **101**, 14742-47 (2004).
34. H. Yuan and N. Khaneja, "Reachable sets of bilinear control system with time varying drift", System and Control Letters, **55**, 501 (2006).
35. S. Helgason *Differential geometry, lie groups, and symmetric spaces* (Academic Press) (1978).
36. M. Nielsen and I. Chuang, *Quantum information and Computation* (Cambridge University Press) (2000).
37. H. Mabuchi and N. Khaneja, "Principles and applications of control in quantum systems", *International Journal of Robust and Nonlinear Control* 15 647-667 (2005).

Chapter 11
Common Threads and Technical Challenges in Controlling Micro- and Nanoscale Systems

Benjamin Shapiro and Jason J. Gorman

The chapters in this book were taken from different groups, each of which has addressed a different topic in the control of miniaturized systems. Yet the broad issues they had to solve turn out to be strikingly similar. We close this book by describing some common threads and technical challenges found throughout the book chapters and in the field in general. This is followed by comments on research directions that we believe are necessary to enable future innovations in this area.

11.1 Common Threads

As editors, one major motivation in assembling this book was to see what unifying themes might emerge when examining a number of different applications involving control at the micro- and nanoscales. It was clear that there would be crosscutting technical challenges, which are discussed in the following section. However, somewhat less expected was the emergence of common threads in how the contributing researchers approached a new problem. These common threads comprise a high-level research philosophy on how to approach new and challenging control problems. Although culled from challenges in controlling micro- and nanoscale systems, the points highlighted below largely apply to any multidisciplinary research that involves control.

B. Shapiro (✉)
Fischell Department of Bioengineering, Institute for Systems Research,
University of Maryland, College Park, MD 20742, USA
e-mail: benshap@umd.edu

J.J. Gorman
Intelligent Systems Division, Engineering Laboratory, National Institute of Standards and Technology, Gaithersburg, MD 20899, USA
e-mail: gorman@nist.gov

J.J. Gorman and B. Shapiro (eds.), *Feedback Control of MEMS to Atoms*,
DOI 10.1007/978-1-4419-5832-7_11, © Springer Science+Business Media, LLC 2012

11.1.1 Picking the Right Problems

All of the chapters in this book present applications where control can have a strong impact such as dramatically improving the performance of atomic force microscopy and enabling fast automated manipulation of nanoscale objects. Thus the first recurring thread through all the chapters is the selection of good problems where control can have a big impact. These problems can either be motivated by current technological needs (e.g., the push to make AFM imaging and manipulation better) or by a vision that control can do something in a new way (e.g., better material growth by *in situ* sensing for online control of reactor parameters).

The problems presented in this book are just a sample. There are other examples of control on small scales, including the control of microchemical reactors (e.g., see [1, 2]), medical implant control (e.g., see [3]), and microscale robots (e.g., see [4]). Control is also emerging in synthetic biology, where the goal is to design novel, robust, and tunable biochemical systems and program them into genes to reprogram living cells [5, 6].

Picking good problems is an art. Success at this stage is dependent on experience, a view of the big picture, and creativity. To build up such a view in a new area, for example, for control theorists interested in learning about miniature systems, we recommend attending high-level seminars far outside one's area of expertise, taking every opportunity to "pick-the-brain" of researchers in disparate fields, and reading widely – doing so is instrumental in uncovering broad needs. Even better is complete immersion in a new environment for a significant period of time – no one we know has ever regretted time spent in this way.

For experts in chemistry, fabrication, optics, microfluidics, or other domains related to miniaturized systems, this book provides some concrete examples of how feedback control can significantly improve capabilities on small scales. In every single case that took significant time. It takes months, sometimes years, for a newcomer to understand what questions to ask. However, once the issues sink in and the creativity starts to work with different tools at hand, the subsequent payoffs can be tremendous.

11.1.2 Model-Based Control

Although the book contributors were selected without any regard for what kind of modeling they use, all of the chapters proceed by either stating a known physical model (e.g., the Schrodinger equation for quantum mechanics) or by deriving one. In each case, control systems were designed using these physical models. Moreover, the modeling was not based on "black-box" system identification (e.g., fitting a neural network to experimental data). Rather, as much as possible, the models were derived from the physics. There are multiple reasons for this, but the simplest one is this: if there is physics-based information (such as Newton's laws or chemical rate equations) that can be translated into "when this actuator is turned on like this, the behavior of the system changes like this," then that information is useful and it permits better design of controllers. Including that known information explicitly through first-principles modeling is more effective than trying to have a computer infer it from limited and noisy data. Simply put, the physical information that is known provides a big advantage for control design, even though it may take months or years to extract, understand, and use that information properly.

Although physical modeling is highly beneficial, the derived models do not have to be perfect. Many of the chapters in this book deal with messy and complex situations where pristine models are not feasible. Yet physical modeling is still a major advantage. Even imperfect models can be used to tell a controller what it should do to make the situation better – feedback then ensures high performance by sensing the errors, choosing actuations to diminish them, and repeating the correction at the next time step to force errors down to near zero. This works even if each correction is imperfect. For dealing with imperfect models, the mathematics of designing effective controllers in the face of modeling errors is known as robust control and is a major field of study within control theory [7].

In some cases, systems exhibit crucial behavior that cannot be fully captured by first-principles modeling, either due to a lack of physical understanding or inherent system complexity (e.g., as in Chaps. 2, 3, 4, 5, 6, and 9). In these cases, it is advantageous to focus fitting and estimation techniques on just the unknown portions of the system behavior and to still codify the remaining known information with a physical first principles model. This leads to a "gray-box" type description, where part of the model is known from the physics and the rest is identified from data.

Regardless of the form of the models created, they must be useful for control design. This means that they must be compatible with available or emerging control design tools. It is of no use to create a system model that contains all the physics and is extremely accurate but is so complex that it requires a super-computer to complete one simulation. More detailed models with a large number of states can be (but are not necessarily) more accurate, but smaller models fit with a wider array of analysis, design, and optimization tools. Roughly speaking, with current computational capabilities and algorithms, linear control design techniques, such as LQR (linear quadratic regulator) control design, can be used on linear or linearized models with up to thousands or tens of thousands of states [7, 8]. But methods for nonlinear control analysis and design [9, 10], such as nonlinear stability analysis and feedback linearization, are only practical for analytical models or models with much fewer states (typically less than tens of states). Molecular dynamic or computational fluid dynamic models, with their millions of states, are too large to be used directly for control analysis or design. The key issue is to strike the right balance. In the words of the 2004 National Science Foundation panel [11], models should be "parsimonious" – the best models include only the most essential elements. When large models originate from physical first principles, for example for computational fluid dynamic or molecular dynamic models, it may be possible to reduce the size of such high-fidelity models while keeping most of their accuracy by using model reduction techniques [12, 13], as was done in Chap. 2. Used wisely, this can be an effective tool for generating small but accurate models of complex systems.

Effective modeling is a necessary part of control design. In new domains, this modeling must be at least initiated by a control expert since they know what types of models are needed for control analysis and design. Once initiated, continued modeling is also usually carried out by the same person. However, information on system physical phenomena is required from a domain expert, such that a tight collaboration between a control scientist and a domain expert is almost always required during the modeling phase. The attitude that control design starts with prerequisite knowledge of a model in the form $\dot{x} = f(x,u)$ and that this model is provided by the domain expert is self defeating. The end result of this thinking is that the control expert will miss a chance to work on something new and interesting and the domain expert will have another example of a situation where control cannot be applied to their problem of interest. Our own experience, and the experience reflected in the book chapters, is that appropriate and high-quality modeling is essential for successful control of micro- and nanoscale systems.

11.1.3 Posing the Right Mathematical Problem for Control

Once a suitable model is available, the next step is to pose the control problem in a form that is tractable using existing, understood, or at least emerging mathematical analyses. Good decisions here can sometimes turn what initially looks like a very difficult task into something solvable. Conversely, it is easy to do this poorly – by not keeping in mind the limitations of control analysis and design tools, or by a desire to attack "the most general problem" – and to arrive at a formulation that cannot be solved. As a concrete example related to control of electrowetting in Chap. 9, consider the task of controlling the shape of liquid droplets by modulating surface tension on their boundaries. Mathematically, this corresponds to control of nonlinear partial differential equations through their moving boundary conditions, and it is beyond the capabilities of current partial-differential-equation (PDE) control methods. But if this problem is rephrased: if it is noted that the dynamic map from the pressure on the boundary to the velocity inside the droplet is linear, that the map from voltage to modulated surface tension pressure is nonlinear but static, and that what is needed is a mapping from a few actuators (those actuators overlaid by the droplet) to a few material points that adequately define the shape, then the problem can be rephrased as a small least squares inversion from surface pressure above the electrodes to velocity at the particles/material points, along with a static inversion of the pressure/voltage nonlinearity. Now the problem can be easily and effectively solved, robustly, and in real-time. These kinds of problem formulation decisions are crucial – by exploiting features of the physics, difficult tasks can turn into tractable problems. So, as in modeling, a careful balance must be struck here between generality (more general problems are harder to solve) and usefulness (a too specific solution will apply to only a small class of problems). Our advice is to initially be motivated by specific applications, as is the case for all the chapters in this book, and then to work up and out from demonstrated specific applications to a wider class of problems.

11.1.4 Experimental Verification

Another common thread that is present in all of the chapters is the essential need for experimental verification. In a new area, like control of micro- and nanoscale phenomena, there are few accepted models and the validity of the existing models is far less certain compared to, for example, models of macroscale mechanical structures. In these cases, it is not sufficient to demonstrate control performance through simulations because one can only have confidence in simulation results when the system models have been experimentally validated over the years. Speaking from personal experience, we have yet to encounter a group of micro- or nanoscale domain experts that will accept simulations as proof of performance. Improved performance must be demonstrated where it counts, in actual working systems.

11.1.5 Communication Across Complementary Fields

The final thread across the chapters is effective cross-disciplinary communication. It is popular to invoke the importance of interdisciplinary collaboration, but the depth of cross-immersion that is necessary to achieve results between two fields is striking. We have found time and again that we can only create working systems once we have deeply understood the needs, physics, numerics, and experiments for that application domain, or, alternately, once each of those areas of expertise is represented in our research team. From reading the chapters of other contributors, we believe this need is universal. Being a controls expert just talking to a MEMS expert or a clinician is almost always not good enough. The clinician must be willing to become part of the team. The reverse is likely also true: being a MEMS expert and talking to a controls expert is insufficient, rather, the microsystems expert or clinician should find the right controls person and make him or her a part of the research team. We hope that this book will add to the impetus to create tight interdisciplinary teams, and to educate the next generation of students, as well as engineers and faculty, in a concrete way – not just in name, but in real effort and significant time spent learning things from new sources and outside disciplines. We know the effort will be worth the rewards.

11.2 Technical Challenges

As seen throughout this book, controlling micro- and nanoscale systems presents some unique and serious challenges in terms of the physics encountered and implementation required. Some of these challenges are a result of scale and are not found in macroscale systems, while others are simply exacerbated by the reduction in size. This section describes some of the most prevalent challenges and limitations.

11.2.1 Noise and Fluctuations

All systems experience noise in one form or another; but at the micro- and nanoscales, noise plays a particularly important role in system behavior. This is because these systems are much closer in scale to the atomic-scale processes that cause noise and they require precision measurements that are more likely to be affected by noise. Thermal noise, which is due to the random interactions between atoms and molecules in gases, liquids, and solids, is one form of noise that is universal. When a particle is suspended in solution, the atoms in the liquid interact with the particle causing Brownian motion [14]. As described in Chap. 6, Brownian motion has a significant effect on the behavior of optically trapped particles. Fluctuations in the particle position limit the manipulation precision and can cause the particle to escape from the trap. Other examples of noise include shot noise in the laser-based beam bounce method used to measure the deflection of AFM cantilevers and Johnson noise in sensor readout electronics. Control issues for noisy systems are well known. Optimal control design tools are appropriate to minimize the influence of noise on the output signal. In most cases, noise in the sensor signal sets the bottom limit on closed-loop resolution. Although there are many optimal control design tools for linear stochastic systems, control theory for nonlinear stochastic systems is currently limited to special cases due to mathematical complexity and remains an open area of research.

11.2.2 Model Uncertainty

Parametric uncertainty is a major issue in micro- and nanoscale systems. As an example, the geometry of micro- and nanofabricated structures generally has much more uncertainty compared to structures machined using conventional macroscale technology. When using a CNC (computer numerically controlled) mill, it is commonplace to machine a 5 mm feature with 25 μm tolerances (tolerance/feature size $= 0.005$). Using standard contact lithography for the microfabrication of MEMS, one can typically fabricate structures with 5 μm features with 0.5 μm tolerances (tolerance/feature size $= 0.1$). Therefore, in this comparison there is 20 times more uncertainty in the microfabricated structures. Similarly, there is far more uncertainty in the material properties of micro- and nanoscale structures compared to bulk properties due to variations in deposition procedures, surface effects, and the breakdown of continuum mechanics as the structures approach nanometer dimensions. As a result of this uncertainty, parameter identification is required in many cases, even when a high-fidelity model of the physics is available. Furthermore, robust control is needed when designing a single controller to be used on many similar devices (e.g., a large batch of MEMS accelerometers) due to such parameter variations.

11.2.3 Precision Sensing

Measuring one or more of a system's state variables is a prerequisite for feedback control at any scale. However, sensing at the micro- and nanoscales is generally more difficult than in macroscale systems. This is partly due to the fact that many of the techniques used to measure macroscale systems do not scale well in terms of dynamic range and resolution. For example, due to the diffraction limit of light, far-field optical measurement techniques often have reduced sensitivity when measuring nanostructures because the focal spot of light is bigger than the measured structure. This is true when using laser-based displacement interferometry for nanoelectromechanical systems [15]. Other issues include difficulty in measuring multiple state variables of a system and coupling between sensors and actuators, both due to the confined area where these measurements take place. Therefore, the challenge is designing a controller that can achieve the desired performance with limited and noncollocated sensing. Given some of the ambitious performance specifications found in micro- and nanosystems (e.g., the MEMS gyroscopes discussed in Chap. 8), this is often not easy.

11.2.4 High-Bandwidth Operation

One straightforward outcome of scaling down to micrometers or nanometers in size is that the mass of the system becomes extremely small. For example, the mass of a 1 μm diameter silica particle is approximately 1 picogram. As a result, systems at these scales are often capable of extremely high accelerations due to their low inertia. From a performance point of view, this means that high-bandwidth operation can be achieved. This has motivated the application of MEMS to a number of problems including hard-disk drive read heads and scanning probe microscopy. This advantage also presents two significant challenges. The root-mean-square (RMS) noise within a system is directly related to the bandwidth of the system when there are white noise sources (e.g., thermal noise, Johnson noise, and shot noise). Therefore, high-bandwidth operation introduces more noise into the system response, which may outweigh the benefits of faster motion. Second, higher bandwidth requires a higher bandwidth control system. As an example, closed-loop nanoscale resonators with resonant frequencies on the order of MHz are so fast that digital signal processing (DSP) and field-programmable gate array (FPGA) controllers are insufficient (e.g., see [16]). Therefore, an all-analog controller implementation is required, which limits tunability and functionality compared to digital controllers.

11.2.5 Surface Forces

Surface forces can dominate at the micro- and nanoscales, which can result in very different behavior than seen at the macroscale. This is because surface forces scale with the area of an object whereas bulk forces, such as gravity and inertia, scale with the volume. Some of the most prevalent surface forces are electrostatic forces, van der Waals forces, surface tension forces, and friction [17, 18]. The action distance of these surface forces varies significantly and is heavily dependent on system geometry and environmental conditions. Surface forces play a major role in the control of micro- and nanoparticle manipulation, as discussed in Chaps.4 and 9. Other examples include micromotor failure due to stiction and the "jump to contact" seen in atomic force microscopy, where the cantilever snaps to a surface during approach because of surface forces. From a controls perspective, surface forces are a challenge due to the speed at which adhesion can occur and the irreversibility of many events driven by surface forces (e.g., a particle stuck to a probe).

11.2.6 Embedded On-Chip Control

It is often desirable for micro- and nanoscale devices, including MEMS/NEMS and micro/nanofluidics, to be stand-alone systems for portability and easy integration into larger systems. As an example, MEMS accelerometers are currently used in a number of hand-held consumer products. Therefore, they must be self-contained and offer functionality that is compatible with existing electronics. This requires embedded control electronics using an application specific integrated circuit (ASIC) based on complementary metal-oxide-semiconductor (CMOS) technology, which presents several additional challenges. Tools for synthesizing controllers under CMOS design constraints and then translating those controllers into CMOS circuit layouts do not yet exist. Also, many micro- and nanoscale actuators require more power and voltage than possible with standard CMOS electronics (e.g., electrostatic MEMS actuators (Chap. 7) require tens of volts but CMOS voltages must typically be below 6 V). Finally, massively parallel MEMS arrays are being used for optical displays (e.g., the Texas Instruments digital light projector (DLP)) and high-density data storage (e.g., the IBM Millipede) and have prospects in many other areas. As these technologies evolve, closed-loop control will be required for every device within an array, resulting in significant complexity in the design and implementation of the controllers. It is expected that as more micro- and nanoscale devices move to market, these system integration issues will become more evident and will require serious research efforts to solve.

11.3 Future Directions

This book has provided an introduction to some of the application domains where control has been demonstrated to have an impact. There are many other examples of control at the micro- and nanoscales and it is safe to say that this field will continue to grow over the next decade. Up to this point most of the research on controlling micro- and nanoscale systems has been application focused, with little cross-fertilization between application domains. However, based on the common threads noted above, there is a large amount of overlap between current and emerging applications. Moving forward it is critical that there be more focus on the big picture that unifies these efforts. We close this book with a few thoughts on where the field of control of miniaturized systems may be going.

At least five of the chapters in this book relate to nanomanufacturing, which closely aligns with current goals in nanotechnology research to move nanoscale research and development from laboratories to the marketplace. Using macroscale manufacturing as an analogy, many manufacturing processes start out as open-loop processes with no methods for correction. However, once a product reaches maturity, there is a much stronger focus on yield, repeatability, reliability, and cost. At this stage, the manufacturing processes are typically reevaluated and closed-loop control is implemented to improve production. Nanomanufacturing is now approaching this phase and, as a result, there are a number of manufacturing applications that can benefit from the control systems perspective. In addition to those already discussed, examples where control can have a large impact include directed self-assembly, dip pen nanolithography, and nanoimprint lithography.

All of the micro- and nanoscale systems discussed in this book have been engineered but there are many such systems found in nature that can equally benefit from control. Research over the last decade in the area of systems biology has striven to provide mathematical formalism to biological sciences. This formalism is a prerequisite to understanding the mechanisms of internal control and finding ways to introduced engineered control into biological systems. With the merging of biotechnology and nanotechnology, and the increasing demand for noninvasive medical diagnostics and treatments, it is clear that controlling micro- and nanoscale biological systems will be a major thrust in the coming decade.

As systems approach the scale of atoms, classical mechanics break down, and quantum mechanics is needed to describe their behavior. In this book, only the nuclear magnetic resonance (NMR) applications discussed in Chap. 10 required a quantum mechanical description. However, it is clear that control engineers will be faced with more and more systems with quantum behavior over the next decade. Quantum computing is one of the biggest drivers because control systems will be needed for preparing quantum bit states and verifying that the proper bits are attained. However, there are many other examples of nanoscale systems that will require quantum-level control, including magnetic resonance force microscopy and quantum communication and encryption. The merging of control theory and implementation with quantum systems is expected to be a growing trend with massive implications in the way we compute, communicate, and further scientific understanding in our world.

This book has largely been application driven because most practical research on controlling micro- and nanoscale systems is still focused on solving specific problems, whether it is improving the performance of an atomic force microscope or an atomic layer deposition process. This is the right way to start. However, based on the common threads and technical challenges discussed earlier and the emerging needs for control in nanomanufacturing, biological systems, and quantum mechanical systems, there is now enough knowledge and momentum to begin to tackle problems at a level higher than a single application. Classes of problems will need to be identified, and then rigorously approached, to maximize such efforts.

For control theorists approaching this subject, a common question is whether new control theory is required to control micro- and nanoscale systems. It is tempting to create grand unified control theory frameworks – these have an air of generality that is satisfying from a mathematical viewpoint. However we know from experience that such frameworks, unless they emerge from real application needs, are rarely useful. Solving specific practical problems is hard enough; control of a general class is even more difficult. The chances that the created framework will be aimed just right (general enough to encompass a variety of applications, simple enough to be tractable, but powerful enough to provide useful results for a class of practical micro- and nanoscale applications) are slim to none unless it is motivated by real-world needs.

A more sensible approach is to first try existing control methods in applications where control is needed, see how they work, and then extend the theory to fill major gaps as necessary. This is the approach that was chosen by the majority of the authors in this book: In Chaps. 3–9 the authors started with existing control theory, adapted and applied it in a micro- and nanoscale setting, and only then began to extend it. In some instances, for example, in Chap. 9, the problems were first recast into a form that allowed standard mathematics to be used.

However, there are cases where it is clear that standard control theory does not suffice. In Chap. 2, existing model reduction and control tools were not sufficient to control nanoparticle synthesis and processing. In Chap. 10, new mathematical structures for infinite dimensional control and control of ensemble systems had to be developed to better manipulate quantum systems. Thus the answer is: new control theory will surely advance control of micro- and nanoscale systems, but its development should be driven by concrete and high-impact applications. We hope this book will help motivate the next generation of researchers who will develop needed theory, and combine it with deep knowledge in applications, to demonstrate the impact that feedback control can have in micro- and nanoscale applications ranging, as the book title says, "from MEMS to atoms."

References

1. M.V. Kothare. Dynamics and control of integrated microchemical systems with applications to micro-scale fuel processing. *Computers and Chemical Engineering*, 30:1725–1734, 2006.
2. B. Chachuat, A. Mitsos, and P.I. Barton. Optimal start-up of microfabricated power generation processes employing fuel cells. *Optimal Control: Applications and Methods*, 31:471–495, 2010.
3. N. Najafi and A. Ludomirsky. Initial animal studies of a wireless, batteryless, MEMS implant for cardiovascular applications. *Biomedical Microdevices*, 6:61–65, 2004.
4. M. Sitti. Miniature devices: Voyage of the microrobots. *Nature*, 458:1121–1122, 2009.
5. E. Andrianantoandro, et al. Synthetic biology: new engineering rules for an emerging discipline. *Molecular Systems Biology*, 2, 2006.0028, 2006.
6. D. Sprinzak and M.B. Elowitz. Reconstruction of genetic circuits. *Nature*, 438:443–448, 2005.
7. K. Zhou, J.C. Doyle, and K. Glover. *Robust and optimal control*, Prentice Hall, Englewood Cliffs, NJ, 1996.
8. P. Dorato, C.T. Abdallah, and V. Cerone. *Linear quadratic control: An introduction*, Krieger Publishing, Malabar, FL, 2000.
9. A. Isidori. *Nonlinear control systems*, Springer, New York, 1995.
10. H.J. Marquez. *Nonlinear control systems: Analysis and design*, Wiley-Interscience, Hoboken, NJ, 2003.
11. B. Shapiro. *NSF workshop on control and system integration of micro- and nano-scale systems*, final report, 2004. Available: http://www.isr.umd.edu/CMN-NSFwkshp/
12. G. Obinata and B.D.O. Anderson. *Model reduction for control system design*, Springer, London, 2001.
13. P.D. Christofides. *Nonlinear and robust control of PDE systems: Methods and applications to transport-reaction processes*, Birkhauser, Boston, MA, 2001.
14. R.M. Mazo. *Brownian motion: Fluctuations, dynamics and applications*, Oxford, New York, 2002.
15. D.W. Carr, L. Sekaric, and H.G. Craighead. Measurement of nanomechanical resonant structures in single-crystal silicon. *Journal of Vacuum Science & Technology B*, 16:3821–3824, 1998.
16. X.L. Feng, et al. A self-sustaining ultrahigh-frequency nanoelectromechanical oscillator. *Nature Nanotechnology*, 3:342–346, 2008.
17. J.N. Israelachvili. *Intermolecular and surface forces*, Academic Press, London, 1991.
18. R.S. Fearing. Survey of sticking effects for micro parts handling. *Proceedings of the IEEE/RSJ International Conference on Intelligent Robots and Systems*, Pittsburgh, PA, 212–217, 1995.

Index

A

Accelerometers
  capacitive, 194–197
  tunneling, 197–199
Acousto-optic deflector (AOD)
  critical performance parameters, 153
  operating principles, 152, 154
Aerosol reactors, nonlinear control
  closed-loop profiles, 31–33
  steady-state profile, 30, 31
  summarization, 29
Aerosol synthesis, 11–12
AFM. *See* Atomic force microscope (AFM)
AFM-based nanomanipulation. *See* Force-based 2-D nanoparticle manipulation
AFM non-contact-mode images, gold nanoparticles, 88, 89
Angle random walk (ARW), 223, 258–259
Angle white noise (AWN), 223, 258
Angular rate estimate, 256
Assembly algorithm, 101
Atomic force microscope (AFM)
  beam-bounce method, 110
  cantilever illustration, 109, 110
  dynamic mode operation
    amplitude modulation, 135–136
    cantilever model parameters, 134
    cantilever-sample system, 136
    challenges, 134–135
    feedback interconnection, 131–133
    transient force atomic force microscopy, 135
  force-balance principle, 112
  forces, 110, 111
  nanopositioning systems, control design
    bandwidth, 120–121
    2DOF control, 128–131
    optimal control framework, 121–125
    performance criteria and limitations, 118–120
    positioning system, 117
    resolution, 120–121
    robustness, 120–121
    ultrahigh resolution, 125–128
  operational principles
  contact mode imaging, 114–115
  dither piezo, 112–113
  dynamic modes, 115–117
  noise sources, 113–114
  position control, MEMS, 179
  scanning tunneling microscope, 109, 110
  uses, 111–112
Automatic gain control (AGC), 237
AWN. *See* Angle white noise (AWN)

B

Back-focal-plane detection method, 155, 156
Batch protein crystallization
  low-order models, particulate processes, 25–27
  particle size distribution, 9–10
Beam-bounce method, 110
Bilinear systems
  coupled spin dynamics, 361
  oscillatory control, 330
Bloch equations
  control input design, 326
  free induction decay, 326
  frequency dispersion, 338–342
  high field NMR instrument, 328
  NMR spectrum, 327
  salient features, 325

J.J. Gorman and B. Shapiro (eds.), *Feedback Control of MEMS to Atoms*,
DOI 10.1007/978-1-4419-5832-7, © Springer Science+Business Media, LLC 2012

Bloch equations (*cont.*)
steering problem, 325
Brownian motion suppression
instrument design, 166–167
scan controller design
linearized closed-loop system, 167, 168
PD controller, 170, 171
PI controller, 170, 172
PID control system, 169
power spectrum, 170

C
Cantilever
AFM, 109, 110
contact mode imaging, 114–115
dynamic mode, 115–117
noise sources, 114
thermally driven scanning, 199–200
Client-server program, 86–87
Closed-loop sense channel, DRG
advantages, 249
block diagram, 250
closed-loop spectra, 254
magnitude function *vs.* detuning frequency, 252, 253
scale factor and optimum demodulation phase, 252
spectral density, 255, 256
transfer function, 250–251
Closed-loop vibratory rate sensor operation, DRG, 220, 221
CMOS-integrated scanning cantilevers, 199–200
Comb-electrode actuator, 182–183
Contact angle saturation, 277
Contact loss detection, 95, 96
Continuous crystallization, 8–9
Coupled spin dynamics
Cartan decomposition, 351
density matrix, 350
description, 348
Hamiltonians, 348, 349
Lie algebra, 351
relaxation
Bloch vector, 358
CNOT operation, 356, 357
decoherence effects, 354
Hamiltonian eigenstates, 355, 356
Lindblad operators, 355
optimal feedback control law, 360
spin ensemble, 355, 356
state-of-the-art pulse sequences, 361
unitary transformation, 357
theorems, 352–354
unitary transformations, 350
Crystallization
batch protein, 9–11
continuous, 8–9

D
Data storage
devices, 200, 201
magnetic HDD
approaches, 204
decoupled sensitivity design, 204, 205
dual-stage actuation, 203, 204
master-slave design, 204, 205
PQ method, 205
VCM suspension, 203, 204
media actuator
LQG regulator, 202, 203
seek-and-settle stage, 201, 202
time-optimal control, 201, 202
track-and-follow, 202, 203
2-D automated micro/nanoparticle manipulation
atomic force microscope (AFM), 70
force-based 2-D nanoparticle manipulation
AFM non-contact-mode images, 89
experimental results, 90–94
force modeling, 90–94
imaging, 88
system description, 86–87
micro/nanoelectromechanical system fabrication process, 70
reference image, 71
vision-based 2-D microparticle manipulation
adhesive forces, 72–73
automated micromanipulation system, structure of, 72
automated single-microparticle manipulation, 81–84
force modeling, 74–78
image processing, 78–79
parameter estimation, 79–81
sliding mode control, 73–74
Disk resonator gyro (DRG)
amplitude and phase coordinates, 215–216
bias terms, analysis of
phase perturbation, 263, 264
rebalance loop gain, 261–262
zero rate bias, 260
closed-loop control architecture
amplitude stability, 237, 238
automatic gain control, 237

excitation and force-to-rebalance loop, 234
loop gain, 234, 235
phase-locked-loop, 236
wideband frequency response, 235, 236
electrostatic biasing
electrical stiffness matrices, 229
electrostatic tuning algorithm, 230–233
model fitting algorithm, 228
equations of motion, 214
noise analysis
angle uncertainty, 256–260
closed-loop sense channel (*see* Closed-loop sense channel, DRG)
Johnson noise, 238
open-loop sense channel (*see* Open-loop sense channel, DRG)
noise power spectrum, 221
operation modes
angular rate sensing mode, 218
closed-loop operation, 219
whole-angle mode, 217, 218
resonant structure, 211, 212
scale factor, 220
skew-symmetric matrix, 213
vibratory sensor model
empirical frequency response, 225–228
gyro transfer function, 225
wideband frequency response, 224
weighting factor, 222
2DOF control architecture
analysis, 130–131
benefits of, 131
control synthesis scheme, 129–130
DOF scheme, 128–129
electrical stiffness matrices, 229
feedforward-feedback scheme, 128, 129
DRG. *See* Disk resonator gyro (DRG)

E

Electro-optic deflectors (EOD)
critical performance parameters, 153
operating principles, 151, 154
Electro-osmotic microflow modes, 299, 300
Electrostatic biasing
electrical stiffness matrices, 229
electrostatic tuning algorithm
eigenvalue problem, 231
empirical frequency response, 232–233
stiffness matrices, 230–231
model fitting algorithm, 228
Electrostatic microactuators
charge-control approach, 191–193
comb-electrode actuator, 182–183
PPA, 182–184
voltage-control approach
closed-loop control, 187, 188
controller gain, 186, 187
controller output waveforms, 186, 187
conventional CMOS process, 184, 185
feedback control algorithm, 189, 190
feedback control system, 184, 185
gimbaled two-axis micro-mirror fabrication, 189
phase margin, 186, 187
silicon-on-insulator wafer, 187, 188
sliding-mode control, 190–191
variable structure control, 191
Electrowetting actuation
applications of, 272
definition, 272
modeling
contact angle saturation, 277
dielectric energy, 274
level-set method, 279
line pinning, 277
Navier–Stokes equations, 274–276
pressure jump, 276
shape change actuation, 273
UCLA EWOD system, 274, 275, 278
variational front-tracking approach, 279
particle steering
algorithm initialization, 282
angular path, 288
closed-loop feedback control architecture, 281, 282
diverging paths, 291, 292
figure 8 path, 287, 288
particle motion, desired direction of, 283
particle position and droplet boundary sensing, 282
particle position updation, 286–287
pressure boundary conditions, least-squares solution of, 283–286
sine wave path, 290
two-particle control, 289
Electrowetting-on-dielectric (EWOD) system
droplet in, 283, 284
particle steering closed-loop feedback control architecture, 281, 282
particle steering control algorithm update, 286, 287
UCLA EWOD system, 274, 275, 278
Emissivity correcting pyrometer (ECP), 47
Ensemble control
multiply rotating frames method, 342–348

Ensemble control (*cont.*)
  parametric inhomogeneities, 335–338
EOD. *See* Electro-optic deflectors (EOD)
Extended Kalman filter (EKF), 49

F
Fault-tolerant control systems, 27–29
Feynman lecture, 5
Force-based 2-D nanoparticle manipulation
  experimental results, 96–99
  flowchart description, 94
  force modeling
    deformation, of cantilever, 92
    friction, 90
    net force *vs.* contact angles, 92, 93
    normal force, 91
    parameters, 90, 91
  imaging, 88, 89
  mechanical pushing/pulling, 89
  principle, 85
  system description, 86–87
Force-to-rebalance controller, 218

G
Gaussian potential, 159, 160
Generalized Lorentz-Mie theory (GLMT), 161

H
Hele-Shaw cell, liquid flow in, 274, 275
Hough transform, 78
HVOF thermal spray coating processes
  microstructure coating, 33–35
  velocity and temperature control
    closed-loop simulations, 38
    feedback controller response, 40, 41
    pressure and fuel/oxygen ratio, 35–37
    profiles of, 38, 39
Hybrid predictive control, low-order models
  continuous crystallizer, 23–25
  limitations, 20, 21
  model predictive control
    implementation, 22–23
    policies, 20

I
Inertia sensors, MEMS
  capacitive accelerometers, 194–197
  closed-loop feedback configuration, 193, 194
  description, 193
  tunneling accelerometers, 197–199

J
Johnson noise, 238

L
Laser cooling, 333–334
Laser intensity modulators, 154–155
Level-set method, 279
Linear quadratic Gaussian (LQG) regulator, 202, 203
Linear transformation, of boundary conditions, 286
Low-order models, particulate processes
  aerosol reactors, nonlinear control
    closed-loop profiles, 31–33
    steady-state profile, 30, 31
    summarization, 29
  batch protein crystallizer, 25–27
  fault-tolerant control systems, 27–29
  hybrid predictive control
    continuous crystallizer, 23–25
    limitations, 20, 21
    model predictive control, 20, 22–23
  nonlinear control
    closed-loop output, 20, 21
    crystal size distribution, 19, 22
    fifth-order moment model, 17
    infinite-order dimensionless system, 16
    manipulated input profiles, 19, 21
    open-loop profiles comparison, 17, 18
    output feedback controller, 19
  particle size distribution evolution, 26, 27

M
Magnetic hard disk drive (HDD), 203–206
Master equations, 331–333
Media actuator
  LQG regulator, 202, 203
  seek-and-settle stage, 201, 202
  time-optimal control, 201, 202
  track-and-follow, 202, 203
MHE. *See* Moving horizon estimation (MHE)
Microfluidics
  boundary conditions, 270
  electrowetting actuation
    applications of, 272
    contact angle saturation, 277
    definition, 272

dielectric energy, 274
level-set method, 279
line pinning, 277
Navier–Stokes equations, 274–276
particle steering (*see* Particle steering, control for)
pressure jump, 276
shape change actuation, 273
UCLA EWOD system, 274, 275, 278
variational front-tracking approach, 279
fabrication methods, 270
geometric uncertainty, 271
object manipulation, flow control/electrophoresis
advantages, 293
feedback control, 298–302
multiple particle control, 302–304
physics and modeling, 294–298
single particle control, 302
single quantum dot control, 305–307
swimming cell control, 304
system setup and device fabrication, 294
three-dimensional control, 308–312
Microscale system. *See* Nanoscale system
Model predictive control (MPC), 51
Moving horizon estimation (MHE)
advantages, 62–63
application of, 59, 61
chemical vapor deposition
experimental apparatus, 52–54
MHE, illustration of, 55, 56
state space model, 54–55
Multiply rotating frames method, 342–348

N

Nanoparticle control, in Brownian motion, 310–311
Nanopositioning systems, control design
bandwidth, 120–121
2DOF control
analysis, 130–131
benefits of, 131
control synthesis scheme, 129–130
DOF scheme, 128–129
feedforward-feedback scheme, 128, 129
optimal control framework
creep effects, 125
hysteresis elimination, 123, 124
MIMO transfer function, 122
parallel-kinematic *xyz*, 121, 122
tracking performance, 123, 124
performance criteria and limitations
block diagram, 118
challenges, 119
finite-waterbed effect, 120
tracking error, 118
waterbed effect, 119
positioning system, 117
resolution, 120–121
robustness, 120–121
ultrahigh resolution
feedback strategies, 125, 126
noise-management scheme, 127
single-axis open-loop system, 126
sub-nanometer positioning resolution, 127
tracking, 128
Nanoscale surface property estimation
optical surface measurement, 45, 46
pyrometry and temperature control
constructive interference, thin transparent film, 47, 48
emissivity correcting pyrometer, 47
reflectometry and film thickness control
inverse problem, 48–49
lithographically patterned surfaces, 49, 50
MHE, 49 (*see also* Moving horizon estimation (MHE))
scatterometry, 51–52
spectroscopic ellipsometry, 50–51
Nanoscale system
length scale, 2
technical challenges
embedded on-chip control, 372
high-bandwidth operation, 371–372
model uncertainty, 370–371
noise and fluctuations, 370
precision sensing, 371
surface forces, 372
threads
cross-disciplinary communication, 369
experimental verification, 369
mathematical problem, 368
model-based control, 366–368
right problems, picking, 366
Noise analysis, DRG
angle uncertainty, 256–260
closed-loop sense channel
advantages, 249
block diagram, 250
closed-loop spectra, 254
magnitude function *vs.* detuning frequency, 252, 253
scale factor and optimum demodulation phase, 252

Noise analysis, DRG (*cont.*)
spectral density, 255, 256
transfer function, 250–251
Johnson noise, 238
open-loop sense channel
coriolis force, 241
cross-coupling, 239
decoupled sensor dynamics, 239, 240
excitation signal, 242
magnitude function *vs.* detuning frequency, 246–247
noise scaling *vs.* detuning frequency, 248
operation of, 243
scale factor, 244–245
Noise-management, 127
Nonlinear control
closed-loop output, 19, 21
low-order models
crystal size distribution, 19, 22
fifth-order moment model, 17
infinite-order dimensionless system, 16
manipulated input profiles, 19, 21
open-loop profiles comparison, 17, 18
output feedback controller, 19
Nonlinear stochastic control theory, 174
Nuclear magnetic resonance (NMR) spectrum
high field instrument, 328
principles, 327

O

Object manipulation, flow control/electrophoresis
advantages, 293
feedback control, 298–302
multiple particle control, 302–304
physics and modeling
Brownian motion, 297
electro-osmotic actuation, 294–295
fluid velocity, 296
single particle control, 302
single quantum dot control, 305–307
swimming cell control, 304
system setup and device fabrication, 294
three-dimensional control, 308–312
Open-loop sense channel, DRG
coriolis force, 241
cross-coupling, 239
decoupled sensor dynamics, 239, 240
excitation signal, 242
magnitude function *vs.* detuning frequency, 246–247
noise scaling *vs.* detuning frequency, 248
operation of, 243
scale factor, 244–245
Optical diagnostics
definition, 46
uses, 59
Optical sensor, 46. *See also* Nanoscale surface property estimation
Optical trapping
actuation and sensing
block diagram, 150
laser intensity modulators, 154–155
particle position sensing, 155–158
performance of, 150
trap scanners, 151–153
in air and vacuum, 173–174
applications, 143–145
Brownian motion suppression
instrument design, 166–167
scan controller design, 167–172
controlled, 173
description, 141
dipole moment, 143
dynamic behavior
empirical closed-form dynamic model, 163–165
finite lifetime, 159–163
method of lines, 160
probability density function, 161
trapping coordinate system, 163, 164
extending trapping lifetime, 173
feedback control
design approaches, 148–149
gradient force, 145
isometric measurements, 146–147
PID control law, 148
trapping configuration, 146
nonlinear stochastic control theory, 174
three-dimensional position control, 172–173
trap formation, 141, 142
trapping forces, 143
Optimal control design
creep effects, 125
hysteresis elimination, 123, 124
MIMO transfer function, 122
parallel-kinematic *xyz*, 121, 122
tracking performance, 123, 124

P

Parallel-plate actuator (PPA)
modelling, 183–184
voltage-control approach
closed-loop control, 187, 188

controller gain, 186, 187
controller output waveforms, 186, 187
conventional CMOS process, 184, 185
feedback control algorithm, 189, 190
feedback control system, 184, 185
gimbaled two-axis micro-mirror fabrication, 189
phase margin, 186, 187
silicon-on-insulator wafer, 187, 188
sliding-mode control, 190–191
variable structure control, 191
Parameter fitting method, 49
Particle-center detection, 94, 95
Particle position sensing
back-focal-plane detection method, 155, 156
microscopic illumination, 157
optical detection mechanism, 158
Particle size distribution
aerosol synthesis, 11–12
batch protein crystallization, 9–10
continuous crystallization, 8–9
HVOF thermal spray coating processes
microstructure coating, 33–35
velocity and temperature control, 35–41
particulate process model
aerosol reactors, nonlinear control, 29–33
batch protein crystallizer, 25–27
description, 7–8, 12–13
fault-tolerant control, 27–29
hybrid predictive control, 20–25
nonlinear control, 16–20
reduction of, 14–16
spatially homogeneous, 13–14
Particle steering, control for
algorithm initialization, 282
angular path, 288
closed-loop feedback control architecture, 281, 282
diverging paths, 291, 292
figure 8 path, 287, 288
particle motion, desired direction of, 283
particle position and droplet boundary sensing, 282
particle position updation, 286–287
pressure boundary conditions, least-squares solution of, 283–286
sine wave path, 290
two-particle control, 289
Particulate process model
description, 7–8, 12–13
low-order models
aerosol reactors, nonlinear control, 29–33
batch protein crystallizer, 25–27
fault-tolerant control, 27–29
hybrid predictive control, 20–25
nonlinear control, 16–20
reduction of, 14–16
spatially homogeneous, 13–14
Pattern-formation planning, 100–101, 103
Phase-locked-loop (PLL), 236
Pietrement's contact mechanics model, 90
Position control, MEMS
AFM system, 179
classical loop-shaping, 181
conventional closed-loop control system, 180
data storage
devices, 200, 201
magnetic HDD, 203–206
media actuator, 201–203
design and implementation issues, 206–207
electrostatic microactuators
charge-control approach, 191–193
comb-electrode actuator, 182–183
PPA, 182–184
voltage-control approach, 184–191
inertia sensors
capacitive accelerometers, 194–197
closed-loop feedback configuration, 193, 194
description, 193
tunneling accelerometers, 197–199
linear time-invariant control systems, 180–181
sigma-delta control, 194–197
SPM, 199–200
types of, 180
PPA. *See* Parallel-plate actuator (PPA)
Proportional-derivative (PD) controller, 170, 171
Proportional-integral (PI) controller, 170, 172
Pyrometer, 46–47

Q

Quantum mechanical system
bloch equations control
control input design, 326
free induction decay, 326
frequency dispersion, 338–342
high field NMR instrument, 328
NMR spectrum, 327
salient features, 325

Quantum mechanical system (*cont.*)
steering problem, 325
coupled spin dynamics
Cartan decomposition, 351
density matrix, 350
description, 348
Hamiltonians, 348, 349
Lie algebra, 351
relaxation, 354–361
theorems, 352–354
unitary transformations, 350
density matrix, 323, 324
description, 3
dispersions, 335–338
ensemble control
multiply rotating frames method, 342–348
parametric inhomogeneities, 335–338
magnetic moment evolution, 322
open quantum systems
laser cooling, 333–334
master equations, 331–333
oscillatory control, 329–331
Schröedinger equation, 321
two level system, 323

R
Real-time optical measurement, 46
Rebalance loop filter, 250
Reflectometer, 48

S
Scan controller
linearized closed-loop system, 167, 168
PD controller, 170, 171
PI controller, 170, 172
PID control system, 169
power spectrum, 169, 170
Scanning probe microscopy (SPM), 3
Scanning tunneling microscope (STM), 109, 110
Scatterometry, 51–52
Shear flow, 310–311
Sigma-delta modulation, 194–197
Simple least squares fitting
application, 60
MHE, 56
Sliding mode control (SMC), 73–74
Spectral reflectometry measurement, 58
Spectroscopic ellipsometry, 50–51
Surface emissivity, 47
Surface forces, 372
Surface roughness, 57–59

T
Thermal spray. *See* HVOF thermal spray coating processes
Three-dimensional particle control, microfluidics, 308, 309
Trap scanners
AODs, 151
critical performance parameters, 153
EODs, 152
operating principles, 151, 152
Tuned gyro, 218
Two-particle separation, 289, 290

V
Variational front-tracking approach, 279
Vibratory gyros, 211, 220
Vibratory sensor model
empirical frequency response, 225–228
gyro transfer function, 225
wideband frequency response, 224
Vision-based 2-D microparticle manipulation
adhesive forces, 72–73
automated micromanipulation system, structure of, 72
automated single-microparticle manipulation
Lyapunov stability criterion, 84
pixel discretization, 82
single polystyrene (PS) particle, 82, 83
force modeling
adhesive force, 74–75
frictional forces, 76
horizontal forces, 77
material properties, 76
spinning friction, 74
image processing, 78–79
parameter estimation
linear transformation, 79
Lyapunov function, 80
sliding-mode iterative parameter estimator, 81
sliding mode control, 73–74

Z
Zero rate bias (ZRB), 260